Selected Titles in This Series

21 **John B. Conway,** A course in operator theory, 2000
20 **Robert E. Gompf and András I. Stipsicz,** 4-manifolds and Kirby calculus, 1999
19 **Lawrence C. Evans,** Partial differential equations, 1998
18 **Winfried Just and Martin Weese,** Discovering modern set theory. II: Set-theoretic tools for every mathematician, 1997
17 **Henryk Iwaniec,** Topics in classical automorphic forms, 1997
16 **Richard V. Kadison and John R. Ringrose,** Fundamentals of the theory of operator algebras. Volume II: Advanced theory, 1997
15 **Richard V. Kadison and John R. Ringrose,** Fundamentals of the theory of operator algebras. Volume I: Elementary theory, 1997
14 **Elliott H. Lieb and Michael Loss,** Analysis, 1997
13 **Paul C. Shields,** The ergodic theory of discrete sample paths, 1996
12 **N. V. Krylov,** Lectures on elliptic and parabolic equations in Hölder spaces, 1996
11 **Jacques Dixmier,** Enveloping algebras, 1996 Printing
10 **Barry Simon,** Representations of finite and compact groups, 1996
9 **Dino Lorenzini,** An invitation to arithmetic geometry, 1996
8 **Winfried Just and Martin Weese,** Discovering modern set theory. I: The basics, 1996
7 **Gerald J. Janusz,** Algebraic number fields, second edition, 1996
6 **Jens Carsten Jantzen,** Lectures on quantum groups, 1996
5 **Rick Miranda,** Algebraic curves and Riemann surfaces, 1995
4 **Russell A. Gordon,** The integrals of Lebesgue, Denjoy, Perron, and Henstock, 1994
3 **William W. Adams and Philippe Loustaunau,** An introduction to Gröbner bases, 1994
2 **Jack Graver, Brigitte Servatius, and Herman Servatius,** Combinatorial rigidity, 1993
1 **Ethan Akin,** The general topology of dynamical systems, 1993

A Course in Operator Theory

GRADUATE STUDIES IN MATHEMATICS 21

A Course in Operator Theory

John B. Conway

AMERICAN MATHEMATICAL SOCIETY
Providence, Rhode Island

1991 *Mathematics Subject Classification.* Primary 47A99.

Library of Congress Cataloging-in-Publication Data

Conway, John B.
A course in operator theory / John B. Conway.
p. cm. — (Graduate studies in mathematics ; v. 21)
Includes bibliographical references and index.
ISBN 0-8218-2065-6 (alk. paper) | Softcover ISBN 1-4704-8061-1
1. Operator theory. I. Title. II. Series.
QA329.C7 1999
515′.724-dc21 99-41229
CIP

Reprinted by the American Mathematical Society, 2025

♾ The paper used in this book is acid-free and falls within the guidelines established to ensure permanence and durability.
Visit the AMS home page at URL: http://www.ams.org/

11 10 9 8 7 6 5 4 3 30 29 28 27 26 25

For Ann, my source for happiness

Contents

Preface xiii

Chapter 1. Introduction to C*-Algebras 1
§1. Definition and examples 1
§2. Abelian C*-algebras and the Functional Calculus 7
§3. The positive elements in a C*-algebra 12
§4. Approximate identities 17
§5. Ideals in a C*-algebra 21
§6. Representations of a C*-algebra 24
§7. Positive linear functionals and the GNS construction 29

Chapter 2. Normal Operators 37
§8. Some topologies on $\mathcal{B}(\mathcal{H})$ 37
§9. Spectral measures 41
§10. The Spectral Theorem 47
§11. Star-cyclic normal operators 51
§12. The commutant 55
§13. Von Neumann algebras 60
§14. Abelian von Neumann algebras 62
§15. The functional calculus for normal operators 65

Chapter 3. Compact Operators 71
§16. C*-algebras of compact operators 71

§17. Ideals of operators 82
§18. Trace class and Hilbert-Schmidt operators 86
§19. The dual spaces of the compact operators and the trace class 93
§20. The weak-star topology 95
§21. Inflation and the topologies 99

Chapter 4. Some Non-Normal Operators 105
§22. Algebras and lattices 105
§23. Isometries 111
§24. Unilateral and bilateral shifts 118
§25. Some results on Hardy spaces 126
§26. The functional calculus for the unilateral shift 132
§27. Weighted shifts 136
§28. The Volterra operator 143
§29. Bergman operators 147
§30. Subnormal operators 157
§31. Essentially normal operators 170

Chapter 5. More on C*-Algebras 181
§32. Irreducible representations 181
§33. Positive maps 187
§34. Completely positive maps 190
§35. An application: Spectral sets and the Sz.-Nagy Dilation Theorem 198
§36. Quasicentral approximate identitites 204

Chapter 6. Compact Perturbations 207
§37. Behavior of the spectrum under a compact perturbation 207
§38. $\mathcal{B}_p$ perturbations of hermitian operators 211
§39. The Weyl-von Neumann-Berg Theorem 214
§40. Voiculescu's Theorem 220
§41. Approximately equivalent representations 229
§42. Some applications 236

Chapter 7. Introduction to Von Neumann Algebras 241
§43. Elementary properties and examples 242
§44. The Kaplansky Density Theorem 250

§45. The Pedersen Up-Down Theorem 253
§46. Normal homomorphisms and ideals 258
§47. Equivalence of projections 265
§48. Classification of projections 270
§49. Properties of projections 278
§50. The structure of Type I algebras 282
§51. The classification of Type I algebras 289
§52. Operator-valued measurable functions 294
§53. Some structure theory for continuous algebras 301
§54. Weak-star continuous linear functionals revisited 305
§55. The center-valued trace 311

Chapter 8. Reflexivity 319
§56. Fundamentals and examples 319
§57. Reflexive operators on finite dimensional spaces 323
§58. Hyperreflexive subspaces 327
§59. Reflexivity and duality 335
§60. Hypereflexive von Neumann algebras 342
§61. Some examples of operators 348

Bibliography 355

Index 367

List of Symbols 371

Preface

The genesis of this book is a pair of courses I taught, one at Indiana University and another at the University of Tennessee. Both of these followed a standard two semester course in functional analysis, though this book is written with only a one semester course in functional analysis as its prerequisite. The aim is to cover with varying depth a variety of subjects that are central to operator theory. Many of these topics have treatises devoted to their explication, and some care has been taken to alert the reader to additional sources for deeper study. So you can think of this book as a sequel to a basic functional analysis course or as a foundation for the study of operator theory.

The prerequisites for this book are a bit fuzzy. The reader is assumed to know fundamental functional analysis. Specifically it will be assumed that the reader knows the material in the first seven chapters of [ACFA]. (Repeated reference to [ACFA] is made in this text and it is referenced in this way rather than the form of other references.) The desire is to make the book as accessible as possible. It is incumbent on me to avoid the arrogance of assuming or requiring familiarity with one of my previous books. But I have to set the standard somewhere, so I will assume the reader knows what I believe to be the material common to all basic courses in functional analysis: the three basic principles of Banach spaces, the definition and elementary properties of locally convex spaces, the foundations of Banach algebras, including the Riesz functional calculus, and the rudiments of operator theory, including the spectral theory of compact operators. This constitutes the first seven chapters of [ACFA].

This book starts with an introduction to C^*-algebras followed by a chapter on normal operators that culminates with the Spectral Theorem and the functional calculus. There is considerable overlap between these first two chapters and Chapters VIII and IX in [ACFA]. If the reader really knows these first nine chapters of [ACFA], he/she will be able to fly through the first two chapters here. Such a reader could realistically begin reading Chapter 3 of the present book. There is, however, material in the first two chapters of this book, specifically §4, that does not appear in [ACFA].

Chapter 3 examines compact operators. Again there is some overlap with [ACFA], but most of this material does not appear there. Chapter 4 begins the study of non-normal operators, which has seen so much significant progress in recent times. In particular the reader is introduced to some rather deep connections between operator theory and analytic functions. This is a hallmark of much that has been done in recent years. The Fredholm index appears here. This is not part of the stated prerequisites and is the subject of Chapter XI in [ACFA]. On the other hand, the need for the index is not that substantial that the reader should be dismayed by encountering it. Indeed, Fredholm theory is discussed later in §37, where the reader will see statements of the pertinent results from this topic, some proofs, and references to [ACFA] for the omitted proofs. The reader who feels some insecurity on this point may examine §37 during the study of Chapter 4.

Chapter 5 returns to the theory of C^*-algebras and examines irreducible representations as well as completely positive maps. As an application, a proof is presented of the Sz.-Nagy Dilation Theorem, which is the basis for a large portion of modern operator theory. The chapter closes with the existence of a quasicentral approximate identity in a C^*-algebra. This is used in the following chapter.

Chapter 6 explores the general topic of compact perturbations. After an abbreviated treatment of Fredholm theory, the Weyl–von Neumann–Berg Theorem is proved as is Voiculescu's Theorem on the approximation of representations of a separable C^*-algebra. The chapter concludes with some applications of these ideas to single operators.

Chapter 7 is a rather extended introduction to von Neumann algebras. The classification scheme is obtained as is complete information on Type I algebras. This is used to recapture the multiplicity theory of normal operators. The chapter includes a proof that a von Neumann algebra is finite if and only if it has a faithful, centered-valued trace.

Chapter 8, the last chapter, gives an introduction to reflexive subspaces of operators. Here the word "reflexive" is used differently from its meaning in Banach space theory. Reflexive subspaces are spaces of operators that are determined by their invariant subspaces. An operator is reflexive if and only if the weakly closed algebra it generates is a reflexive subspace. This, together with the related notion of a hyperreflexive subspace, is a still developing area of research. In many ways this subject is one of the more successful episodes in the modern exploration of asymmetric algebras.

There are many topics in operator theory that are not included here. Given the vastness of the subject, that is no surprise. Many important topics have been omitted because good treatments already exist in the literature. For example, nothing is in this book on the Brown–Douglas–Fillmore theory, which would have been a natural sequel to the chapter on compact perturbations. But whatever approach I would have used would not have differed in any substantial way from that in Davidson [1996]. Also the theory of dual algebras is not presented, but this is due more to space and time limitations. So much has happened since Bercovici, Foias, and Pearcy [1985] that the area is ripe for a further exposition. Such an exposition would have almost doubled the size of the present book. This list of omissions could continue.

Like all books, especially more advanced ones, the material is not linearly dependent. The reader can skip around, especially after the first three chapters. For example, Chapter 8 on reflexivity does not depend on the preceding three chapters in any substantial way. Developing a dependency chart was a temptation, but instead I'll encourage readers to skip around, covering the topics that interest them and filling in the gaps as necessary.

One final caveat. Be aware that I am a bit schizophrenic about separability. On the one hand, I don't wish to present anything in that setting as though it depends on the assumption of separability. On the other hand, in Hilbert space this is where the interest lies. There are also parts of operator theory that really only hold in a reasonable way when the underlying Hilbert space is separable. There are others that are connected to measure theory and I did not want to get into discussing non-separable measure spaces. So I start out with no assumption of separability, but occasionally giving a result that does depend on this. Later, in §51, all Hilbert spaces are assumed to be separable for the remainder of the book.

Throughout the text I give references for further study. I frequently also will cite the source of some, but not all, of the results. I have confidence in the attributions I give, but I certainly am not infallible. I'll maintain a list of corrections and updates on this book linked to my web page (http://www.math.utk.edu/~conway), and any corrections or changes in attribution will be found there as well in future printings should they come to be.

I have many people to thank for their assistance during the preparation of this book. My former student Nathan Feldman read various editions of the manuscript and made many helpful suggestions. Similar help came from my current students Gabriel Prajitura and Sherwin Kouchekian, who, with Nathan, were in one of the courses that eventually led to this book. In the final analysis, of course, I am the one who is responsible for any errors.

Chapter 1

Introduction To C*-Algebras

§1. Definition and examples

1.1 Definition. If $\mathcal{A}$ is a Banach algebra, an *involution* is a map $a \to a^*$ of $\mathcal{A}$ into itself such that for all a and b in $\mathcal{A}$ and all scalars α the following conditions hold: (i) $(a^*)^* = a$; (ii) $(ab)^* = b^*a^*$; (iii) $(\alpha a + b)^* = \bar{\alpha}a^* + b^*$.

If $\mathcal{A}$ has an involution as well as an identity, then $1^*a = (1^*a)^{**} = (a^*1)^* = a$. Thus $1^* = 1$ by the uniqueness of the identity. It also follows that $\alpha^* = \bar{\alpha}$ for any multiple of the identity. An algebra that has an identity is called *unital*.

1.2 Definition. A C*-*algebra* $\mathcal{A}$ is a Banach algebra with involution such that

$$\|a^*a\| = \|a\|^2$$

for every a in $\mathcal{A}$.

This seems a rather simple requirement and it has always impressed the author how much follows from this condition. Here are a few examples.

1.3 Examples. (a) $\mathcal{B} = \mathcal{B}(\mathcal{H})$, the algebra of bounded operators on a Hilbert space, is a C*-algebra, where for each operator A, A^* is the adjoint of A. (See [ACFA], Proposition II.2.7.) In many ways this algebra is the subject of study of this book, though not just as an algebra.

(b) For any Hilbert space $\mathcal{H}$, $\mathcal{B}_0 = \mathcal{B}_0(\mathcal{H})$, the algebra of compact operators on the Hilbert space $\mathcal{H}$, is a C*-algebra, though it lacks an identity when $\mathcal{H}$ is not finite dimensional.

(c) If X is a compact space, $C(X)$, the collection of all continuous functions from X into $\mathbb{C}$, is a C*-algebra if for each continuous function f on X, $f^*(x) \equiv \overline{f(x)}$. Here $C(X)$ is an abelian C*-algebra with an identity. In the next section it is shown that every abelian C*-algebra with identity is isomorphic to one of this type.

(d) If X is locally compact but not compact, then $C_0(X)$, the algebra of continuous functions on X that vanish at infinity, is an abelian C*-algebra without identity. The next section shows that every abelian C*-algebra without identity is isomorphic to one of this type.

The starting point is a collection of basic properties of the norm for a C*-algebra. Throughout this section, $\mathcal{A}$ will be a C*-algebra. Recall that for any Banach space $\mathcal{X}$, $\operatorname{ball}\mathcal{X}$ denotes the closed unit ball.

1.4 Proposition. *If $a \in \mathcal{A}$, then:*

(a) $\|a^*\| = \|a\|$;

(b) $\|aa^*\| = \|a\|^2$;

(c) $\|a\| = \sup\{\, \|ax\| : x \in \operatorname{ball}\mathcal{A}\,\} = \sup\{\, \|xa\| : x \in \operatorname{ball}\mathcal{A}\,\}$.

Proof. (a) $\|a\|^2 = \|a^*a\| \leq \|a^*\|\,\|a\|$. So $\|a^*\| \leq \|a\|$. Since $a^{**} = a$, substituting a^* for a in this inequality gives the reverse inequality.

(b) Combine (a) with the definition.

(c) Let α be the first supremum in (c). Clearly $\alpha \leq \|a\|$ with equality if $a = 0$. If $a \neq 0$, let $x = a^*/\|a\|$. So $\|x\| = 1$ by part (a) and so $\alpha \geq \|a\|$. The proof of the other equality is similar. ∎

When $\mathcal{A}$ is a C*-algebra without an identity, there is a unique way of adjoining an identity. For the algebras $\mathcal{B}_0(\mathcal{H})$ and $C_0(X)$ above, the way to do this is clear. Doing this in general takes a bit more work. If $\mathcal{A}$ and $\mathcal{C}$ are two C*-algebras, a map $\rho : \mathcal{A} \to \mathcal{C}$ is called a *$*$-homomorphism* or *homomorphism* if ρ is an algebraic homomorphism such that $\rho(a^*) = \rho(a)^*$ for all a in $\mathcal{A}$. A *$*$-isomorphism* or *isomorphism* between C*-algebras is a bijective homomorphism. A monomorphism or epimorphism between C*-algebras is defined similarly. It is easy to check that if ρ is an isomorphism, so is ρ^{-1}. This is the natural means of identification in the category of C*-algebras and the word "unique," as used in the first sentence of this paragraph, means unique up to an isomorphism. Similarly when, as in the next proposition, it is stated that the algebra $\mathcal{A}$ is contained in the algebra $\mathcal{A}_1$, this means that $\mathcal{A}$ is isomorphic to a subalgebra of $\mathcal{A}_1$.

1.5 Proposition. *If $\mathcal{A}$ is a C^*-algebra without identity, then there is a unique C^*-algebra with identity $\mathcal{A}_1$ containing $\mathcal{A}$ as an ideal and such that $\mathcal{A}_1/\mathcal{A}$ is one dimensional. If $\rho\colon \mathcal{A} \to \mathcal{C}$ is a $*$-homomorphism, then $\rho_1(a+\alpha) = \rho(a) + \alpha$ defines a $*$-homomorphism $\rho_1\colon \mathcal{A}_1 \to \mathcal{C}_1$.*

Proof. Let $\mathcal{B}(\mathcal{A})$ be all the bounded operators on the algebra $\mathcal{A}$ considered as a Banach space. For a in $\mathcal{A}$ define L_a in $\mathcal{B}(\mathcal{A})$ by $L_a(x) = ax$. The reader is asked to verify that the map $\lambda\colon \mathcal{A} \to \mathcal{B}(\mathcal{A})$ defined by $\lambda(a) = L_a$ is an algebraic homomorphism. By part (c) of the preceding proposition λ is an isometry. Define $\mathcal{A}_1 = \lambda(\mathcal{A}) + \mathbb{C}$; so $\mathcal{A}_1$ is a Banach subalgebra of $\mathcal{B}(\mathcal{A})$. On $\mathcal{A}_1$ define $[L_a + \alpha]^* \equiv L_{a^*} + \bar{\alpha}$. It is left to the reader to verify that this defines an involution on $\mathcal{A}_1$. The claim is that with this involution $\mathcal{A}_1$ is a C*-algebra.

If $a \in \mathcal{A}$ and α is a scalar, for any $\epsilon > 0$, there is an x in ball $\mathcal{A}$ such that $\|L_a + \alpha\|^2 - \epsilon < \|ax + \alpha x\|^2$. Thus

$$\begin{aligned}\|L_a + \alpha\|^2 - \epsilon &< \|(x^*a^* + \bar{\alpha}x^*)(ax + \alpha x)\| \\ &= \|x^*(L_a + \alpha)^*(L_a + \alpha)x\| \\ &\le \|(L_a + \alpha)^*(L_a + \alpha)\|.\end{aligned}$$

Therefore $\|L_a + \alpha\|^2 \le \|(L_a + \alpha)^*(L_a + \alpha)\|$. On the other hand, because $\mathcal{A}_1$ is a Banach algebra, $\|(L_a + \alpha)^*(L_a + \alpha)\| \le \|(L_a + \alpha)^*\|\, \|L_a + \alpha\|$. So the proof will be complete if it can be shown that $\|(L_a + \alpha)^*\| \le \|L_a + \alpha\|$. But using Proposition 1.4.c twice gives

$$\begin{aligned}\|(L_a + \alpha)^*\| &= \sup\{\, \|a^*x + \bar{\alpha}x\| : x \in \operatorname{ball}\mathcal{A}\,\} \\ &= \sup\{\, \|y^*a^*x + \bar{\alpha}y^*x\| : x, y \in \operatorname{ball}\mathcal{A}\,\} \\ &= \sup\{\, \|x^*ay + \alpha x^*y\| : x, y \in \operatorname{ball}\mathcal{A}\,\} \\ &= \sup\{\, \|x^*(L_a + \alpha)y\| : x, y \in \operatorname{ball}\mathcal{A}\,\} \\ &\le \|L_a + \alpha\|.\end{aligned}$$

Thus $\mathcal{A}_1$ is a C*-algebra and λ is a $*$-monomorphism. Identify $\mathcal{A}$ and its image $\lambda(\mathcal{A})$. Clearly $\mathcal{A}$ is an ideal in $\mathcal{A}_1$ that has codimension 1.

The proof of the statement about the homomorphism ρ is left to the reader. It remains to prove the uniqueness of $\mathcal{A}_1$. But if $\mathcal{A}_1'$ is another C*-algebra with identity that contains $\mathcal{A}$ as an ideal of codimension 1, it follows that $\mathcal{A}_1' = \mathcal{A} + \mathbb{C}$. It is easy to verify that $\mathcal{A}_1$ and $\mathcal{A}_1'$ are isomorphic by means of an isomorphism that fixes the elements of $\mathcal{A}$. ∎

If $\mathcal{A}$ is a C*-algebra with identity and $a \in \mathcal{A}$, then the spectrum of a, $\sigma(a)$, is defined in the usual way. If $\mathcal{A}$ does not have an identity, the spectrum of a is defined as its spectrum when it is considered as an element of the algebra $\mathcal{A}_1$ obtained by adjoining an identity to $\mathcal{A}$.

1.6 Definition. If $\mathcal{A}$ is a C*-algebra and $a \in \mathcal{A}$, then: (a) a is *hermitian* if $a^* = a$; (b) a is *normal* if $a^*a = aa^*$; (c) when $\mathcal{A}$ has an identity, a is *unitary* if $a^*a = aa^* = 1$.

The reader can check the examples in (1.3) to see when an element of these C*-algebras is hermitian, normal, or unitary. Of course every element of an abelian C*-algebra is normal. If $\mathcal{A}$ is any C*-algebra and a is a normal element, then the C*-algebra generated by a is abelian.

1.7 Proposition. *If $a \in \mathcal{A}$, the following hold.*

(a) *If a is invertible, then a^* is invertible and $(a^*)^{-1} = (a^{-1})^*$.*

(b) *$a = x + iy$, where x and y are hermitian elements of $\mathcal{A}$.*

(c) *If u is a unitary in $\mathcal{A}$, then $\|u\| = 1$.*

(d) *If a is hermitian, then $\|a\| = r(a)$, the spectral radius of a.*

(e) *If $\mathcal{B}$ is another C*-algebra and $\rho\colon \mathcal{A} \to \mathcal{B}$ is a $*$-homomorphism, then $\|\rho(a)\| \le \|a\|$ for all a in $\mathcal{A}$.*

Proof. The proofs of (a), (b), and (c) are left as exercises. To prove (d), it suffices to assume that $\mathcal{A}$ is unital. If a is a hermitian element of $\mathcal{A}$, then $\|a\|^2 = \|a^*a\| = \|a^2\|$. By induction, $\|a\|^{2n} = \|a^{2n}\|$ for all $n \ge 1$. Thus $r(a) = \lim_n \|a^{2n}\|^{1/2n} = \|a\|$.

(e) First note that by adjoining an identity if necessary, it can be assumed that $\mathcal{A}$ is unital. Now the definition of a $*$-homomorphism does not assume that $\rho(1)$ is the identity of $\mathcal{B}$. However it is easy to see that $\rho(1)$ is an identity for $\rho(\mathcal{A})$. So there is no loss of generality in assuming that $\mathcal{B}$ has an identity and $\rho(1) = 1$.

For a in $\mathcal{A}$, it is easy to check that $\sigma(\rho(a)) \subseteq \sigma(a)$. From (d) it follows that $\|\rho(a)\|^2 = \|\rho(a^*a)\| = r(\rho(a^*a)) \le r(a^*a) \le \|a\|^2$. ■

Since the inverse of a $*$-isomorphism is also a $*$-isomorphism, the following corollary is immediate.

1.8 Corollary. *If $\mathcal{A}$ and $\mathcal{B}$ are C*-algebras and $\rho : \mathcal{A} \to \mathcal{B}$ is a $*$-isomorphism, then ρ is an isometry.*

For any C*-algebra $\mathcal{A}$, $\operatorname{Re}\mathcal{A}$ will denote the collection of hermitian elements of $\mathcal{A}$. It is worth noting that $\operatorname{Re}\mathcal{A}$ is a real Banach space.

1.9 Proposition. *If $h : \mathcal{A} \to \mathbb{C}$ is an algebraic homomorphism, then the following hold.*

(a) *If $a \in \operatorname{Re} \mathcal{A}$, $h(a) \in \mathbb{R}$.*

(b) *For an arbitrary element a of $\mathcal{A}$, $h(a^*) = \overline{h(a)}$.*

(c) *$h(a^*a) \geq 0$ for all a in $\mathcal{A}$.*

(d) *If $\mathcal{A}$ is unital and u is a unitary in $\mathcal{A}$, then $|h(u)| = 1$.*

Proof. By Proposition 1.5 it can be assumed that $\mathcal{A}$ has an identity and $h(1) = 1$. By Proposition VII.8.4 of [ACFA], the restriction of h to any abelian subalgebra of $\mathcal{A}$ has norm 1. Let $a \in \operatorname{Re} \mathcal{A}$ and note that for any real number t, $a + it$ is normal; hence the C*-algebra generated by $a + it$ and the identity is abelian. Thus

$$\begin{aligned} |h(a+it)|^2 &\leq \|a+it\|^2 \\ &= \|(a+it)^*(a+it)\| \\ &= \|(a-it)(a+it)\| \\ &= \|a^2+t^2\| \\ &\leq \|a\|^2 + t^2. \end{aligned}$$

Suppose $h(a) = \alpha + i\beta$, $\alpha, \beta \in \mathbb{R}$. So the above inequalities yield

$$\begin{aligned} \|a\|^2 + t^2 &\geq |\alpha + i(\beta+t)|^2 \\ &= \alpha^2 + (\beta+t)^2 \\ &= \alpha^2 + \beta^2 + 2\beta t + t^2; \end{aligned}$$

hence $\|a\|^2 \geq \alpha^2 + \beta^2 + 2\beta t$ for every real number t. Therefore $\beta = 0$, proving (a).

Let $a = x + iy$, where x and y are hermitian. Because $h(x), h(y) \in \mathbb{R}$ and $a^* = x - iy$, part (b) follows. Also $h(a^*a) = h(a^*)h(a) = |h(a)|^2 \geq 0$, yielding (c). Finally, if u is unitary, $|h(u)|^2 = h(u^*u) = h(1) = 1$. ■

The next corollary is a restatement of part (b) of the preceding proposition.

1.10 Corollary. *Every algebraic homomorphism from a C^*-algebra into the real numbers is a $*$-homomorphism.*

Suppose $\mathcal{A}$ is an abelian C*-algebra and $a \in \mathcal{A}$. From basic Banach algebra theory (see [ACFA], Theorem VII.8.6) it is known that the spectrum of a is $\{\, h(a) : h \text{ is a homomorphism from } \mathcal{A} \text{ into } \mathbb{C} \,\}$. In light of the preceding proposition, this implies that $\sigma(a) \subseteq \mathbb{R}$ whenever $\mathcal{A}$ is abelian and $a \in \operatorname{Re}\mathcal{A}$. Now suppose $\mathcal{A}$ is not abelian, but $a \in \operatorname{Re}\mathcal{A}$. Let $\mathcal{C}$ be the C*-algebra generated by a and the identity; so $\mathcal{C}$ is abelian. Thus $\sigma_{\mathcal{C}}(a)$, the spectrum of a as an element of the algebra $\mathcal{C}$, is a subset of $\mathbb{R}$. According to Theorem VII.5.4 in [ACFA], $\sigma_{\mathcal{A}}(a) \subseteq \sigma_{\mathcal{C}}(a)$ and $\partial\sigma_{\mathcal{C}}(a) \subseteq \partial\sigma_{\mathcal{A}}(a)$. But because $\sigma_{\mathcal{C}}(a) \subseteq \mathbb{R}$, it equals its boundary. Thus $\sigma_{\mathcal{A}}(a) = \sigma_{\mathcal{C}}(a) \subseteq \mathbb{R}$ for a in $\operatorname{Re}\mathcal{A}$. This proves another corollary of Proposition 1.9 and sets the stage for the final theorem of this section.

1.11 Corollary. *If $a \in \operatorname{Re}\mathcal{A}$, $\sigma(a) \subseteq \mathbb{R}$.*

1.12 Theorem. *If $\mathcal{A}$ and $\mathcal{B}$ are two C^*-algebras with a common identity such that $\mathcal{B} \subseteq \mathcal{A}$, then $\sigma_{\mathcal{B}}(a) = \sigma_{\mathcal{A}}(a)$ for any a in $\mathcal{B}$.*

Proof. From the argument preceding the last corollary, it follows that $\sigma_{\mathcal{B}}(a) = \sigma_{\mathcal{A}}(a)$ if $a \in \operatorname{Re}\mathcal{B}$. Now let a be an arbitrary element of $\mathcal{B}$. Since $\sigma_{\mathcal{B}}(a) \supseteq \sigma_{\mathcal{A}}(a)$ by Theorem VII.5.4 of [ACFA], it suffices to show that if a is invertible in $\mathcal{A}$, it is invertible in $\mathcal{B}$. So suppose there is an x in $\mathcal{A}$ such that $ax = xa = 1$. Thus $(a^*a)(xx^*) = (xx^*)(a^*a) = 1$. But $a^*a \in \operatorname{Re}\mathcal{B}$, so the first step in the proof implies that a^*a is invertible in $\mathcal{B}$. Since inverses are unique, $xx^* \in \mathcal{B}$. Hence $x = x(x^*a^*) = (xx^*)a^* \in \mathcal{B}$. ∎

Exercises.

$\operatorname{cl} A$ and $\operatorname{int} A$ are used to denote the closure and interior of a set A in a topological space.

1. Let $\mathcal{A} = \{ f \in C(\operatorname{cl}\mathbb{D}) : f \text{ is analytic on } \mathbb{D} \}$ and for f in $\mathcal{A}$ define $f^*(z) = \overline{f(\bar{z})}$. Show that $f \to f^*$ is an involution on $\mathcal{A}$ and $\|f^*\| = \|f\|$ for all f in $\mathcal{A}$, but $\mathcal{A}$ is not a C*-algebra.

2. Let $\mathcal{A}$ be a C*-algebra and let X be a locally compact space. Show that $C_0(X, \mathcal{A})$, the set of all continuous functions $f \colon X \to \mathcal{A}$ such that for every $\epsilon > 0$, $\{x : \|f(x)\| \geq \epsilon\}$ is compact, is a C*-algebra, where all the operations are defined pointwise and the norm is the sup norm: $\|f\| = \sup\{ \|f(x)\| : x \in X \}$.

3. With the notation of the preceding exercise, show that if X and Y are locally compact spaces, then there is a natural $*$-isomorphism from $C_0(X, C_0(Y))$ onto $C_0(X \times Y)$.

4. Let $\{ \mathcal{A}_i : i \in I \}$ be a collection of C*-algebras. Show that $\bigoplus\{\mathcal{A}_i : i \in I\} \equiv \{ \{a_i\} \in \prod_i \mathcal{A}_i \text{ with } \sup_i \|a_i\| < \infty \}$ is a C*-algebra if the operations are defined coordinatewise. Show that if $\bigoplus_0\{ \mathcal{A}_i : i \in I \}$ is the set of those $\{a_i\}$ in $\prod_i \mathcal{A}_i$ such that for every $\epsilon > 0$, $\{ i : \|a_i\| \geq \epsilon \}$ is finite, then $\bigoplus_0\{ \mathcal{A}_i : i \in I \}$ is a C*-algebra.

5. Show that the C*-algebra obtained by adjoining an identity to $C_0(X)$ is isomorphic to $C(\widehat{X})$, where $\widehat{X}$ is the one-point compactification of X.

§2. Abelian C*-algebras and the Functional Calculus

Recall a few facts concerning the Gelfand transform. Let $\mathcal{A}$ be an abelian Banach algebra with identity and let Σ be its maximal ideal space. Σ is identified with the set of non-zero homomorphisms from $\mathcal{A}$ into the complex numbers. Each such homomorphism has norm 1 ([ACFA], VII.8.4) and so $\Sigma \subseteq \operatorname{ball} \mathcal{A}^*$. Furnish Σ with the relative weak* topology, and Σ becomes a compact Hausdorff space. If $a \in \mathcal{A}$, define $\widehat{a} : \Sigma \to \mathbb{C}$ by $\widehat{a}(h) = h(a)$. Thus $\widehat{a} \in C(\Sigma)$ and this function $\widehat{a}$ is called the *Gelfand transform* of a. The map $\gamma : \mathcal{A} \to C(\Sigma)$ defined by $\gamma(a) = \widehat{a}$ is the *Gelfand transform* for the algebra.

It is known that γ is an algebraic homomorphism having norm 1. When $\mathcal{A}$ is an abelian C*-algebra, much more is true.

2.1 Theorem. *If $\mathcal{A}$ is an abelian unital C*-algebra and Σ its maximal ideal space, then the Gelfand transform $\gamma : \mathcal{A} \to C(\Sigma)$ is a $*$-isomorphism.*

Proof. For each a in $\mathcal{A}$, $\|\widehat{a}\|_\infty \leq \|a\|$. By Proposition 1.7.d, $\|\gamma(a)\|_\infty = \|a\|$ whenever $a \in \operatorname{Re} \mathcal{A}$. In particular, $\|\gamma(a^*a)\|_\infty = \|a^*a\|$ for any a in $\mathcal{A}$. By Proposition 1.9.b, $\widehat{a^*} = \gamma(a^*) = \overline{\widehat{a}}$ for all a in $\mathcal{A}$. That is, $\gamma(a^*) = \gamma(a)^*$ and γ is a $*$-homomorphism. Therefore $\|\gamma(a)\|_\infty^2 = \|\gamma(a)^*\gamma(a)\|_\infty = \|\gamma(a^*a)\|_\infty = \|a^*a\| = \|a\|^2$. Therefore γ is an isometry.

It remains to show that γ is surjective; this is done with the aid of the Stone–Weierstrass Theorem. Because γ is a $*$-homomorphism and an isometry, $\gamma(\mathcal{A})$ is a closed subalgebra of $C(\Sigma)$ containing the constants and closed under complex conjugation. So to show that γ is surjective, it suffices to prove that $\gamma(\mathcal{A})$ separates points of Σ. But if h_1 and h_2 are distinct homomorphisms on $\mathcal{A}$, the definition of homomorphism implies there is an a in $\mathcal{A}$ such that $h_1(a) \neq h_2(a)$. Thus $\widehat{a} = \gamma(a)$ separates the points h_1 and h_2. ■

Now let $\mathcal{A}$ be an abelian C*-algebra without an identity and let $\mathcal{A}_1$ be the algebra obtained by adjoining an identity to $\mathcal{A}$. Let Σ and Σ_1 be the two maximal ideal spaces. Now $\mathcal{A}$ is a maximal ideal in $\mathcal{A}_1$ so if $h_0 : \mathcal{A}_1 \to \mathbb{C}$ is the unique homomorphism with $\ker h_0 = \mathcal{A}$, then $\Sigma = \Sigma_1 \setminus \{h_0\}$ in a natural manner. Combining this observation with the preceding theorem and a little work produces the following corollary.

2.2 Corollary. *If $\mathcal{A}$ is an abelian C*-algebra without an identity and Σ is its maximal ideal space, then the Gelfand transform $\gamma : \mathcal{A} \to C_0(\Sigma)$ is a $*$-isomorphism.*

If (X, Ω, μ) is a σ-finite measure space, $L^\infty(\mu)$ is an abelian C*-algebra. Thus $L^\infty(\mu)$ is $*$-isomorphic to $C(\Sigma)$, the algebra of continuous functions on its maximal ideal space. See Exercise 5 for more.

A bit of notation will prove useful here and later. If $\mathcal{A}$ is a C*-algebra and $S \subseteq \mathcal{A}$, let $C_0^*(S)$ be the C*-algebra generated by S. That is, $C_0^*(S)$ is the intersection of all the C*-algebras contained in $\mathcal{A}$ that contain S. It is trivial to see that $C_0^*(S)$ is a C*-algebra, though it may not have an identity. Let $C^*(S) = C_0^*(S \cup \{1\})$. In particular, if $a \in \mathcal{A}$, $C^*(a)$ is the closure of all words in a, a^*, and 1. The case of an abelian C*-algebra with a single generator is worthy of special consideration. I like to call the next result the "a to z theorem."

2.3 Theorem. *If $\mathcal{A}$ is an abelian unital C^*-algebra such that there is an element a with $\mathcal{A} = C^*(a)$, then there is a unique $*$-isomorphism $\rho : \mathcal{A} \to C(\sigma(a))$ such that $\rho(a) = z$.*

Proof. The hypothesis of the proposition implies that $\mathcal{A}$ is the closure of $\{\, p(a, a^*) : p(a, a^*) \text{ is a polynomial in } a \text{ and } a^* \,\}$. Define $\tau \colon \Sigma \to \sigma(a)$ by $\tau(h) = h(a)$. Clearly τ is continuous and Theorem VII.8.6 in [ACFA] implies τ is surjective. If $h_1, h_2 \in \Sigma$ and $\tau(h_1) = \tau(h_2)$, then algebraic manipulations show that $h_1(p(a, a^*)) = h_2(p(a, a^*))$ for every polynomial $p(a, a^*)$ in a and a^*. By the observation at the beginning of the proof, this implies $h_1 = h_2$. Thus τ is injective and must be a homeomorphism.

Define $\rho : \mathcal{A} \to C(\sigma(a))$ by $\rho(x) = \widehat{x} \circ \tau^{-1}$. It is a routine exercise to check that ρ is a $*$-isomorphism. The fact that ρ is the unique $*$-isomorphism mapping a to z follows directly from the fact that a generates the C*-algebra $C^*(a)$. ∎

Now to exploit this and define $f(a)$ for a normal element a and a continuous function f on $\sigma(a)$. Let $\mathcal{B}$ be any C*-algebra with identity and let a be a normal element of $\mathcal{B}$. Put $\mathcal{A} = C^*(a)$. If $\tau \colon \Sigma \to \sigma(a)$ is the homeomorphism defined in the last proposition, let $\tau^\# : C(\sigma(a)) \to C(\Sigma)$ be defined by $\tau^\#(f) = f \circ \tau$. It is left to the reader to check that $\tau^\#$ is a $*$-homomorphism. In fact, since τ is a homeomorphism, $\tau^\#$ is a $*$-isomorphism.

Thus $\gamma^{-1} \circ \tau^\# : C(\sigma(a)) \to \mathcal{A}$ is a $*$-isomorphism. If $p(z, \bar{z})$ is a polynomial, $\gamma^{-1} \circ \tau^\# = p(a, a^*)$. This is the functional calculus for the normal element a. Since this map $f \to f(a)$ takes the identity to the identity and the function z to a, it extends the Riesz Functional Calculus by the uniqueness of that functional calculus. (See Theorem VII.4.7 in [ACFA].) Summarize this as follows.

2.4 Theorem. *If $\mathcal{B}$ is any C^*-algebra and a is a normal element of $\mathcal{B}$, then there is a $*$-monomorphism $f \to f(a)$ from $C(\sigma(a))$ into $\mathcal{B}$ having the following properties:*

(a) $\|f(a)\| = \|f\|_\infty$ *for all f in $C(\sigma(a))$;*

(b) $f \to f(a)$ *extends the Riesz Functional Calculus for a;*

(c) *the range of this map is $C^*(a)$.*

Moreover this map is unique in the following sense. If $\rho : C(\sigma(a)) \to C^*(a)$ is a $*$-homomorphism such that $\rho(1) = 1$ and $\rho(z) = a$, then $\rho(f) = f(a)$ for all f in $C(\sigma(a))$.

Proof. The only part not proved yet is the uniqueness statement. Let ρ be as described. Algebra implies that $\rho(p(z, \bar{z})) = p(a, a^*)$ for any polynomial in z and $\bar{z}$. If $f \in C(\sigma(a))$, there is a sequence of polynomials $\{p_n(z, \bar{z})\}$ that converges to f uniformly on $\sigma(a)$. But $p_n(a, a^*) = \rho(p_n) \to \rho(f)$ and also $p_n(a, a^*) \to f(a)$. ■

This referred to as the *functional calculus of the normal element a*. Because of the uniqueness of the functional calculus, it is not necessary to remember its exact form but only that it constitutes a $*$-monomorphism that extends the Riesz Functional calculus. This will be illustrated shortly, but the stage needs to be set.

The next two results might be considered examples, but they are truly more than that. The first is an example of a normal operator that is the building block for all normal operators on a separable Hilbert space. The second explores an abelian C*-algebra that is the prototype for all maximal abelian C*-algebras. The second result is actually a generalization of the first, so only that one will be proved. Deriving the first from the second is left as an exercise.

The *support* of a measure μ is defined as the complement of $\bigcup\{U : U \text{ is open and } \mu(U) = 0\}$ and is denoted by $\operatorname{supp}\mu$. Another standard notion from measure theory is the *essential range* of a measurable function,

$$\text{ess-ran}\,(\phi) = \bigcap\{\,\text{cl}\,(\phi(\Delta)) : \Delta \in \Omega \text{ and } \mu(X \setminus \Delta) = 0\,\}.$$

2.5 Proposition. *If μ is a compactly supported, regular Borel measure on $\mathbb{C}$, and $N_\mu : L^2(\mu) \to L^2(\mu)$ is defined by $N_\mu f(z) = zf(z)$ for all f in $L^2(\mu)$, then N_μ is a normal operator with the following properties:*

(a) ${N_\mu}^* f(z) = \bar{z} f(z)$ *for all f in $L^2(\mu)$;*

(b) $\sigma(N_\mu)$ *is the support of the measure μ.*

2.6 Proposition. *If (X, Ω, μ) is a σ-finite measure space, for each ϕ in $L^\infty(\mu)$, define M_ϕ on $L^2(\mu)$ as multiplication by ϕ. If $\phi \in L^\infty(\mu)$, the following statements hold:*

(a) *the operator M_ϕ is normal and $M_\phi{}^* = M_{\bar{\phi}}$;*

(b) *$\phi \to M_\phi$ is a $*$-homomorphism from $L^\infty(\mu)$ into $\mathcal{B}(L^2(\mu))$;*

(c) *$\|M_\phi\| = \|\phi\|_\infty$;*

(d) *$\sigma(M_\phi) =$ ess-ran ϕ.*

Proof. The proofs of (a) and (b) are routine. If $f \in L^2(\mu)$, $\|M_\phi f\|^2 = \int |\phi f|^2 \, d\mu \le \|\phi\|_\infty^2 \, \|f\|^2$, so $\|M_\phi\| \le \|\phi\|_\infty$. If $\epsilon > 0$, let $\Delta \in \Omega$ such that Δ has full measure and $\sup\{ |\phi(x)| : x \in \Delta \} > \|\phi\|_\infty - \epsilon$. There is a Borel subset Λ of Δ such that $0 < \mu(\Lambda) < \infty$ and $|\phi(x)| > \|\phi\|_\infty - \epsilon$ for all x in Λ. If $f = [\mu(\Lambda)]^{-1/2}\chi_\Lambda$, then f is a unit vector in $L^2(\mu)$ and $\|M_\phi f\|^2 = [\mu(\Lambda)]^{-1} \int_\Lambda |\phi|^2 \, d\mu \ge (\|\phi\|_\infty - \epsilon)^2$. This proves (c).

(d) Assume that $\lambda \notin$ ess-ran (ϕ). So there is a set Δ in Ω having full measure such that $\lambda \notin \text{cl}\,[\phi(\Delta)]$. Thus there is a $\delta > 0$ with $|\phi(x) - \lambda| > \delta$ for all x in Δ. Hence $\psi = (\phi - \lambda)^{-1} \in L^\infty(\mu)$. From (b) it follows that $M_\psi = (M_\phi - \lambda)^{-1}$.

Conversely, assume that $\lambda \in$ ess-ran (ϕ). Thus for every integer n there is a set Δ_n in Ω with $0 < \mu(\Delta_n) < \infty$ and $|\phi(x) - \lambda| < n^{-1}$ for all x in Δ_n. If $f_n = [\mu(\Delta_n)]^{-1/2}\chi_{\Delta_n}$, then $\{f_n\}$ is a sequence of unit vectors in $L^2(\mu)$ and $\| [M_\phi - \lambda] f_n \| \le 1/n$. Hence $\lambda \in \sigma(M_\phi)$. (Actually this shows that λ belongs to the approximate point spectrum of M_ϕ. But for a normal operator, the spectrum and approximate point spectrum coincide. A review of these notions is warranted.) ■

In the preceding proof, where is use made of the fact that the measure space is σ-finite?

2.7 Example. Continue with the notation as in the last proposition. If $\phi \in L^\infty(\mu)$ and $f \in C(\sigma(M_\phi))$, then $f(M_\phi) = M_{f \circ \phi}$. In fact, since $\sigma(M_\phi)$ is the essential range of ϕ, $f \circ \phi$ makes sense and is seen to also belong to $L^\infty(\mu)$. Thus $M_{f \circ \phi}$ is well defined. The reader is asked to check that the map $f \to M_{f \circ \phi}$ is a $*$-monomorphism. Clearly the function z is mapped to M_ϕ. Thus by the uniqueness of the functional calculus, $f(M_\phi) = M_{f \circ \phi}$.

2.8 Example. If X is a compact space and $g \in C(X)$, then $\sigma(g) = g(X)$ and the functional calculus for g is given by $f \to f \circ g$ for all f in $C(g(X))$.

These examples illustrate that computing the functional calculus is more a matter of making an educated guess and then verifying that the uniqueness criteria are satisfied for the guess. Also logic says that anything that must be proved about the functional calculus can be proved by using its determining characteristics. There may be an occasion when the exact form of the functional calculus is useful in executing a proof, but my experience is that usually this precise form just gets in the way of understanding.

The next, and final, result of the section is the answer to a fundamental question that should be addressed whenever there is a functional calculus. What is $\sigma(f(a))$?

2.9 Spectral Mapping Theorem. *If $\mathcal{A}$ is a C^*-algebra, a is a normal element in $\mathcal{A}$, and $f \in C(\sigma(a))$, then*

$$\sigma(f(a)) = f(\sigma(a)).$$

Proof. This is so simple it may confuse you. The map $f \to f(a)$ is a $*$-isomorphism from $C(\sigma(a))$ onto $C^*(a)$. Thus $\sigma(f(a)) = \sigma_{C(\sigma(a))}(f) = f(\sigma(a))$. ∎

Exercises.

1. Prove a converse to Theorem 2.3 by showing that when X is a compact subset of the plane, $C(X)$ is a C*-algebra with a single generator.

2. If $\mathcal{A}$ is an abelian C*-algebra with a finite number of generators $a_1, \ldots, a_n$, then there is a compact subset X of $\mathbb{C}^n$ and a $*$-isomorphism $\rho : \mathcal{A} \to C(X)$ such that $\rho(a_k) = z_k$, the k-th coordinate function.

3. If $\mathcal{A}$ is an abelian C*-algebra, show that $\mathcal{A}$ is separable if and only if its maximal ideal space is metrizable. (Theorem V.6.6 in [ACFA] has a more general result.)

4. A projection in a C*-algebra is a hermitian idempotent. If X is compact, show that X is totally disconnected if and only if $C(X)$, as a C*-algebra, is generated by its projections.

5. Using the preceding exercise, show that for a σ-finite measure space (X, Ω, μ), the maximal ideal space of $L^\infty(\mu)$ is totally disconnected.

6. If $\mathcal{A}$ is a unital C*-algebra and $a \in \operatorname{Re} \mathcal{A}$, show that $u = \exp(ia)$ is unitary. Is the converse true?

7. Assume that (X, Ω, μ) is a finite measure space and ϕ is a bounded measurable function. Using the notation of Proposition 2.6, show that the essential range of ϕ is the support of the measure $\mu \circ \phi^{-1}$.

8. Verify the statements in Example 2.8.

9. Let X be a compact space, $x_0 \in X$, and define $\mathcal{A} = \{\{f_n\} : f_n \in C(X), \sup_n \|f_n\| < \infty$, and $\{f_n(x_0)\}$ is a convergent sequence$\}$. If $\|\{f_n\}\|$ is defined as $\sup_n \|f_n\|$ and the operations on $\mathcal{A}$ are defined entrywise, show that $\mathcal{A}$ is an abelian C*-algebra and find its maximal ideal space.

10. For a completely regular space X, let $C_b(X)$ be the C*-algebra of bounded continuous functions from X into $\mathbb{C}$. Show that the maximal ideal space of $C_b(X)$ is homeomorphic to βX, the Stone–Čech compactification of X.

§3. The positive elements in a C*-algebra

In this section the functional calculus developed in the last section will be exploited to study a fundamental concept, the positive elements in a C*-algebra.

3.1 Definition. If $\mathcal{A}$ is a C*-algebra, an element a of $\mathcal{A}$ is *positive* if $a \in \operatorname{Re}\mathcal{A}$ and $\sigma(a) \subseteq \mathbb{R}_+$, the set of non-negative real numbers. This is denoted by $a \geq 0$ and $\mathcal{A}_+$ denotes the collection of all positive elements in $\mathcal{A}$. Say that an element a is *negative* if $-a \in \mathcal{A}_+$. Write this as $a \leq 0$ and let $\mathcal{A}_-$ be the collection of all negative elements in $\mathcal{A}$.

It is easy to see that f is a positive element of the C*-algebra $C(X)$ if and only if $f(x) \geq 0$ for all x in X. A function ϕ in $L^\infty(\mu)$ is positive if and only if $\phi(x) \geq 0$ a.e. $[\mu]$. Clearly the identity is a positive element of any C*-algebra.

Also $\mathcal{A}_+ \cap \mathcal{A}_- = (0)$. In fact, if $a \in \mathcal{A}_+ \cap \mathcal{A}_-$, then $\sigma(a) = \{0\}$. But a is hermitian so that $\|a\| = r(a) = 0$.

3.2 Proposition. *If $a \in \operatorname{Re}\mathcal{A}$, then there are unique positive elements u and v in $\mathcal{A}$ such that $a = u - v$ and $uv = vu = 0$.*

Proof. Let $f(t) = \max(t, 0)$ and $g(t) = -\min(t, 0)$. So $f, g \in C(\mathbb{R})_+$ and $f(t) - g(t) = t$. Using the functional calculus for a, let $u = f(a)$ and $v = g(a)$. From the Spectral Mapping Theorem it follows that $u, v \in \mathcal{A}_+$. Also $a = u - v$ and $uv = vu = 0$ since $fg = 0$ in $C(\sigma(a))$. This proves existence.

To prove uniqueness, assume $u_1, v_1 \in \mathcal{A}_+$ satisfy $a = u_1 - v_1$ and $u_1v_1 = v_1u_1 = 0$. Note that this implies that u_1 and v_1 commute with a. Let $\{p_n\}$ be a sequence of polynomials such that $p_n(0) = 0$ and $p_n(t) \to f(t)$ uniformly on the spectrum of a. So $p_n(a) \to f(a) = u$ in $\mathcal{A}$. Because $u_1a = au_1$, $u_1p_n(a) = p_n(a)u_1$ for all n, so that $u_1u = uu_1$. Similar reasoning shows that the elements a, u, v, u_1, v_1 are pairwise commuting hermitian elements of $\mathcal{A}$. let $\mathcal{B} = C^*(a, u, v, u_1, v_1)$. So $\mathcal{B}$ is abelian and, therefore, $\mathcal{B} \cong C(X)$ for some compact space X. Thus the proof of uniqueness for the general C*-algebra will be proved if it is proved for $C(X)$.

So assume that a, u, v are continuous functions on X having the relations listed above. It is left to the reader to show that for any x in X, $u(x) = \max\{a(x), 0\}$ and is, therefore, unique. Similarly, $v(x) = -\min\{a(x), 0\}$. ∎

The elements u and v in the preceding proposition are called the *positive part* and the *negative part* of the hermitian element a and are denoted by $u = a_+$ and $v = a_-$.

The proof of the next result is similar and is left to the reader as an exercise.

3.3 Proposition. *If $a \in \mathcal{A}_+$ and $n \in \mathbb{N}$, there is a unique b in $\mathcal{A}_+$ such that $a = b^n$.*

The positive element b in this proposition is called the *n-th root* of the positive element a and is denoted by $b = a^{1/n}$.

Now for some alternate characterizations of positive elements of a C*-algebra. Some of these are useful for verifying that examples are positive and others are only useful at a theoretical level. Observe that if $f \in C(X)_+$, then $|f(x) - t| \leq t$ for all real numbers $t \geq \|f\|$. Conversely if $f \in C(X)$ and $|f(x) - t| \leq t$ for some $t \geq \|f\|$, then $f(x) \geq 0$ for all x in X. These observations will be used in the next proof.

3.4 Theorem. *If $\mathcal{A}$ is a C^*-algebra, the following statements are equivalent.*

(a) $a \geq 0$.

(b) $a = b^2$ *for some* b *in* $\mathcal{A}_+$.

(c) $a = x^*x$ *for some* x *in* $\mathcal{A}$.

(d) $a \in \operatorname{Re}\mathcal{A}$ *and* $\|t - a\| \leq t$ *for all* $t \geq \|a\|$.

(e) $a \in \operatorname{Re}\mathcal{A}$ *and* $\|t - a\| \leq t$ *for some* $t \geq \|a\|$.

Proof. Clearly (b) implies (c), and (d) implies (e). Proposition 3.3 says that (a) implies (b).

(e) *implies* (a). Since a is hermitian, $C^*(a)$ is abelian. Let $X = \sigma(a)$, so that $X \subseteq \mathbb{R}$. Transferring the inequality in (e) to $C(X)$, there is a $t \geq \|a\| = \sup|X|$ such that $|t - x| \leq t$ for all x in X. As was pointed out before the statement of the theorem, this implies $X \subseteq \mathbb{R}_+$.

(a) *implies* (d). This follows along the same line as the proof that (e) implies (a) and is left to the reader.

Note that at this point, conditions (a), (d), and (e) are equivalent. This is used to prove the next proposition, which, in turn, is used to establish the remaining implication. No slight of hand here, just an economy of expression.

(c) *implies* (a). The condition that $a = x^*x$ for some x in $\mathcal{A}$ implies that $a \in \operatorname{Re}\mathcal{A}$. Let $a = u - v$ as in Proposition 3.2. It must be shown that $v = 0$.

Put $xv^{1/2} = b + ic$, $b, c \in \operatorname{Re}\mathcal{A}$. So $(xv^{1/2})^*(xv^{1/2}) = (b - ic)(b + ic) = b^2 + c^2 + i(bc - cb)$. But performing the computation the other way shows that $(xv^{1/2})^*(xv^{1/2}) = v^{1/2}x^*xv^{1/2} = v^{1/2}(u - v)v^{1/2} = -v^2$, since $uv = vu = 0$. Hence $i(bc - cb) = -v^2 - b^2 - c^2$. By the next proposition, $i(bc - cb) \leq 0$.

Also $(xv^{1/2})^*(xv^{1/2}) = -v^2 \leq 0$ by the Spectral Mapping Theorem. But the spectrum of $(xv^{1/2})^*(xv^{1/2})$ and $(xv^{1/2})(xv^{1/2})^*$ differ by at most 0 (Exercise VII.3.7 in [ACFA]). Thus $(xv^{1/2})(xv^{1/2})^* \leq 0$. Put $(xv^{1/2})(xv^{1/2})^* = -y$, where $y \in \mathcal{A}_+$. Hence $-y = (b + ic)(b - ic) = b^2 + c^2 - i(bc - cb)$, so that $i(bc - cb) = y + b^2 + c^2 \geq 0$ by the next proposition. Therefore $i(bc - cb) \in \mathcal{A}_+ \cap \mathcal{A}_- = (0)$. But this implies that $-v^2 = (xv^{1/2})^*(xv^{1/2}) = b^2 + c^2 \in \mathcal{A}_+ \cap \mathcal{A}_- = (0)$, so that $v^2 = 0$. Since $v^2 \geq 0$, the uniqueness of the square root says that $v = 0$. ■

3.5 Proposition. *$\mathcal{A}_+$ is a closed cone in $\mathcal{A}$.*

Proof. First show that $\mathcal{A}_+$ is closed. Let $\{a_n\}$ be a sequence of positive elements and suppose $\|a_n - a\| \to 0$. Clearly $a \in \operatorname{Re}\mathcal{A}$. By condition (d) in the preceding theorem, $\|a_n - \|a_n\|\,\| \leq \|a_n\|$ for all $n \geq 1$. Taking limits shows that $\|a - \|a\|a\,\| \leq \|a\|$. By (3.4.e), $a \geq 0$.

From the definition, $\alpha a \geq 0$ when $a \in \mathcal{A}_+$ and $\alpha \in \mathbb{R}_+$. Now let $a, b \in \mathcal{A}_+$. To show that $a + b \in \mathcal{A}_+$, it suffices to also assume that $\|a\|, \|b\| \leq 1$. Applying (3.4.d) shows that $\|1 - \frac{1}{2}(a + b)\| = \frac{1}{2}\|(1 - a) + (1 - b)\| \leq 1$. By (3.4.e), $\frac{1}{2}(a + b) \geq 0$, so that $a + b \in \mathcal{A}_+$. ■

Because $\mathcal{A}_+$ is a cone, $\operatorname{Re}\mathcal{A}$ is a real ordered vector space. If $a, b \in \operatorname{Re}\mathcal{A}$, say that $a \leq b$ if $b - a \geq 0$. This has a pleasing aspect in that the theory of ordered spaces can be applied. But usually it is easier to just prove what is needed about the ordering directly rather than appealing to the theory of ordered spaces. It is useful, especially when making certain estimates, to note that if $a \in \operatorname{Re}\mathcal{A}$, then $-\|a\| \leq a \leq \|a\|$. In particular, if a is any element of $\mathcal{A}$, then $0 \leq a^*a \leq \|a\|^2$.

The remainder of the section is devoted to an examination of positivity for the C*-algebra $\mathcal{B}(\mathcal{H})$ and to giving an important application to operator theory.

3.6 Proposition. *If $\mathcal{H}$ is a Hilbert space and $A \in \mathcal{B}(\mathcal{H})$, then A is positive if and only if $\langle Ah, h\rangle \geq 0$ for every vector h.*

Proof. If $A \geq 0$, then Theorem 3.4.c implies there is a T in $\mathcal{B}(\mathcal{H})$ such that $A = T^*T$. Thus $\langle Ah, h\rangle = \|Th\|^2 \geq 0$. Conversely, assume that $\langle Ah, h\rangle \geq 0$ for all h in $\mathcal{H}$. By Proposition II.2.12 of [ACFA], $A = A^*$. If $\lambda < 0$ and $h \in \mathcal{H}$, the fact that $\langle Ah, h\rangle \geq 0$ implies

$$\begin{aligned}\|(A-\lambda)h\|^2 &= \|Ah\|^2 - 2\lambda\langle Ah, h\rangle + \lambda^2\|h\|^2 \\ &\geq -2\lambda\langle Ah, h\rangle + \lambda^2\|h\|^2 \\ &\geq \lambda^2\|h\|^2.\end{aligned}$$

Thus $\lambda \notin \sigma_{ap}(A)$, the approximate point spectrum of A. This says that $A - \lambda$ is left invertible. (See Exercise VII.6.5 of [ACFA].) Since $A - \lambda$ is hermitian, it must also be right invertible. That is, $\lambda \notin \sigma(A)$ and so $A \geq 0$. ■

3.7 Definition. For any element a in a C*-algebra, define the *absolute value* of a by $|a| = (a^*a)^{1/2}$.

If a is hermitian, then it is not difficult to show that $|a| = a_+ + a_-$. See Exercise 4.

3.8 Definition. A *partial isometry* is an operator W on a Hilbert space $\mathcal{H}$ such that $\|Wh\| = \|h\|$ for all vectors h in $(\ker W)^\perp$. The space $(\ker W)^\perp$ is called the *initial space* of W and $\operatorname{ran} W$ is called the *final space*.

The exercises at the end of this section contain a lot more information about partial isometries. Later, when the discussion turns to von Neumann algebras (Chapter 7), they will play a crucial role.

3.9 Polar Decomposition. *If $A \in \mathcal{B}(\mathcal{H})$, then there is a partial isometry W with initial space $(\ker A)^\perp$ and final space $\operatorname{cl}(\operatorname{ran} A)$ such that $A = W|A|$. Moreover, if $A = UP$, where $P \geq 0$ and U is a partial isometry with $\ker U = \ker P$, then $P = |A|$ and $U = W$.*

Proof. If $h \in \mathcal{H}$, then $\|Ah\|^2 = \langle A^*Ah, h\rangle = \langle |A|h, |A|h\rangle$. Thus

$$\|Ah\| = \|\,|A|h\,\|$$

for all vectors h. Thus permits the definition of $W : \operatorname{ran}|A| \to \operatorname{ran} A$ by

$$W(|A|h) = Ah.$$

The preceding equation shows that W is an isometry on $\operatorname{ran} A$ and, as such, can be extended to map $\operatorname{cl}[\operatorname{ran}|A|]$ onto $\operatorname{cl}[\operatorname{ran} A]$. Extend W to all of $\mathcal{H}$ by letting it be 0 on $(\operatorname{ran}|A|)^\perp$. This makes W a partial isometry and, clearly $W|A| = A$. It remains to check that it has the correct initial space.

That is, it is necessary to show that $\text{cl}\,[\text{ran}\,|A|] = (\ker A)^{\perp}$. If $f \in \text{ran}\, A^*$, then $f = A^*g$ for some g in $(\ker A^*)^{\perp} = \text{cl}\,[\text{ran}\, A]$. Therefore $\text{ran}\, A^*A$ is dense in $\text{cl}\,[\text{ran}\, A^*] = (\ker A)^{\perp}$. But $A^*Ak = |A|^2k = |A|h$, where $h = |A|k$. Thus $\text{ran}\,|A|$ is dense $(\ker A)^{\perp}$ as asserted.

To show uniqueness, note that $A^*A = PU^*UP$. But $U^*U = E$, the projection onto the initial space of U, $(\ker U)^{\perp} = (\ker P)^{\perp} = \text{cl}\,[\text{ran}\, P]$. (See Exercise 10.) Thus $A^*A = P^2$, so that $P = |A|$ by the uniqueness part of Proposition 3.3. If $h \in \mathcal{H}$, $W|A|h = Ah = U|A|h$. That is, U and W agree on a dense manifold in their common initial space. Hence $U = W$. ∎

When reference is made to *the* polar decomposition of an operator, this means the unique decomposition obtained in the preceding theorem.

Exercises.

1. Using the notation of Proposition 2.6 give necessary and sufficient conditions that M_ϕ be positive. If $M_\phi \geq 0$, find ${M_\phi}^{1/n}$. If M_ϕ is hermitian, find its positive and negative parts.

2. Find an example of a positive operator that has a non-hermitian square root. Can you find one on a 2-dimensional space?

3. If $\{\mathcal{A}_i : i \in I\}$ is a collection of C*-algebras and $\mathcal{A} = \bigoplus_i \mathcal{A}_i$ as in Exercise 1.4, show that an element $\{a_i\}$ in $\mathcal{A}$ is positive if and only if $a_i \geq 0$ for all i.

4. Show that if $a \in \text{Re}\,\mathcal{A}$, then $|a| = a_+ + a_-$.

5. Show that each element a in $\text{ball}\,\text{Re}\,\mathcal{A}$ is the sum of two unitaries in $\mathcal{A}$. (Hint: First show this when $\mathcal{A} = \mathbb{C}$.)

6. If $a \in \mathcal{A}_+$, show that a has a positive logarithm if and only if a is invertible. (a has a logarithm if there is an element b in $\mathcal{A}$ such that $e^b = a$.) Show that an invertible hermitian element has a logarithm. Is this logarithm hermitian?

7. Give an example of a C*-algebra $\mathcal{A}$ and two positive elements a and b in $\mathcal{A}$ such that ab is not positive. If $ab = ba$, show that $ab \in \mathcal{A}_+$.

8. Show that for a hermitian operator, the spectrum and the approximate spectrum are the same.

9. Show that the following statements are equivalent for an operator W in $\mathcal{B}(\mathcal{H})$. (a) W is a partial isometry. (b) W^* is a partial isometry. (c) W^*W is a projection. (d) WW^* is a projection. (e) $WW^*W = W$. (f) $W^*WW^* = W^*$.

10. If W is a partial isometry, show that W^*W is the projection onto the initial space of W and WW^* is the projection onto the final space of W.

11. For partial isometries W_1 and W_2, define $W_1 \lesssim W_2$ to mean that $W_1^* W_1 \le W_2^* W_2$, $W_1 W_1^* \le W_2 W_2^*$, and $W_1 h = W_2 h$ for all vectors h in the initial space of W_1. Show that $\lesssim$ is a partial ordering on the set of partial isometries and that a partial isometry is a maximal element relative to this ordering if and only if either W or W^* is an isometry.

12. Using the terminology of the preceding exercise, show that the extreme points of ball $\mathcal{B}(\mathcal{H})$ are the maximal partial isometries.

13. Find the polar decomposition of the operator M_ϕ from Proposition 2.6.

14. If the operator A has polar decomposition $A = W|A|$ and $\alpha \in \mathbb{C}$, find the polar decomposition of $A \oplus \alpha$.

15. For a Hilbert space $\mathcal{H}$, define $\mathcal{K} = \mathcal{H} \oplus \mathcal{H} \cdots$ and define $S : \mathcal{K} \to \mathcal{K}$ by $S(h_1, h_2, \dots) = (0, h_1, h_2, \dots)$. What is the polar decomposition of S? If A is a positive operator on $\mathcal{H}$ and T is defined on $\mathcal{K}$ by $T(h_1, h_2, \dots) = (0, Ah_1, Ah_2, \dots)$, find the polar decomposition of T.

16. Define $T : \mathbb{C}^n \to \mathbb{C}^n$ by $Te_k = e_{k+1}$ for $1 \le k \le n-1$ and $Te_n = 0$, where $\{e_1, \dots, e_n\}$ is the standard basis for $\mathbb{C}^n$. Find the polar decomposition of T.

17. Show that the parts of the polar decomposition of a normal operator commute.

18. If $A \in \mathcal{B}(\mathcal{H})$, show that there is a positive operator P and a partial isometry W such that $A = PW$. Discuss the uniqueness of P and W.

19. If A is an invertible normal operator, show that the parts of the polar decomposition of A belong to $C^*(A)$. Give an example of a normal operator A for which the parts of the polar decomposition do not belong to $C^*(A)$.

§4. Approximate identities

Most of the C*-algebras encountered in this book will have an identity. There are many notable exceptions, the most prominent of which is the C*-algebras of compact operators examined in detail in §16 below. But it is necessary to consider ideals in a C*-algebra, and an ideal constitutes a C*-algebra without identity. (See §5.) Of course, as seen, it is always possible to adjoin an identity, and this helps in many cases. In this section approximate identities are examined, and this often provides a better remedy than adjoining an identity.

Let $\mathcal{A}$ be a C*-algebra and, in this section, assume that $\mathcal{A}$ does not have an identity. Let $\mathcal{A}_1$ be the C*-algebra obtained by adjoining an identity to

$\mathcal{A}$. A consideration of $\mathcal{A}_1 = \mathcal{A} + \mathbb{C}$ is not required for what is done below, but it is a notational convenience.

4.1 Definition. An *approximate identity* for $\mathcal{A}$ is a net $\{e_i\}$ in $\mathcal{A}$ such that: (i) $0 \leq e_i \leq 1$ for all i; (ii) $e_i \leq e_j$ when $i \leq j$; (iii) $\lim_i \|xe_i - x\| = 0$ for all x in $\mathcal{A}$. If the net is a sequence, call it a *sequential approximate identity.*

First note that by taking adjoints, condition (iii) in the definition implies that the symmetric condition $\lim_i \|e_i x - x\| = 0$ also holds for all x in $\mathcal{A}$. A second fact to note is that if a net $\{ e_i \}$ satisfies conditions (i) and (ii), then, due to its boundedness, to prove it is an approximate identity it suffices to show that it satisfies (iii) for x belonging to a total subset of $\mathcal{A}$.

If $\mathcal{H}$ is separable and $\{P_n\}$ is an increasing sequence of finite rank projections that increases to the identity, then $\{P_n\}$ is a sequential approximate identity for $\mathcal{B}_0(\mathcal{H})$. (Note that for a non-separable Hilbert space, $\mathcal{B}_0(\mathcal{H})$ cannot have a sequential approximate identity.)

Start with a simple lemma. (It might be mentioned that the displayed inequalities in this lemma illustrate the convenience of having an identity lurking in the background.)

4.2 Lemma. *If $\mathcal{A}$ is a C^*-algebra and $a, b \in \mathcal{A}$ with $0 \leq a \leq b \leq 1$, then*

$$\|x - bx\|^2 \leq \|x^*(1 - a)x\|$$

and

$$\|x - xb\|^2 \leq \|x^*(1 - a)x\|$$

for all x in $\mathcal{A}$.

Proof. If $x \in \mathcal{A}$, $\|x - bx\|^2 = \|(1 - b)x\|^2 = \|x^*(1 - b)^2 x\| \leq \|x^*(1 - b)x\| \leq \|x^*(1 - a)x\|$. The other inequality is proved similarly. ∎

For certain technical reasons, it is helpful to prove the main result of this section for a dense ideal of the C*-algebra $\mathcal{A}$.

4.3 Theorem. *Let $\mathcal{I}$ be a dense ideal of the C^*-algebra $\mathcal{A}$. If $\mathcal{E} = \{ e \in \mathcal{I} : 0 \leq e < 1 \}$ has the natural ordering, then $\mathcal{E}$ is a directed set and, when considered as a net, is an approximate identity for $\mathcal{A}$.*

Proof. The proof begins by establishing that $\mathcal{E}$ is directed. Let $a, b \in \mathcal{E}$ and define the functions ϕ and ψ by $\phi(t) = t/(1 - t)$ for $0 \leq t < 1$ and $\psi(t) = t/(1 + t) = 1 - (1 + t)^{-1}$ for $t \geq 0$. Put $c = \phi(a) + \phi(b)$ and $d = \psi(c)$. Now for any e in $\mathcal{E}$, $\phi(e) = e(1 - e)^{-1} \in \mathcal{I}_+$. Thus $c \in \mathcal{I}_+$. Similarly, $d = c(1 + c)^{-1} \in \mathcal{I}$. Because $0 \leq \psi(t) < 1$ for all positive t, $d \in \mathcal{E}$. Now $x = \phi(a) \leq c$, so $(1 + x)^{-1} \geq (1 + c)^{-1}$. Since $\psi(\phi(t)) = t$, this implies that

$a = \psi(x) = 1 - (1+x)^{-1} \leq \psi(c) = d$. Similarly, $b \leq d$. Thus d is the sought for element of $\mathcal{E}$ that dominates both a and b, and $\mathcal{E}$ is directed.

If $x \in \mathcal{I}_+$, let $a_n = \psi(nx)$; so $a_n \in \mathcal{E}$. Define $\eta(t) = t^2(1 - \psi(nt)) = t^2(1+nt)^{-1} \leq t/n$. Thus $\eta(x) = x^2(1 - \psi(nx)) = x(1 - a_n)x$ and so

$$\|x(1 - a_n)x\| \leq \|\eta\|_{\sigma(x)} \leq \|x\|/n.$$

If $\epsilon > 0$ and n is chosen larger than $\epsilon^{-1}\|x\|$, the preceding lemma implies that for e in $\mathcal{E}$ and $e \geq a_n$, $\|x - xe\|^2 \leq \epsilon$. That is

$$\lim_{e \in \mathcal{E}} \|xe - x\| = 0$$

whenever $x \in \mathcal{I}_+$. If $x \in \mathcal{A}_+$, let $\{y_n\}$ be a sequence in $\mathcal{I}$ such that $\|y_n - x^{1/2}\| \to 0$. Thus $x_n = y_n^* y_n \to x$ and $x_n \in \mathcal{I}_+$. Since $\mathcal{E}$ is bounded, the above limit holds for every x in $\mathcal{A}_+$. But each element of $\mathcal{A}$ can be written as a linear combination of four elements in $\mathcal{A}_+$, so this limit is 0 for all x in $\mathcal{A}$. ■

4.4 Corollary. *A separable C^*-algebra has a sequential approximate identity.*

Proof. Let $\{a_n\}$ be a countable dense subset of $\mathcal{A}$ and let $\{e_i\}$ be an approximate identity. A simple induction argument shows that there are $i_1 \leq i_2 \leq \cdots$ such that $\|a_k e_i - a_k\| < 2^{-n}$ for $1 \leq k \leq n$ and $i \geq i_n$. Letting $x_n = e_{i_n}$, it follows that for all $k \geq 1$, $\|a_k x_n - a_k\| \to 0$ as $n \to \infty$. Since $\{x_n\}$ is bounded, $\|a x_n - a\| \to 0$ for all a in $\mathcal{A}$. ■

In general, there is no polar decomposition in a C*-algebra. For example, if $\mathcal{A} = \mathcal{B}_0(\mathcal{H}) + \mathbb{C}$, every projection in $\mathcal{A}$ has finite dimensional range or kernel. It is left to the reader to find a compact operator K that cannot be expressed as $K = W|K|$, where W is a partial isometry in $\mathcal{A}$. What happens if the requirement that one of the factors be a partial isometry is relaxed? In this case such a factorization is possible using the techniques developed in this section.

4.5 Lemma. *Let α and β be positive scalars with $\alpha + \beta > 1$. If $\mathcal{A}$ is a C^*-algebra, $x, y \in \mathcal{A}$, $a \in \mathcal{A}_+$ such that $x^*x \leq a^\alpha$ and $y^*y \leq a^\beta$, and $u_n = x(n^{-1} + a)^{-1/2} y$, then $\{u_n\}$ converges to an element u in $\mathcal{A}$ with $\|u\| \leq \|a^{\frac{1}{2}(\alpha+\beta-1)}\|$.*

Proof. If $d_{nm} = (n^{-1} + a)^{-1/2} - (m^{-1} + a)^{-1/2}$, then

$$\begin{aligned}\|u_n - u_m\|^2 &= \|xd_{nm}y\|^2 = \|y^* d_{nm} x^* x d_{nm} y\| \le \|y^* d_{nm} a^\alpha d_{nm} y\| \\ &= \|a^{\frac{\alpha}{2}} d_{nm} y\|^2 = \|a^{\frac{\alpha}{2}} d_{nm} y y^* d_{nm} a^{\frac{\alpha}{2}}\| \\ &\le \|a^{\frac{\alpha}{2}} d_{nm} a^\beta d_{nm} a^{\frac{\alpha}{2}}\| = \|d_{nm}^2 a^{\alpha+\beta}\| = \|d_{nm} a^{\frac{\alpha+\beta}{2}}\|^2.\end{aligned}$$

But $\{(n^{-1} + t)^{-1/2} t^{\frac{\alpha+\beta}{2}}\}$ is an increasing sequence of functions on the positive real axis. Thus $\{(n^{-1} + a)^{-1/2} a^{\frac{\alpha+\beta}{2}}\}$ is an increasing sequence in $\mathcal{A}_+$. Applying Dini's Theorem from real variables shows that

$$(n^{-1} + a)^{-1/2} a^{\frac{1}{2}(\alpha+\beta)} \to a^{\frac{1}{2}(\alpha+\beta-1)}$$

in $\mathcal{A}$. Applying this to the above inequalities it follows that $\{u_n\}$ is a Cauchy sequence in $\mathcal{A}$. Let $u = \lim_n u_n \in \mathcal{A}$. Using the same techniques as above gives that $\|u_n\| = \|x(n^{-1} + a)^{-1/2} y\| \le \|(n^{-1} + a)^{-1/2} a^{\frac{1}{2}(\alpha+\beta)}\| \nearrow \|a^{\frac{1}{2}(\alpha+\beta-1)}\|$. ■

4.6 Proposition. *If $a \in \mathcal{A}_+$, $x \in \mathcal{A}$ such that $x^* x \le a$, and $0 < \alpha < \frac{1}{2}$, then there is an element u in $\mathcal{A}$ with $\|u\| \le \|a^{\frac{1}{2}-\alpha}\|$ and $x = ua^\alpha$.*

Proof. By the preceding lemma, $u_n \equiv x(n^{-1} + a)^{-1/2} a^{\frac{1}{2}-\alpha} \to u$ in $\mathcal{A}$ and $\|u\| \le \|a^{\frac{1}{2}(1+1-2\alpha-1)}\| = \|a^{\frac{1}{2}-\alpha}\|$. Since $u_n a^\alpha = x(n^{-1} + a)^{-1} a^{1/2}$,

$$\begin{aligned}\|x - u_n a^\alpha\|^2 &= \left\| x\left[1 - (n^{-1} + a)^{-1/2} a^{1/2}\right]\right\|^2 \\ &= \left\|\left[1 - (n^{-1} + a)^{-1/2}\right] x^* x \left[1 - (n^{-1} + a)^{-1/2}\right]\right\| \\ &\le \left\|\left[1 - (n^{-1} + a)^{-1/2}\right] a \left[1 - (n^{-1} + a)^{-1/2}\right]\right\| \\ &= \left\|a^{1/2}\left[1 - (n^{-1} + a)^{-1/2}\right]\right\|^2 \\ &= \left\|a^{1/2} - (n^{-1} + a)^{-1/2} a^{1/2}\right\|^2.\end{aligned}$$

Again Dini's Theorem implies that $u_n a^\alpha \to x$ so that $x = ua^\alpha$. ■

4.7 Corollary. *If $\mathcal{A}$ is a unital C^*-algebra, $x \in \mathcal{A}$, and $0 < \beta < 1$, then there is a u in $\mathcal{A}$ such that*

$$x = u|x|^\beta.$$

Proof. Apply the preceding proposition with $a = x^* x$ and $\alpha = \beta/2$. ■

Exercises.

1. If X is a locally compact space, find an approximate identity for $C_0(X)$. What is a necessary and sufficient condition that $C_0(X)$ have a sequential approximate identity?

2. For any Hilbert space $\mathcal{H}$, if $\mathcal{F}$ is the collection of finite rank projections ordered by inclusion, show that $\mathcal{F}$ is an approximate identity for $\mathcal{B}_0(\mathcal{H})$.

3. Let $\{E_i\}$ be an approximate identity for $\mathcal{B}_0(\mathcal{H})$. Show that for every operator A in $\mathcal{B}(\mathcal{H})$ and each vector h in $\mathcal{H}$, $\lim_i \|AE_ih - Ah\| = \lim_i \|E_iAh - Ah\| = 0$. If $A \geq 0$, show that $\lim_i \|A^{1/2}E_iA^{1/2}h - Ah\| = 0$ for all h in $\mathcal{H}$.

§5. Ideals in a C*-algebra

Start by examining the behavior of the functional calculus relative to a closed one-sided ideal. $\mathcal{A}$ will always denote a C*-algebra in this section.

5.1 Proposition. *If $\mathcal{I}$ is a closed left or right ideal of $\mathcal{A}$, $a \in \mathcal{I}$ with $a = a^*$, and $f \in C(\sigma(a))$ such that $f(0) = 0$, then $f(a) \in \mathcal{I}$. Consequently, for a hermitian element a of $\mathcal{I}$, $a_+, a_-, |a|$, and $|a|^{1/2} \in \mathcal{I}$.*

Proof. Note that if $\mathcal{I}$ is proper, $0 \in \sigma(a)$. Since $\sigma(a) \subseteq \mathbb{R}$, if $f \in C(\sigma(a))$ and $f(0) = 0$, there is a sequence of polynomials p_n such that $p_n(0) = 0$ and $p_n(t) \to f(t)$ uniformly in $\sigma(a)$. Clearly $p_n(a) \in \mathcal{I}$ for each $n \geq 1$, so $f(a) \in \mathcal{I}$. ■

5.2 Definition. If $\mathcal{A}$ is a C*-algebra, then a *hereditary subalgebra* of $\mathcal{A}$ is a C*-subalgebra $\mathcal{B}$ of $\mathcal{A}$ such that if $a \in \mathcal{B}_+$ and $x \in \mathcal{A}$ with $0 \leq x \leq a$, then $x \in \mathcal{B}$.

The compact operators form a hereditary subalgebra of $\mathcal{B}(\mathcal{H})$. If P is a projection on $\mathcal{H}$ and $\mathcal{B} = \{T \in \mathcal{B}(\mathcal{H}) : T = PT = TP\}$, then $\mathcal{B}$ is a hereditary subalgebra of $\mathcal{B}(\mathcal{H})$. The next proposition defines an order preserving bijection between the collection of all hereditary subalgebras of $\mathcal{A}$ and the collection of closed left ideals.

5.3 Proposition. *Let $\mathcal{A}$ be a C*-algebra.*

(a) *If $\mathcal{L}$ is a closed left ideal of $\mathcal{A}$ and $\mathcal{B} = \mathcal{L} \cap \mathcal{L}^*$, then $\mathcal{B}$ is a hereditary subalgebra of $\mathcal{A}$.*

(b) *If $\mathcal{B}$ is a hereditary subalgebra of $\mathcal{A}$ and $\mathcal{L} = \{x \in \mathcal{A} : x^*x \in \mathcal{B}\}$, then $\mathcal{L}$ is a closed left ideal of $\mathcal{A}$.*

(c) *If $\mathcal{L}$ is a closed left ideal of $\mathcal{A}$ and $\mathcal{B} = \mathcal{L} \cap \mathcal{L}^*$, then $\mathcal{L} = \{x : x^*x \in \mathcal{B}\}$.*

(d) *If $\mathcal{B}$ is a hereditary subalgebra of $\mathcal{A}$ and $\mathcal{L} = \{x : x^*x \in \mathcal{B}\}$, then $\mathcal{B} = \mathcal{L} \cap \mathcal{L}^*$.*

Proof. (a) Clearly $\mathcal{B}$ is a C*-algebra. Let $a \in \mathcal{B}_+$ and suppose $x \in \mathcal{A}$ with $0 \leq x \leq a$. So $x^{1/2}x^{1/2} \leq a$ and Proposition 4.6 implies there is an element u in $\mathcal{A}$ such that $x^{1/2} = ua^{1/3}$. But $a^{1/3} \in \mathcal{B}_+ \subseteq \mathcal{L}$. Thus $x^{1/2} \in \mathcal{L}$ from which it follows that $x \in \mathcal{L}$. Since x is hermitian, $x \in \mathcal{B}$ and $\mathcal{B}$ is hereditary.

(b) If $x \in \mathcal{L}$ and $a \in \mathcal{A}$, then $(ax)^*(ax) = x^*a^*ax \leq \|a\|^2x^*x \in \mathcal{B}$. Hence $ax \in \mathcal{L}$ whenever $x \in \mathcal{L}$. If $x, y \in \mathcal{L}$, then $(x+y)^*(x+y) \leq (x+y)^*(x+y) + (x-y)^*(x-y) = 2(x^*x + y^*y) \in \mathcal{B}$. Thus $\mathcal{L}$ is a left ideal. It is easy to see that $\mathcal{L}$ is closed.

(c) If $x \in \mathcal{L}$, then $x^*x \in \mathcal{L}$ since $\mathcal{L}$ is a left ideal. Thus $x^*x \in \mathcal{B}$. On the other hand, if x is an arbitrary element of $\mathcal{A}$ and $x^*x \in \mathcal{B}$, then $x^*x \in \mathcal{L}$ and so $|x|^{1/2} \in \mathcal{L}$ by Proposition 5.1. According to Proposition 4.6 there is an element u in $\mathcal{A}$ such that $x = u|x|^{1/2}$. Therefore $x \in \mathcal{L}$.

(d) To show that $\mathcal{B} = \mathcal{L} \cap \mathcal{L}^*$, Proposition 3.2 implies it suffices to show that $\mathcal{B}_+ = (\mathcal{L} \cap \mathcal{L}^*)_+ = \mathcal{L}_+$. If $x \in \mathcal{L}_+$, then $x^2 \in \mathcal{B}$ by the definition of $\mathcal{L}$, so $x = (x^2)^{1/2} \in \mathcal{B}_+$. Now assume that $x \in \mathcal{B}_+$. So $(x^{1/2})^*(x^{1/2}) = x \in \mathcal{B}$, and, by definition, this implies that $x^{1/2} \in \mathcal{L}$. Thus $x \in \mathcal{L}$. ∎

An ideal of $\mathcal{A}$ always means a two-sided ideal, though it is not always assumed that ideals are closed. The next result says that closed ideals constitute C*-algebras. This is an easy consequence of Corollary 4.7.

5.4 Proposition. *If $\mathcal{I}$ is a closed ideal of $\mathcal{A}$, then $a^* \in \mathcal{I}$ whenever $a \in \mathcal{I}$.*

Proof. If $x \in \mathcal{I}$, then Corollary 4.7 implies there is a u in $\mathcal{A}$ such that $x = u|x|^{1/2}$. But $|x|^{1/2} \in \mathcal{I}$ by Proposition 5.1. Thus $x^* = |x|^{1/2}u^* \in \mathcal{I}$. ∎

Now for one of the standard rituals of mathematics. Whenever presented with an algebra and an ideal, form the quotient. This must be carried out in the C*-algebra setting.

5.5 Lemma. *If $\mathcal{I}$ is a closed ideal in $\mathcal{A}$ and $a \in \mathcal{A}$, then $\|a + \mathcal{I}\| = \inf\{\, \|a - ax\| : x \in (\operatorname{ball}\mathcal{I})_+ \,\}$.*

Proof. If α denotes the infimum, clearly $\|a + \mathcal{I}\| \leq \alpha$ since $a\mathcal{I} \subseteq \mathcal{I}$. Let $\{e_i\}$ be an approximate identity for $\mathcal{I}$. Since $0 \leq 1 - e_i \leq 1$, if $y \in \mathcal{I}$, then $\|(a+y)(1-e_i)\| \leq \|a+y\|$. Thus

$$\begin{aligned}\|a+y\| &\geq \liminf_i \|(a+y)(1-e_i)\| \\ &= \liminf_i \|(a - ae_i) + (y - ye_i)\| \\ &= \liminf_i \|a - ae_i\|\end{aligned}$$

since $\|y - ye_i\| \to 0$. Therefore $\|a + I\| \geq \alpha$. ∎

5.6 Theorem. *Let* $\mathcal{A}$ *be a* C^**-algebra and let* $\mathcal{I}$ *be a closed ideal of* $\mathcal{A}$. *For* a *in* $\mathcal{A}$ *define* $(a+\mathcal{I})^* = a^* + \mathcal{I}$. *With this definition and the quotient norm,* $\mathcal{A}/\mathcal{I}$ *is a* C^**-algebra.*

Proof. From the general theory of Banach algebras, it is known that $\mathcal{A}/\mathcal{I}$ is a Banach algebra. It must be shown that $\mathcal{A}/\mathcal{I}$ satisfies the C* identity. That is, show that $\|a+\mathcal{I}\|^2 = \|a^*a+\mathcal{I}\|$ for all a in $\mathcal{A}$. By Proposition 5.4, $\|a^*+\mathcal{I}\| = \|a+\mathcal{I}\|$ for all a in $\mathcal{A}$. Hence,

$$\begin{aligned} \|a^*a+\mathcal{I}\| &= \|(a^*+\mathcal{I})(a+\mathcal{I})\| \\ &\le \|a^*+\mathcal{I}\|\,\|a+\mathcal{I}\| \\ &= \|a+\mathcal{I}\|^2. \end{aligned}$$

On the other hand, the preceding lemma implies that

$$\begin{aligned} \|a+\mathcal{I}\|^2 &= \inf\{\, \|a-ax\|^2 : x \in (\operatorname{ball}\mathcal{I})_+ \,\} \\ &= \inf\{\, \|a(1-x)\|^2 : x \in (\operatorname{ball}\mathcal{I})_+ \,\} \\ &= \inf\{\, \|(1-x)a^*a(1-x)\| : x \in (\operatorname{ball}\mathcal{I})_+ \,\} \\ &\le \inf\{\, \|a^*a(1-x)\| : x \in (\operatorname{ball}\mathcal{I})_+ \,\} \\ &= \inf\{\, \|a^*a-a^*ax\| : x \in (\operatorname{ball}\mathcal{I})_+ \,\} \\ &= \|a^*a+\mathcal{I}\|. \quad \blacksquare \end{aligned}$$

5.7 Corollary. *If* $\rho : \mathcal{A} \to \mathcal{B}$ *is a* $*$*-homomorphism, then* $\operatorname{ran}\rho$ *is a* C^**-algebra, and the induced map* $\widetilde{\rho} : \mathcal{A}/\ker\rho \to \operatorname{ran}\rho$ *is a* $*$*-isomorphism.*

Proof. The only thing that needs to be proved is that $\operatorname{ran}\rho$ is closed. By Proposition 1.7, ρ is bounded. Thus $\mathcal{I} = \ker\rho$ is a closed ideal of $\mathcal{A}$. Thus the induced map $\widetilde{\rho} : \mathcal{I} \to \mathcal{B}$ is a $*$-monomorphism. By Corollary 1.8, $\widetilde{\rho}$ is an isometry, so that $\operatorname{ran}\widetilde{\rho} = \operatorname{ran}\rho$ is closed. ■

This section closes with an illustration of the preceding results for the case of an abelian C*-algebra.

5.8 Proposition. *If* X *is a compact space and* $\mathcal{I}$ *is a closed ideal of* $C(X)$, *then there is a unique closed subset* F *of* X *such that* $\mathcal{I} = \{f \in C(X) : f(x) = 0$ *for all* x *in* $F\}$. *The* C^**-algebra* $C(X)/\mathcal{I}$ *is isomorphic to* $C(F)$ *by means of the map induced by the restriction map from* $C(X)$ *to* $C(F)$.

Proof. Let $F = \{x \in X : f(x) = 0$ for all f in $\mathcal{I}\}$; clearly F is a closed subset of X. If $\mu \in M(X)$ and $\mu \perp \mathcal{I}$, then for each f in $\mathcal{I}$, $\int |f|^2\, d\mu = 0$ since $|f|^2 \in \mathcal{I}$. Thus each f in $\mathcal{I}$ must vanish on the support of μ. Hence

$\operatorname{supp}\mu \subseteq F$ or $|\mu|(X \setminus F) = 0$. Conversely, if $\mu \in M(X)$ and $|\mu|(X \setminus F) = 0$, then $\int f\,d\mu = 0$ for all f in $\mathcal{I}$. Therefore $\mathcal{I}^\perp = \{\mu \in M(X) : |\mu|(X\setminus F) = 0\}$. Duality in Banach spaces now implies that $\mathcal{I} = (\mathcal{I}^\perp)_\perp = \{\mu \in M(X) : |\mu|(X \setminus F) = 0\}_\perp = \{f \in C(X) : f(x) = 0 \text{ for all } x \text{ in } F\}$.

Let $\rho : C(X) \to C(F)$ be the restriction map: $\rho(f) = f|F$. Clearly ρ is a $*$-homomorphism and $\ker \rho = \mathcal{I}$. By the Tietze Extension Theorem ρ is surjective. Thus the induced map $\widetilde{\rho}$: $C(X)/\mathcal{I} \to C(F)$ is a $*$-isomorphism. ■

Exercises.

1. Show that M_n, the C*-algebra of all $n \times n$ matrices over the complex numbers, has no proper ideals.

2. If $\mathcal{A}$ is a C*-algebra, I is a closed ideal of $\mathcal{A}$, and $\mathcal{B}$ is a C*-subalgebra of $\mathcal{A}$, show that the C*-algebra generated by $I \cup \mathcal{B}$ is $I + \mathcal{B}$.

3. If $\mathcal{A}$ is a C*-algebra, let $\widetilde{L}$ be the collection of all closed left ideals of $\mathcal{A}$ and let $\widetilde{B}$ be the collection of hereditary subalgebras of $\mathcal{A}$. Define $L : \widetilde{B} \to \widetilde{L}$ by $L(\mathcal{B}) = \{x \in \mathcal{A} : x^*x \in \mathcal{B}\}$ and define $B : \widetilde{L} \to \widetilde{B}$ by $B(\mathcal{L}) = \mathcal{L} \cap \mathcal{L}^*$. Show that L and B are order preserving bijections and $B = L^{-1}$. What are the fixed points of L and B?

4. If $\mathcal{A}$ is a C*-algebra and I and J are closed ideals of $\mathcal{A}$, show that $I + J$ is a closed ideal of $\mathcal{A}$.

5. Give an example of a non-closed ideal of $C(\operatorname{cl}\mathbb{D})$ that is not self-adjoint.

6. Find all the left ideals and hereditary subalgebras of M_n.

§6. Representations of a C*-algebra

A representation is just a $*$-homomorphism with a specific range.

6.1 Definition. If $\mathcal{A}$ is a C*-algebra, a *representation* of $\mathcal{A}$ is a pair $(\pi, \mathcal{H})$, where $\mathcal{H}$ is a Hilbert space and $\pi : \mathcal{A} \to \mathcal{B}(\mathcal{H})$ is a $*$-homomorphism. If $\mathcal{A}$ is unital, then it is required that $\pi(1) = 1$.

The role here of $\mathcal{H}$ will often be suppressed and it will be stated that π is a representation. If $\mathcal{A}$ does not have an identity and $\mathcal{A}_1 = \mathcal{A} + \mathbb{C}$, then whenever $\pi : \mathcal{A} \to \mathcal{B}(\mathcal{H})$ is a representation, it can be extended to a representation $\widetilde{\pi}$ of $\mathcal{A}_1$ by letting $\widetilde{\pi}(a+\alpha) = \pi(a) + \alpha$ for a in $\mathcal{A}$ and α in $\mathbb{C}$. Note that in light of Proposition 1.7.e, every representation is contractive. By Corollary 5.7, the range of a representation is closed.

6.2 Examples. (a) If $\mathcal{A}$ is a C*-subalgebra of $\mathcal{B}(\mathcal{H})$, then the inclusion map $\mathcal{A} \to \mathcal{B}(\mathcal{H})$ is a representation.

(b) If (X, Ω, μ) is a σ-finite measure space, then $\pi : L^\infty(\mu) \to \mathcal{B}(L^2(\mu))$ defined by $\pi(\phi) = M_\phi$ is a representation.

(c) If X is compact and μ is a positive Borel measure on X, then $\pi_\mu : C(X) \to \mathcal{B}(L^2(\mu))$ defined by $\pi_\mu(f) = M_f$ is a representation of $C(X)$.

The notation set in part (c) of the last example will be used in the sequel.

6.3 Definition. Let $\mathcal{A}$ be a C*-algebra.

(a) If d is a cardinal number and $\mathcal{H}$ is a Hilbert space, let $\mathcal{H}^{(d)}$ denote the direct sum of $\mathcal{H}$ with itself d times. If $A \in \mathcal{B}(\mathcal{H})$, $A^{(d)}$ is the direct sum of A with itself d times. The operator $A^{(d)}$ is called the *d-fold inflation* of A, or simply the *inflation* of A. If $\pi : \mathcal{A} \to \mathcal{B}(\mathcal{H})$ is a representation, then $\pi^{(d)} : \mathcal{A} \to \mathcal{B}(\mathcal{H}^{(d)})$ defined by $\pi^{(d)}(a) = \pi(a)^{(d)}$ is a representation. The representation $\pi^{(d)}$ is called the *inflation* of π. For convenience, the inflations $\mathcal{H}^{(\aleph_0)}$ and $\pi^{(\aleph_0)}$ are denoted by $\mathcal{H}^{(\infty)}$ and $\pi^{(\infty)}$.

(b) If $\{(\pi_i, \mathcal{H}_i)\}$ is a collection of representations of $\mathcal{A}$, then the *direct sum* of these representations is the representation $\pi = \bigoplus_i \pi_i : \mathcal{A} \to \mathcal{B}(\bigoplus_i \mathcal{H}_i)$ defined by $\pi(a) = \bigoplus_i \pi_i(a)$.

Note that in the definition of the direct sum of representations the fact that $\|\pi_i(a)\| \le \|a\|$ for all i implies that the direct sum of operators $\bigoplus_i \pi_i(a)$ is a well-defined bounded operator. It is easily checked that if π is an isometry, then so is $\pi^{(d)}$. Also $\pi : \mathcal{B}(\mathcal{H}) \to \mathcal{B}(\mathcal{H}^{(d)})$ defined by $\pi(T) = T^{(d)}$ is an isometric representation of $\mathcal{B}(\mathcal{H})$.

6.4 Example. Let X be a compact space and let $\{\mu_n\}$ be a sequence of positive Borel measures on X. For each $n \ge 1$ define $\pi_n : C(X) \to \mathcal{B}(L^2(\mu_n))$ as in Example 6.2.c. Then $\bigoplus_n \pi_n$ is a representation.

6.5 Definition. Two representations of the C*-algebra $\mathcal{A}$, $(\pi_1, \mathcal{H}_1)$ and $(\pi_2, \mathcal{H}_2)$, are *equivalent* if there is an isomorphism $U : \mathcal{H}_1 \to \mathcal{H}_2$ such that $\pi_2(a) = U\pi_1(a)U^{-1}$ for all a in $\mathcal{A}$. In symbols this is denoted by $\pi_1 \cong \pi_2$. (The terminology of [ACFA] is maintained here. An isomorphism between Hilbert spaces is an isomorphism in that category. That is, an isomorphism is an inner product preserving linear bijection. The term "unitary" is reserved for an isomorphism from a Hilbert space onto itself.)

6.6 Example. (a) In Example 6.4, if the measures $\{\mu_n\}$ are pairwise singular and $\mu = \sum_n (2^n\|\mu_n\|)^{-1}\mu_n$, then the representation $\pi_\mu : C(X) \to \mathcal{B}(L^2(\mu))$ is equivalent to the direct sum of the representations $\{\pi_{\mu_n}\}$.

(b) Let μ and ν be two positive measures on the compact space X and π_μ and π_ν the two representations defined in Example 6.2.c. Then $\pi_\mu \cong \pi_\nu$ if and only if μ and ν are mutually absolutely continuous; that is, if and only if the two measures have the same sets of measure zero.

To see part (b), first assume that that the measures are mutually absolutely continuous. If ϕ is the Radon–Nikodym derivative $d\mu/d\nu$, then define $U : L^2(\mu) \to L^2(\nu)$ by $Uh = h\sqrt{\phi}$. It is left to the reader to check that U is the required isomorphism. For the converse, assume that an isomorphism $U : L^2(\mu) \to L^2(\nu)$ is given such that for each f in $C(X)$ and each h in $L^2(\mu)$, $U(fh) = fUh$. Letting $\psi = U(1)$, this implies that $U(f) = U(f \cdot 1) = f\psi$ for each f in $C(X)$. Since U is an isometry, $\int |f|^2|\psi|^2\, d\nu = \int |f|^2\, d\mu$. But this implies that $\int f|\psi|^2\, d\nu = \int f\, d\mu$ for every positive continuous function f on X. Thus $\mu = |\psi|^2\nu$ and so $\mu << \nu$. The same argument applied to the isomorphism U^{-1} shows that $\nu << \mu$.

This high ratio of definitions to results must continue a bit longer.

6.7 Definition. A representation π of a C*-algebra $\mathcal{A}$ is *cyclic* if there is a vector e in $\mathcal{H}$ such that $\{\pi(a)e : a \in \mathcal{A}\}$ is dense in $\mathcal{H}$. Any vector e that satisfies this condition is called a *cyclic vector* for the representation π. The representation π is called *faithful* if it is injective. A *subrepresentation* of π is a representation of $\mathcal{A}$ of the form $a \to \pi(a)|\mathcal{M}$, where $\mathcal{M}$ is a closed subspace of $\mathcal{H}$ that reduces $\pi(\mathcal{A})$.

It is easy to see that if X is compact and μ is a positive Borel measure on X, then π_μ, the representation of $C(X)$ defined in (6.2.c), is cyclic with cyclic vector 1. In fact, any function in $L^2(\mu)$ that does not vanish on a set of positive μ measure is cyclic for this representation. (See Exercise 1.) Also the identity representation of $\mathcal{B}(\mathcal{H})$ is cyclic with any non-zero vector as a cyclic vector.

It is also easy to see that the representation π_μ is faithful if and only if the support of μ is all of X. Note that a faithful representation is an isometry. There is more about faithful representations in the next section. The reader might find some subrepresentations of the representations in Examples 6.2.b and 6.2.c.

6.8 Theorem. *Every representation is equivalent to the direct sum of cyclic representations. If the representation acts on a separable Hilbert space, the representation is equivalent to the direct sum of a sequence of cyclic representations.*

Proof. The proof is a rather simple extrapolation from an elementary observation. Fix a representation $\pi : \mathcal{A} \to \mathcal{B}(\mathcal{H})$. For any vector e in $\mathcal{H}$, put $\mathcal{M} = [\pi(\mathcal{A})e]$. (For any collection $\mathcal{S}$ of vectors in a Hilbert space, $[\mathcal{S}]$ denotes the closed linear span of the set $\mathcal{S}$.) Since $\mathcal{A}$ is a C*-algebra, $\mathcal{M}$ is invariant for every operator $\pi(a)$ as well as its adjoint. That is, $\mathcal{M}$ reduces

$\pi(\mathcal{A})$. So if $\pi_1 \colon \mathcal{A} \to \mathcal{B}(\mathcal{M})$ is the subrepresentation $\pi_1(a) = \pi(a)|\mathcal{M}$ and $\pi_2 : \mathcal{A} \to \mathcal{B}(\mathcal{M}^\perp)$ is defined by $\pi_2(a) = \pi(a)|\mathcal{M}^\perp$, π_1 and π_2 are representations and π_1 is cyclic.

Now use Zorn's Lemma to obtain a maximal family of vectors $\{e_i\}$ such that $[\pi(\mathcal{A})e_i] \perp [\pi(\mathcal{A})e_j]$ for $i \neq j$. If $\mathcal{H}_i = [\pi(\mathcal{A})e_i]$, then maximality implies that $\mathcal{H} = \bigoplus_i \mathcal{H}_i$. If $\pi_i(a) = \pi(a)|\mathcal{H}_i$, then $\pi = \bigoplus_i \pi_i$. If $\mathcal{H}$ is separable, this family must be countable. ■

A natural question to ask is whether a representation can be extended from a C*-subalgebra of the C*-algebra $\mathcal{A}$. The answer is no. For example consider the C*-algebra of all $n \times n$ matrices, M_n. Let $\mathcal{B}$ be the subalgebra consisting of all diagonal matrices and define $\rho : \mathcal{B} \to \mathbb{C} = \mathcal{B}(\mathbb{C})$ by $\rho(\operatorname{diag}(\alpha_1, \ldots, \alpha_n)) = \alpha_1$. It is left to the reader to show that ρ is a representation. But there can be no representation $\kappa : M_n \to \mathbb{C}$ other than the 0 representation. In fact, if there were, then $\ker \kappa$ is a proper ideal in M_n, a contradiction (Exercise 5.1). Therefore ρ has no extension. If the representation is defined on an ideal of $\mathcal{A}$, however, the story has a different conclusion.

6.9 Definition. If $\mathcal{A}$ is a C*-algebra, say that a representation $\rho : \mathcal{A} \to \mathcal{B}(\mathcal{H})$ is *non-degenerate* if $[\rho(\mathcal{A})\mathcal{H}] = \mathcal{H}$.

For any representation ρ, $[\rho(\mathcal{A})\mathcal{H}]$ is a reducing subspace for $\rho(\mathcal{A})$ and this is where the representation "lives." So to say that a representation is non-degenerate is to say that it is living on the entire Hilbert space and not in some corner. Equivalently, a representation $\rho : \mathcal{A} \to \mathcal{B}(\mathcal{H})$ is non-degenerate if the only vector g in $\mathcal{H}$ satisfying $Ag = 0$ for every A in $\mathcal{A}$ is $g = 0$. Also note that if ρ is a representation of $\mathcal{A}$ and $\rho = \bigoplus_i \rho_i$, then ρ is non-degenerate if and only if each ρ_i is non-degenerate.

6.10 Proposition. *If $\mathcal{A}$ is a C^*-algebra and $\mathcal{I}$ is an ideal of $\mathcal{A}$, then every representation $\rho : \mathcal{I} \to \mathcal{B}(\mathcal{H})$ can be extended to a representation $\widetilde{\rho} : \mathcal{A} \to \mathcal{B}(\mathcal{H})$. If the representation ρ is non-degenerate, the extension $\widetilde{\rho}$ is unique.*

Proof. First assume that ρ is non-degenerate.

Claim. For every a in $\mathcal{A}$ there is a unique $\widetilde{\rho}(a)$ in $\mathcal{B}(\mathcal{H})$ such that for every x in $\mathcal{I}$, $\widetilde{\rho}(a)\rho(x) = \rho(ax)$.

By means of Theorem 6.8, it suffices to prove the claim for a cyclic representation. So assume ρ is cyclic with cyclic vector h_0. Fix a in $\mathcal{A}$ and let $\{e_i\}$ be an approximate identity for the ideal $\mathcal{I}$. Thus for every x in $\mathcal{I}$

$$\begin{aligned}\|\rho(ax)h_0\| &= \lim_i \|\rho(ae_ix)h_0\| \\ &= \lim_i \|\rho(ae_i)\rho(x)h_0\| \\ &\le \sup_n \|ae_i\| \, \|\rho(x)h_0\| \\ &\le \|a\| \, \|\rho(x)h_0\|.\end{aligned}$$

This shows that $\rho(x)h_0 \to \rho(ax)h_0$ is a well defined bounded operator on a dense manifold in $\mathcal{H}$. Extend this to a bounded operator on all of $\mathcal{H}$ and denote this operator by $\widetilde{\rho}(a)$. If $x, y \in \mathcal{I}$, then $\widetilde{\rho}(a)\rho(x)\rho(y)h_0 = \widetilde{\rho}(a)\rho(xy)h_0 = \rho(axy)h_0 = \rho(ax)\rho(y)h_0$. Since $\mathcal{H} = [\rho(\mathcal{I})h_0]$, this shows that $\widetilde{\rho}(a)\rho(x) = \rho(ax)$ for all x in $\mathcal{I}$ and establishes the existence part of the claim.

To show the uniqueness statement in the claim, observe that if T is any other operator on $\mathcal{H}$ such that $T\rho(x) = \rho(ax)$ for all x in $\mathcal{I}$, then $T\rho(x)h = \widetilde{\rho}(a)\rho(x)h$ for all x in $\mathcal{I}$ and all h in $\mathcal{H}$. Since ρ is non-degenerate, $T = \widetilde{\rho}(a)$.

Therefore there is a well defined map $\widetilde{\rho}\colon \mathcal{A} \to \mathcal{B}(\mathcal{H})$. If $a, b \in \mathcal{A}$, $x \in \mathcal{I}$, and $h \in \mathcal{H}$, then $\widetilde{\rho}(ab)\rho(x)h = \rho(abx)h = \widetilde{\rho}(a)\rho(bx)h = \widetilde{\rho}(a)\widetilde{\rho}(b)\rho(x)h$. Thus $\widetilde{\rho}(ab) = \widetilde{\rho}(a)\widetilde{\rho}(b)$. Similar arguments show that $\widetilde{\rho}$ is a representation of $\mathcal{A}$ and it is clear that it extends ρ.

Now assume that ρ is degenerate. Put $\mathcal{H}_0 = [\rho(\mathcal{I})\mathcal{H}]$ and let $\rho_0\colon \mathcal{I} \to \mathcal{B}(\mathcal{H}_0)$ be the subrepresentation $\rho_0(x) = \rho(x)|\mathcal{H}_0$. Since ρ_0 is non-degenerate, there is a unique extension $\widetilde{\rho_0} : \mathcal{A} \to \mathcal{B}(\mathcal{H}_0)$. If $\kappa : \mathcal{A} \to \mathcal{B}(\mathcal{H}_0^\perp)$ is any representation (like the 0 representation, for example), then $\widetilde{\rho_0} \oplus \kappa$ is an extension of ρ. ∎

It's easy to see from the preceding proof why uniqueness fails unless the representation is non-degenerate. The proof also identifies all the possible representations in the degenerate case. Since cyclic representations are non-degenerate, the following is immediate.

6.11 Corollary. *If $\mathcal{A}$ is a C^*-algebra and $\mathcal{I}$ is an ideal of $\mathcal{A}$, then every cyclic representation $\rho : \mathcal{I} \to \mathcal{B}(\mathcal{H})$ has a unique extension to a cyclic representation of $\mathcal{A}$*

6.12 Corollary. *If $\mathcal{A}$ is a C^*-algebra, $\mathcal{I}$ is an ideal of $\mathcal{A}$, and $(\rho, \mathcal{H})$ and $(\kappa, \mathcal{K})$ are two representations of $\mathcal{A}$ such that:*

(a) *$\rho|\mathcal{I}$ is a non-degenerate representation of the ideal $\mathcal{I}$;*

(b) *$\rho|\mathcal{I} \cong \kappa|\mathcal{I}$;*

then $\rho \cong \kappa$.

Proof. The hypothesis implies there is an isomorphism $U : \mathcal{H} \to \mathcal{K}$ such that $U^*\kappa(x)U = \rho(x)$ for all x in $\mathcal{I}$. Now $\rho(\cdot)$ and $U^*\kappa(\cdot)U$ are both extensions of $\rho|\mathcal{I}$. By Proposition 6.10 and condition (a), $\rho|\mathcal{I}$ has a unique extension to $\mathcal{A}$. Thus $U^*\kappa(a)U = \rho(a)$ for all a in $\mathcal{A}$, so that $\rho \cong \kappa$. ■

Exercises.

1. Show that the representation in Example 6.2.b is cyclic and that a function f in $L^2(\mu)$ is a cyclic vector for this representation if and only if it does not vanish on a set of positive μ measure.

2. Let X be a compact space, μ a positive Borel measure on X, and $(\pi_\mu, L^2(\mu))$ the representation from Example 6.2.c. Determine the kernel of π_μ. If X is metrizable, show that the measure μ can be chosen so that π_μ is injective.

3. Consider the two positive Borel measures μ and ν on the compact space X and the representations π_μ and π_ν defined in Example 6.2.c. Show that $\pi_\mu \oplus \pi_\nu$ is cyclic if and only if $\mu \perp \nu$. If $\mu \perp \nu$, show that $\pi_\mu \oplus \pi_\nu \cong \pi_{\mu+\nu}$. Note that this implies that $\pi_\mu^{(d)}$ is not cyclic for $d \geq 2$. Can you show this directly?

4. Verify the statements in Example 6.6.

5. Let $\mathcal{A}$ be any C*-algebra contained in $\mathcal{B}(\mathcal{H})$ that contains the compact operators. If $\pi : \mathcal{A} \to \mathcal{B}(\mathcal{H})$ is the identity representation, show that $\pi^{(\infty)}$ is a cyclic representation.

6. Let $\rho : \mathcal{A} \to \mathcal{B}(\mathcal{H})$ be a representation and let $\mathcal{I}$ be an ideal of $\mathcal{A}$. Define $\mathcal{H}_\mathcal{I} = [\rho(\mathcal{I})\mathcal{H}]$ and let $\rho_\mathcal{I} : \mathcal{I} \to \mathcal{B}(\mathcal{H}_\mathcal{I})$ by $\rho_\mathcal{I}(x) = \rho(x)|\mathcal{H}_\mathcal{I}$. Show that $\rho_\mathcal{I}$ is a non-degenerate representation of $\mathcal{I}$. What is its extension?

§7. Positive linear functionals and the GNS construction

Suppose $\mathcal{A}$ is a unital C*-algebra and $\pi : \mathcal{A} \to \mathcal{B}(\mathcal{H})$ is a representation. If e is a unit vector in $\mathcal{H}$ and $\phi : \mathcal{A} \to \mathbb{C}$ is defined by $\phi(a) = \langle \pi(a)e, e \rangle$, then ϕ is a positive linear functional. That is, $\phi(a) \geq 0$ when $a \in \mathcal{A}_+$. Also $\phi(1) = 1$ since $\|e\| = 1$.

7.1 Definition. A *state* on a C*-algebra is a positive linear functional of norm 1. Let $\Sigma = \Sigma(\mathcal{A})$ be the collection of all states on $\mathcal{A}$. $\Sigma(\mathcal{A})$ is called the *state space* of $\mathcal{A}$.

A state on a C*-algebra without identity is extended to the algebra obtained by adjoining an identity if we set $\phi(a + \alpha) = \phi(a) + \alpha$.

The paragraph preceding the definition shows that whenever there is a representation, this produces a state. The main result of this section is the converse. Whenever a state is present, it is possible to manufacture a representation. Indeed this is done in such a manner that these two processes are the inverses of each other. The first step in carrying out this plan is deriving some elementary properties of positive linear functionals. The first is that the Cauchy–Schwarz–Bunyakowskii Inequality holds.

7.2 Proposition. *If ϕ is a positive linear functional on a C^*-algebra $\mathcal{A}$, then for any x, y in $\mathcal{A}$*

$$|\phi(y^*x)|^2 \leq \phi(y^*y)\phi(x^*x).$$

Proof. If $[x, y] \equiv \phi(y^*x)$ for x, y in $\mathcal{A}$, then it is easily checked that $[\cdot, \cdot]$ is a semi-inner product on $\mathcal{A}$. Thus the inequality follows from the CBS Inequality. ∎

The preceding inequality is also called the CBS Inequality.

7.3 Corollary. *If ϕ is a positive linear functional on $\mathcal{A}$, then ϕ is bounded. If $\{e_i\}$ is an approximate identity for $\mathcal{A}$, then $\|\phi\| = \lim_i \phi(e_i)$.*

Proof. The proof of the case where $\mathcal{A}$ is unital is left to the reader; it follows from the CBS Inequality. Note that in this case $\|\phi\| = \phi(1)$. Assume $\mathcal{A}$ does not have an identity and let $\{e_i\}$ be an approximate identity. First show that ϕ is bounded on $(\text{ball}\,\mathcal{A})_+$. If not, then for every integer $k \geq 1$ there is an a_k in $(\text{ball}\,\mathcal{A})_+$ with $\phi(a_k) > 2^k$. If $a = \sum_k 2^{-k} a_k$, then $\phi(a) \geq \phi(\sum_{k=1}^n 2^{-k} a_k) = \sum_{k=1}^n \phi(2^{-k} a_k) > n$, an impossibility. Thus $\alpha \equiv \sup\{\, \phi(a) : a \in (\text{ball}\,\mathcal{A})_+ \,\} < \infty$. Since each element of $\mathcal{A}$ can be written as the linear combination of 4 positive elements, ϕ is bounded with $\|\phi\| \leq 4\alpha$.

Note also that $\alpha \geq \sup_i \phi(e_i) = \lim_i \phi(e_i) \equiv \beta$, since the approximate identity is increasing. Now if $a \in \text{ball}\,\mathcal{A}$, then $0 \leq a^*a \leq 1$ and so

$$|\phi(a)|^2 = \lim |\phi(e_i a)|^2 \leq \lim_i \phi(e_i^2)\, \phi(a^*a) \leq \beta \|\phi\|.$$

This implies that $\|\phi\| \leq \beta \leq \alpha \leq \|\phi\|$. ∎

7.4 Example. If X is a compact space, then the positive linear functionals on $C(X)$ correspond to the positive measures. The states correspond to the probability measures.

Recall from measure theory that if μ is a finite, complex-valued measure on (X, Ω) and $\|\mu\| = \mu(X)$, then μ is positive. The next result extends this to the C*-algebra setting.

7.5 Proposition. *If $\mathcal{A}$ is a unital C^*-algebra and $\phi : \mathcal{A} \to \mathbb{C}$ is a bounded linear functional such that $\|\phi\| = \phi(1)$, then $\phi \geq 0$.*

Proof. If $\mathcal{A} = C(X)$, then ϕ corresponds to a measure μ on X and the hypothesis is the statement that $\mu(X) = \|\mu\|$. This implies that $\mu \geq 0$. For an arbitrary C*-algebra $\mathcal{A}$, let $a \in \mathcal{A}_+$ and consider $\mathcal{C} = C^*(a)$. So $\mathcal{C} \cong C(\sigma(a))$. If $\phi_0 = \phi|\mathcal{C}$, then $\phi_0(1) \leq \|\phi_0\| \leq \|\phi\| = \phi(1) = \phi_0(1)$. So the result from measure theory implies that $\phi(a) \geq 0$. Hence ϕ is positive. ∎

When mention is made of a "subalgebra" of $\mathcal{A}$ and the subalgebra has an identity, then it is assumed that it has the same identity as $\mathcal{A}$. However every subalgebra of an algebra with identity need not have one.

7.6 Corollary. *If $\mathcal{A}$ is a C^*-algebra and $\mathcal{B}$ is a C^*-subalgebra of $\mathcal{A}$, then any state on $\mathcal{B}$ has an extension to a state on $\mathcal{A}$.*

Proof. Without loss of generality it may be assumed that $\mathcal{A}$ has an identity. Let $\psi : \mathcal{B} \to \mathbb{C}$ be a state. If $\mathcal{B}$ has an identity, let ϕ be a norm preserving extension to $\mathcal{A}$. Then $1 = \|\phi\| = \|\psi\| = \psi(1) = \phi(1)$. Thus ϕ is a state on $\mathcal{A}$ by the preceding proposition. If $\mathcal{B}$ does not have an identity, let $\mathcal{B}_1 = \mathcal{B} + \mathbb{C}$, a subalgebra of $\mathcal{A}$. If $\psi_1 : \mathcal{B}_1 \to \mathbb{C}$ is defined by $\psi_1(b + \beta) = \psi(b) + \beta$, then ψ_1 is a state on $\mathcal{B}_1$, and the first part of the proof applies. ∎

Now for the promised converse to the observation made at the beginning of the section. Before stating and proving the result, examine what happens when $\mathcal{A}$ is abelian. If $\phi : C(X) \to \mathbb{C}$ is a state, then there is a probability measure μ such that $\phi(f) = \int f \, d\mu$. It seems clear that the representation that should be associated with the state ϕ is the representation $\pi_\mu : C(X) \to \mathcal{B}(L^2(\mu))$. Note that $L^2(\mu)$ is the completion of $C(X)$ with respect to the norm defined by the inner product $\langle f, g \rangle = \int f\bar{g} \, d\mu = \phi(f\bar{g})$. This furnishes the idea of how to prove the next theorem. The reader's comprehension might benefit by examining for the abelian case each step of the next proof.

7.7 The Gelfand–Naimark–Segal Construction. *Let $\mathcal{A}$ be a C^*-algebra with identity.*

(a) *If ϕ is a state on $\mathcal{A}$, then there is a cyclic representation $(\pi_\phi, \mathcal{H}_\phi)$ of $\mathcal{A}$ with unit cyclic vector e such that $\phi(a) = \langle \pi_\phi(a)e, e \rangle$ for all a in $\mathcal{A}$.*

(b) *If $(\pi, \mathcal{H})$ is a cyclic representation with unit cyclic vector e, ϕ is the state on $\mathcal{A}$ defined by $\phi(a) = \langle \pi(a)e, e \rangle$, and $(\pi_\phi, \mathcal{H}_\phi)$ is the representation obtained in* (a), *then $\pi \cong \pi_\phi$.*

Proof. Let $\mathcal{L} = \{x \in \mathcal{A} : \phi(x^*x) = 0\}$. It is easy to see that $\mathcal{L}$ is closed in $\mathcal{A}$. Also if $a \in \mathcal{A}$ and $x \in \mathcal{L}$, then $0 \leq \phi((ax)^*(ax)) = \phi(x^*a^*ax) \leq$

$\phi(x^*\|a\|^2 x) = \|a\|^2\phi(x^*x) = 0$. Thus $ax \in \mathcal{L}$ whenever $x \in \mathcal{L}$ and a is arbitrary. It follows that $\mathcal{L}$ is a closed left ideal in $\mathcal{A}$. Consider the vector space $\mathcal{A}/\mathcal{L}$ and for x, y in $\mathcal{A}$ define

$$\langle x + \mathcal{L}, y + \mathcal{L}\rangle = \phi(y^*x).$$

It is a routine exercise for the reader to show that $\langle\cdot,\cdot\rangle$ defines an inner product on $\mathcal{A}/\mathcal{L}$. Let $\mathcal{H}$ denote the completion of $\mathcal{A}/\mathcal{L}$ with respect to the norm defined by this inner product.

If $a, x \in \mathcal{A}$, then $\|ax + \mathcal{L}\|^2 = \langle ax + \mathcal{L}, ax + \mathcal{L}\rangle = \phi(x^*a^*ax) \leq \|a\|^2\phi(x^*x) = \|a\|^2 \|x + \mathcal{L}\|^2$. Thus if $\pi_\phi(a) : \mathcal{A}/\mathcal{L} \to \mathcal{A}/\mathcal{L}$ is defined by $\pi_\phi(a)(x + \mathcal{L}) = ax + \mathcal{L}$, $\pi_\phi(a)$ extends to a bounded operator on $\mathcal{H}$ with $\|\pi_\phi(a)\| \leq \|a\|$. It is left to the reader to verify that $\pi_\phi : \mathcal{A} \to \mathcal{B}(\mathcal{H})$ is a representation. If $e = 1 + \mathcal{L}$, then e is a unit vector in $\mathcal{H}$ and $\pi_\phi(\mathcal{A})e = \{a + \mathcal{L} : a \in \mathcal{A}\}$, which is dense in $\mathcal{H}$ by definition. Thus e is a cyclic vector for the representation $(\pi_\phi, \mathcal{H})$. Note that $\phi(a) = \langle\pi_\phi(a)e, e\rangle$. This proves (a).

Now let $(\pi, \mathcal{H})$, e, and ϕ be as in (b) and let $(\pi_\phi, \mathcal{H}_\phi)$ be the representation constructed in (a). Let e_ϕ be the cyclic vector for the representation π_ϕ. So for each a in $\mathcal{A}$, $\langle\pi_\phi(a)e_\phi, e_\phi\rangle = \phi(a) = \langle\pi(a)e, e\rangle$. Define U on the dense manifold $\pi_\phi(\mathcal{A})e_\phi$ by $U(\pi_\phi(x)e_\phi) = \pi(x)e$. Now $\|U(\pi_\phi(x)e_\phi)\|^2 = \|\pi(x)e\|^2 = \langle\pi(x)e, \pi(x)e\rangle = \phi(x^*x) = \|\pi_\phi(x)e_\phi\|^2$. Hence U extends to a well defined isometry $U : \mathcal{H}_\phi \to \mathcal{H}$. Since $U(\pi_\phi(\mathcal{A})e_\phi)$ is dense in $\mathcal{H}$, U is an isomorphism. But if $a, x \in \mathcal{A}$, $U\pi_\phi(a)(\pi_\phi(x)e_\phi) = U\pi_\phi(ax)e_\phi = \pi(a)\pi(x)e = \pi(a)U(\pi_\phi(x)e_\phi)$. That is, $U\pi_\phi(a) = \pi(a)U$ and so $\pi_\phi \cong \pi$. ■

If, in the preceding theorem, $\mathcal{A}$ does not have an identity, the result still holds with a suitable modification. If $\mathcal{A}_1 = \mathcal{A} + \mathbb{C}$, then a state or a representation can be extended from $\mathcal{A}$ to $\mathcal{A}_1$ and the preceding theorem invoked. Also the perspicacious reader probably felt a sense of *deja vu* when seeing that $\mathcal{L}$ in the preceding proof is a closed left ideal. In fact there is a relation with Proposition 5.3. In the proof of (5.3) note that the only thing about the hereditary subalgebra $\mathcal{B}$ that was used to show that $\mathcal{L}$ is a closed left ideal is the fact that $\mathcal{B}_+$ is a closed cone with the hereditary property. In the preceding proof it could have been noted that for a state ϕ, $(\ker\phi)_+$ is a closed hereditary cone and use the same proof as in (5.3) to show that $\mathcal{L}$ is a closed left ideal.

The construction of the cyclic representation from the state on $\mathcal{A}$ is called the *GNS construction*. It can now be shown that every C*-algebra can be realized as a C*-algebra of operators on a Hilbert space. The first step is to prove a result that will also be used later in another context.

7.8 Proposition. *If $\mathcal{A}$ is a C^*-algebra, a is a hermitian element of $\mathcal{A}$, $\alpha = \min \sigma(a)$, and $\beta = \max \sigma(a)$, then* $[\alpha, \beta] = \{\phi(a) : \phi \text{ is a state on } \mathcal{A}\}$.

Proof. Let $\mathcal{C} = C^*(a)$. So $\mathcal{C} = \{f(a) : f \in C(\sigma(a))\}$. If ϕ is a state on $\mathcal{A}$ and $\phi_0 = \phi|\mathcal{C}$, there is a probability measure μ on $\sigma(a)$ such that $\phi_0(f(a)) = \int f d\mu$ for all f in $C(\sigma(a))$. In particular, $\phi(a) = \int t\, d\mu(t) \in [\alpha, \beta]$.

Conversely, if $\alpha \le t \le \beta$, define $\phi_0 : \mathcal{C} \to \mathbb{C}$ by

$$\phi_0(f(a)) = \left(\frac{t-\alpha}{\beta-\alpha}\right) f(\beta) + \left(\frac{\beta-t}{\beta-\alpha}\right) f(\alpha).$$

Note that ϕ_0 is positive and $\phi_0(a) = t$. If ϕ is a state on $\mathcal{A}$ that is an extension of ϕ_0, $t = \phi(a)$. ■

7.9 Corollary. *If $a \in \mathcal{A}$ and Σ_1 is a weak* dense subset of $\Sigma(\mathcal{A})$, then* $\|a\|^2 = \sup\{\phi(a^*a) : \phi \in \Sigma_1\}$.

The next result says that every abstract C*-algebra is isomorphic to a C*-algebra of operators.

7.10 Theorem. *Every C^*-algebra $\mathcal{A}$ has a faithful representation. $\mathcal{A}$ has a faithful representation $(\pi, \mathcal{H})$ with $\mathcal{H}$ separable if and only if there are a countable number of states on $\mathcal{A}$ that separate the points of $\mathcal{A}_+$, each of which defines a separable representation. In particular, every separable C^*-algebra has a faithful, separable representation.*

Proof. Let Σ_1 be a weak* dense subset of $\Sigma = \Sigma(\mathcal{A})$. Define $\mathcal{H} = \bigoplus\{\mathcal{H}_\phi : \phi \in \Sigma_1\}$ and $\pi = \bigoplus\{\pi_\phi : \phi \in \Sigma_1\}$. An application of Corollary 7.9 shows that $(\pi, \mathcal{H})$ is a faithful representation.

Now suppose that $\pi : \mathcal{A} \to \mathcal{B}(\mathcal{H})$ is faithful and $\mathcal{H}$ is separable. Let $\{e_n\}$ be a dense sequence in $\{h \in \mathcal{H} : \|h\| = 1\}$ and define $\phi_n(a) = \langle \pi(a)e_n, e_n\rangle$ and $\mathcal{H}_n = [\pi(\mathcal{A})e_n]$. The representation defined by ϕ_n via the GNS construction is equivalent to $\pi_n(a) = \pi(a)|\mathcal{H}_n$ and is, therefore, separable. If $a \in \mathcal{A}_+$ and $\phi_n(a) = 0$ for all n, put $a = b^*b$ for some b in $\mathcal{A}$. So $0 = \phi_n(a) = \langle \pi(b^*b)e_n, e_n\rangle = \|\pi(b)e_n\|^2$ for all n. Thus $\pi(b) = 0$ and so $a = 0$. Thus $\{\phi_n\}$ separates the points of $\mathcal{A}_+$.

Now assume that $\{\phi_n\}$ is a sequence of states on $\mathcal{A}$ that separates the points of $\mathcal{A}_+$ and that each representation defined by ϕ_n by the GNS construction is separable. It is left to the reader to check that $\pi = \bigoplus_n \pi_n$ is a faithful, separable representation of $\mathcal{A}$.

If $\mathcal{A}$ is separable, then the weak* topology on ball $\mathcal{A}^*$ is metrizable. Hence there is a countable weak* dense subset of Σ. By Corollary 7.9, this countable set separates the points of $\mathcal{A}_+$. Since $\mathcal{A}$ is separable, every cyclic representation is separable. ■

For a C*-algebra $\mathcal{A}$, consider the representation $\pi = \bigoplus\{ \pi_\phi : \phi \in \Sigma \}$. As pointed out in the preceding theorem, this representation is faithful. This is called the *universal* or *all state* representation of $\mathcal{A}$. This has an interesting and useful property. But to prove that, it is necessary to prove an additional result that has interest in its own right. Say that a linear functional $L : \mathcal{A} \to \mathbb{C}$ is *hermitian* if $L(a^*) = \overline{L(a)}$ for all a in $\mathcal{A}$. Equivalently, L is hermitian if $L(a) \in \mathbb{R}$ for every a in $\operatorname{Re}\mathcal{A}$.

7.11 Lemma. *If $\mathcal{A}$ is a C^*-algebra and $L : \mathcal{A} \to \mathbb{C}$ is a bounded hermitian linear functional, then* $\|L\| = \sup\{ L(a) : a \in \operatorname{ball}\operatorname{Re}\mathcal{A} \}$.

Proof. Let $\epsilon > 0$ and let $x \in \operatorname{ball}\mathcal{A}$ such that $|L(x)| > \|L\| - \epsilon$. Replacing x by a suitable unimodular multiple of x, it can be assumed that $L(x) \geq 0$. In particular, $L(x) = \overline{L(x)} = L(x^*)$. Therefore, if $a = \operatorname{Re} x$, $\|L\| - \epsilon < L(a)$. This proves half the desired inequality, the other half being clear. ∎

7.12 Theorem. *If $\mathcal{A}$ is a C^*-algebra and $L : \mathcal{A} \to \mathbb{C}$ is a bounded hermitian linear functional, then there are unique positive linear functionals $\phi_\pm$ such that $L = \phi_+ - \phi_-$ and $\|L\| = \|\phi_+\| + \|\phi_-\|$.*

Proof. Let Λ be the set of all hermitian linear functionals in $\operatorname{ball}\mathcal{A}$. If Σ is the collection of states, then $\Sigma \cup (-\Sigma) \subseteq \Lambda$, so $\widetilde{\Lambda}$, the weak* closed convex hull of $\Sigma \cup (-\Sigma)$ in $\mathcal{A}^*$, is contained in Λ. The first task is to show equality. Suppose this is not the case. If $L_0 \in \Lambda \setminus \widetilde{\Lambda}$, then the Hahn–Banach Theorem implies there is an x in $\mathcal{A}$ and a real number α such that $\operatorname{Re} L_0(x) > \alpha$ and $\operatorname{Re} R(x) \leq \alpha$ for all R in $\widetilde{\Lambda}$. If $a = \operatorname{Re} x$, then the fact that the functionals are hermitian implies that $L_0(a) > \alpha$ and $R(a) \leq \alpha$ for all R in $\widetilde{\Lambda}$. Thus $\alpha \geq \sup\{ |\phi(a)| : \phi \in \Sigma \} = \|a\|$ while $\alpha < L_0(a) \leq \|a\|$ since $\|L_0\| \leq 1$, a contradiction.

Claim. $\Lambda = \{ s\phi - t\psi : \phi, \psi \in \Sigma, s, t \geq 0, s + t = 1 \}$.

In fact the set in question is clearly convex and contains $\Sigma \cup (-\Sigma)$. Since it is the image of a compact set under the continuous function $(s, t, \phi, \psi) \to s\phi - t\psi$, this set is weak* compact. By the first paragraph of the proof, the claim is established.

Now to directly attack the theorem. Without loss of generality it can be assumed that $\|L\| = 1$. According to the claim, there are states ϕ and ψ such that $L = s\phi - t\psi$, where $s, t \geq 0$ and $s + t = 1$. Let $\phi_+ = s\phi$ and $\phi_- = t\psi$. Since $\|\phi_-\| + \|\phi_-\| = s + t = 1 = \|L\|$, establishing existence.

To prove uniqueness, assume that ψ_+ and ψ_- are states with $L = \psi_+ - \psi_-$ and $\|\psi_+\| + \|\psi_-\| = 1$. Let $\epsilon > 0$ and choose a in $\operatorname{ball}\operatorname{Re}\mathcal{A}$ such that

$L(a) > 1 - \epsilon^2/2$. Put $b = (1-a)/2$; so $0 \le b \le 1$. Thus

$$\phi_+(b) + \phi_-(1-b) = \frac{1}{2}\left[\phi_+(1-a) + \phi_-(1+a)\right] < \frac{1}{4}\epsilon^2.$$

(Before going further, notice that the same inequality holds for $\psi_+(b) + \psi_-(1-b)$.) But $\phi_\pm$ is positive and $0 \le b \le 1$, so that $\phi_+(b) < \epsilon^2/4$ and $\phi_-(1-b) < \epsilon^2/4$. For any x in $\mathcal{A}$, the CBS Inequality implies that $|\phi_+(bx)|^2 = |\phi_+(b^{1/2}b^{1/2}x)|^2 \le \phi_+(b)\,|\phi_+(x^*bx)| < \epsilon^2\|x\|^2/4$. Similarly $|\phi_-((1-b)x)|^2 \le \epsilon^2\|x\|^2/4$.

Applying the same reasoning to $\psi_\pm$ shows that for all x in $\mathcal{A}$

$$\begin{aligned} |\phi_+(bx)| &< \frac{\epsilon}{2}\|x\| & |\psi_+(bx)| &< \frac{\epsilon}{2}\|x\| \\ |\phi_-((1-b)x)| &< \frac{\epsilon}{2}\|x\| & |\psi_-((1-b)x)| &< \frac{\epsilon}{2}\|x\|. \end{aligned}$$

But $\phi_+ - \psi_+ = \phi_- - \psi_-$, so

$$\begin{aligned} \phi_+(x) - \psi_+(x) &= (\phi_+ - \psi_+)(bx) + (\phi_+ - \psi_+)((1-b)x) \\ &= (\phi_+ - \psi_+)(bx) + (\phi_- - \psi_-)((1-b)x). \end{aligned}$$

Applying the inequalities just obtained, this shows that $|\phi_+(x) - \psi_+(x)| < 2\epsilon\|x\|$ for all x in $\mathcal{A}$. Since ϵ was arbitrary, $\phi_+ = \psi_+$. It also follows that $\phi_- = \psi_-$. ∎

The decomposition of a hermitian linear functional as the difference of two positive functionals whose norms add appropriately is called the *Jordan decomposition* of the hermitian functional. If this is applied to a real-valued measure considered as a linear functional on $C(X)$, it yields the standard Jordan decomposition of the measure as the difference of two positive measures with disjoint carriers.

7.13 Corollary. *Every bounded linear functional on a C^*-algebra is the linear combination of four positive linear functionals.*

Proof. Define $\widetilde{L}(a) = \overline{L(a^*)}$; clearly $\widetilde{L}$ is a bounded linear functional on $\mathcal{A}$. If $L_1 = (L + \widetilde{L})/2$ and $L_2 = (L = \widetilde{L})/2i$, then $L = L_1 + iL_2$ and the functionals L_j are hermitian. Now apply the theorem. ∎

7.14 Corollary. *If $\pi : \mathcal{A} \to \mathcal{B}(\mathcal{H})$ is the universal representation of $\mathcal{A}$ and L is a bounded linear functional on $\mathcal{A}$, then there are vectors g, h in $\mathcal{H}$ such that $L(a) = \langle \pi(a)g, h\rangle$ for all a in $\mathcal{A}$.*

Proof. According to the preceding corollary, there are states ϕ_j and complex scalars α_j, $1 \leq j \leq 4$, such that $L = \alpha_1\phi_1 + \alpha_2\phi_2 + \alpha_3\phi_3 + \alpha_4\phi_4$. It can be assumed that the states ϕ_j are all distinct. Let e_j be the unit vector such that $\phi_j(a) = \langle \pi(a)e_j, e_j\rangle$. Because these states are distinct, $[\pi(\mathcal{A})e_j] \perp [\pi(\mathcal{A})e_i]$ for $i \neq j$ so that $\langle \pi(a)e_j, e_i\rangle = 0$ for all a in $\mathcal{A}$ as long as $i \neq j$. If $g = \sum_j \alpha_j e_j$ and $h = \sum_j e_j$, it follows that $L(a) = \langle \pi(a)g, h\rangle$. ∎

Exercises.

1. Show that $\phi_0 : \mathcal{B}(\mathcal{H})/\mathcal{B}_0(\mathcal{H}) \to \mathbb{C}$ is a state if and only if there is a state $\phi : \mathcal{B}(\mathcal{H}) \to \mathbb{C}$ such that $\phi_0(T + \mathcal{B}_0(\mathcal{H})) = \phi(T)$ for every T in $\mathcal{B}(\mathcal{H})$.

2. Fix a Banach limit LIM on $\ell^\infty = \ell^\infty(\mathbb{N})$ and let $\{e_n\}$ be an orthonormal basis for the separable Hilbert space $\mathcal{H}$. Define $\phi : \mathcal{B}(\mathcal{H}) \to \mathbb{C}$ by $\phi(T) = \text{LIM}\,\langle Te_n, e_n\rangle$. Prove that ϕ is a state on $\mathcal{B}(\mathcal{H})$. If $(\pi_\phi, \mathcal{H}_\phi)$ is the representation defined by the GNS construction, show that $\ker \pi_\phi$ is the set of compact operators.

3. Show that an element a of a C*-algebra is positive if and only if $\phi(a) \geq 0$ for every state ϕ.

4. If $a \in \mathcal{A}$ and $a \neq 0$, show that there is a state ϕ such that $\phi(a) \neq 0$.

5. If $\mathcal{A}$ is a separable C*-algebra and $\{\phi_n\}$ a weak* dense sequence in the state space, put $\phi = \sum_n 2^{-n}\phi_n$. Show that ϕ is a state and the representation π_ϕ is faithful. How does this refine Theorem 7.10?

6. If ϕ is a state on a C*-algebra and $\mathcal{L}$ is the closed left ideal $\mathcal{L} = \{\, x : \phi(x^*x) = 0 \,\}$, identify the hereditary subalgebra $\mathcal{B} = \mathcal{L} \cap \mathcal{L}^*$ in terms of ϕ.

7. Generalize Proposition 7.8 as follows. Show that if $\mathcal{A}$ is a C*-algebra and x is a normal element of $\mathcal{A}$, then $\{\, \phi(x) : \phi \in \Sigma(\mathcal{A}) \,\}$ is the closed convex hull of the spectrum of x.

8. Does the algebra M_n have any representations that are not faithful?

9. If $\mathcal{A}$ is a C*-algebra and L is a bounded linear functional on $\mathcal{A}$, show that there is a state ϕ on $\mathcal{A}$ and a vector h in $\mathcal{H}_\phi$ such that $L(a) = \langle \pi_\phi(a)e_\phi, h\rangle$ for all a in $\mathcal{A}$. (Hint: Start with Corollary 7.14.)

Chapter 2

Normal Operators

There is considerable overlap between Chapter IX of [ACFA] and the present chapter. The reader who is familiar with that material should certainly feel comfortable skipping this. This chapter, basically the Spectral Theorem and the characterization of abelian von Neumann algebras, is included for the convenience of the veterans and the instruction of the novitiate.

§8. Some topologies on $\mathcal{B}(\mathcal{H})$

In this section a few definitions and results concerning two topologies on $\mathcal{B}(\mathcal{H})$ are collected. $\mathcal{B}(\mathcal{H})$ is, of course, a normed space, but this topology is not the focus of this section. The *weak operator topology* (WOT) is the topology on $\mathcal{B}(\mathcal{H})$ defined by the collection of seminorms $\{p_{f,g} : f, g \in \mathcal{H}\}$, where $p_{f,g}(T) = |\langle Tf, g\rangle|$ for T in $\mathcal{B}(\mathcal{H})$. The *strong operator topology* (SOT) is defined by the seminorms $\{p_f : f \in \mathcal{H}\}$, where $p_f(T) = \|Tf\|$ for T in $\mathcal{B}(\mathcal{H})$. If $\mathcal{H}$ is separable, both topologies, when restricted to a bounded subset of $\mathcal{B}(\mathcal{H})$, are metrizable (Exercise 1).

8.1 Proposition. *If $\Phi : \mathcal{B}(\mathcal{H}) \to \mathbb{C}$ is a linear functional, the following statements are equivalent.*

(a) *Φ is* SOT-*continuous.*

(b) *Φ is* WOT-*continuous.*

(c) *There are vectors $f_1, \ldots, f_n, g_1, \ldots, g_n$ in $\mathcal{H}$ such that*

$$\Phi(T) = \sum_{k=1}^{n} \langle Tf_k, g_k\rangle$$

for all T in $\mathcal{B}(\mathcal{H})$.

Proof. Clearly (c) implies (b) and (b) implies (a). Assume (a) holds. There are vectors $g_1, \ldots, g_n$ and positive scalars $\alpha_1, \ldots, \alpha_n$ such that $|L(A)| \leq \sum_k \alpha_k \|Ag_k\|$ for all A in $\mathcal{B}(\mathcal{H})$ (see [ACFA], Theorem IV.3.1.f). Replacing g_k by $\alpha_k g_k$, it follows that

$$|L(A)| \leq \sum_{k=1}^{n} \|Ag_k\| \leq \sqrt{n} \left[\sum_{k=1} \|Ag_k\|^2 \right]^{1/2}$$

for all A in $\mathcal{B}(\mathcal{H})$. Replacing each g_k by $\sqrt{n}\, g_k$ gives

$$|L(A)| \leq \left[\sum_{k=1}^{n} \|Ag_k\|^2 \right]^{1/2} \equiv p(A).$$

Note that p is a seminorm. Let $\mathcal{K} = \text{cl}\{ Ag_1 \oplus \cdots \oplus Ag_n : A \in \mathcal{B}(\mathcal{H}) \}$; so $\mathcal{K} \subseteq \mathcal{H}^{(n)}$. If $Ag_1 \oplus \cdots \oplus Ag_n = 0$, $p(A) = 0$ and therefore $L(A) = 0$. Thus $F(Ag_1 \oplus \cdots \oplus Ag_n) = L(A)$ is a well defined linear functional on a dense manifold of $\mathcal{K}$. But $|F(Ag_1 \oplus \cdots \oplus Ag_n)| \leq p(A) = \|Ag_1 \oplus \cdots \oplus Ag_n\|$, so F can be extended to a bounded linear functional F_1 on $\mathcal{H}^{(n)}$. Hence there are vectors $h_1, \ldots, h_n$ in $\mathcal{H}$ such that $F_1(f_1 \oplus \cdots \oplus f_n) = \langle f_1 \oplus \cdots \oplus f_n, h_1 \oplus \cdots \oplus h_n \rangle = \sum_k \langle f_k, h_k \rangle$. In particular, $L(A) = F(Ag_1 \oplus \cdots \oplus Ag_n) = \sum_k \langle Ag_k, h_k \rangle$. ∎

The following corollary combines this proposition and the fact that if a vector space has two locally convex topologies with identical collections of continuous linear functionals, then a convex subset has the same closure in both topologies.

8.2 Corollary. *If $\mathcal{S}$ is a convex subset of $\mathcal{B}(\mathcal{H})$, then the* WOT *closure of $\mathcal{S}$ equals the* SOT *closure of $\mathcal{S}$.*

There is another topology on $\mathcal{B}(\mathcal{H})$ that must be discussed later. It will be shown that $\mathcal{B}(\mathcal{H})$ is the dual of a Banach space (§19); as such it has a weak* topology. This weak* topology agrees with the WOT on bounded sets, but has other distinctive attributes.

8.3 Proposition. *The closed unit ball of $\mathcal{B}(\mathcal{H})$ with the* WOT *is a compact space.*

Proof. The proof is like the proof of Alaoglu's Theorem ([ACFA], Theorem V.3.1). For each vector g in ball $\mathcal{H}$, let B_g be a copy of ball $\mathcal{H}$ equipped with the weak topology. Let $X = \prod\{ B_g : g \in \text{ball}\, \mathcal{H} \}$ and define $\phi : \text{ball}\, \mathcal{B}(\mathcal{H}) \to X$ by $\phi(T) = \{ Tg : g \in \text{ball}\, \mathcal{H} \}$. If ball $\mathcal{B}(\mathcal{H})$ has the WOT, it is easy to see that ϕ is continuous. It is even easier to see that ϕ is one-to-one. If it can be shown that $\phi(\text{ball}\, \mathcal{B}(\mathcal{H}))$ is closed in X, the proof will be complete.

Let $\{T_i\}$ be a net in ball $\mathcal{B}(\mathcal{H})$ and assume that $\phi(T_i) \to F$ in X. So for each g in ball $\mathcal{H}$, $T_i g \to F(g)$ weakly. It is left to the reader to check that there is a linear transformation $T : \mathcal{H} \to \mathcal{H}$ such that $Tg = F(g)$ for all g in ball $\mathcal{H}$. But if $g \in$ ball $\mathcal{H}$, $\|Tg\| = \|F(g)\| \leq 1$. Thus $T \in$ ball $\mathcal{H}$ and $F = \phi(T)$. ■

There are some interesting and unsuspected relationships between the various topologies that have been introduced and the inflation of operators. The first illustration of such a relationship requires a concept that will be more fully discussed in Chapter 8. It is introduced here because it affords an algebraic way to characterize the strong closure of a linear manifold of operators.

8.4 Definition. For any linear manifold $\mathcal{S}$ of $\mathcal{B}(\mathcal{H})$, define

$$\operatorname{Ref}\mathcal{S} = \{T \in \mathcal{B}(\mathcal{H}) : Th \in [\mathcal{S}h] \text{ for all } h \text{ in } \mathcal{H}\}.$$

Call $\operatorname{Ref}\mathcal{S}$ the *attached space* for $\mathcal{S}$.

This definition first appeared in Loginov and Sulman [1975]. At first this may seem an odd concept. In §22 and §56 the attached space is explored in more detail when it is related to the study of invariant subspaces. In the meantime, the reader has the author's apology for not explaining in more detail the significance of the concept or the choice of "Ref" for the notation. But be patient, gentle reader, all will become clear to those who persevere.

8.5 Proposition. *Let $\mathcal{S}$ be a linear manifold in $\mathcal{B}(\mathcal{H})$.*

(a) $\operatorname{Ref}\mathcal{S}$ *is a strongly closed subspace of* $\mathcal{B}(\mathcal{H})$.

(b) $\operatorname{Ref}\operatorname{Ref}\mathcal{S} = \operatorname{Ref}\mathcal{S}$.

(c) *If $\mathcal{S}^*$ is the linear manifold consisting of the adjoints of the operators in $\mathcal{S}$,* $\operatorname{Ref}\mathcal{S}^* = (\operatorname{Ref}\mathcal{S})^*$.

Proof. The proofs of (a) and (c) are left as easy exercises. To prove (b) first observe that $\mathcal{S} \subseteq \operatorname{Ref}\mathcal{S}$ always holds; thus $\operatorname{Ref}\mathcal{S} \subseteq \operatorname{Ref}\operatorname{Ref}\mathcal{S}$. Now let $T \in \operatorname{Ref}\operatorname{Ref}\mathcal{S}$. So for any vector h, $Th \in [\operatorname{Ref}\mathcal{S}h]$. Therefore there is a sequence $\{A_n\}$ in $\operatorname{Ref}\mathcal{S}$ such that $A_n h \to Th$. But since each A_n belongs to $\operatorname{Ref}\mathcal{S}$, $A_n h \in [\mathcal{S}h]$. Therefore $Th \in [\mathcal{S}h]$. Since h was arbitrary, $T \in \operatorname{Ref}\mathcal{S}$. ■

Here is the promised characterization of the strong closure of a linear manifold. The fact that it shows an equivalence between the topological condition of "closure" and what is nearly an algebraic condition indicates that this is potentially very useful.

8.6 Proposition. *If $\mathcal{S}$ is a linear manifold in $\mathcal{B}(\mathcal{H})$, then the* SOT *closure of $\mathcal{S}$ equals*

$$\{T \in \mathcal{B}(\mathcal{H}) : T^{(n)} \in \operatorname{Ref} \mathcal{S}^{(n)} \text{ for } 1 \leq n < \infty\}.$$

Proof. Let $\mathcal{T}$ denote the set in the display. Note that $\mathcal{T}$ is SOT closed (exercise). Clearly $\mathcal{T}$ contains $\mathcal{S}$, and hence its strong closure. Now take an arbitrary operator T from $\mathcal{T}$. If $h_1, \ldots, h_n \in \mathcal{H}$ and $\epsilon > 0$, it must be shown that there is an operator S in $\mathcal{S}$ such that $\|Th_j - Sh_j\| < \epsilon$ for $1 \leq j \leq n$. But if $h = (h_1, \ldots, h_n)$ in $\mathcal{H}^{(n)}$, then the fact that $T \in \mathcal{T}$ implies that $T^{(n)}h \in [\mathcal{S}^{(n)}h]$. But this says exactly what was required. ∎

Even though the WOT closed linear spaces are the same as the SOT closed ones, there is an advantage to working with the strong operator topology. It is, after all, a stronger topology. The preceding proposition shows another advantage.

Exercises.

1. If $\mathcal{H}$ is separable, show that both the relative WOT and relative SOT on ball $\mathcal{B}(\mathcal{H})$ are metrizable. (See Proposition IV.2.1 in [ACFA].)

2. Show that the WOT is the weakest locally convex topology on $\mathcal{B}(\mathcal{H})$ such that every linear functional defined in Proposition 8.1.c is continuous.

3. Let $\mathcal{B}_{00}(\mathcal{H})(= \mathcal{B}_{00}$ if the Hilbert space is understood) denote the algebra of finite rank operators on $\mathcal{H}$. Show that ball $\mathcal{B}_{00}(\mathcal{H})$ is WOT (respectively, SOT) dense in ball $\mathcal{B}(\mathcal{H})$.

4. If $\{E_i\}$ is a net of projections and E is a projection, show that $E_i \to E$ (SOT) if and only if $E_i \to E$ (WOT).

5. Let $\{T_i\}$ be a bounded net in $\mathcal{B}(\mathcal{H})$ and fix an orthonormal basis $\{e_n\}$. (a) Prove that $T_i \to 0$ (WOT) if and only if for all e_n and e_m, $\langle T_i e_n, e_m\rangle \to 0$ with i. (b) Prove that $T_i \to 0$ (SOT) if and only if for all e_n, $\|T_i e_n\| \to 0$ with i.

6. Let $\{T_i\}$ and $\{S_i\}$ be nets in $\mathcal{B}(\mathcal{H})$ and assume that $\{T_i\}$ is uniformly bounded. (a) If $T_i \to 0$ (WOT) and $S_i \to 0$ (SOT), then $T_i S_i \to 0$ (WOT). (b) If, in addition, $T_i \to 0$ (SOT), then $T_i S_i \to 0$ (SOT).

7. Let $\mathcal{H} = \ell^2$ and define S on $\mathcal{H}$ by $Se_n = e_{n+1}$, where $\{e_n\}$ is the standard basis for ℓ^2. (S is called the *unilateral shift*; it will be studied in some detail in §24 through §26.) Examine the sequences $\{S^k\}$, $\{S^{*k}\}$, $\{S^k S^{*k}\}$, and $\{S^{*k} S^k\}$ for their convergence properties in the WOT and SOT. (Hint: use Exercise 5.) Compare with Exercise 6.

8. (Halmos) Let $\mathcal{H}$ be a separable Hilbert space and fix an orthonormal basis $\{e_n\}$ for $\mathcal{H}$. (a) Show that 0 belongs to the weak closure of $\{\sqrt{n}e_n : n \geq 1\}$. (b) Let $\{n_i\}$ be a net of integers such that $\sqrt{n_i}e_{n_i} \to 0$ weakly and define $T_i f = \sqrt{n_i}\langle f, e_{n_i}\rangle e_{n_i}$ for all f in $\mathcal{H}$. Show that $T_i \to 0$ (SOT) but $\{T_i^2\}$ does not converge to 0 (SOT).

9. Let $\{A_i\}$ be a net of hermitian operators on $\mathcal{H}$ such that there is a hermitian operator T with $A_i \leq T$ for all i. If for each h in $\mathcal{H}$, $\{\langle A_i h, h\rangle\}$ is an increasing net of real numbers, show that there is a hermitian operator A such that $A_i \to A$ (WOT).

10. Define the strong* topology on $\mathcal{B}(\mathcal{H})$ using the family of seminorms $\{ q_h : h \in \mathcal{H}\}$, where $q_h(T) = \|Th\| + \|T^*h\|$. Show that a linear functional on $\mathcal{B}(\mathcal{H})$ is strong* continuous if and only if it is strongly continuous.

§9. Spectral measures

Though almost all examples and applications only involve Borel measures defined on locally compact spaces, the properties of a spectral measure be defined and established in the abstract setting of a σ-algebra of subsets of an arbitrary set.

Recall that the infinite sum of a sequence of pairwise orthogonal projections is the projection onto the closed linear span of the ranges of the projections in the sequence. Equivalently, the sequence of finite partial sums of the projections converges in the strong operator topology to this projection.

9.1 Definition. If X is a set, Ω a σ-algebra of subsets of X, and $\mathcal{H}$ a Hilbert space, a *spectral measure* for $(X, \Omega, \mathcal{H})$ is a function $E : \Omega \to \mathcal{B}(\mathcal{H})$ having the following properties.

(a) $E(\Delta)$ is a projection.

(b) $E(\emptyset) = 0$ and $E(X) = 1$.

(c) If $\Delta_1, \Delta_2 \in \Omega$, $E(\Delta_1 \cap \Delta_2) = E(\Delta_1)\, E(\Delta_2)$.

(d) If $\{\Delta_n\}_{n=1}^{\infty}$ is a sequence of pairwise disjoint sets from Ω, then

$$E\left(\bigcup_{n=1}^{\infty} \Delta_n\right) = \sum_{n=1}^{\infty} E(\Delta_n).$$

By condition (c) in the definition, the projections $E(\Delta_n)$ in (d) are pairwise orthogonal since the sets Δ_n are pairwise disjoint. Thus the convergence in part (d) is in the strong operator topology, as discussed before the statement of the definition.

9.2 Examples. (a) Let (X, Ω, μ) be a σ-finite measure space, and, for each Δ in Ω, let $E(\Delta) : L^2(\mu) \to L^2(\mu)$ be defined by $E(\Delta)f = \chi_\Delta f$. That is, $E(\Delta) = M_{\chi_\Delta}$, the operator defined as multiplication by the characteristic function of Δ. (See Proposition 2.6.) E defines a spectral measure for $(X, \Omega, L^2(\mu))$.

(b) Let X be any set, Ω the collection of all subsets of X, and $\mathcal{H}$ a Hilbert space. Fix an orthonormal basis $\mathcal{E}$ for $\mathcal{H}$ and a subset $\{ x_e \in X : e \in \mathcal{E} \}$ of X, indexed by the orthonormal basis. For any set Δ contained in X, let $E(\Delta)$ be the projection onto $[\{ e : x_e \in \Delta \}]$. Then E is a spectral measure for $(X, \Omega, \mathcal{H})$.

Actually, when the Hilbert space is separable and, therefore, the basis is countable, the spectral measure in part (b) above is a special case of the one in part (a) . If $\mu = \sum_n \delta_{x_n}$, the sum of the unit point masses at the points $\{x_n\}$, then there is a natural identification of $\mathcal{H}$ with $L^2(\mu)$, where e_n and $\chi_{\{x_n\}}$ are identified.

The next lemma is crucial to the study of spectral measures. It helps to reduce almost every question concerning spectral measures to a question about scalar-valued measures.

9.3 Lemma. *If E is a spectral measure for $(X, \Omega, \mathcal{H})$ and $g, h \in \mathcal{H}$, then*

$$E_{g,h}(\Delta) \equiv \langle E(\Delta)g, h\rangle$$

defines a countably additive, complex-valued measure on Ω with total variation at most $\|g\|\,\|h\|$.

Proof. Put $\mu = E_{g,h}$; it is left to the reader to verify that μ is a countably additive measure. Now to estimate $\|\mu\|$. Let $\Delta_1, \ldots, \Delta_n$ be pairwise disjoint sets in Ω and let $\alpha_1, \ldots, \alpha_n$ be complex numbers such that $|\alpha_j| = 1$ and $|\langle E(\Delta_j)g, h\rangle| = \alpha_j\langle E(\Delta_j)g, h\rangle$. So $\sum_j |\mu(\Delta_j)| = \sum_j \alpha_j\langle E(\Delta_j)g, h\rangle = \langle\sum_j \alpha_j E(\Delta_j)g, h\rangle \le \|\sum_j \alpha_j E(\Delta_j)g\|\,\|h\|$. But the vectors $\{\alpha_j E(\Delta_j)g : 1 \le j \le n\}$ are pairwise orthogonal. Therefore,

$$\left\|\sum_j \alpha_j E(\Delta_j)g\right\|^2 = \sum_j \|E(\Delta_j)g\|^2 = \|E(\cup_j \Delta_j)g\|^2 \le \|g\|^2.$$

Hence $\sum_j |\mu(\Delta_j)| \le \|g\|\,\|h\|$. ∎

Now to define what is meant by the integral of a bounded measurable function with respect to a spectral measure.

9.4 Proposition. *If E is a spectral measure for $(X, \Omega, \mathcal{H})$ and $\phi: X \to \mathbb{C}$ is a bounded Ω-measurable function, then there is a unique operator A in $\mathcal{B}(\mathcal{H})$ such that if $\epsilon > 0$ and $\{\Delta_1, \ldots, \Delta_n\}$ is an Ω-partition of X with $\sup\{|\phi(x) - \phi(y)| : x, y \in \Delta_k\} < \epsilon$ for $1 \leq k \leq n$, then for any choice of x_k in Δ_k,*

$$\left\| A - \sum_{k=1}^{n} \phi(x_k) E(\Delta_k) \right\| < \epsilon.$$

Proof. For each pair of vectors g, h in $\mathcal{H}$, define $B(g,h) = \int \phi \, dE_{g,h}$. Using Lemma 9.3 it is easy to see that $B : \mathcal{H} \times \mathcal{H} \to \mathbb{C}$ is a sesquilinear form and $|B(g,h)| \leq \|\phi\|_\infty \|g\| \, \|h\|$. Therefore there is a bounded operator A such that $B(g,h) = \langle Ag, h\rangle$ for all vectors g, h in $\mathcal{H}$ and $\|A\| \leq \|\phi\|_\infty$.

Let $\epsilon > 0$ and let $\{\Delta_1, \ldots, \Delta_n\}$ and $x_1, \ldots, x_n$ be as in the statement of the proposition. For arbitrary vectors g, h,

$$\begin{aligned} \left| \langle Ag, h\rangle - \sum_{k=1}^{n} \phi(x_k) \langle E(\Delta_k) g, h\rangle \right| &= \left| \sum_{k=1}^{n} \int_{\Delta_k} [\phi(x) - \phi(x_k)] \, dE_{g,h}(x) \right| \\ &\leq \sum_{k=1}^{n} \int_{\Delta_k} |\phi(x) - \phi(x_k)| \, d|E_{g,h}|(x) \\ &\leq \epsilon \int d|E_{g,h}|(x) \\ &\leq \epsilon \|g\| \, \|h\|. \end{aligned}$$

By taking the supremum over all unit vectors g and h, the desired inequality is established. ■

The operator A obtained for the bounded function ϕ in the preceding proposition is called the *integral* of ϕ with respect to E and is written

$$A = \int \phi \, dE.$$

So for any vectors g, h,

9.5
$$\left\langle \left(\int \phi \, dE \right) g, h \right\rangle = \int \phi \, dE_{g,h}.$$

If $B(X, \Omega)$ denotes the collection of all bounded Ω-measurable, complex-valued functions and it is furnished with the supremum norm, then it is clear that $B(X, \Omega)$ is a Banach space. In fact, it is an abelian C*-algebra, where the star is complex conjugation.

9.6 Proposition. *If E is a spectral measure for $(X,\Omega,\mathcal{H})$ and $\rho : B(X,\Omega) \to \mathcal{B}(\mathcal{H})$ is defined by $\rho(\phi) = \int \phi\, dE$, then ρ is a representation of $B(X,\Omega)$ and $\rho(\phi)$ is a normal operator for every ϕ in $B(X,\Omega)$.*

Proof. The only property of a representation that will be verified here is that ρ is multiplicative. Showing the remaining properties of a representation is a routine exercise.

Fix ϕ, ψ in $B(X,\Omega)$. If $\epsilon > 0$, let $\{\Delta_1, \dots, \Delta_n\}$ be a partition such that for $\omega = \phi, \psi$ or $\phi\psi$, the oscillation of ω on each Δ_k is less than ϵ. Hence by Proposition 9.4, for any choice of points x_k in Δ_k and for $\omega = \phi, \psi, \phi\psi$, $\|\int \omega\, dE - \sum_{k=1}^n \omega(x_k)E(\Delta_k)\| < \epsilon$. Thus for any choice of points x_k in Δ_k,

$$\begin{aligned}
&\left\| \int \phi\psi\, dE - \left(\int \phi\, dE\right)\left(\int \psi\, dE\right) \right\| \\
&\le \epsilon + \left\| \sum_{k=1}^n \phi(x_k)\psi(x_k)E(\Delta_k) - \left[\sum_{i=1}^n \phi(x_i)E(\Delta_i)\right]\left[\sum_{j=1}^n \psi(x_j)E(\Delta_j)\right] \right\| \\
&\quad + \left\| \left[\sum_{i=1}^n \phi(x_i)E(\Delta_i)\right]\left[\sum_{j=1}^n \psi(x_j)E(\Delta_j)\right] - \left(\int \phi\, dE\right)\left(\int \psi\, dE\right) \right\|.
\end{aligned}$$

But $E(\Delta_i)E(\Delta_j) = E(\Delta_i \cap \Delta_j)$ and $\{\Delta_1, \dots, \Delta_n\}$ is a partition. Therefore

$$\begin{aligned}
&\left\| \int \phi\psi\, dE - \left(\int \phi\, dE\right)\left(\int \psi\, dE\right) \right\| \\
&\le \epsilon + \left\| \left[\sum_{i=1}^n \phi(x_i)E(\Delta_i)\right]\left[\sum_{j=1}^n \psi(x_j)E(\Delta_j) - \int \psi\, dE\right] \right\| \\
&\quad + \left\| \left[\sum_{i=1}^n \phi(x_i)E(\Delta_i) - \int \phi\, dE\right]\left[\int \psi\, dE\right] \right\| \\
&\le \epsilon[1 + \|\phi\|_\infty + \|\psi\|_\infty].
\end{aligned}$$

Since ϵ was arbitrary, ρ is multiplicative. ■

If X is a compact space and Ω is the collection of Borel sets, any spectral measure defines a representation of $C(X)$ since $C(X) \subseteq B(X,\Omega) \equiv B(X)$. The next theorem gives a converse of this. But first a lemma that will be used here as well as in the subsequent treatment of spectral measures and normal operators.

9.7 Lemma. *If X is a compact space and ϕ is a bounded Borel function on X, then there is a net $\{u_i\}$ of continuous functions on X such that $\|u_i\|_\infty \le \|\phi\|_\infty$ for all i and $\int u_i\, d\mu \to \int \phi\, d\mu$ for every μ in $M(X)$.*

Proof. This lemma is just a specialization of a general Banach space result to the situation at hand. The general result (Proposition V.4.1 in [ACFA]) is that if $\mathcal{X}$ is a Banach space and $\mathcal{X}$ is considered as a subspace of $\mathcal{X}^{**}$, then ball $\mathcal{X}$ is weak* dense in ball $\mathcal{X}^{**}$. Take $C(X)$ for the Banach space $\mathcal{X}$; so ball $C(X)$ is weak* dense in ball $M(X)^*$. Identifying $B(X)$ with a subspace of $M(X)^* = C(X)^{**}$, the stated proposition says precisely what is wanted to prove this lemma. ■

9.8 Theorem. *If X is a compact space and $\rho : C(X) \to \mathcal{B}(\mathcal{H})$ is a representation, then there is a unique spectral measure E defined on the Borel subsets of X such that for all vectors g, h in $\mathcal{H}$, $E_{g,h}$ is a regular, complex-valued Borel measure and*

$$\rho(u) = \int u\, dE$$

for every continuous function u on X.

Proof. For each pair of vectors g, h in $\mathcal{H}$, $u \to \langle \rho(u)g, h\rangle$ is a bounded linear functional on $C(X)$ with norm at most $\|g\|\,\|h\|$. Thus there is a unique measure $\mu_{g,h}$ in $M(X)$ with $\|\mu_{g,h}\| \le \|g\|\,\|h\|$ such that

9.9 $$\langle \rho(u)g, h\rangle = \int u\, d\mu_{g,h}$$

for all u in $C(X)$. If $\phi \in B(X)$, the bounded Borel functions on X, then $[g, h] \equiv \int \phi\, d\mu_{g,h}$ is easily seen to be a bounded sesquilinear form on $\mathcal{H} \times \mathcal{H}$ with norm at most $\|\phi\|_\infty$. Thus there is a bounded operator $\widetilde{\rho}(\phi)$ such that

9.10 $$\langle \widetilde{\rho}(\phi)g, h\rangle = \int \phi\, d\mu_{g,h}$$

for all g, h.

9.11 Claim. $\widetilde{\rho} : B(X) \to \mathcal{B}(\mathcal{H})$ is a representation and $\widetilde{\rho}(u) = \rho(u)$ for all continuous functions u.

The fact that $\widetilde{\rho}(u) = \rho(u)$ for all u in $C(X)$ follows from (9.9). The first task is to show that $\widetilde{\rho}$ is multiplicative. If $\phi \in B(X)$, the preceding lemma implies there is a net $\{u_i\}$ in $C(X)$ with $\sup_i \|u_i\| \le \|\phi\|$ and such that $\int u_i\, d\mu \to \int \phi\, d\mu$ for all μ in $M(X)$. If $\psi \in B(X)$ and $\mu \in M(X)$, consider $\psi\mu$ as a measure in the usual way. Thus $\int u_i\psi\, d\mu \to \int \phi\psi\, d\mu$ for every ψ in $B(X)$ and μ in $M(X)$. By (9.10), $\widetilde{\rho}(u_i\psi) \to \widetilde{\rho}(\phi\psi)$ (WOT) for every ψ in $B(X)$. In particular, if $\psi \in C(X)$,

$$\widetilde{\rho}(\phi\psi) = \lim_i \widetilde{\rho}(u_i\psi) = \lim_i \rho(u_i)\rho(\psi) = \widetilde{\rho}(\phi)\rho(\psi),$$

where the limits are in the WOT. Hence $\widetilde{\rho}(u_i\psi) = \rho(u_i)\widetilde{\rho}(\psi)$ for all ψ in $B(X)$ and for all u_i. Since $\rho(u_i) \to \widetilde{\rho}(\phi)$ (WOT), this shows that $\widetilde{\rho}$ is multiplicative. The remainder of the proof of Claim 9.11 is left to the reader.

To define the spectral measure E, let $E(\Delta) = \widetilde{\rho}(\chi_\Delta)$ for all Borel sets Δ. It is necessary to check that E is, indeed, a spectral measure. Since every characteristic function is a hermitian idempotent in $B(X)$, the claim implies that $E(\Delta)$ is a projection. That $E(\emptyset) = 0$ and $E(X) = 1$ also follows from the fact that $\widetilde{\rho}$ is a representation. For any pair of Borel sets Δ_1, Δ_2, $E(\Delta_1 \cap \Delta_2) = \widetilde{\rho}(\chi_{\Delta_1\cap\Delta_2}) = \widetilde{\rho}(\chi_{\Delta_1}\chi_{\Delta_2}) = \widetilde{\rho}(\chi_{\Delta_1})\widetilde{\rho}(\chi_{\Delta_2}) = E(\Delta_1)E(\Delta_2)$.

Now let $\{\Delta_n\}$ be a sequence of pairwise disjoint Borel sets. Put $\Delta = \bigcup_n \Delta_n$ and $\Lambda_n = \bigcup_{k=n+1}^\infty \Delta_k$. The fact that E is finitely additive follows easily from the linearity of $\widetilde{\rho}$. Therefore for each h in $\mathcal{H}$

$$\begin{aligned}\left\| E(\Delta)h - \sum_{k=1}^n E(\Delta_k)h \right\|^2 &= \langle E(\Lambda_n)h, E(\Lambda_n)h\rangle \\ &= \langle E(\Lambda_n)h, h\rangle \\ &= \langle \widetilde{\rho}(\chi_{\Lambda_n})h, h\rangle \\ &= \int \chi_{\Lambda_n}\, d\mu_{h,h} \\ &= \mu_{h,h}(\Lambda_n) \to 0,\end{aligned}$$

as $n \to \infty$. Therefore E is a spectral measure.

Now let $\phi \in B(X)$, $\epsilon > 0$, and let $\{\Delta_1, \ldots, \Delta_n\}$ be a partition of X such that the oscillation of ϕ over each Δ_k is less than ϵ. For any choice of points x_k in Δ_k, $\|\phi - \sum_k \phi(x_k)\chi_{\Delta_k}\| < \epsilon$. Thus $\|\widetilde{\rho}(\phi) - \sum_k \phi(x_k)E(\Delta_k)\| < \epsilon$. From the definition of the integral with respect to a spectral measure, it follows that $\widetilde{\rho}(\phi) = \int \phi\, dE$. In particular, $\rho(u) = \int u\, dE$ for every continuous function u.

The proof of uniqueness is an exercise. ∎

Exercises.

1. For the spectral measures in (9.2) give the corresponding representations of $B(X, \Omega)$.

2. If $\rho : C(X) \to \mathcal{B}(\mathcal{H})$ is a representation and E its spectral measure, show that ρ is injective if and only if $E(G) \neq 0$ for every non-empty open subset G of X.

3. Show that a representation $\rho : C(X) \to \mathcal{B}(\mathcal{H})$ has the property that there are no proper subspaces of $\mathcal{H}$ that are invariant for all the operators $\rho(f)$, $f \in C(X)$, if and only if $\dim \mathcal{H} = 1$. (Such a representation is called *irreducible*. They will be studied in detail in §32.) What is the spectral measure?

§10. The Spectral Theorem

If N is a normal operator on the Hilbert space $\mathcal{H}$, then $C^*(N)$, the C*-algebra generated by N and the identity, is abelian. Therefore there is the functional calculus $f \to f(N) \equiv \rho(f)$ that takes $C(\sigma(N))$ isometrically into $\mathcal{B}(\mathcal{H})$, such that the function z is taken to the operator N (2.4). This functional calculus $\rho : C(\sigma(N)) \to \mathcal{B}(\mathcal{H})$ is, therefore, a representation of the C*-algebra $C(\sigma(N))$ and Theorem 9.8 applies to give a spectral measure E on $\sigma(N)$ such that

10.1 $$f(N) = \int_{\sigma(N)} f(z)\, dE(z)$$

for every continuous function f on the spectrum of N. This is the first step in the proof of the Spectral Theorem.

10.2 The Spectral Theorem. *If N is a normal operator, then there is a spectral measure E defined on the Borel subsets of $\sigma(N)$ such that* (10.1) *holds and the following are satisfied:*

(a) *if G is a non-empty relatively open subset of $\sigma(N)$, $E(G) \neq 0$;*

(b) *If $T \in \mathcal{B}(\mathcal{H})$, then $TN = NT$ and $TN^* = N^*T$ if and only if $TE(\Delta) = E(\Delta)T$ for every Borel set Δ.*

Proof. The spectral measure has already been shown to exist such that Equation 10.1 holds. As in the proof of Theorem 9.8, extend ρ to the representation $\widetilde{\rho} : B(\sigma(N)) \to \mathcal{B}(\mathcal{H})$.

Let G be a non-empty relatively open subset of $\sigma(N)$ and let f be a continuous function on $\sigma(N)$ such that $0 \leq f \leq \chi_G$, $f \neq 0$. So $0 \leq \rho(f) = f(N) \leq \widetilde{\rho}(\chi_G)$. Since ρ is an isometry, $\rho(f) \neq 0$, establishing (a).

Now let T be a bounded operator on $\mathcal{H}$ such that $TN = NT$ and $TN^* = N^*T$. An algebraic argument shows that $Tp(N, N^*) = p(N, N^*)T$ for every polynomial in N and N^*. That is, $T\rho(p) = \rho(p)T$ for every polynomial p in z and $\bar{z}$. By the Stone–Weierstrass Theorem, $T\rho(f) = \rho(f)T$ for all f in $C(\sigma(N))$. Let $\Omega \equiv \{\Delta : \Delta \text{ is a Borel subset of } \sigma(N) \text{ and } TE(\Delta) = E(\Delta)T\}$. Clearly Ω contains $\sigma(N)$ as well as the empty set. It is left to the reader to show that Ω is a σ-algebra of sets. If it can be shown that Ω contains the open sets, then Ω must be the σ-algebra of all Borel sets.

If G is open in $\sigma(N)$, there is a sequence of continuous functions $\{ f_n \}$ on $\sigma(N)$ such that $0 \leq f_n \leq \chi_G$ and $f_n(z) \nearrow \chi_G(z)$ for all z in $\sigma(N)$. Thus for all vectors g, h in $\mathcal{H}$

$$
\begin{aligned}
\langle TE(G)g,h\rangle &= \langle E(G)g,T^*h\rangle \\
&= E_{g,T^*h}(G) \\
&= \lim \int f_n \, dE_{g,T^*h} \\
&= \lim\langle f_n(N)g,T^*h\rangle \\
&= \lim\langle Tf_n(N)g,h\rangle \\
&= \lim\langle f_n(N)Tg,h\rangle \\
&= \langle E(G)Tg,h\rangle.
\end{aligned}
$$

To prove the converse of (b), note that the hypothesis implies that $T\widetilde{\rho}(f) = \widetilde{\rho}(f)T$ for every simple function. Now approximate the functions z and $\bar{z}$ by simple functions. The details are left for the reader. ■

Once Fuglede's Theorem in §12 is proved, the condition that $TN^* = N^*T$ in part (b) of the Spectral Theorem can be omitted.

Now define $\phi(N)$ for every bounded Borel function ϕ on $\sigma(N)$ by

$$\phi(N) \equiv \int \phi \, dE,$$

where E is the spectral measure associated with the normal operator N as in the preceding theorem. In fact, $\phi(N) = \widetilde{\rho}(\phi)$, where $\widetilde{\rho}$ is the extension of ρ that appears in the proof of Theorem 9.8. The proof of the next theorem is a routine assembly of the pertinent facts from the material that has passed in front of our eyes in the last two sections. This assemblage of spectral visions is left to the reader. It is a good exercise to test whether this material has only passed in front of the reader's eyes or has gone somewhat further into the cranial cavity.

10.3 Theorem. *If N is a normal operator with spectral measure E and $B(\sigma(N))$ is the C^*-algebra of bounded Borel functions on $\sigma(N)$, then the map*

$$\phi \to \phi(N)$$

is a representation of $B(\sigma(N))$. If $\{\phi_i\}$ is a net in $B(\sigma(N))$ such that $\int \phi_i \, d\mu \to 0$ for every μ in $M(X)$, then $\phi_i(N) \to 0$ (WOT). This map is unique in the sense that if $\tau : B(\sigma(N)) \to \mathcal{B}(\mathcal{H})$ is a representation such that $\tau(z) = N$ and $\tau(\phi_i) \to 0$ (WOT) whenever $\{\phi_i\}$ is a net in $B(\sigma(N))$ such that $\int \phi_i \, d\mu \to 0$ for every μ in $M(\sigma(N))$, then $\tau(\phi) = \phi(N)$ for every ϕ in $B(\sigma(N))$.

If ϕ is a Borel function on a subset of $\mathbb{C}$ that contains $\sigma(N)$ and ϕ is bounded on the spectrum, $\phi(N)$ can still be defined by considering the restriction of ϕ to $\sigma(N)$. This simple statement will have technical advantages later.

10.4 Proposition. *With the notation from Proposition 2.6, fix a particular representative of the element ϕ in $L^\infty(\mu)$ and also denote it by ϕ.*

(a) *The spectral measure for M_ϕ is given by $\Delta \to M_{\chi_{\phi^{-1}(\Delta)}}$.*

(b) *If ψ is a bounded Borel function on $\sigma(M_\phi)$, then $\psi(M_\phi) = M_{\psi\circ\phi}$.*

Proof. If $E(\Delta) = M_{\chi_{\phi^{-1}(\Delta)}}$ for every Borel set Δ in the plane, E is a spectral measure (exercise). Now if $\epsilon > 0$, there is a Borel partition $\{\Delta_1, \ldots, \Delta_n\}$ of $\phi(X)$ such that for any choice of points x_k in $\phi^{-1}(\Delta_k)$, $\|\phi - \sum_k \phi(x_k)\chi_{\phi^{-1}(\Delta_k)}\|_\infty < \epsilon$. If $z_k = \phi(x_k)$, then (2.6) implies that $\epsilon > \|M_\phi - \sum_k \phi(x_k) M_{\chi_{\phi^{-1}(\Delta_k)}}\| = \|M_\phi - \sum_k z_k E(\Delta_k)\|$. By uniqueness, E is the spectral measure for M_ϕ.

Part (a) says that $\psi(M_\phi) = M_{\psi\circ\phi}$ whenever ψ is a characteristic function. An algebraic manipulation shows that this is also true when ψ is a simple function. Now uniformly approximate an arbitrary bounded Borel function by a simple function and complete the proof. ■

Though the next proposition is not connected with the Spectral Theorem, this seems an appropriate place to state it. Recall that $L^\infty(\mu)$ is the dual space of $L^1(\mu)$ and therefore has a weak* topology.

10.5 Proposition. *If (X, Ω, μ) is any σ-finite measure space and $\{\phi_i\}$ is a net in $L^\infty(\mu)$, then $\phi_i \to 0$ weak* in $L^\infty(\mu)$ if and only if $M_{\phi_i} \to 0$ WOT in $\mathcal{B}(L^2(\mu))$.*

Proof. Assume that $\phi_i \to 0$ weak* in $L^\infty(\mu)$. If $f, g \in L^2(\mu)$, $\langle M_{\phi_i} f, g\rangle = \int \phi_i f\bar{g}\, d\mu \to 0$ since $f\bar{g} \in L^1(\mu)$. Conversely, assume $M_{\phi_i} \to 0$ WOT in $\mathcal{B}(L^2(\mu))$. If $h \in L^1(\mu)$, factor h as $h = f\bar{g}$, where $f, g \in L^2(\mu)$. Thus $\int \phi_i h\, d\mu = \langle M_{\phi_i} f, g\rangle \to 0$. ■

Exercises.

1. For each $k \geq 1$, let N_k be a normal operator on $\mathcal{H}_k$ with spectral measure E_k. If $\sup_k \|N_k\| < \infty$ and $N = \bigoplus_k N_k$, show that $\sigma(N) = \operatorname{cl}[\bigcup_k \sigma(N_k)]$ and the spectral measure E for N is $E(\Delta) = \bigoplus_k E_k(\Delta)$ for every Borel subset Δ of $\sigma(N)$.

In the exercises that follow, N is always a normal operator on a Hilbert space $\mathcal{H}$ and E is its spectral measure.

2. Show that λ is an eigenvalue for N if and only if $E(\{\lambda\}) \neq 0$, in which case the range of $E(\{\lambda\})$ is the eigenspace $\ker(N - \lambda)$.

3. If Δ is a subset of $\sigma(N)$ that is both closed and relatively open, show that $E(\Delta)$ coincides with the Riesz idempotent for Δ.

4. Show that the parts of the polar decomposition of N commute. Show that it is always possible to write $N = U|N|$, where U is a unitary that is a function of N. (This may not be the polar decomposition of N. Why?)

5. If A is a positive contraction with spectral decomposition $A = \int_0^1 t\,dE(t)$, and for each $k \geq 1$

$$P_k = E\left(\bigcup_{j=1}^{2^k-1}\left(\frac{j}{2^k}, \frac{j+1}{2^k}\right]\right),$$

show that $A = \sum_{k=1}^{\infty} 2^{-k} P_k$, where the convergence is in the norm.

6. Show that N is compact if and only if for every $\epsilon > 0$, $E(\{\, z : |z| > \epsilon \,\})$ has finite rank.

7. This exercise interprets the Spectral Theorem for compact normal operators. Assume N is compact and $\{\,\lambda_1, \lambda_2, \dots\}$ are the distinct non-zero points in $\sigma(N)$. Let $E_n \equiv E(\{\lambda_n\})$. (a) Show that each E_n is a finite rank, non-zero projection and $E_n \perp E_m$ for $n \neq m$. (b) Show that $N = \sum_n \lambda_n E_n$, where convergence is in norm. (c) Show that $[\sum_n E_n]^\perp = \ker N$. (d) Let $\{e_k\}$ be an orthonormal sequence obtained by taking a basis for each space E_n and then taking the union of these. Let $\zeta_k = \lambda_n$ if $e_k \in E_n$. Show that $N = \sum_k \zeta_k P_k$, where P_k is the projection onto the one dimensional space $[e_k]$. (e) If ϕ is a bounded Borel function on the plane, show that $\phi(N) = \sum_n \phi(\lambda_n) E_n = \sum_k \phi(\zeta_k) P_k$. (f) Give a necessary and sufficient condition on a Borel function ϕ such that $\phi(N)$ is compact. (g) Give a necessary and sufficient condition on a continuous function f such that $f(N)$ is compact.

8. Show that if A is hermitian, then $U = \exp(iA)$ is unitary. Show that every unitary can be so written. Find the spectral measure of the unitary in terms of that of the hermitian operator.

9. Prove that the set of unitaries on Hilbert space when furnished with its norm topology is an arcwise connected space. Is it simply connected?

10. Show that every normal operator on a separable Hilbert space has a square root that commutes with it. Uniqueness?

11. Show that for any normal operator N there is a sequence of invertible normal operators $\{N_k\}$ such that $\|N_k - N\| \to 0$. Show that the normal operators N_k can be chosen such that $\sigma(N_k)$ is either: (i) finite; (ii) a set bounded by a finite number of smooth curves; or (iii) a finite collection of

line segments. Can the spectrum of each N_k be chosen to be connected? What if $\sigma(N)$ is connected?

12. If $N_1, \ldots, N_p$ are commuting normal operators on $\mathcal{H}$ such that $N_j N_k^* = N_k^* N_j$ for $1 \leq j, k \leq p$, show that there is a unique spectral measure E defined on a compact subset X of $\mathbb{C}^p$ such that for $1 \leq k \leq p$, $N_k = \int z_k \, dE(z)$. (Here z_k denotes the k-th coordinate of the point z in $\mathbb{C}^p$.) How do you recapture the spectral measure for each N_k from the spectral measure E?

13. Show that if $N_1, \ldots, N_p$ are compact operators as in Exercise 12, then there is a basis for $\mathcal{H}$ that simultaneously diagonalizes each N_k.

14. Show that the operator U defined on $L^2(\mathbb{R})$ by $Uf(t) = f(t-1)$ is a unitary. Find its spectral measure.

§11. Star-cyclic normal operators.

Recall that a subspace $\mathcal{M}$ of $\mathcal{H}$ is reducing for a collection of operators $\mathcal{S}$ if both $\mathcal{M}$ and $\mathcal{M}^\perp$ are invariant for every operator in $\mathcal{S}$. This is equivalent to the condition that $S\mathcal{M} \subseteq \mathcal{M}$ and $S^*\mathcal{M} \subseteq \mathcal{M}$ for every S in $\mathcal{S}$. Thus if $S^* \in \mathcal{S}$ whenever $S \in \mathcal{S}$, then every invariant subspace for $\mathcal{S}$ is reducing. (See [ACFA], Proposition II.3.7.)

11.1 Definition. If $\mathcal{S} \subseteq \mathcal{B}(\mathcal{H})$, a vector e_0 is *star-cyclic* for $\mathcal{S}$ if $\mathcal{H}$ is the smallest reducing subspace for $\mathcal{S}$ that contains e_0. The vector e_0 is *cyclic* for $\mathcal{S}$ if $\mathcal{H}$ is the smallest invariant subspace for $\mathcal{S}$ that contains e_0. $\mathcal{S}$ is *star-cyclic* or *cyclic* if it has a star-cyclic or a cyclic vector.

Note that the vector e_0 is star-cyclic for $\mathcal{S}$ if and only if $\mathcal{H} = \text{cl}\{ Ae_0 : A \in C^*(\mathcal{S}) \}$, where $C^*(\mathcal{S})$ is the C*-algebra generated by $\mathcal{S}$ and the identity. Similarly e_0 is cyclic for $\mathcal{S}$ if $\mathcal{H} = \text{cl}\{ p(A)e_0 : p \text{ is a polynomial and } A \in \mathcal{S} \}$. So e_0 is star-cyclic for $\mathcal{S}$ if and only if it is cyclic for $C^*(\mathcal{S})$. In this section interest will focus on the case where $\mathcal{S}$ is a single operator. Observe that if an operator T is a star-cyclic operator, the underlying Hilbert space must be separable. Now to characterize the normal operators that appear in Proposition 2.5.

11.2 Theorem. *A normal operator N is star-cyclic if and only if it is unitarily equivalent to N_μ for some compactly supported measure μ on $\mathbb{C}$. If e_0 is a star-cyclic vector for N, then the unitary $U\colon \mathcal{H} \to L^2(\mu)$ such that $N_\mu = UNU^{-1}$ can be chosen such that $Ue_0 = 1$, in which case U and μ are unique.*

Proof. It is an easy consequence of the Stone–Weierstrass Theorem that N_μ is star-cyclic with the constant function 1 as the vector. So assume that N is a normal operator with star-cyclic vector e_0. If E is the spectral measure for N, let $\mu(\Delta) = \langle E(\Delta)e_0, e_0\rangle = \|E(\Delta)e_0\|^2$. If $K = \operatorname{supp}\mu$ and $\phi \in B(K)$, then

$$\begin{aligned}\|\phi(N)e_0\|^2 &= \langle \phi(N)e_0, \phi(N)e_0\rangle \\ &= \langle |\phi|^2(N)e_0, e_0\rangle \\ &= \int |\phi|^2\, d\mu.\end{aligned}$$

If $B(K)$ is considered as a submanifold of $L^2(\mu)$, then the above equations show that $U(\phi(N)e_0) = \phi$ is a well defined isometry from a dense manifold in $\mathcal{H}$ onto $B(K)$, a dense manifold in $L^2(\mu)$. Extend U to an isomorphism $U : \mathcal{H} \to L^2(\mu)$. Note that by taking ϕ to be the constantly 1 function or the function z it follows that $Ue_0 = 1$ and $UNe_0 = z$. If $\phi \in B(K)$, then

$$UNU^{-1}(\phi) = UN(\phi(N)e_0) = U[(z\phi)(N)e_0] = z\phi = N_\mu\phi.$$

Therefore $UNU^{-1} = N_\mu$.

The proof of uniqueness is left as an exercise. ■

11.3 Definition. Two measures μ and ν are *mutually absolutely continuous* if they have the same sets of measure zero. This is denoted by $[\mu] = [\nu]$. In this case it is possible to form the Radon–Nikodym derivatives $d\mu/d\nu$ and $d\nu/d\mu$. Say that μ and ν are *boundedly mutually absolutely continuous* if $[\mu] = [\nu]$ and the two Radon–Nikodym derivatives are bounded functions.

11.4 Theorem. *The star-cyclic normal operators N_μ and N_ν are unitarily equivalent if and only if $[\mu] = [\nu]$.*

Proof. If $[\mu] = [\nu]$ and $\phi = d\mu/d\nu$, define $U : L^2(\mu) \to L^2(\nu)$ by $Uf = f\sqrt{\phi}$. Now $\|Uf\|^2 = \int |f|^2\phi\, d\nu = \int |f|^2\, d\mu = \|f\|^2$. Thus U is an isometry. Since $\phi^{-1} = d\nu/d\mu$, if $g \in L^2(\nu)$, then $f = \sqrt{\phi^{-1}}g \in L^2(\mu)$ and $Uf = g$. Thus U is an isomorphism. Finally, for each g in $L^2(\nu)$

$$UN_\mu U^{-1}g = UN_\mu(\sqrt{\phi^{-1}}g) = U(z\sqrt{\phi^{-1}}g) = zg = N_\nu g.$$

Conversely, assume there is an isomorphism $V\colon L^2(\mu) \to L^2(\nu)$ such that $VN_\mu V^{-1} = N_\nu$. Put $\psi = V(1)$; so $\psi \in L^2(\nu)$. Since it is also true that $VN_\mu{}^*V^{-1} = N_\nu{}^*$, algebraic manipulation shows that $Vp(N_\mu.N_\mu{}^*)V^{-1} = p(N_\nu, N_\nu{}^*)$ for every polynomial p in z and $\bar{z}$. Hence for any such polynomial, $V(p) = Vp(N_\mu, N_\mu{}^*)(1) = p(N_\nu.N_\nu{}^*)V(1) = \psi p$ in $L^2(\nu)$. Because V is an isometry, this implies that $\int |p|^2\, d\mu = \int |p|^2|\psi|^2\, d\nu$. Now because $N_\mu \cong N_\nu$, $\operatorname{supp}\mu = \sigma(N_\mu) = \sigma(N_\nu) = \operatorname{supp}\nu \equiv K$. Hence, taking uniform limits of polynomials in z and $\bar{z}$ in this last equation, it follows that $\int |f|^2\, d\mu = \int |f|^2|\psi|^2\, d\nu$ for every continuous function f on K. But this says that the two measures μ and $|\psi|^2\, d\nu$ integrate every positive continuous function on K the same way. Thus $\mu = |\psi|^2\, d\nu$ by the uniqueness part of the Riesz Representation Theorem. Therefore, $\mu << \nu$. Applying the same argument to the isomorphism V^{-1} shows that $\nu << \mu$. Thus μ and ν are mutually absolutely continuous. ∎

It should be mentioned that a star-cyclic normal operator is also cyclic. This deeper result is from Bram [1955]. (Also see Theorem V.14.21 in Conway [1991].)

This section closes with an application of the characterization of star-cyclic normal operators to derive what some call the Spectral Theorem. It was already observed that all multiplication operators on L^2 spaces are normal. The converse is also true.

11.5 Theorem. *If N is a normal operator, then there is a measure space (X, Ω, μ) and a function ϕ in $L^\infty(\mu)$ such that N is unitarily equivalent to M_ϕ acting on $L^2(\mu)$.*

Proof. If $e \in \mathcal{H}$, let $\mathcal{H}_e$ be the smallest reducing subspace for N that contains e. Using Zorn's Lemma, find a maximal family of unit vectors $\mathcal{E}$ such that the spaces $\{\, \mathcal{H}_e : e \in \mathcal{E}\,\}$ are pairwise orthogonal. By maximality, $\mathcal{H} = \bigoplus\{\, \mathcal{H}_e : e \in \mathcal{E}\,\}$. For each e in $\mathcal{E}$, let $N_e = N|\mathcal{H}_e$. So N_e is a star-cyclic normal operator and there is a measure supported on $X_e = \sigma(N_e)$ and an isomorphism $U_e : \mathcal{H}_e \to L^2(\mu_e)$ such that $U^{-1}N_{\mu_e}U = N_e$. Let X be the disjoint union of the sets $\{\, X_e : e \in \mathcal{E}\,\}$ and define Ω to be the collection of all subsets Δ of X such that for every e in $\mathcal{E}$, $\Delta \cap X_e$ is a Borel set. It is for the reader to check that Ω is a σ-algebra of sets. Note that a function on X is Ω-measurable if and only if its restriction to each X_e is a Borel function.

Now define a measure μ on Ω by setting

$$\mu(\Delta) = \sum \{\mu_e(\Delta \cap X_e) : e \in \mathcal{E}\}.$$

(Of course, μ may not be a σ-finite measure. A measurable subset Δ of X, however, has $\mu(\Delta) = 0$ if and only if $\mu_e(\Delta \cap X_e) = 0$ for every e in $\mathcal{E}$.) Also define a function $\phi : X \to \mathbb{C}$ by letting $\phi(z) = z$ when $z \in X_e$. It is easy to check from the definitions that ϕ is Ω-measurable. Also note that for z in X_e, $|z| \le \|N_e\| \le \|N\|$. Thus $\phi \in L^\infty(\mu)$.

Each space $L^2(\mu_e)$ can be identified with a subspace of $L^2(\mu)$ by extending each function in $L^2(\mu_e)$ to X by setting it to be 0 off X_e. With this identification, $L^2(\mu) = \bigoplus_e L^2(\mu_e)$. Finally define $U : \mathcal{H} \to L^2(\mu)$ by

$U = \bigoplus_e U_e$. Clearly U is an isomorphism and a routine exercise shows that $U^{-1}M_\phi U = \bigoplus_e N_e = N$. ∎

As pointed out, the measure space (X, Ω, μ) is not necessarily σ-finite. It is, however, *decomposable*. That is, it has the property that a set Δ in Ω has measure 0 if $\mu(\Delta \cap \Lambda) = 0$ for every Ω-measurable set Λ that has finite μ-measure. It turns out the decomposable measure spaces are the proper setting for measure theory and enjoy the properties that courses in measure theory usually only prove under the assumption of σ-finiteness. For example the Radon–Nikodym Theorem holds for decomposable measure spaces as does the fact that the Banach space dual of the L^1 space is the L^∞ space. Such properties are false without some type of assumption. The interested reader can look at Kelley [1966], where these measure theoretic results are shown to be equivalent to decomposability. This book will continue to assume the σ-finiteness of all measure spaces to avoid the technical details required to handle the case of a decomposable measure space. It is also important to remain as deep in the comfort zone of the reader as possible.

Later the focus of this book will be on separable Hilbert spaces, the locale of all interesting operator theory. Note that if the underlying Hilbert space $\mathcal{H}$ is separable, then the measure space obtained in the preceding theorem is σ-finite. In fact, the maximal family of vectors $\mathcal{E}$ obtained in the proof must be countable and each of the sets X_e has finite μ measure.

Say that a measure space (X, Ω, μ) is *separable* if the space $L^2(\mu)$ is separable. It follows that a separable measure space must be σ-finite (Exercise 5).

Exercises.

1. If μ is a compactly supported measure on the plane, show that a function f in $L^2(\mu)$ is star-cyclic for N_μ if and only if $\mu(\{\, z : f(z) = 0 \,\}) = 0$.

2. Give an example of a star-cyclic normal operator and a star-cyclic vector that is not cyclic.

3. Prove the uniqueness statement in Theorem 11.2.

4. If μ and ν are compactly supported measures on $\mathbb{C}$, show that the following statements are equivalent. (a) μ and ν are boundedly mutually absolutely continuous. (b) There is an isomorphism $U\colon L^2(\mu) \to L^2(\nu)$ such that $UN_\mu U^{-1} = N_\nu$ and $U[L^\infty(\mu)] \subseteq L^\infty(\nu)$. (c) The identity map $Rp(z, \bar{z}) = p(z, \bar{z})$ on polynomials in z and $\bar{z}$ extends to a bounded bijection $R\colon L^2(\mu) \to L^2(\nu)$.

5. Show that a separable measure space is σ-finite. If (X, Ω, μ) is a separable measure space, are the spaces $L^p(\mu)$, $1 \le p < \infty$, separable?

6. If (X, Ω, μ) is a separable measure space and $\phi \in L^\infty(\mu)$, show that M_ϕ on $L^2(\mu)$ is star-cyclic if and only if there is a measurable subset Y of X having full measure on which ϕ is one-to-one.

7. Let N be a *diagonalizable* normal operator. That is, N is a normal operator such that there is an orthonormal basis consisting of eigenvectors for N. (a) Show that N is star-cyclic if and only if there are points $\lambda_1, \lambda_2, \ldots$ in the plane such that $N \cong N_\mu$, where $\mu = \sum_n 2^{-n}\delta_{\lambda_n}$. In this case, each λ_n is an eigenvalue for N and the eigenspace of each eigenvalue is one dimensional. (b) Characterize the star-cyclic vectors in terms of their expansion with respect to the basis of eigenvectors. (c) If N is compact, show that every cyclic vector is star-cyclic.

8. Let $N_1, \ldots, N_p$ be commuting normal operators on $\mathcal{H}$ such that $N_j N_k^* = N_k^* N_j$ for $1 \leq j, k \leq p$. (See Exercise 10.12.) Show that the set $\{ N_1, \ldots, N_p \}$ has a star-cyclic vector e_0 if and only if there is a compactly supported measure μ on $\mathbb{C}^p$ and an isomorphism $U : \mathcal{H} \to L^2(\mu)$ such that for $1 \leq k \leq p$ and each f in $L^2(\mu)$, $UN_kU^{-1}f = z_k f$, where z_k is the k-th coordinate function on $\mathbb{C}^p$. If the star-cyclic vector is given, then the isomorphism U can be chosen such that $Ue_0 = 1$. In this case U and μ are unique.

9. Using the preceding exercise, show that if $N_1, \ldots, N_p$ are commuting normal operators on $\mathcal{H}$ such that $N_j N_k^* = N_k^* N_j$ for $1 \leq j, k \leq p$, then there is a measure space (X, Ω, μ), functions $\phi_1, \ldots, \phi_p$ in $L^\infty(\mu)$, and an isomorphism $U : \mathcal{H} \to L^2(\mu)$ such that $U^{-1} M_{\phi_k} U = N_k$ for $1 \leq k \leq p$.

§12. The commutant

Interest here is primarily in looking at operators that commute with a normal operator. Some degree of generality, however, is desirable so that results here will also apply to non-normal operators.

12.1 Definition. If $\mathcal{S} \subseteq \mathcal{B}(\mathcal{H})$, the *commutant* of $\mathcal{S}$ is denoted by $\mathcal{S}'$ and is defined to be the set of all operators that commute with each operator in $\mathcal{S}$. That is,

$$\mathcal{S}' \equiv \{ T \in \mathcal{B}(\mathcal{H}) : TS = ST \text{ for all } S \text{ in } \mathcal{S} \}.$$

The *double commutant* of $\mathcal{S}$ is defined by $\mathcal{S}'' \equiv \{\mathcal{S}'\}'$.

12.2 Proposition. *Let $\mathcal{S}$ be a subset of $\mathcal{B}(\mathcal{H})$.*

(a) *If $\mathcal{S} \subseteq \mathcal{T}$, then $\mathcal{T}' \subseteq \mathcal{S}'$ and $\{\mathcal{S}''\}' = \mathcal{S}'$.*

(b) *$\mathcal{S}'$ is a strongly closed algebra that contains the identity.*

(c) *If* $A = A_1 \oplus A_2 \oplus \cdots$ *is a bounded operator on* $\mathcal{H} = \mathcal{H}_1 \oplus \mathcal{H}_2 \oplus \cdots$ *and* $T = [T_{ij}] \in \mathcal{B}(\mathcal{H})$, *then* $T \in \{A\}'$ *if and only if* $T_{ij}A_j = A_iT_{ij}$ *for all* i, j.

(d) *If* $T = [T_{ij}] \in \mathcal{B}(\mathcal{H}^{(n)})$ *and* $A \in \mathcal{B}(\mathcal{H})$, *then* $T \in \{A^{(n)}\}'$ *if and only if* $T_{ij}A = AT_{ij}$ *for all* i, j.

(e) *If* $\mathcal{S} \subseteq \mathcal{B}(\mathcal{H})$, $[\mathcal{S}^{(n)}]'' = [\mathcal{S}'']^{(n)}$.

Proof. (a) The first part of (a) is routine. For the second part, note that since $\mathcal{S} \subseteq \mathcal{S}''$, $\{\mathcal{S}''\}' \subseteq \mathcal{S}'$. If $A \in \mathcal{S}'$ and $B \in \mathcal{S}''$, $AB = BA$. Thus $A \in \{\mathcal{S}\}''$.

The proof of (b) is a straightforward exercise in applying the definitions. The proof of (c) follows by elementary matrix computations. Part (d) is a special case of (c).

Part (e) will be proved under the additional assumptions that $\mathcal{S}$ consists of a single operator A and $n = 2$. The reader can then supply the bookkeeping and notation to prove the full statement. Let $T = [T_{ij}] \in \{A \oplus A\}''$. Since $\begin{bmatrix} 0 & 1 \\ 0 & 0 \end{bmatrix} \in \{A \oplus A\}'$, a matrix computation shows that $T_{11} = T_{22}$ and $T_{21} = 0$. A similar computation with the adjoint of this matrix shows that $T_{12} = 0$. If $S = T_{11} = T_{22}$, $T = S \oplus S$. If $X \in \{A\}'$, then $X \oplus X \in \{A \oplus A\}'$ and so $T(X \oplus X) = (X \oplus X)T$. It follows that $S \in \{A\}''$. Thus, $\{A \oplus A\}'' \subseteq [\{A\}'']^{(2)}$. Conversely, if $S \in \{A\}''$, then an application of part (d) shows that $S \oplus S \in \{A \oplus A\}''$. ■

One important way in which the commutant surfaces is the observation ([ACFA], Proposition II.3.7) that a subspace reduces a set of operators if and only if the corresponding projection belongs to the commutant of the set. This observation and an application of Proposition 8.6 permit a proof a deep result of von Neumann [1929].

12.3 The Double Commutant Theorem. *If* $\mathcal{A}$ *is a* C^**-subalgebra of* $\mathcal{B}(\mathcal{H})$ *that contains the identity, then the strong closure of* $\mathcal{A}$ *is* $\mathcal{A}''$.

Proof. Let $\mathcal{S}$ denote the strong closure of $\mathcal{A}$. Since $\mathcal{A} \subseteq \mathcal{A}''$ and $\mathcal{A}''$ is strongly closed, $\mathcal{S} \subseteq \mathcal{A}''$. Now fix an operator T in $\mathcal{A}''$ and fix $n \geq 1$. If $h = (h_1, \ldots, h_n) \in \mathcal{H}^{(n)}$, $\mathcal{M} \equiv [\mathcal{A}^{(n)}h]$ is an invariant subspace for $\mathcal{A}^{(n)}$. But $\mathcal{A}^{(n)}$ is a C*-algebra, so $\mathcal{M}$ is reducing for $\mathcal{A}^{(n)}$. So if P is the projection of $\mathcal{H}^{(n)}$ onto $\mathcal{M}$, $P \in \{\mathcal{A}^{(n)}\}'$. Therefore $T^{(n)}P = PT^{(n)}$ by Proposition 12.2.e. That is, $\mathcal{M}$ reduces $T^{(n)}$. In particular, $\mathcal{M}$ is invariant for $T^{(n)}$ and so $T^{(n)}h \in [\mathcal{A}^{(n)}h]$. Thus $T^{(n)} \in \operatorname{Ref} \mathcal{A}^{(n)}$ for any $n \geq 1$. By Proposition 8.6, $T \in \mathcal{S}$. ■

Often a theorem can be seen to be important when, like the Double Commutant Theorem, it shows that two statements whose natures are distinct are equivalent. In the present case an algebraic statement is equivalent to a topological one. This theorem forms the motivation in the next section for defining a von Neumann algebra. But a few results remain to be obtained about the commutant of a normal operator before introducing this topic. The first of these will be regarded by some as an example, while, in fact, it's much more. These algebras form the fundamental building blocks for all abelian von Neumann algebras. Recall the definition of the multiplication operator M_ϕ on $L^2(\mu)$ (2.6).

12.4 Proposition. *If (X, Ω, μ) is a σ-finite measure space and $\mathcal{A}_\mu = \{ M_\phi : \phi \in L^\infty(\mu) \}$, then*

$$\mathcal{A}_\mu = \mathcal{A}_\mu' = \mathcal{A}_\mu''.$$

Proof. It is shown that $\mathcal{A}_\mu = \mathcal{A}_\mu'$, which implies that the algebra equals its double commutant since it is abelian. Let $T \in \mathcal{A}_\mu'$. Consider two cases.

Case 1. $\mu(X) < \infty$. Here $1 \in L^2(\mu)$. Let $\phi = T(1)$; so $\phi \in L^2(\mu)$. If $\psi \in L^\infty(\mu)$, then $\psi \in L^2(\mu)$ and $T(\psi) = TM_\psi 1 = M_\psi T(1) = \psi\phi$. Also $\|\phi\psi\|_2 = \|T\psi\|_2 \le \|T\| \, \|\psi\|_2$.

Let $\Delta_n = \{x \in X : |\phi(x)| \ge n\}$. Putting $\psi = \chi_{\Delta_n}$ in the preceding paragraph gives that

$$\|T\|^2 \mu(\Delta_n) = \|T\|^2 \|\psi\|^2 \ge \|\phi\psi\|^2 = \int_{\Delta_n} |\phi|^2 \, d\mu \ge n^2 \mu(\Delta_n).$$

Since T is bounded, $\mu(\Delta_n) = 0$ for sufficiently large n. That is, $\phi \in L^\infty(\mu)$. But $T = M_\phi$ on $L^\infty(\mu)$, which is dense in $L^2(\mu)$. Thus $T = M_\phi$.

Case 2. $\mu(X) = \infty$. For each measurable set Δ having finite measure, let $L^2(\mu|\Delta) = \{ f \in L^2(\mu) : f = 0 \text{ off } \Delta \}$. So $L^2(\mu|\Delta)$ is the range of the projection M_{χ_Δ} and, hence, is left invariant by T. Let $T_\Delta = T|L^2(\mu|\Delta)$ and apply Case 1 to T_Δ. Thus for every measurable set Δ having finite measure, there is a function ϕ_Δ in $L^\infty(\mu|\Delta)$ such that $Tf = \phi_\Delta f$ for all f in $L^2(\mu|\Delta)$. If Δ_1 and Δ_2 are two such measurable sets and $f \in L^2(\mu|(\Delta_1 \cap \Delta_2))$, then $\phi_{\Delta_1} f = Tf = \phi_{\Delta_2} f$. Thus $\phi_{\Delta_1} = \phi_{\Delta_2}$ on $\Delta_1 \cap \Delta_2$.

Because (X, Ω, μ) is σ-finite, $X = \bigcup_n \Delta_n$, where $\Delta_n \in \Omega$ and $\mu(\Delta_n) < \infty$ for each $n \ge 1$. Since the different functions ϕ_{Δ_n} agree on the overlap of the corresponding sets, setting $\phi(x) = \phi_{\Delta_n}(x)$ when $x \in \Delta_n$ gives a well defined function. It is seen that ϕ is measurable since $\phi|\Delta_n = \phi_n$ is measurable for each $n \ge 1$. Also, $\|\phi|\Delta_n\|_\infty = \|T|L^2(\mu|\Delta_n)\| \le \|T\|$, so $\phi \in L^\infty(\mu)$. Moreover $Tf = M_\phi f$ whenever $n \ge 1$ and $f \in L^2(\mu|\Delta_n)$.

Because $X = \bigcup_n \Delta_n$, the linear span of the spaces $\{L^2(\mu|\Delta_n)\}$ is dense in $L^2(\mu)$. Therefore $T = M_\phi$. ∎

The definition $\mathcal{A}_\mu$, introduced in the preceding proposition, will be used frequently in the sequel.

12.5 The Fuglede–Putnam Theorem. *If N and M are normal operators and B is an operator such that $NB = BM$, then $N^*B = BM^*$.*

Proof. From the hypothesis it follows that if $p(z)$ is any polynomial in z, $p(N)B = Bp(M)$. But for any fixed complex number w, there is one sequence of polynomials $\{p_n\}$ such that $p_n(M) \to \exp(i\overline{w}N)$ and $p_n(N) \to \exp(i\overline{w}M)$. Therefore $\exp(i\overline{w}N)B = B\exp(i\overline{w}M)$ for all w in $\mathbb{C}$. Equivalently, $B = e^{-i\overline{w}N}Be^{i\overline{w}M}$ for all complex w.

Because $e^Xe^Y = e^{X+Y}$ for all commuting operators X and Y, the normality of N and M implies that

$$\begin{aligned} f(w) &\equiv e^{-iwN^*}Be^{iwM^*} \\ &= e^{-iwN^*}e^{-i\overline{w}N}Be^{i\overline{w}M}e^{iwM^*} \\ &= e^{-i(wN^*+\bar{w}N)}Be^{i(wM^*+\bar{w}M)}. \end{aligned}$$

Note that f is an entire, operator-valued function. But $wN^* + \bar{w}N$ and $wM^* + \bar{w}M$ are hermitian operators, so $e^{-i(wN^*+\bar{w}N)}$ and $e^{i(wM^*+\bar{w}M)}$ are unitaries. Hence f is bounded. By Liouville's Theorem, f is constant. Therefore

$$0 = f'(w) = -iN^*e^{-iwN^*}Be^{iwM^*} + ie^{-iwN^*}BM^*e^{iwM^*}.$$

Setting $w = 0$ gives that $0 = -iN^*B + iBM^*$. ∎

This was originally proved in Fuglede [1950] under the assumption that $N = M$. Putnam [1951] established the stated version. The proof of the theorem presented here is due to Rosenblum [1958]. Another proof is in Radjavi and Rosenthal [1973]. Berberian [1959] observed that the version for two normal operators follows from the original version of Fuglede by using the following trick. If

$$L = \begin{bmatrix} N & 0 \\ 0 & M \end{bmatrix} \quad \text{and} \quad A = \begin{bmatrix} 0 & B \\ 0 & 0 \end{bmatrix},$$

then L is normal and $LA = AL$. Hence $L^*A = AL^*$ by Fuglede's version. An examination of the entries from this last matrix equation yields Putnam's version.

Now for a pair of corollaries that can be regarded as technical improvements of previous results. The first is an improvement of the Spectral Theorem.

12.6 Corollary. *If $N = \int z\,dE(z)$ is normal and $BN = NB$, then $BE(\Delta) = E(\Delta)B$ for all Borel sets Δ.*

Proof. If $BN = NB$, then $BN^* = N^*B$ and now the conclusion follows from the Spectral Theorem. ■

12.7 Corollary. *If μ is any compactly supported measure on $\mathbb{C}$, then*

$$\{N_\mu\}' = \mathcal{A}_\mu.$$

Proof. Clearly $\mathcal{A}_\mu \subseteq \{N_\mu\}'$. If $T \in \{N_\mu\}'$, then Theorem 12.5 implies $TN_\mu{}^* = N_\mu{}^*T$. Algebraic massage shows that $TM_p = M_pT$ whenever p is a polynomial in z and $\bar{z}$. Taking weak* limits of such polynomials shows that $T \in \mathcal{A}_\mu'$, which equals $\mathcal{A}_\mu$ by (12.4). ■

Putnam [1951] used his generalization of Fuglede's Theorem to show that similar normal operators must be unitarily equivalent. This is generalized in a somewhat formal way below, but this version has useful additional applications.

12.8 Proposition. *If N_1 and N_2 are normal operators on Hilbert spaces $\mathcal{H}_1$ and $\mathcal{H}_2$ and $X : \mathcal{H}_1 \to \mathcal{H}_2$ is a bounded operator such that $XN_1 = N_2X$, then:*

(a) $\text{cl}\,[\text{ran}\,X]$ *reduces* N_2;

(b) $\ker X$ *reduces* N_1;

(c) *if* $M_1 \equiv N_1|(\ker X)^\perp$ *and* $M_2 \equiv N_2|\text{cl}\,[\text{ran}\,X]$, *then* $M_1 \cong M_2$.

Proof. (a) If $h_1 \in \mathcal{H}_1$, then $N_2Xh_1 = XN_1h_1 \in \text{ran}\,X$; so $\text{cl}\,[\text{ran}\,X]$ is invariant for N_2. By the Fuglede–Putnam Theorem, $N_2^*X = XN_1^*$, so the analogous argument shows that $\text{cl}\,[\text{ran}\,X]$ is also invariant for N_2^*.

(b) Exercise.

(c) With the additional assumption that X is injective and has dense range, it will be shown that $N_1 \cong N_2$. Deriving the general case from this one is left to the reader.

Let $X = W|X|$ be the polar decomposition of X. Because $\ker X = (0)$ and $\text{ran}\,X$ is dense, W is an isomorphism. Also the Fuglede–Putnam Theorem implies $X^*N_2 = N_1X^*$, so $|X|^2 = X^*X \in \{N_1\}'$. Thus $|X| \in \{N_1\}'$. (Why?) Hence $N_2W|X| = N_2X = W|X|N_1 = WN_1|X|$. That is $N_2W = WN_1$ on $\text{ran}\,|X|$. But $\ker|X| = \ker X = (0)$, so $\text{ran}\,|X|$ is dense. Therefore $N_2W = WN_1$. ■

12.9 Corollary. *Similar normal operators are unitarily equivalent.*

The corollary appears in Putnam [1951], while the proposition appears in Douglas [1969]. A non-abelian version of this corollary appears below in Exercise 6. The proof is like the proof of the preceding proposition.

Exercises.

1. Let $\mathcal{A}$ be a C*-algebra contained in $\mathcal{B}(\mathcal{H})$, but do not assume that $\mathcal{A}$ contains the identity. Put $\mathcal{M} = \bigvee\{\operatorname{ran} A : A \in \mathcal{A}\}$ and let $P = P_{\mathcal{M}}$. Show that the strong closure of $\mathcal{A}$ is $P\mathcal{A}'' = \mathcal{A}''P$.

2. What is $\{N_\mu \oplus N_\mu\}'$? $\{N_\mu \oplus N_\mu\}''$?

3. Find a non-normal 2×2 matrix that is similar to a normal matrix.

4. Let $\{e_n\}$ be a basis for the separable Hilbert space $\mathcal{H}$ and define the diagonalizable operator A on $\mathcal{H}$ by $Ae_n = \lambda_n e_n$, where $\sup_n |\lambda_n| < \infty$. Determine $\{A\}'$ and $\{A\}''$ and give a necessary and sufficient condition that they are equal.

5. Find an $n \times n$ matrix whose commutant is the set of all $n \times n$ matrices.

6. If $\mathcal{A}$ is any C*-algebra, show that two representations π_1 and π_2 are equivalent if they are similar; that is, if there is an invertible operator $R : \mathcal{H}_1 \to \mathcal{H}_2$ such that $R^{-1}\pi_2(a)R = \pi_1(a)$ for all a in $\mathcal{A}$.

§13. Von Neumann algebras

13.1 Definition. A *von Neumann algebra* is a strongly closed C*-subalgebra of $\mathcal{B}(\mathcal{H})$ that contains the identity.

So, in light of the Double Commutant Theorem, a C*-algebra of operators is a von Neumann algebra if and only if it contains the identity and equals its double commutant. This section establishes a few elementary facts about von Neumann algebras. The next section explores abelian von Neumann algebras, and later Chapter 7 explores a more thorough investigation of the subject.

Before starting this introduction, observe that, unlike the definition of a C*-algebra, it is required that a von Neumann algebra be contained in $\mathcal{B}(\mathcal{H})$. That is, an abstract, space free definition of a von Neumann algebra has not been given. Such a space free definition is possible. Indeed, every unital C*-algebra that is the dual of a Banach space has a faithful representation as a von Neumann algebra. The theory of von Neumann algebras, moreover, can be developed from this starting point. This is done in Sakai [1971].

In order to increase familiarity with some of the ideas, the proof of the first proposition is left to the reader.

13.2 Proposition. (a) *If $\mathcal{A}$ is any $*$-subalgebra of $\mathcal{B}(\mathcal{H})$, $\mathcal{A}'$ is a von Neumann algebra.*

(b) *If $\{\mathcal{A}_i\}$ is a collection of von Neumann algebras, then $\bigcap_i \mathcal{A}_i$ is a von Neumann algebra.*

(c) *If $\{\mathcal{A}_i\}$ is a collection of $*$-subalgebras of $\mathcal{B}(\mathcal{H})$, then the smallest von Neumann algebra containing $\bigcup_i \mathcal{A}_i$ is $[\bigcup_i \mathcal{A}_i]''$ and its commutant is $\bigcap_i \mathcal{A}_i'$.*

(d) *If $\mathcal{A}$ is a von Neumann algebra, then the center of $\mathcal{A}$, $\mathcal{A} \cap \mathcal{A}'$, is an abelian von Neumann algebra.*

If $\mathcal{S}$ is any subset of $\mathcal{B}(\mathcal{H})$, $W^*(\mathcal{S})$ is the intersection of all von Neumann algebras that contain $\mathcal{S}$. Since $\mathcal{B}(\mathcal{H})$ is one such algebra, this is well defined. By the preceding proposition, $W^*(\mathcal{S})$ is a von Neumann algebra. $W^*(\mathcal{S})$ is called the *von Neumann algebra generated by $\mathcal{S}$.*

13.3 Proposition. *Let $\mathcal{A}$ be a von Neumann algebra contained in $\mathcal{B}(\mathcal{H})$ and let $A \in \mathcal{A}$.*

(a) *If A is normal and ϕ is a bounded Borel function on $\sigma(A)$, then $\phi(A) \in \mathcal{A}$.*

(b) *The operator A is the linear combination of four unitary operators that belong to $\mathcal{A}$.*

(c) *If E and F are the projections onto* $\text{cl}\,[\text{ran}\, A]$ *and* $\ker A$, *respectively, then $E, F \in \mathcal{A}$.*

(d) *If $A = W|A|$ is the polar decomposition of A, then W and $|A|$ belong to $\mathcal{A}$.*

(e) *A von Neumann algebra is the norm closed linear span of its projections.*

Proof. (a) In fact, if $\phi \in B(\sigma(A))$, there is a net $\{f_i\}$ of continuous functions on $\sigma(A)$ such that $\|f_i\|_\infty \le \|\phi\|_\infty$ and $\int f_i\, d\mu \to \int \phi\, d\mu$ for every μ in $M(\sigma(A))$ (9.7). By Theorem 10.3, $f_i(A) \to \phi(N)$ (WOT). Since each $f_i(A)$ belongs to $\mathcal{A}$, $\phi(A) \in \mathcal{A}$.

(b) This is true for C*-algebras (Exercise 3.5). It's clear that the real and imaginary parts of A belong to $\mathcal{A}$, so there is no loss of generality in assuming that $A \in \text{ball}\,(\text{Re}\,\mathcal{A})$ and showing that A is the linear combination of two unitaries. But for such an A, it is easy to check that $U_\pm = A \pm i(1 - A^2)^{1/2}$ are unitaries in $\mathcal{A}$ and $A = (U_+ + U_-)/2$.

(c) If $h \in \ker A$ and $A' \in \mathcal{A}'$, then $AA'h = A'Ah = 0$. That is, F is left invariant under every operator in $\mathcal{A}'$. Since $\mathcal{A}'$ is a C*-algebra, $F \in \mathcal{A}'' = \mathcal{A}$. Since $E = \ker A^*$, this also shows that $E \in \mathcal{A}$.

(d) The fact that $|A| = (A^*A)^{1/2} \in \mathcal{A}$ is a consequence of the fact that $\mathcal{A}$ is a C*-algebra. Recall that when it is said that $A = W|A|$ is *the* polar decomposition, this means that W is a specific partial isometry with $W^*W = E = \text{cl}\,[\text{ran}\,|A|]$ and $WW^* = \text{cl}\,[\text{ran}\,A]$. In light of part (b), to show that W belongs to $\mathcal{A}$ it suffices to show that $WU = UW$ for every unitary U in $\mathcal{A}'$. Fix such a unitary U and observe that $A = U^*AU = (U^*WU)(U^*|A|U) = V|A|$, where $V = U^*WU$. It must be shown that $V = U$. This is done using the uniqueness of the polar decomposition.

Since $E, F \in \mathcal{A}$ and $U, U^* \in \mathcal{A}'$, part (c) implies that U maps $\text{cl}\,[\text{ran}\,|A|]$ and $\text{cl}\,[\text{ran}\,A]$ onto themselves. Thus V is a partial isometry with the same initial and final space as W. Therefore $V = W$.

(e) According to part (a), the spectral projections of a hermitian operator in $\mathcal{A}$ also belong to $\mathcal{A}$. By the Spectral Theorem, every hermitian operator is the norm limit of a linear combination of its spectral projections. By writing each element of $\mathcal{A}$ in its Cartesian decomposition, part (e) follows. ∎

13.4 Corollary. *If $\mathcal{A}$ is a von Neumann algebra and $A \in \mathcal{B}(\mathcal{H})$, then $A \in \mathcal{A}$ if and only if $AP = PA$ for every projection P in $\mathcal{A}'$.*

Proof. Since $\mathcal{A}'$ is a von Neumann algebra, it is the norm closed linear span of its projections (13.3.e). So if $AP = PA$ for every projection P in $\mathcal{A}'$, then $A \in \mathcal{A}'' = \mathcal{A}$. The converse follows by the definition of the commutant. ∎

§14. Abelian von Neumann algebras

With algebras of operators, as opposed to abstract algebras, it is necessary to consider a special type of isomorphism.

14.1 Definition. If $\mathcal{S}_1$ and $\mathcal{S}_2$ are sets of operators on Hilbert spaces $\mathcal{H}_1$ and $\mathcal{H}_2$, then $\mathcal{S}_1$ and $\mathcal{S}_2$ are said to be *spatially isomorphic* if there is an isomorphism $U : \mathcal{H}_1 \to \mathcal{H}_2$ such that $U\mathcal{S}_1U^{-1} = \mathcal{S}_2$. In notation this is denoted by $\mathcal{S}_1 \cong \mathcal{S}_2$.

Isomorphic algebras need not be spatially isomorphic. For example, if μ is a compactly supported measure on the plane, $\mathcal{A}_\mu$ and its double inflation, $\mathcal{A}_\mu^{(2)}$, are isomorphic under the map $M_\phi \to {M_\phi}^{(2)}$. However they are not spatially isomorphic. Indeed, it is easy to see that if $\mathcal{S}_1$ and $\mathcal{S}_2$ are spatially isomorphic, then so are their commutants (using the same isomorphism). However $\mathcal{A}_\mu' = \mathcal{A}_\mu$, an abelian algebra. But Proposition 12.2 implies that $\{\mathcal{A}_\mu^{(2)}\}'$ is not abelian.

14.2 Definition. A vector e_0 is said to be *separating* for a collection of operators $\mathcal{S}$ if there is no non-zero operator S in $\mathcal{S}$ such that $Se_0 = 0$.

For a separable measure space (X, Ω, μ), any function in $L^2(\mu)$ that does not vanish on a set of positive measure is a separating vector for $\mathcal{A}_\mu$. The reader may have noticed that such vectors are also cyclic for $\mathcal{A}_\mu$ (Exercise 11.11). This is no accident; see the next proposition. But if f is a function in $L^2(\mu)$ that does not vanish on a set of positive measure, $f \oplus f$ is also separating for $\mathcal{A}_\mu^{(2)}$, but it is not a cyclic vector for this algebra. Now $\mathcal{B}(\mathcal{H})$ has no separating vectors (if $\dim \mathcal{H} > 1$), while every non-zero vector is separating for $\mathcal{B}(\mathbb{C})$.

14.3 Proposition. *If $\mathcal{S}$ is a linear manifold of operators, every cyclic vector for $\mathcal{S}$ is separating for $\mathcal{S}'$. If $\mathcal{A}$ is a unital C^*-algebra of operators, a vector is cyclic for $\mathcal{A}$ if and only if it is separating for $\mathcal{A}'$.*

Proof. If e_0 is cyclic for $\mathcal{S}$ and $T \in \mathcal{S}'$ such that $Te_0 = 0$, then for each S in $\mathcal{S}$, $TSe_0 = STe_0 = 0$. So $T[\mathcal{S}e_0] = (0)$, and hence $T = 0$. If $\mathcal{A}$ is a C*-algebra of operators containing the identity and e_0 is separating for $\mathcal{A}'$, let P be the projection onto $\mathcal{M} = [\mathcal{A}e_0]^\perp$. Since $\mathcal{M}$ reduces $\mathcal{A}$, $P \in \mathcal{A}'$. But $e_0 \perp \mathcal{M}$, so $Pe_0 = 0$. Since e_0 is separating for $\mathcal{A}'$, $P = 0$ and, therefore, e_0 is a cyclic vector for $\mathcal{A}$. ∎

When $\mathcal{A}$ is an abelian algebra, $\mathcal{A} \subseteq \mathcal{A}'$. This yields the following corollary.

14.4 Corollary. *If $\mathcal{A}$ is an abelian algebra of operators, every cyclic vector for $\mathcal{A}$ is also a separating vector for $\mathcal{A}$.*

As was pointed out with the examples above, an abelian von Neumann algebra can have separating vectors that are not cyclic. The abelian von Neumann algebras that are cyclic are quite special. In the next theorem the assumption that $\mathcal{H}$ is a separable Hilbert space is needed. For the non-separable case, see Dixmier [1956], §I.7. Exercise 2 contains a version of this result where the separability of the Hilbert space is replaced by an additional assumption about $\mathcal{A}$.

14.5 Theorem. *If $\mathcal{H}$ is separable and $\mathcal{A}$ is an abelian C^*-algebra contained in $\mathcal{B}(\mathcal{H})$ that contains the identity, then the following statements are equivalent.*

(a) *$\mathcal{A}$ is a maximal abelian von Neumann algebra.*

(b) $\mathcal{A} = \mathcal{A}'$.

(c) *$\mathcal{A}$ is strongly closed and has a cyclic vector.*

(d) *There is a compact metric space X, a regular Borel measure μ with support X, and an isomorphism $U : L^2(\mu) \to \mathcal{H}$ such that $U\mathcal{A}_\mu U^{-1} = \mathcal{A}$.*

Proof. The proof that (a) and (b) are equivalent is an exercise in mental calisthenics and holds even when the algebra is not a C*-algebra. See Exercise 1.

(b) *implies* (c). The fact that $\mathcal{A}$ is strongly closed and contains the identity is trivial. Let $\{e_n\}$ be a maximal sequence of unit vectors such that $[\mathcal{A}e_n] \perp [\mathcal{A}e_m]$ for $n \neq m$. Thus $\mathcal{H} = \bigoplus_n [\mathcal{A}e_n]$. Let $P_n = [\mathcal{A}e_n]$ and set $e_0 = \sum_n 2^{-n} e_n$. Now each P_n reduces $\mathcal{A}$, so $P_n \in \mathcal{A}'$. By (b), $P_n \in \mathcal{A}$. Therefore, $e_n = 2^n P_n e_0 \in [\mathcal{A}e_0]$. Hence $[\mathcal{A}e_n] \subseteq [\mathcal{A}e_0]$ for each $n \geq 1$. Therefore e_0 is cyclic for $\mathcal{A}$.

(c) *implies* (d). Since $\mathcal{H}$ is separable, ball $\mathcal{A}$ with the WOT is a compact metric space. (See Proposition 8.3 and Exercise 8.1.) Hence there is a countable weakly dense set. If $\mathcal{A}_1$ is the C*-algebra generated by that weakly dense sequence, then $\mathcal{A}_1$ is a separable C*-algebra that is WOT dense in $\mathcal{A}$. Let X be the maximal ideal space of $\mathcal{A}_1$; because $\mathcal{A}_1$ is separable, X is metrizable (Exercise 2.3). Let $\rho\colon C(X) \to \mathcal{A}_1 \subseteq \mathcal{A} \subseteq \mathcal{B}(\mathcal{H})$ be the inverse of the Gelfand map. Thus ρ is a representation of $C(X)$. By Theorem 9.8 there is a spectral measure E on X such that $\rho(f) = \int f\, dE$ for all f in $C(X)$. Using previous approximation properties of functions, for every bounded Borel function ϕ on X, $\widetilde{\rho}(\phi) \equiv \int \phi\, dE \in \{\mathcal{A}_1\}'' = \mathcal{A}'' = \mathcal{A}$ by the Double Commutant Theorem.

Let e_0 be a cyclic vector for $\mathcal{A}$ and put $\mu(\Delta) = \langle E(\Delta)e_0, e_0\rangle$. Thus $\langle \widetilde{\rho}(\phi)e_0, e_0\rangle = \int \phi\, d\mu$ for every ϕ in $B(X)$. Therefore

$$\begin{aligned} \|\widetilde{\rho}(\phi)e_0\|^2 &= \langle \widetilde{\rho}(\phi)^* \widetilde{\rho}(\phi)e_0, e_0\rangle \\ &= \int |\phi|^2\, d\mu \end{aligned}$$

This implies that if $B(X)$ is considered as a dense manifold in $L^2(\mu)$ and $U : B(X) \to \mathcal{H}$ is defined by $U\phi = \widetilde{\rho}(\phi)e_0$, U is a well defined isometry. Extend U to be an isometry on all of $L^2(\mu)$.

If $\phi \in B(X)$ and $\psi \in L^\infty(\mu)$, then $UM_\psi\phi = U(\psi\phi) = \widetilde{\rho}(\psi\phi)e_0 = \widetilde{\rho}(\psi)\widetilde{\rho}(\phi)e_0 = \widetilde{\rho}(\psi)U\phi$. That is, $U\mathcal{A}_\mu U^{-1} = \widetilde{\rho}(L^\infty(\mu))$. But $\mathcal{A}_\mu$ is weakly closed in $\mathcal{B}(L^2(\mu))$, so $\widetilde{\rho}(L^\infty(\mu))$ is weakly closed in $\mathcal{B}(\mathcal{H})$. Also $\widetilde{\rho}(L^\infty(\mu)) \supseteq \rho(C(X)) = \mathcal{A}_1$, which is weakly dense in $\mathcal{A}$. Therefore $U\mathcal{A}_\mu U^{-1} = \mathcal{A}$.

(d) *implies* (b). This follows by (12.4). ■

Mercer [1986] shows that if $\mathcal{A}$ is a maximal abelian von Neumann algebra, there is a basis for the Hilbert space consisting of vectors that are cyclic and separating for $\mathcal{A}$.

14.6 Corollary. *If $\mathcal{H}$ is separable, every abelian C^*-algebra contained in $\mathcal{B}(\mathcal{H})$ has a separating vector.*

Proof. By Zorn's Lemma, there is a maximal abelian C*-algebra that contains $\mathcal{A}$. By the preceding Theorem this maximal algebra has a cyclic vector, which must also be separating by Proposition 14.4. This vector is, therefore, also separating for $\mathcal{A}$. ∎

Exercises.

1. Show that $\mathcal{A}$ is a maximal abelian subalgebra of $\mathcal{B}(\mathcal{H})$ if and only if $\mathcal{A} = \mathcal{A}'$.

2. This exercise gives a version of Theorem 14.5 that does not assume the underlying Hilbert space is separable, but substitutes another assumption. Suppose $\mathcal{A}$ is a C*-subalgebra of $\mathcal{B}(\mathcal{H})$ that contains the identity and has the property that its commutant contains no uncountable family of pairwise orthogonal projections. If (X, Ω, μ) is any σ-finite measure space, separable or not, then $\mathcal{A}_\mu = \{ M_\phi : \phi \in L^\infty(\mu) \}$ is such an algebra. Show that the following statements are equivalent. (a) $\mathcal{A}$ is a maximal abelian von Neumann algebra. (b) $\mathcal{A} = \mathcal{A}'$. (c) $\mathcal{A}$ has a cyclic vector and is strongly closed. (d) There is a σ-finite measure space (X, Ω, μ) and an isomorphism $U \colon L^2(\mu) \to \mathcal{H}$ such that $U\mathcal{A}_\mu U^{-1} = \mathcal{A}$. Show that the measure space in (d) can be chosen to be finite.

§15. The functional calculus for normal operators

For the remainder of this chapter, N will always denote a normal operator with spectral decomposition $N = \int z \, dE(z)$. This section gives the final, definitive exposé of the functional calculus for a normal operator on a separable Hilbert space.

Recall that for any set of operators $\mathcal{S}$, $W^*(\mathcal{S})$ is the von Neumann algebra generated by $\mathcal{S}$. The next result has implicitly appeared, but needs to be made explicit.

15.1 Proposition. *If N is a normal operator,*

$$W^*(N) = \{N\}'' \supseteq \{ \phi(N) : \phi \in B(\sigma(N)) \}.$$

Proof. The first equality is a consequence of the Double Commutant Theorem and the Fuglede–Putnam Theorem. If $\phi \in B(\sigma(N))$, then the Spectral Theorem implies that $\phi(N) = \int \phi \, dE$ commutes with any operator that commutes with N. ∎

It will be shown that when the underlying Hilbert space is separable, there is always a measure μ supported on the spectrum of N such that

$\phi(N)$ is well defined for any ϕ in $L^\infty(\mu)$ and the map $\phi \to \phi(N)$ defines a $*$-isomorphism of $L^\infty(\mu)$ onto $W^*(N)$. Without the assumption that the Hilbert space is separable, the result is false. (See Exercise 4.) Therefore, throughout this section it is assumed that $\mathcal{H}$ is separable.

A moment's reflection shows that the first difficulty in obtaining this $*$-isomorphism is in defining $\phi(N)$. Specifically, μ must be chosen so that $\phi(N) = 0$ whenever ϕ is a bounded Borel function that vanishes a.e. $[\mu]$. By Corollary 14.6, $W^*(N)$ has a separating vector e_0. Define the measure μ by

15.2 $$\mu(\Delta) = \langle E(\Delta)e_0, e_0\rangle = \|E(\Delta)e_0\|^2.$$

15.3 Proposition. *If Δ is a Borel set, $\mu(\Delta) = 0$ if and only if $E(\Delta) = 0$.*

Proof. If $\mu(\Delta) = 0$, then $E(\Delta)e_0 = 0$. Since e_0 is separating for $W^*(N)$, Proposition 15.1 implies $E(\Delta) = 0$. The converse is clear. ∎

15.4 Definition. A *scalar-valued spectral measure* for N is a positive measure ν on $\sigma(N)$ such that ν and E have the same sets of measure zero. That is, ν and E are mutually absolutely continuous.

Proposition 15.3 shows that scalar-valued spectral measures exist. Note that for a scalar-valued spectral measure μ, if ϕ is a bounded Borel function on $\sigma(N)$ and $\phi = 0$ a.e. $[\mu]$, then $\phi(N) = 0$. So this completes the first step in the process of finalizing the functional calculus for normal operators.

Now for some additional notation and the localization of normal operators. If $h \in \mathcal{H}$, define $\mu_h(\cdot) = \langle E(\cdot)h, h\rangle$; that is, $\mu_h = E_{h,h}$. So each μ_h is a positive measure on $\sigma(N)$ and μ_h is a scalar-valued spectral measure for N if h is a separating vector for $W^*(N)$. Let $\mathcal{H}_h = [W^*(N)h]$ and put $N_h = N|\mathcal{H}_h$. So N_h is a $*$-cyclic normal operator with $*$-cyclic vector h. The uniqueness of the spectral measure for a normal operator implies that the spectral measure for N_h is $E(\cdot)|\mathcal{H}_h$. Thus Theorem 11.2 implies there is an isomorphism $U_h : \mathcal{H}_h \to L^2(\mu_h)$ such that $U_h N_h U_h^{-1} f = zf$ for all f in $L^2(\mu_h)$ and $U_h h = 1$.

The operators N_h are the localizations of N. In light of Theorem 11.2, the local operators are thoroughly understood. Indeed, all the information about the localization N_h is contained in the measure μ_h. The task that remains is to assemble all this local information to get a global picture of the normal operator. The first task is to show that for the functional calculus all the information is contained in the local operator N_{e_0}, where e_0 is a separating vector.

Note that if $A \in W^*(N)$, then each space $\mathcal{H}_h$ reduces A, and so it is possible to consider $A|\mathcal{H}_h$. Also there is a net $\{p_i\}$ of polynomials in z and

$\bar{z}$, such that $p_i(N, N^*) \to A$ (WOT). Looking at the restrictions of these operators to $\mathcal{H}_h$ it follows that $A|\mathcal{H}_h \in W^*(N_h)$.

15.5 Lemma. *If $h \in \mathcal{H}$ and $\rho_h : W^*(N) \to W^*(N_h)$ is defined by $\rho_h(A) = A|\mathcal{H}_h$, then ρ_h is a $*$-epimorphism that is WOT continuous. Moreover, if $\psi \in B(\sigma(N))$, then $\rho_h(\psi(N)) = \psi(N_h)$; if $A \in W^*(N)$, then there is a function ϕ in $B(\sigma(N))$ such that $\rho_h(A) = \phi(N_h)$.*

Proof. The fact that ρ_h is a $*$-homomorphism is left to the reader. If $\{A_i\}$ is a net in $W^*(N)$ such that $A_i \to 0$ (WOT), then for all g, k in $\mathcal{H}_h$, $\langle A_i g, k\rangle \to 0$. Thus ρ_h is WOT continuous.

If $p(z, \bar{z})$ is a polynomial in z and $\bar{z}$, then algebraic manipulation shows that $\rho_h(p(N, N^*)) = p(N_h, N_h^*)$. Because $\sigma(N_h) \subseteq \sigma(N)$ (Why?), it is possible to take uniform limits of such polynomials and conclude that $\rho_h(u(N)) = u(N_h)$ for every u in $C(\sigma(N))$. If $\psi \in B(\sigma(N))$, let $\{u_i\}$ be a net in $C(\sigma(N))$ such that $\|u_i\| \leq \|\psi\|$ for all i and $\int u_i \, d\mu \to \int \psi \, d\mu$ for every measure μ in $M(\sigma(N))$ (9.7). Thus $u_i(N) \to \psi(N)$ (WOT) and so $u_i(N_h) \to \psi(N_h)$ (WOT). It follows that $\rho_h(\psi(N)) = \psi(N_h)$.

Now fix an operator A in $W^*(N)$ and put $A_h = \rho_h(A)$. If U_h is the isomorphism $U_h : \mathcal{H}_h \to L^2(\mu_h)$ such that $U_h h = 1$ and $U_h N_h U_h^{-1} = N_{\mu_h}$, it follows that $A_h N_h = N_h A_h$ and so $U_h A_h U_h^{-1} \in \{N_{\mu_h}\}'$. By Proposition 12.4 there is a function ϕ in $L^\infty(\mu_h)$ such that $U_h A_h U_h^{-1} = M_\phi$. It is left to the reader to chase through the various mappings and finish the argument that $A_h = \phi(N_h)$.

It remains to show that ρ_h is surjective. If $B \in W^*(N_h)$, then, as in the preceding paragraph, there is a ψ in $L^\infty(\mu_h)$ such that $B = \psi(N_h)$. Thus $B = \rho_h(\psi(N))$. ∎

15.6 Lemma. *If $e \in \mathcal{H}$ such that μ_e is a scalar-valued spectral measure for N and ν is a measure on $\sigma(N)$ such that $\nu << \mu_e$, then there is a vector h in $\mathcal{H}_e$ such that $\nu = \mu_h$.*

Proof. Let $f = [d\nu/d\mu]^{1/2}$; so $f \in L^2(\mu_e)$. Put $h = U_e^{-1} f$; so $h \in \mathcal{H}_e$. For any Borel set Δ, $\nu(\Delta) = \int \chi_\Delta \, d\nu = \int \chi_\Delta f^2 \, d\mu = \langle M_{\chi_\Delta} f, f\rangle = \langle U_e^{-1} M_{\chi_\Delta} f, U_e^{-1} f\rangle = \langle E(\Delta)h, h\rangle = \mu_h(\Delta)$. ∎

15.7 Lemma. $W^*(N) = \{\, \phi(N) : \phi \in B(\sigma(N))\,\}$.

Proof. Let $\mathcal{A} = \{\, \phi(N) : \phi \in B(\sigma(N))\,\}$. So $\mathcal{A}$ is a C*-algebra and $\mathcal{A} \subseteq W^*(N)$ by Proposition 15.1. Since $N \in \mathcal{A}$, to complete the proof it suffices to show that $\mathcal{A}$ is weakly closed. So let $\{\phi_i\}$ be a net in $B(\sigma(N))$ such that $\phi_i(N) \to A$ (WOT) in $\mathcal{B}(\mathcal{H})$. So $A \in W^*(N)$ and $\phi_i(N_h) \to A_h = A|\mathcal{H}_h$ for every h in $\mathcal{H}$. By Lemma 15.5, for each vector h there is a ϕ_h in $B(\sigma(N))$

such that $A_h = \phi_h(N_h)$. Fix a separating vector e so that μ_e is a scalar-valued spectral measure for N.

Because $\phi_i(N_h) \to \phi_h(N_h)$ (WOT), $\phi_i \to \phi_h$ weak* in $L^\infty(\mu_h)$. Also $\phi_i \to \phi_e$ weak* in $L^\infty(\mu_e)$. But $\mu_h << \mu_e$, so $d\mu_h/d\mu_e \in L^1(\mu_e)$. Hence for any Borel set Δ,

$$\int_\Delta \phi_i \, d\mu_h = \int_\Delta \phi_i \frac{d\mu_h}{d\mu_e} \, d\mu_e \to \int_\Delta \phi_e \, d\mu_e.$$

But also

$$\int_\Delta \phi_i \, d\mu_h \to \int_\Delta \phi_h \, d\mu_h.$$

Thus for any Borel set Δ, $\int_\Delta (\phi_h - \phi_e) \, d\mu_h = 0$. Hence $\phi_h = \phi_e$ $[\mu_h]$ a.e.

Now let $g \in \mathcal{H}_h$. So $\langle \phi_h(N_h)g, g\rangle = \langle \phi_h(N)g, g\rangle = \int \phi_h \, d\mu_g = \int \phi_e \, d\mu_g$, since $\mu_g << \mu_h$. Thus $\langle \phi_h(N)g, g\rangle = \langle \phi_e(N)g, g\rangle$; that is, $\phi_h(N_h) = \phi_e(N_h)$. In particular, $Ah = \phi_h(N_h)h = \phi_e(N_h)h = \phi_e(N)h$. Since h was arbitrary, $A = \phi_e(N)$. ∎

It seems to me that lemmas should never have corollaries. Nevertheless, the next result is an immediate consequence of the preceding lemma. The details are left to the reader.

15.8 Lemma. *If $h \in \mathcal{H}$ and $\rho_h : W^*(N) \to W^*(N_h)$ is defined as in Lemma* 15.5, *then* $\ker \rho_h = \{\phi(N) : \phi = 0 \text{ a.e. } [\mu_h]\}$.

The next theorem is not the main result of this section, though it is important. Its proof is a collage of the preceding material from this section.

15.9 Theorem. *For a vector e in $\mathcal{H}$, the following statements are equivalent.*

(a) *e is a separating vector for $W^*(N)$.*

(b) *μ_e is a scalar-valued spectral measure for N.*

(c) *The map $\rho_e : W^*(N) \to W^*(N_e)$ is a $*$-isomorphism.*

(d) $\{\phi \in B(\sigma(N)) : \phi(N) = 0\} = \{\phi \in B(\sigma(N)) : \phi = 0 \text{ a.e. } [\mu_e]\}$.

Proof. (a) *implies* (b). Proposition 15.3.

(b) *implies* (c). It is already established that ρ_e is a $*$-epimorphism. According to the preceding lemma, $\ker \rho_e = \{\phi(N) : \phi = 0 \text{ a.e. } [\mu_e]\}$. But using (b) it follows that these are the operators t$\phi(N)$ with $\phi = 0$ off a set with spectral measure zero. For such a ϕ, $\phi(N) = 0$ so that ρ_e is injective.

(c) *implies* (d). Combine (c) with Lemma 15.8.

(d) *implies* (a). Suppose $A \in W^*(N)$ and $Ae = 0$. By Lemma 15.7 there is a bounded Borel function ϕ on $\sigma(N)$ such that $A = \phi(N)$. Thus $0 = \|Ae\| = \langle A^*Ae, e\rangle = \int |\phi|^2 \, d\mu_e$. So $\phi = 0$ a.e. $[\mu_e]$. By (d), $A = \phi(N) = 0$. ■

Now for the main theorem in this section. In many ways, it is the culmination of the entire chapter. Most of the work for the proof has already been done.

15.10 The Functional Calculus for Normal Operators. *If N is a normal operator on the separable Hilbert space $\mathcal{H}$ and μ is a scalar-valued spectral measure for N, then there is a well defined map $\rho : L^\infty(\mu) \to W^*(N)$ given by $\rho(\phi) = \phi(N)$ such that:*

(a) *ρ is a $*$-isomorphism;*

(b) *$\rho : (L^\infty(\mu), \text{weak}^*) \to (W^*(N), \text{WOT})$ is a homeomorphism.*

Proof. By (15.6) there is a vector e such that $\mu = \mu_e$. By (15.9), e is a separating vector for $W^*(N)$. Theorem 15.9.d also implies that if $\phi \in B(\sigma(N))$ and $\phi = 0$ a.e. $[\mu]$, then $\phi(N) = 0$. Hence $\phi(N)$ is well defined for each ϕ in $L^\infty(\mu)$. So it is possible to define $\rho : L^\infty(\mu) \to W^*(N)$ by $\rho(\phi) = \phi(N)$. It is left to the reader to show that ρ is a $*$-homomorphism. Theorem 15.9.d implies ρ is injective and Lemma 15.7 implies ρ is surjective. Thus ρ is a $*$-isomorphism.

It remains to prove (b). Let $\{\phi_i\}$ be a net in $L^\infty(\mu)$ such that $\phi_i(N) \to 0$ (WOT). If $f \in L^1(\mu)$, Lemma 15.6 implies there is a vector h in $\mathcal{H}_e$ such that $f\mu = \mu_h$. Thus $\int \phi_i f \, d\mu = \int \phi_i \, d\mu_h = \langle \phi_i(N)h, h\rangle \to 0$. Therefore $\phi_i \to 0$ (weak*). The proof of the other half of (b) is left to the reader. ■

15.11 The Spectral Mapping Theorem. *If N is a normal operator on a separable Hilbert space, μ is a scalar-valued spectral measure for N, and $\phi \in L^\infty(\mu)$, then the spectrum of $\phi(N)$ is the essential range of ϕ.*

Proof. Use (15.10) and the fact that the spectrum of ϕ as an element of $L^\infty(\mu)$ is the essential range of ϕ (2.6). ■

At this point the reader can consult §IX.10 in [ACFA] for the complete classification of normal operators. This book will present this classification in §50 as part of the discussion of Type I von Neumann algebras.

Exercises.

1. What is a scalar-valued spectral measure for a diagonalizable normal operator?

2. If N_k, $k = 1, 2$, are normal operators with scalar-valued spectral measures μ_k, what is a scalar-valued spectral measure for $N_1 \oplus N_2$?

3. If $\{e_n\}$ is a basis for $\mathcal{H}$, show that $\mu(\Delta) = \sum_n \|E(\Delta)e_n\|^2$ is a scalar-valued spectral measure for N.

4. Show that there is a normal operator on a non-separable Hilbert space with no scalar-valued spectral measure.

5. For compactly supported measures on the plane, μ and ν, show that the following statements are equivalent. (a) $N_\mu \oplus N_\nu$ is $*$-cyclic. (b) $W^*(N_\mu \oplus N_\nu) = W^*(N_\mu) \oplus W^*(N_\nu)$. (c) $\mu \perp \nu$.

6. If M and N are normal operators with scalar-valued spectral measures μ and ν, respectively, show that the following statements are equivalent. (a) $W^*(M \oplus N) = W^*(M) \oplus W^*(N)$. (b) $\{M \oplus N\}' = \{M\}' \oplus \{N\}'$. (c) There is no non-zero operator A such that $MA = AN$. (d) $\mu \perp \nu$.

7. If M and N are normal operators, show that $C^*(M \oplus N) = C^*(M) \oplus C^*(N)$ if and only if $\sigma(M) \cap \sigma(N) = \emptyset$.

8. Give an example of two normal operators M and N such that $C^*(M \oplus N) \neq C^*(M) \oplus C^*(N)$ but $W^*(M \oplus N) = W^*(M) \oplus W^*(N)$.

9. Let N be a normal operator with μ as a scalar-valued spectral measure. If $\phi \in L^\infty(\mu)$, show that $\mu \circ \phi^{-1}$ is a scalar-valued spectral measure for $\phi(N)$.

Chapter 3

Compact Operators

This chapter discusses a number of topics, some of which may be known to the reader. There is almost no overlap between this material and [ACFA]. It will characterize the C*-algebras of compact operators and the ideals of $\mathcal{B}(\mathcal{H})$, and explore the trace class and the Hilbert–Schmidt operators.

§16. C*-algebras of compact operators

Consider the C*-algebras $\mathcal{A}$ that are contained in the algebra of compact operators, $\mathcal{B}_0 = \mathcal{B}_0(\mathcal{H})$. Assume here that $\mathcal{A}$ is not *degenerate* in $\mathcal{B}_0(\mathcal{H})$. That is, assume that $\mathcal{H} = [\mathcal{A}\mathcal{H}]$. Equivalently, if $g \in \mathcal{H}$ and $Ag = 0$ for every A in $\mathcal{A}$, then $g = 0$. Understanding the general case reduces to understanding this one. Indeed, if $\mathcal{A}$ is degenerate, let $\mathcal{H}_1 = [\mathcal{A}\mathcal{H}]$ and $\mathcal{H}_0 = \mathcal{H}^{\perp}$. Now each of these subspaces reduces $\mathcal{A}$, $\mathcal{A} = \mathcal{A}_1 \oplus (0)$ on $\mathcal{H} = \mathcal{H}_1 \oplus \mathcal{H}_0$, and $\mathcal{A}_1$ is non-degenerate.

Throughout this section $\mathcal{A}$ will be a non-degenerate C*-subalgebra of $\mathcal{B}_0$. Since $\mathcal{A}$ does not have an identity, a bit of caution must be exercised. If $\mathcal{A}_1$ is the algebra $\mathcal{A} + \mathbb{C}$, then $\mathcal{A}$ is an ideal in $\mathcal{A}_1$. Thus, according to Proposition 5.1, if A is a hermitian element of $\mathcal{A}$ and f is a continuous function on the spectrum of A such that $f(0) = 0$, then $f(A) \in \mathcal{A}$. If it is not assumed that $f(0) = 0$, then $f(A) \in \mathcal{A}_1$. Also remain cognizant that because the operators in $\mathcal{A}$ are compact, the spectrum is a sequence $\{\alpha_n\}$ converging to 0. So the stipulation that f is a continuous function on $\sigma(A)$ and vanishes at 0 is only the requirement that $\lim_n f(\alpha_n) = 0$.

It is necessary to set some terminology. If $\{\alpha_i : i \in I\}$ is a collection of scalars, say that $\alpha_i \to 0$ if for every $\epsilon > 0$, there is a finite subset I_0 of I such that $|\alpha_i| < \epsilon$ when $i \notin I_0$. Here is another way to say the same thing. Let $\mathcal{F}$ be the collection of all finite subsets of I and order $\mathcal{F}$ by inclusion. Thus

$\mathcal{F}$ is a directed set. If for each F in $\mathcal{F}$, $\alpha_F = \max\{|\alpha_i| : i \in F\}$, then the statement that $\alpha_i \to 0$ is precisely the statement that the net $\{\alpha_F : F \in \mathcal{F}\}$ converges to 0. Note that this implies that at most a countable number of the numbers α_i are not 0. Similarly, a discussion of the convergence of the uncountable sum, $\sum_i \alpha_i$, is a discussion about the convergence of the net $\{\sum_{i\in F} \alpha_i : F \in \mathcal{F}\}$.

16.1 Examples. (a) Let $\{E_n\}$ be a sequence of finite rank projections on a separable space that are pairwise orthogonal and satisfy $\sum_n E_n = 1$. Put $\mathcal{A} = \{\sum_{n=1}^{\infty} \lambda_n E_n : \{\lambda_n\} \subseteq \mathbb{C} \text{ and } \lambda_n \to 0\}$. Note that as a C*-algebra, $\mathcal{A}$ is isomorphic to c_0, irrespective of the dimensions of the projections $\{E_n\}$. If the condition $\sum_n E_n = 1$ is deleted, the resulting C*-algebra is degenerate.

(b) Let $\{\mathcal{H}_i : i \in I\}$ be a collection of Hilbert spaces and define $\mathcal{A} = \{\bigoplus_i T_i : T_i \in \mathcal{B}_0(\mathcal{H}_i) \text{ and } \|T_i\| \to 0\}$. So $\mathcal{A} \subseteq \mathcal{B}_0(\bigoplus_i \mathcal{H}_i)$. If each space $\mathcal{H}_i$ has the same dimension, say d, and I is given the discrete topology, then this algebra is isomorphic to $C_0(I, \mathcal{B}_n(\mathcal{K}))$, where $\mathcal{K}$ is a Hilbert space of dimension d. (See Exercise 1.2.) If $d < \infty$, then this algebra becomes $C_0(I, M_d)$.

(c) If $1 \leq d < \infty$ and $\mathcal{A}$ is a C*-subalgebra of $\mathcal{B}_0(\mathcal{H})$, then $\mathcal{A}^{(d)}$ is a C*-subalgebra of $\mathcal{B}_0(\mathcal{H}^{(d)})$.

(d) Fix a Hilbert space $\mathcal{H}$ and let U be a unitary on $\mathcal{H}$. Let $\mathcal{A} = \{T \oplus U^*TU : T \in \mathcal{B}_0(\mathcal{H})\}$. Note that if $W = 1 \oplus U$, W is a unitary on $\mathcal{H} \oplus \mathcal{H}$ and $\mathcal{A} = W^*\mathcal{B}_0(\mathcal{H})^{(2)}W$.

(e) Take the direct sum of any finite number of the preceding examples; or take the direct sum of an infinite number of them, but insist that the norms of the direct summands converge to 0.

When dealing with compact operators, the following notation is advantageous. For vectors g, h in $\mathcal{H}$ define the rank one operator

16.2 $$g \otimes h(f) = \langle f, h\rangle g.$$

The properties of $g \otimes h$ are contained in the next proposition, whose proof is left to the reader.

16.3 Proposition. *If $g, h \in \mathcal{H}$, the following hold.*

(a) $\operatorname{ran}(g \otimes h) = \mathbb{C}g$ *and* $\ker(g \otimes h) = [h]^{\perp}$ *if* $g \neq 0$.

(b) $(g \otimes h)^* = h \otimes g$.

(c) *The map* $(g, h) \to g \otimes h$ *from* $\mathcal{H} \times \mathcal{H}$ *into* $\mathcal{B}(\mathcal{H})$ *is sesquilinear.*

(d) $\|g \otimes h\| = \|g\|\,\|h\|$.

(e) *If* $T \in \mathcal{B}(\mathcal{H})$, *then* $T(g \otimes h) = (Tg) \otimes h$ *and* $(g \otimes h)T = g \otimes (T^*h)$.

16.4 Proposition. *If $A \in \operatorname{Re} \mathcal{A}$ and λ is a non-zero element of the spectrum of A, then the spectral projection for A corresponding to the singleton $\{\lambda\}$ belongs to $\mathcal{A}$.*

Proof. Let $A = \int_a^b t\, dE(t)$. Since non-zero points in the spectrum of a compact operator are isolated, χ, the characteristic function of the singleton $\{\lambda\}$, is a continuous function on $\sigma(A)$ that vanishes at 0. Thus $E(\{\lambda\}) = \chi(A) \in \mathcal{A}$. ■

A projection E in $\mathcal{A}$ is *minimal* if $E \neq 0$ and there are no non-zero projections in $\mathcal{A}$ that are dominated by E. Amongst the examples in (16.1) it is easy to see several minimal projections.

16.5 Proposition. *Let E be a projection in $\mathcal{A}$.*

(a) *E is minimal if and only if $E\mathcal{A}E = \{\lambda E : \lambda \in \mathbb{C}\}$.*

(b) *Every projection in $\mathcal{A}$ is the sum of a finite number of pairwise orthogonal minimal projections in $\mathcal{A}$.*

Proof. Clearly if $E\mathcal{A}E = \{\lambda E : \lambda \in \mathbb{C}\}$, then E is minimal. Assume that E is minimal. Note that $E\mathcal{A}E \subseteq \mathcal{A}$ and $E\mathcal{A}E$ is a C*-algebra of compact operators on $E(\mathcal{H})$. By Proposition 16.2, $E\mathcal{A}E$ contains the spectral projections corresponding to the non-zero eigenvalues of the hermitian elements of $E\mathcal{A}E$. All such projections are dominated by E. Thus (a) follows.

Let E be a projection in $\mathcal{A}$. Since E is compact, it has finite rank. Examine all the projections in $\mathcal{A}$ that are dominated by E. A simple argument constructs a descending chain of such projections. Because E has finite rank, this construction process must terminate after a finite number of steps; the last projection is a minimal projection of $\mathcal{A}$. What this argument has shown is that every projection in $\mathcal{A}$ dominates a minimal projection. Let E_1 be a minimal projection such that $E_1 \leq E$. Now let E_2 be a minimal projection such that $E_2 \leq E - E_1$. Continuing proves (b). ■

An algebra of operators is *irreducible* if it has no proper reducing subspaces. If $\mathcal{A} \subseteq \mathcal{B}(\mathcal{H})$, the fact that $\mathcal{A}$ is a C*-algebra, means this is equivalent to the condition that $\mathcal{A}$ has no proper invariant subspace.

16.6 Theorem. *If $\mathcal{A}$ is an irreducible C^*-subalgebra of $\mathcal{B}_0(\mathcal{H})$, then $\mathcal{A} = \mathcal{B}_0(\mathcal{H})$.*

Proof. If E is a minimal projection and its rank is greater than 1, then there are orthogonal unit vectors g, h in E. Let A be an arbitrary operator in $\mathcal{A}$ and let $\lambda \in \mathbb{C}$ such that $EAE = \lambda E$. Thus $\langle Ag, h\rangle = \langle EAEg, h\rangle = \lambda\langle g, h\rangle = 0$. That is $h \perp [\mathcal{A}g]$. Because $\mathcal{A}$ is not degenerate, this shows that

$[\mathcal{A}g]$ is a non-trivial reducing subspace for $\mathcal{A}$, a contradiction. Thus every minimal projection in $\mathcal{A}$ must have rank 1.

Now let $E = e \otimes e$ be a minimal projection in $\mathcal{A}$ and let $P = p \otimes p$ be any rank one projection. Again, $[\mathcal{A}e]$ is a non-zero reducing subspace for the algebra, so $[\mathcal{A}e] = \mathcal{H}$. In particular, there is an operator A in $\mathcal{A}$ such that $\|Ae - p\| < \epsilon$. Now $Ae \otimes Ae = AEA^* \in \mathcal{A}$. If h is any unit vector, $(p \otimes p - Ae \otimes Ae)h = \langle h, p\rangle p - \langle h, Ae\rangle Ae = \langle h, p\rangle (p - Ae) + \langle h, p - Ae\rangle Ae$. Thus $\|p \otimes p - Ae \otimes Ae\| \leq \epsilon + \epsilon(1 + \epsilon)$. Since ϵ was arbitrary, $p \otimes p \in \mathcal{A}$. That is, $\mathcal{A}$ contains every rank one projection and so $\mathcal{A} = \mathcal{B}_0(\mathcal{H})$. ■

There are a few corollaries of this theorem. The first will be improved in §17 and the proof of that result will be independent of the the preceding theorem. It appears here for illustration. A C*-algebra is *simple* if it has no proper closed ideals.

16.7 Corollary. *For every Hilbert space $\mathcal{H}$, $\mathcal{B}_0(\mathcal{H})$ is simple.*

Proof. Let $\mathcal{I}$ be a non-zero closed ideal in $\mathcal{B}_0(\mathcal{H})$. For a non-zero vector e, $[\mathcal{I}e]$ is a non-zero reducing subspace for $\mathcal{B}_0(\mathcal{H})$ and hence $[\mathcal{I}e] = \mathcal{H}$. Thus $\mathcal{I}$ is an irreducible C*-algebra (see Exercise 6) of compact operators, so that $\mathcal{I} = \mathcal{B}_0(\mathcal{H})$ by the theorem. ■

16.8 Corollary. *If $\mathcal{C}$ is an irreducible C*-subalgebra of $\mathcal{B}(\mathcal{H})$ and $\mathcal{C} \cap \mathcal{B}_0(\mathcal{H}) \neq (0)$, then $\mathcal{C} \supseteq \mathcal{B}_0(\mathcal{H})$.*

Proof. Let $\mathcal{A} = \mathcal{C} \cap \mathcal{B}_0(\mathcal{H})$ and note that $\mathcal{A}$ is a non-zero ideal of $\mathcal{C}$. Hence $\mathcal{A}$ is an irreducible C*-algebra of compact operators. By the theorem, $\mathcal{A} = \mathcal{B}_0(\mathcal{H})$. ■

The goal is now to assemble the ingredients that will lead to the characterization of the C*-algebras of compact operators. Theorem 16.6 is one of those ingredients; a fundamental one. The next proposition is another. Before giving this result, adopt a convention that will be followed in this book: identify projections with their ranges. So if E is a projection, the statement "Let e be a non-zero vector in E," means "Let e be a non-zero vector in $E\mathcal{H}$."

16.9 Proposition. *If E is a minimal projection in $\mathcal{A}$, e is any non-zero vector in E, and $\mathcal{H}_0 = [\mathcal{A}e]$, then $\mathcal{A}|\mathcal{H}_0 = \mathcal{B}_0(\mathcal{H}_0)$.*

Proof. Put $P_0 = \mathcal{H}_0$; since $\mathcal{H}_0$ reduces $\mathcal{A}$, P_0 commutes with every operator in $\mathcal{A}$. Also $A \to A|\mathcal{H}_0$ is a $*$-homomorphism of $\mathcal{A}$ onto a non-degenerate C*-subalgebra $\mathcal{A}_0$ of $\mathcal{B}_0(\mathcal{H}_0)$. The proof concludes by showing that $\mathcal{A}_0$ is irreducible and invoking Theorem 16.6.

Indeed, suppose that $\mathcal{L}$ is a subspace of $\mathcal{H}_0$ that reduces $\mathcal{A}_0$ and let P be the projection of $\mathcal{H}_0$ onto $\mathcal{L}$. So P commutes with every operator in $\mathcal{A}_0$. Let $T = P - \langle Pe, e\rangle 1$; so T commutes with every operator in $\mathcal{A}_0$. Therefore if A, B are arbitrary operators in $\mathcal{A}$,

$$\begin{aligned}\langle TAe, Be\rangle &= \langle TB^*AEe, Ee\rangle \\ &= \langle TEB^*AEe, e\rangle.\end{aligned}$$

But $EB^*AE \in E\mathcal{A}E$. Since E is minimal, there is a scalar λ such that $EB^*AE = \lambda E$. Thus

$$\langle TAe, Be\rangle = \lambda\langle Te, e\rangle = \lambda[\langle Pe, e\rangle - \langle Pe, e\rangle] = 0.$$

Since A and B were arbitrary, $T = 0$. That is $P = \langle Pe, e\rangle 1$. Because P is a projection, this says that $P = 0$ or $P = 1$. That is, $\mathcal{A}_0$ is irreducible. ■

In order to achieve the characterization of the C*-subalgebras of $\mathcal{B}_0(\mathcal{H})$, it is necessary to have a small interlude to obtain a result about the representations of a C*-algebra of compact operators. If $(\rho, \mathcal{K})$ is a representation of $\mathcal{A}$, it need not be the case that $\rho(\mathcal{A})$ be a C*-algebra of compact operators. Indeed, the infinite inflation of the identity representation fails to satisfy this condition.

16.10 Definition. If $\mathcal{C}$ is any C*-algebra and $\rho : \mathcal{C} \to \mathcal{B}(\mathcal{K})$ is a representation, then ρ is *irreducible* if $\rho(\mathcal{C})$ is an irreducible subalgebra of $\mathcal{B}(\mathcal{K})$.

Recall the definition of a subrepresentation (6.7). So an irreducible representation is one that has no proper subrepresentations. §32 examines irreducible representations of an arbitrary C*-algebra in detail. At present the concentration is on the irreducible representations of a C*-algebra of compact operators. The next theorem shows that every irreducible representation of a C*-algebra of compact operators has range inside the compact operators of the host space. Thus, by Theorem 16.6, if ρ is an irreducible representation of such an algebra, the range equals all the compact operators on the host space.

16.11 Theorem. *If $(\rho, \mathcal{K})$ is a representation of a C*-algebra $\mathcal{A}$ of compact operators, then $\rho = \bigoplus_i \rho_i$, where each ρ_i is an irreducible representation that is equivalent to a subrepresentation of the identity representation $i \colon \mathcal{A} \to \mathcal{B}(\mathcal{H})$.*

Proof. First observe that there is a minimal projection E in $\mathcal{A}$ such that $\rho(E) \neq 0$. In fact, there must be a hermitian element A in $\mathcal{A}$ with $\rho(A) \neq 0$. By Proposition 16.4 and the Spectral Theorem, there must be a spectral projection F for A such that $\rho(F) \neq 0$. But then F is the sum of a finite number of minimal projections (16.5). So ρ must not vanish on at least one minimal projection in $\mathcal{A}$.

Momentarily fix a minimal projection E such that $\rho(E) \neq 0$. Choose a unit vector p in $\rho(E) \leq \mathcal{K}$ and define $\mathcal{K}_0 = [\rho(\mathcal{A})p]$ and $\rho_0(A) = \rho(A)|\mathcal{K}_0$.

Claim. $(\rho_0, \mathcal{K}_0)$ is equivalent to an irreducible subrepresentation of the identity representation of $\mathcal{A}$.

It is known (16.5) that for every A in $\mathcal{A}$ there is a unique scalar λ such that $EAE = \lambda E$. Letting $\phi(A) = \lambda$ defines a function $\phi : \mathcal{A} \to \mathbb{C}$, which is easily seen to be linear. Let e be a unit vector in E. If $A \in \mathcal{A}$, then

$$\begin{aligned}\|\rho(A)p\|^2 &= \|\rho(AE)p\|^2 = \langle \rho(EA^*AE)p, p\rangle \\ &= \phi(A^*A) = \langle EA^*AEe, e\rangle \\ &= \|Ae\|^2.\end{aligned}$$

Thus setting $U(\rho(A)p) = Ae$ gives a well defined isometry from $\rho(\mathcal{A})p$ onto $\mathcal{A}e$. Extend U to an isomorphism $U : \mathcal{K}_0 \to \mathcal{H}_0 \equiv [\mathcal{A}e]$.

Note that $Up = e$. Also for A and B in $\mathcal{A}$, $U\rho(A)\rho(B)p = ABe$. So $U\rho_0(A)U^{-1} = A|\mathcal{H}_0$. Thus ρ_0 is equivalent to the subrepresentation of the identity representation of $\mathcal{A}$ defined by the reducing subspace $\mathcal{H}_0$. By Proposition 16.9 this subrepresentation is irreducible, so ρ_0 is irreducible. This establishes the claim.

Now apply Zorn's Lemma to obtain a maximal family $\{(\rho_i, \mathcal{K}_i)\}$ of irreducible subrepresentations of $(\rho, \mathcal{K})$, each of which is equivalent to a subrepresentation of the identity representation of $\mathcal{A}$ and with the spaces $\{\mathcal{K}_i\}$ pairwise orthogonal. Assume that $\mathcal{K}_1 = \mathcal{K} \ominus \bigoplus_i \mathcal{K}_i \neq (0)$. Let $\rho_1(A) = \rho(A)|\mathcal{K}_1$. It follows that $(\rho_1, \mathcal{K}_1)$ is a non-degenerate representation of $\mathcal{A}$ and the first part of the proof yields a contradiction to the maximality of $\{(\rho_i, \mathcal{K}_i)\}$. ∎

16.12 Corollary. *If $\mathcal{A}$ is a C^*-algebra of compact operators and $(\rho, \mathcal{K})$ is an irreducible representation of $\mathcal{A}$, then $\rho(\mathcal{A}) = \mathcal{B}_0(\mathcal{K})$.*

16.13 Corollary. *If $\mathcal{H}$ and $\mathcal{K}$ are Hilbert spaces and $\rho\colon \mathcal{B}_0(\mathcal{H}) \to \mathcal{B}_0(\mathcal{K})$ is an isomorphism, then there is an isomorphism $U : \mathcal{H} \to \mathcal{K}$ such that $\rho(T) = UTU^*$ for every T in $\mathcal{B}_0(\mathcal{H})$.*

Proof. Such an isomorphism ρ is an irreducible representation of $\mathcal{B}_0(\mathcal{H})$. By the theorem, ρ is equivalent to a subrepresentation of the identity representation of $\mathcal{B}_0(\mathcal{H})$. But the identity representation of $\mathcal{B}_0(\mathcal{H})$ has no proper subrepresentations. ∎

16.14 Corollary. *If $\mathcal{H}$ and $\mathcal{K}$ are Hilbert spaces and $\rho : \mathcal{B}(\mathcal{H}) \to \mathcal{B}(\mathcal{K})$ is an isomorphism, then there is an isomorphism $U : \mathcal{H} \to \mathcal{K}$ such that $\rho(A) = UAU^*$ for every A in $\mathcal{B}(\mathcal{H})$.*

Proof. Since $\rho(\mathcal{B}_0(\mathcal{H}))$ is an ideal of $\mathcal{B}(\mathcal{K})$, $\rho(\mathcal{B}_0(\mathcal{H}))$ is irreducible (Exercise 6). By Corollary 16.12, $\rho(\mathcal{B}_0(\mathcal{H})) = \mathcal{B}_0(\mathcal{K})$. By the preceding corollary there is an isomorphism $U : \mathcal{H} \to \mathcal{K}$ such that $\rho(A) = UAU^*$ for every compact operator A on $\mathcal{H}$. Now let A be any positive operator on $\mathcal{H}$ and let $\{E_i\}$ be an approximate identity for $\mathcal{B}_0(\mathcal{H})$ consisting of finite rank projections (Exercise 4.2). Thus $A_i \equiv A^{1/2}E_iA^{1/2} \uparrow A$ (SOT) (Exercise 4.3). Now $\rho(A_i) = U^*A_iU \uparrow T = U^*AU$ (SOT). Since $A_i \leq A$, $\rho(A_i) \leq \rho(A)$. Hence $T \leq \rho(A)$. On the other hand, each $\rho(A_i)$ is dominated by T, so $A_i \leq \rho^{-1}(T)$. Thus $A \leq \rho^{-1}(T)$ so that $\rho(A) \leq T$. Therefore $\rho(A) = T = UAU^*$. ■

Return to the problem of characterizing the C*-algebras of compact operators and fix some notation. Let $\mathbb{H} = \{\, \mathcal{L}_i : i \in I \,\}$ be a maximal collection of pairwise orthogonal reducing subspaces for $\mathcal{A}$ such that $\mathcal{A}|\mathcal{L}_i = \mathcal{B}_0(\mathcal{L}_i)$. By Proposition 16.14 and the maximality of $\mathbb{H}$, $\mathcal{H} = \bigoplus_i \mathcal{L}_i$. If $\mathcal{L}_i$ and $\mathcal{L}_j$ are two of these spaces, say that $\mathcal{L}_i$ *does not depend on* $\mathcal{L}_j$ if there is an operator A in $\mathcal{A}$ such that $A|\mathcal{L}_j = 0$ and $A|\mathcal{L}_i \neq 0$.

16.15 Lemma. *If $\mathcal{L}_i$ does not depend on $\mathcal{L}_j$, then $\mathcal{A}|(\mathcal{L}_i \oplus \mathcal{L}_j) = \mathcal{B}_0(\mathcal{L}_i) \oplus \mathcal{B}_0(\mathcal{L}_j)$.*

Proof. Let $\mathcal{J} = \{\, A \in \mathcal{A} : A|\mathcal{L}_j = 0 \,\}$; so $\mathcal{J}$ is a closed ideal in $\mathcal{A}$. Since $A \to A|\mathcal{L}_i$ is a $*$-homomorphism, $\mathcal{J}|\mathcal{L}_i$ is a closed ideal of $\mathcal{A}|\mathcal{L}_i = \mathcal{B}_0(\mathcal{L}_i)$. The hypothesis of the lemma says that $\mathcal{J}|\mathcal{L}_i$ is not (0), so $\mathcal{J}|\mathcal{L}_i = \mathcal{B}_0(\mathcal{L}_i)$.

Now let K_i, K_j be arbitrary compact operators on $\mathcal{L}_i, \mathcal{L}_j$. Let $A \in \mathcal{A}$ such that $A|\mathcal{L}_j = K_j$. Pick B in $\mathcal{J}$ such that $B|\mathcal{L}_i = K_i - A|\mathcal{L}_i$. So $(A + B)|\mathcal{L}_i = K_i$ and $(A + B)|\mathcal{L}_j = K_j$. ■

In light of this lemma, if $\mathcal{L}_i$ does not depend on $\mathcal{L}_j$, then $\mathcal{L}_j$ does not depend on $\mathcal{L}_i$. Say that $\mathcal{L}_i$ and $\mathcal{L}_j$ are *independent* if neither space depends on the other. Two of these spaces are *dependent* if they are not independent.

16.16 Lemma. *If the spaces $\mathcal{L}_i$ and $\mathcal{L}_j$ are dependent, then $\|A|\mathcal{L}_i\| = \|A|\mathcal{L}_j\|$ for every A in $\mathcal{A}$. Thus two dependent spaces have the same dimension and each dependency class is finite.*

Proof. The hypothesis says that if $A \in \mathcal{A}$ and $A|\mathcal{L}_j = 0$, then $A|\mathcal{L}_i = 0$. Therefore it is possible to define a map $\rho : \mathcal{B}_0(\mathcal{L}_j) \to \mathcal{B}_0(\mathcal{L}_i)$ as follows. If $K \in \mathcal{B}_0(\mathcal{L}_j)$, let $A \in \mathcal{A}$ such that $A|\mathcal{L}_j = K$. Define $\rho(K) = A|\mathcal{L}_i$. If there is also a B in $\mathcal{A}$ such that $B|\mathcal{L}_j = K$, then $(A - B)|\mathcal{L}_j = 0$, so

$(A - B)|\mathcal{L}_i = 0$. Thus ρ is well defined. It is left to the reader to check that ρ is a $*$-homomorphism.

In fact, ρ is injective. Indeed, if $K \in \mathcal{B}_0(\mathcal{L}_j)$ and $\rho(K) = 0$ and $A \in \mathcal{A}$ such that $A|\mathcal{L}_j = K$, then $A|\mathcal{L}_i = 0$. If it were the case that $A|\mathcal{L}_j = K \neq 0$, this would say that the spaces are independent, a contradiction. But a $*$-monomorphism is an isometry (1.8), proving the first statement of the lemma.

That $\dim \mathcal{L}_i = \dim \mathcal{L}_j$ is clear from the fact that the algebras of compact operators on the spaces are isomorphic. By compactness of the operators in $\mathcal{A}$, for any infinite collection of the spaces $\{\mathcal{L}_k\}$, $\|A|\mathcal{L}_k\| \to 0$. Thus each dependency class must be finite. ∎

Consider one dependency class. For convenience, change the notation and assume that $\mathcal{L}_1, \ldots, \mathcal{L}_n$ are the spaces belonging to this dependency class. Lemma 16.16 implies that for $1 \leq k \leq n$, $A|\mathcal{L}_1 \to A|\mathcal{L}_k$, A in $\mathcal{A}$, defines a $*$-isomorphism from $\mathcal{B}_0(\mathcal{L}_1)$ onto $\mathcal{B}_0(\mathcal{L}_k)$. By Corollary 16.13 there is an isomorphism $U_k : \mathcal{L}_k \to \mathcal{L}_1$ such that $U_k^*(A|\mathcal{L}_1)U_k = A|\mathcal{L}_k$ for every A in $\mathcal{A}$. Let $U_1 = 1$ and put $U = U_1 \oplus \cdots \oplus U_n : \mathcal{L}_1 \oplus \cdots \oplus \mathcal{L}_n \to \mathcal{L}_1^{(n)}$; U is an isomorphism. Also if K is any compact operator on $\mathcal{L}_1$ and $A \in \mathcal{A}$ such that $A|\mathcal{L}_1 = K$, then $U^*K^{(n)}U = A|\mathcal{L}_1 \oplus \cdots \oplus A|\mathcal{L}_n$. Thus

$$\mathcal{A}|(\mathcal{L}_1 \oplus \cdots \oplus \mathcal{L}_n) \cong \mathcal{B}_0(\mathcal{L}_1)^{(n)}.$$

Pick one member of each dependency class and relabel these spaces $\{\mathcal{H}_d : d \in D\}$. Let n_d be the number of spaces belonging to the dependency class of $\mathcal{H}_d$. The discussion of the preceding paragraph shows that by replacing $\mathcal{A}$ by $U^*\mathcal{A}U$ for a suitable isomorphism U, it can be assumed that

$$\mathcal{H} = \bigoplus_d \mathcal{H}_d^{(n_d)}$$

in such a way that

16.17 $$\mathcal{A}|\mathcal{H}_d^{(n_d)} = \mathcal{B}_0(\mathcal{H}_d)^{(n_d)}.$$

This gives the promised characterization. The notation is as above.

16.18 Theorem. *With the preceding notation, if $\mathcal{A}$ is a non-degenerate C^*-algebra of compact operators, then there is a unitary operator U such that*

$$U^*\mathcal{A}U = \left\{ \bigoplus_d K_d^{(n_d)} : K_d \in \mathcal{B}_0(\mathcal{H}_d) \text{ and } \|K_d\| \to 0 \right\}.$$

Proof. As discussed prior to the theorem, it can be assumed that (16.17) holds; the goal is to show that $\mathcal{A}$ equals the restricted direct sum appearing in the statement of the theorem. The fact that $\mathcal{A}$ is contained in this restricted direct sum is clear. To show equality, it suffices to show that each $\bigoplus_d K_d^{(n_d)}$ with K_d a positive compact contraction can be interpolated.

Let $\epsilon > 0$ and choose a finite subset D_1 of D such that $\|K_d\| < \epsilon/2$ for $d \notin D_1$. An application of Lemma 16.15 produces an operator A in $\mathcal{A}$ such that $A|\mathcal{H}_d = K_d^{1/2}$ for all d in D_1. Replacing A by $|A|$ it may be assumed that $A \in \mathcal{A}_+$. If $f(t) = \min\{1, t\}$, then $f(K_d) = K_d$, so that replacing A by $f(A)$ it follows that it can also be assumed that $A \in (\operatorname{ball} \mathcal{A})_+$.

Now let D_2 be a finite subset of D that contains D_1 such that $\|A|\mathcal{H}_d\| < \epsilon/2$ when $d \notin D_2$. Arguing as in the preceding paragraph, there is a B in $(\operatorname{ball} \mathcal{A})_+$ such that $B|\mathcal{H}_d = K_d^{1/2}$ for d in D_1 and $B|\mathcal{H}_d = 0$ for d in $D_2 \setminus D_1$. Consider $T = AB$ in $\mathcal{A}$. If $d \in D_1$, $T|\mathcal{H}_d = K_d$; if $d \in D_2 \setminus D_1$, $T|\mathcal{H}_d = 0$; if $d \notin D_2$, $\|T|\mathcal{H}_d\| \leq \|A|\mathcal{H}_d\| \, \|B|\mathcal{H}_d\| < \epsilon/2$. Therefore,

$$T - \bigoplus_d K_d^{(n_d)} = - \bigoplus_{d \in D_2 \setminus D_1} K_d^{(n_d)} + \bigoplus_{d \notin D_2} \left[T|\mathcal{H}_d^{(n_d)} - K_d^{(n_d)} \right],$$

and this operator has norm at most ϵ. Thus $\bigoplus_d K_d^{(n_d)}$ can be approximated by operators from $\mathcal{A}$, achieving equality. ∎

If the underlying Hilbert space is finite dimensional, the preceding theorem is a special case of the Wedderburn decomposition of a semisimple algebra of matrices.

To complete this section, return to Theorem 16.11 and examine representations of a C*-algebra $\mathcal{A}$ contained in the compact operators. For any C*-algebra, not just one contained in $\mathcal{B}_0$, let $\widehat{\mathcal{A}}$ be the set of equivalence classes of irreducible representations of the C*-algebra $\mathcal{A}$. $\widehat{\mathcal{A}}$ is called the *spectrum* of $\mathcal{A}$. For each ζ in $\widehat{\mathcal{A}}$, fix a ρ_ζ in ζ. If $\mathcal{A}$ is a C*-algebra of compact operators as in this section, for each representation ρ of $\mathcal{A}$ define the cardinal number-valued function m_ρ on $\widehat{\mathcal{A}}$ as follows. By Theorem 16.11, $\rho \cong \bigoplus_i \rho_i$, where each ρ_i is irreducible. For each ζ in $\widehat{\mathcal{A}}$, let $m_\rho(\zeta)$ to be the cardinality of the set $\{ i : \rho_i \in \zeta \}$. The function m_ρ is called the *multiplicity function* of the representation ρ. For each ζ in $\widehat{\mathcal{A}}$ pick one element ρ_ζ for ζ. Thus Theorem 16.11 can be rephrased as follows.

16.19 Theorem. *If ρ is a representation of a C*-algebra of compact operators and m_ρ is its multiplicity function, then*

$$\rho \cong \bigoplus \left\{ \rho_\zeta^{(m_\rho(\zeta))} : \zeta \in \widehat{\mathcal{A}} \right\}.$$

If the reader is nervous about set theory and logical difficulties connected with this theorem, realize that since every irreducible representation of $\mathcal{A}$ is equivalent to a subrepresentation of the identity representation of $\mathcal{A}$, this imposes a restriction on the cardinality of $\widehat{\mathcal{A}}$. The next result is a routine application of elementary facts from operator theory.

16.20 Corollary. *If $\mathcal{A}$ is a C^*-algebra that is contained in $\mathcal{B}_0(\mathcal{H})$, then two representations of $\mathcal{A}$ are equivalent if and only if they have the same multiplicity function.*

16.21 Corollary. *If ρ and κ are two representations of $\mathcal{A}$ with multiplicity functions m_ρ and m_κ, then ρ is equivalent to a subrepresentation of κ if and only if $m_\rho \leq m_\kappa$.*

16.22 Corollary. *If $\mathcal{A}$ is a C^*-algebra of compact operators and ρ and κ are two representations of $\mathcal{A}$ such that each is equivalent to a subrepresentation of the other, then $\rho \cong \kappa$.*

Now to specialize to the case of a separable Hilbert space. Say that the *rank* of an operator T on a separable Hilbert space is the dimension of $\operatorname{ran} T$ if this range is finite dimensional; otherwise the rank of T is ∞. Thus $\operatorname{rank} T$ is the Hilbert space dimension of $\dim \operatorname{cl}[\operatorname{ran} T]$.

16.23 Proposition. *If $\mathcal{A}$ is a C^*-algebra of compact operators and $(\rho, \mathcal{R})$ and $(\kappa, \mathcal{K})$ are two non-degenerate separable representations of $\mathcal{A}$, then ρ and κ are unitarily equivalent if and only if for each A in $\mathcal{A}$, $\rho(A)$ and $\kappa(A)$ have the same rank.*

Proof. Clearly if $\rho \cong \kappa$, then $\operatorname{rank} \rho(A) = \operatorname{rank} \kappa(A)$ for every A in $\mathcal{A}$. For the converse, interpret Theorem 16.19 for the case of separable representations. Thus there are r_j, k_j in $\{0, 1, \ldots, \aleph_0\}$ and subrepresentations of the identity representation of $\mathcal{A}$, $\{\eta_1, \eta_2, \ldots\}$, such that

$$\rho \cong \bigoplus_j \eta_j^{(r_j)} \qquad \text{and} \qquad \kappa \cong \bigoplus_j \eta_j^{(k_j)}.$$

To show that $\rho \cong \kappa$ it is necessary to show that $r_j = k_j$ for all $j \geq 1$. Let $\mathcal{H}_j$ be the reducing subspace for $\mathcal{A}$ such that $\eta_j(A) = A|\mathcal{H}_j$ for all A in $\mathcal{A}$. So $\mathcal{A}|\mathcal{H}_i = \mathcal{B}_0(\mathcal{H}_j)$. By Theorem 16.18, for each $j \geq 1$ there is a projection E_j in $\mathcal{A}$ such that $\eta_j(E_j)$ is a rank 1 projection and $\eta_i(E_j) = 0$ for $i \neq j$. Thus $r_j = \operatorname{rank} \rho(E_j) = \operatorname{rank} \kappa(E_j) = k_j$. ∎

Exercises.

1. Verify the statements in Example 16.1.

2. Compute the polar decomposition of $g \otimes h$.

3. Show that every finite rank operator can be written as $\sum_{k=1}^{n} g_k \otimes h_k$ for some vectors $g_1, \ldots, g_n, h_1, \ldots, h_n$ in $\mathcal{H}$.

4. If X is a compact space, characterize the positive measures μ such that π_μ is the direct sum of irreducible representations. Characterize the compact spaces X such that every representation of $C(X)$ is the direct sum of irreducible representations.

5. If $\{T_n\}$ is a sequence of operators such that $T_n \to 0$ (SOT), show that for every compact operator K, $\|T_nK\| \to 0$ and $\|KT_n\| \to 0$. (Hint: First show this when K has finite rank.) If $T_n \to 0$ (WOT), then $KT_n \to 0$ (SOT). If S is the unilateral shift, show that there is a compact operator K such that $\{S^nK\}$ does not converge to 0 (SOT) even though $S^n \to 0$ (WOT).

6. If $\mathcal{C}$ is a C*-algebra contained in $\mathcal{B}(\mathcal{H})$, show that $\mathcal{C}$ is irreducible if and only if $[\mathcal{C}e] = \mathcal{H}$ for each non-zero vector e in $\mathcal{H}$. Show that if $\mathcal{C}$ is an irreducible C*-algebra contained in $\mathcal{B}(\mathcal{H})$ and $\mathcal{I}$ is an ideal of $\mathcal{C}$, then $\mathcal{I}$ is irreducible. (It is actually the case that $\mathcal{C}$ is irreducible if and only if $\mathcal{C}e = \mathcal{H}$ for each non-zero vector e — no closure required. See Kadison and Ringrose [1983], Corollary 5.4.4.)

7. For each of the representations in Example 16.1, find the multiplicity function for the identity representation.

8. Let N be a normal compact operator on $\mathcal{H}$ and let $\mathcal{A} = C^*(N) \cap \mathcal{B}_0(\mathcal{H})$. (So $\mathcal{A}$ is the smallest C*-algebra containing N and contained in $\mathcal{B}_0(\mathcal{H})$.) Find the multiplicity function of the identity representation of $\mathcal{A}$ and compare this with the spectral decomposition of N.

9. Let T be any compact operator on $\mathcal{H}$ and let $\mathcal{A} = C^*(T) \cap \mathcal{B}_0(\mathcal{H})$. If Theorem 16.11 is applied to the identity representation of $\mathcal{A}$, what does this say about the operator T?

10. For $1 \le d < \infty$, let $\mathcal{H}_d = \mathbb{C}^d$ and for $d = \aleph_0$ let $\mathcal{H}_d = \ell^2 = \ell^2(\mathbb{N})$. If $D \subseteq \{1, 2, \ldots, \aleph_0\}$ and $\{n_d : d \in D\}$ is any collection of positive integers, show that there is a separable Hilbert space and a C*-algebra contained in $\mathcal{B}_0(\mathcal{H})$ such that $\mathcal{A}$ is unitarily equivalent to

$$\left\{ \bigoplus_{d \in D} K_d^{(n_d)} : K_d \in \mathcal{B}_0(\mathcal{H}_d) \text{ and } \|K_d\| \to 0 \right\}.$$

11. What happens to the results of this section if $\mathcal{A}$ is degenerate?

§17. Ideals of operators.

This section will characterize the closed ideals of $\mathcal{B}(\mathcal{H})$ for an arbitrary, possibly non-separable, Hilbert space. This is a result that has long been known, but seldom discussed in books on operator theory. For a separable Hilbert space there is only one proper closed ideal, $\mathcal{B}_0(\mathcal{H})$. In general, the ideals correspond to the infinite cardinal numbers less than or equal to $\dim \mathcal{H}$.

The next result is one whose use in this section is not essential, but the knowledge of which is required of all budding operator theorists. This theorem first appeared in Douglas [1966] and is sometimes called Douglas's Range Inclusion Theorem.

17.1 Theorem. *If A and B are bounded operators on the Hilbert space $\mathcal{H}$, the following statements are equivalent.*

(a) $\operatorname{ran} A \subseteq \operatorname{ran} B$.

(b) *There is a positive constant c such that $AA^* \leq c^2 BB^*$.*

(c) *There is a bounded operator C such that $A = BC$.*

Proof. It's trivial that (c) implies (a) and almost trivial that (c) implies (b): $AA^* = BCC^*B^* \leq c^2 BB^*$, where $c = \|C\|$.

(a) *implies* (c). If B_1 is the restriction of B to $(\ker B)^\perp$, then B_1 is injective and the (possibly unbounded) inverse $B_1^{-1} : \operatorname{ran} B \to (\ker B)^\perp$ has closed graph. Since (a) is valid, it is possible to define $C = B_1^{-1}A : \mathcal{H} \to (\ker B)^\perp \subseteq \mathcal{H}$. It is easy to check that C has closed graph since B_1^{-1} does and A is bounded. Thus C is a bounded operator by the Closed Graph Theorem. Clearly $A = BC$.

(b) *implies* (c). Condition (b) quickly translates to the the inequality $\|A^*h\| \leq c\|B^*h\|$ for all h in $\mathcal{H}$. Thus $D(B^*h) = A^*h$ gives a well defined operator $D : \operatorname{ran} B^* \to \operatorname{ran} A^*$ that is bounded. Extend D to the closure of the range of B^* and set it equal to 0 on $(\operatorname{ran} B^*)^\perp$. Taking $C = D^*$ demonstrates (c). ∎

The study of operator ranges is a subject in itself. See Fillmore and Williams [1971] for a survey and additional references. Now to consider the closed ideals of $\mathcal{B}(\mathcal{H})$. Here is an an easy but useful fact about ideals of operators.

17.2 Lemma. *If $\mathcal{I}$ is a closed ideal of $\mathcal{B}(\mathcal{H})$, $A \in \mathcal{I}$, P is a projection such that $P \subseteq \operatorname{ran} A$, and Q is any projection with $\dim Q = \dim P$, then $Q \in \mathcal{I}$.*

Proof. If $P \subseteq \operatorname{ran} A$, then the preceding theorem implies there is a bounded operator C such that $P = AC$, so $P \in \mathcal{I}$. If Q is another projection such that $\dim Q = \dim P$, then there is a partial isometry W with initial space P and final space Q. Thus $W^*W = P$ and $WW^* = Q$. Thus $W = WP \in \mathcal{I}$, so that $W^* \in \mathcal{I}$ (5.4). But then $Q = WW^* \in \mathcal{I}$. ■

17.3 Lemma. *If $\mathcal{H}$ is infinite dimensional, $A, B \in \mathcal{B}(\mathcal{H})$, and $A + B = 1$, then either* $\operatorname{ran} A$ *or* $\operatorname{ran} B$ *contains a subspace with the same dimension as $\mathcal{H}$.*

Proof. If $Af = 0$, then $f = Bf$. Thus $\operatorname{ran} B \supseteq \ker A = \ker |A|$. Let $A = W|A|$ be the polar decomposition of A. If $a = \|A\|$ and $|A| = \int_0^a t\, dE(t)$ is the spectral decomposition, then $(\ker A)^\perp = E(a/2, a] \oplus E(a/3, a/2] \oplus \cdots$, so that $\mathcal{H} = \ker A \oplus E(a/2, a] \oplus E(a/3, a/2] \oplus \cdots$. It is necessary to look at two cases separately. If $\mathcal{H}$ is not separable, then at least one of the spaces in this direct sum must have the same dimension as $\mathcal{H}$. Now W is isometric on each of the spaces $E(a/n+1, a/n]$, $n \geq 1$. Since $\ker A \subseteq \operatorname{ran} B$ and $WE(a/n+1, a/n] \subseteq \operatorname{ran} A$, the lemma is proved in the non-separable case. If $\mathcal{H}$ is separable, it may be that each of the spaces $E(a/n+1, a/n]$ as well as $\ker A$ is finite dimensional. But in this case, $|A|$ is compact. Since $B = 1 - A$, the Fredholm Alternative ([ACFA], VII.7.9) implies $\operatorname{ran} B$ is closed. Of course it must also be infinite dimensional. ■

Now fix the Hilbert space $\mathcal{H}$ and put $\delta = \dim \mathcal{H}$. If α is an infinite cardinal number with $\alpha \leq \delta$, define

17.4
$$\mathcal{I}_\alpha \equiv \{\, A \in \mathcal{B}(\mathcal{H})\text{: if } \mathcal{M} \text{ is a closed subspace of } \mathcal{H} \text{ and } \mathcal{M} \subseteq \operatorname{ran} A, \text{then } \dim \mathcal{M} < \alpha \,\}.$$

So $\mathcal{I}_\alpha$ is the collection of all operators whose range does not contain a closed subspace whose Hilbert space dimension is at least α. If $\delta = \aleph_0$, then $\aleph_0$ is the only possible choice for α. In this case the ideal I_α becomes the set of all operators A such that the only closed subspaces of $\operatorname{ran} A$ are finite dimensional. Such operators are compact (Exercise 1) so that the ideal is $\mathcal{B}_0(\mathcal{H})$.

17.5 Theorem. *For each infinite cardinal number α, $\mathcal{I}_\alpha$ is a closed ideal of $\mathcal{B}(\mathcal{H})$ and every closed ideal of $\mathcal{B}(\mathcal{H})$ has this form for some cardinal number α.*

Proof. First it will be shown that $\mathcal{I}_\alpha$ is a closed ideal. Suppose $A \in \mathcal{I}_\alpha$ and $T \in \mathcal{B}(\mathcal{H})$. If P is a projection and $P \subseteq \operatorname{ran} AT \subseteq \operatorname{ran} A$, then $\dim P < \alpha$. So $\mathcal{I}_\alpha$ is invariant under right multiplication by bounded operators.

Claim 1. If $A \in \mathcal{I}_\alpha$, then $|A| \in \mathcal{I}_\alpha$.

Suppose $P \subseteq \operatorname{ran}|A|$. If $A = W|A|$ is the polar decomposition, $P \subseteq \operatorname{ran} W^*A \subseteq \operatorname{ran} W^* = (\ker W)^\perp$, the initial space of W. Thus W is isometric on P and, hence, $\dim P = \dim W(P)$. But $W(P) \subseteq W(\operatorname{ran}|A|) = \operatorname{ran} A$. Thus $\dim P < \alpha$ and Claim 1 is true.

Claim 2. If $A \in \mathcal{I}_\alpha$, then $A^* \in \mathcal{I}_\alpha$.

In fact if $A = W|A|$ is the polar decomposition, $A^* = |A|W^*$. By Claim 1, $|A| \in \mathcal{I}_\alpha$. Since $\mathcal{I}_\alpha$ is invariant under right multiplication, Claim 2 is true.

To see that $\mathcal{I}_\alpha$ is also invariant under left multiplication, note that if $A \in \mathcal{I}_\alpha$, then $A^*T^* \in \mathcal{I}_\alpha$ and so $TA \in \mathcal{I}_\alpha$ by Claim 2.

Now let $A, B \in \mathcal{I}_\alpha$ and show that $A + B \in \mathcal{I}_\alpha$. (Here is where it is necessary for α to be infinite.) Assume P is a projection and $P \subseteq \operatorname{ran}(A + B)$. It must be shown that $\dim P < \alpha$. Theorem 17.1 implies there is a bounded operator C such that $P = (A+B)C = AC + BC$. Note that $AC, BC \in \mathcal{I}_\alpha$. Also $P = PACP + PBCP$, and $PACP, PBCP \in \mathcal{I}_\alpha$. But $PACP, PBCP$ are operators in $\mathcal{B}(P)$ that add up to the identity. By Lemma 17.3, either $\operatorname{ran} PACP$ or $\operatorname{ran} PBCP$ contains a closed subspace with the same dimension as P. Thus $\dim P < \alpha$.

It remains to show that $\mathcal{I}_\alpha$ is topologically closed. Let $A \in \operatorname{cl}[\mathcal{I}_\alpha]$ and let P be a projection with $P \subseteq \operatorname{ran} A$. Put $Q = A^{-1}(P) \cap (\ker A)^\perp$. So Q is a closed subspace and A maps Q injectively onto P. Thus A is bounded below on Q. That is, there is an $\epsilon > 0$ such that $\|Ah\| \geq \epsilon\|h\|$ for every h in Q. Choose a B in $\mathcal{I}_\alpha$ such that $\|A - B\| < \epsilon/2$. It's easy to see that $\|Bh\| \geq (\epsilon/2)\|h\|$ for all h in Q. So $B(Q)$ is a subspace contained in $\operatorname{ran} B$. Hence $\dim B(Q) < \alpha$. But $\dim P = \dim Q = \dim B(Q)$, so $\dim P < \alpha$. Since P was arbitrary, $A \in \mathcal{I}_\alpha$.

For the converse, let $\mathcal{I}$ be any closed ideal in $\mathcal{B}(\mathcal{H})$.

Claim 3. If $A \in \mathcal{I}_+$, $A = \int_0^1 t\, dE(t)$, $\epsilon > 0$, and Δ is any Borel subset of $[\epsilon, 1]$, then $E(\Delta) \in \mathcal{I}$.

In fact, if $B = \int_\Delta t^{-1}\, dE(t)$, then $E(\Delta) = BA \in \mathcal{I}$. Thus it is possible to define the cardinal number α by

$$\alpha \equiv \sup\{\dim P : P \text{ a projection in } \mathcal{I}\}.$$

By Lemma 17.2 any projection in $\mathcal{B}(\mathcal{H})$ with dimension strictly smaller than α belongs to $\mathcal{I}$.

Now to show that $\mathcal{I} = \mathcal{I}_\alpha$. It suffices to show that $\operatorname{ball}\mathcal{I}_+ = (\operatorname{ball}\mathcal{I}_\alpha)_+$. Let $A \in \operatorname{ball}\mathcal{I}_+$ with $A = \int_0^1 t\, dE(t)$. By Claim 3 and the definition of α, $E[\epsilon, 1] \in \mathcal{I}_\alpha$. Thus $AE[\epsilon, 1] \in \mathcal{I}_\alpha$, but $AE[\epsilon, 1] \to A$ in norm. Thus $\mathcal{I} \subseteq \mathcal{I}_\alpha$.

Now let $T \in (\operatorname{ball}\mathcal{I}_\alpha)_+$ with $T = \int_0^1 t\,dP(t)$. Since $P(\Delta) \subseteq \operatorname{ran} T$ whenever Δ is a Borel subset of $[0,1]$ with $0 \notin \operatorname{cl}\Delta$, it follows that $\dim P(\Delta) < \alpha$ for every such set Δ. By the observations above, this gives that every such spectral projection for T belongs to $\mathcal{I}$. Now the Spectral Theorem implies that $T \in \mathcal{I}$. ■

17.6 Corollary. *For any Hilbert space $\mathcal{H}$, $\mathcal{B}_0(\mathcal{H})$ is the smallest closed ideal of $\mathcal{B}(\mathcal{H})$.*

Proof. Each ideal $\mathcal{I}_\alpha$ clearly contains the finite rank operators, and thus contains $\mathcal{B}_0(\mathcal{H})$. ■

This corollary could have been proved immediately after Lemma 17.2. Note that if $\alpha = \delta = \dim\mathcal{H}$, $\mathcal{I}_\alpha$ is a maximal ideal in $\mathcal{B}(\mathcal{H})$. If $\alpha = \aleph_0$, then $\mathcal{I}_\alpha = \mathcal{B}_0$, the minimal ideal of compact operators. Since the cardinal numbers are linearly ordered, the collection of proper ideals in $\mathcal{B}(\mathcal{H})$ is a chain with a largest and smallest element.

17.7 Proposition. *If $\mathcal{H}$ is separable, an operator T is not compact if and only if there are operators A, B such that $ATB = 1$.*

Proof. If T is compact, ATB is compact for all operators A and B and, hence, can never be the identity. Now assume that T is not compact; for convenience, also assume that $\|T\| = 1$. Consider the positive operator $|T|$ and its spectral decomposition, $|T| = \int t\,dE(t)$. Because T is not compact there is an $\epsilon > 0$ such that $\mathcal{M} = E[\epsilon, 1]\mathcal{H}$ is infinite dimensional. Since $\mathcal{H}$ is separable, there is an injective operator B such that $B\mathcal{H} = \mathcal{M}$. But for any h in $\mathcal{M}$, $\|Th\|^2 = \langle T^*Th, h\rangle = \int t^2\,d\langle E(t)h, h\rangle \geq \epsilon^2\|h\|^2$. Thus $T\mathcal{M} = \mathcal{N}$ is a closed infinite dimensional space. Thus there is a bounded operator A such that $A\mathcal{N}^\perp = 0$ and $ATBh = h$ for every h in $\mathcal{H}$. ■

17.8 Corollary. *If $\mathcal{H}$ is separable and $\mathcal{I}$ is any non-trivial ideal (not necessarily closed) of $\mathcal{B}(\mathcal{H})$, then $\mathcal{I} \subseteq \mathcal{B}_0(\mathcal{H})$, the ideal of compact operators.*

Exercises.

1. Show that if $A \in \mathcal{B}(\mathcal{H})$, then A is compact if and only if every closed subspace of $\operatorname{ran} A$ is finite dimensional.

2. If $\delta = \dim\mathcal{H}$, show that the condition in Proposition 17.7 characterizes the ideal $\mathcal{I}_\delta$.

3. If A is a hermitian operator with $A = \int t\,dE(t)$, show that $A \in \mathcal{I}_\alpha$ (17.4) if and only if for every $\epsilon > 0$, $\dim E(\{t : |t| \geq \epsilon\}) < \alpha$.

§18. Trace class and Hilbert–Schmidt operators

Recall the definition of an uncountable sum discussed at the beginning of §16.

18.1 Proposition. *If $\mathcal{E}$ and $\mathcal{F}$ are orthonormal bases for $\mathcal{H}$, then for every bounded operator A,*

$$\sum_{\mathcal{E}} \|Ae\|^2 = \sum_{\mathcal{F}} \|A^* f\|^2 = \sum_{\mathcal{E}} \sum_{\mathcal{F}} |\langle Ae, f\rangle|^2.$$

(Note: All the terms in these sums are positive and this result is to be interpreted as stating that one of these infinite sums converges if and only if all of them do, in which case the three sums are equal.)

Proof. From Parseval's Identity, $\|Ae\|^2 = \sum_{\mathcal{F}} |\langle Ae, f\rangle|^2$ for each e in $\mathcal{E}$. Also for every f in $\mathcal{F}$, $\|A^* f\|^2 = \sum_{\mathcal{E}} |\langle e, A^* f\rangle|^2$. The result follows. ∎

Applying the preceding proposition to the square root of $|A|$ yields the following.

18.2 Corollary. *The sum $\sum_{e\in\mathcal{E}} \langle |A|e, e\rangle$ is independent of the choice of basis $\mathcal{E}$.*

18.3 Definition. An operator $A : \mathcal{H} \to \mathcal{H}$ is *trace class* if there is a basis $\mathcal{E}$ such that $\sum_{e\in\mathcal{E}} \langle |A|e, e\rangle < \infty$. The set of trace class operators on $\mathcal{H}$ is denoted by $\mathcal{B}_1(\mathcal{H})$, or $\mathcal{B}_1$ if the space $\mathcal{H}$ is understood.

In light of Corollary 18.2, an operator A is trace class if and only if $\sum_{\mathcal{E}} \langle |A|e, e\rangle$ is finite for every choice of orthonormal basis $\mathcal{E}$ and

$$\|A\|_1 \equiv \sum_{\mathcal{E}} \langle |A|e, e\rangle$$

is well defined, depending only on the operator A. The number $\|A\|_1$ is called the *trace norm* of A. It will be shown below that $\mathcal{B}_1$ is a vector space, the trace norm does indeed define a norm on $\mathcal{B}_1$, and, with respect to this norm, $\mathcal{B}_1$ is complete.

18.4 Definition. An operator A on $\mathcal{H}$ is called a *Hilbert–Schmidt operator* if $|A|^2$ is trace class. The set of all Hilbert–Schmidt operators is denoted by $\mathcal{B}_2(\mathcal{H})$, or $\mathcal{B}_2$ if the space $\mathcal{H}$ is understood. For any Hilbert–Schmidt operator A and any basis $\mathcal{E}$ define

$$\|A\|_2 \equiv \left[\sum_{\mathcal{E}} \langle |A|^2 e, e\rangle\right]^{1/2} = \|\,|A|^2\|_1^{1/2}.$$

18.5 Example. Fix an orthonormal basis $\mathcal{E}$ for $\mathcal{H}$ and consider the diagonal operator $Ae = a(e)e$, where $a \in \ell^\infty(\mathcal{E})$. Then $A \in \mathcal{B}_p(\mathcal{H})$ $(p = 1, 2)$ if and only if $a \in \ell^p(\mathcal{E})$.

It turns out that it is somewhat easier to verify certain elementary properties for $\mathcal{B}_2$ than the analogous properties for $\mathcal{B}_1$.

18.6 Proposition. *Fix an operator A in $\mathcal{B}_2(\mathcal{H})$.*

(a) $\|A\|_2 = \left[\sum_{\mathcal{E}} \|Ae\|^2\right]^{1/2}$ *for any basis $\mathcal{E}$.*

(b) $\|A^*\|_2 = \|A\|_2$.

(c) $\|A\| \leq \|A\|_2$.

(d) *If $T \in \mathcal{B}(\mathcal{H})$, then $AT, TA \in \mathcal{B}_2(\mathcal{H})$ and $\|AT\|_2$ and $\|TA\|_2 \leq \|T\|\,\|A\|_2$.*

(e) *$\mathcal{B}_2(\mathcal{H})$ is an ideal of $\mathcal{B}(\mathcal{H})$ and $\|\cdot\|_2$ is a norm on $\mathcal{B}_2(\mathcal{H})$.*

Proof. The proof of part (a) is easy. The proof of (b) follows by combining (a) with Proposition 18.1. To prove (c), fix a unit vector h and choose an orthonormal basis $\mathcal{E}$ that contains h. From (a) it follows that $\|Ah\| \leq \|A\|_2$. Since h was arbitrary, part (c) follows. If $\mathcal{E}$ is any basis and $T \in \mathcal{B}(\mathcal{H})$, then for each e in $\mathcal{E}$, $\|TAe\|^2 \leq \|T\|^2\|Ae\|^2$. Thus $TA \in \mathcal{B}_2$ and $\|TA\|_2 \leq \|T\|\,\|A\|_2$. Condition (b) implies that $T^*A^* \in \mathcal{B}_2$ and so $AT \in \mathcal{B}_2$ with $\|AT\|_2 \leq \|T\|\,\|A\|_2$. This proves (d).

Now fix a basis $\mathcal{E}$ and let $A, B \in \mathcal{B}_2$; so $\{\|Ae\| : e \in \mathcal{E}\}$ and $\{\|Be\| : e \in \mathcal{E}\} \in \ell^2(\mathcal{E})$. Using the triangle inequality for $\ell^2(\mathcal{E})$, $[\sum_{\mathcal{E}}(\|Ae\|+\|Be\|)^2]^{1/2} \leq \|A\|_2 + \|B\|_2$. Hence $\|A + B\|_2^2 = \sum_{\mathcal{E}} \|Ae + Be\|^2 \leq \sum_{\mathcal{E}}(\|Ae\| + \|Be\|)^2 \leq (\|A\|_2 + \|B\|_2)^2 < \infty$. Thus $A+B \in \mathcal{B}_2$ and $\|A+B\|_2 \leq \|A\|_2+\|B\|_2$. This essentially completes the proof of (e), the remainder of the proof being left to the reader. ∎

18.7 Corollary. *If $A \in \mathcal{B}_2(\mathcal{H})$ and $\epsilon > 0$, there is a finite rank operator B with $\|A - B\|_2 < \epsilon$. Consequently, every Hilbert–Schmidt operator is compact.*

Proof. First understand that the second statement does follow from the first. In fact, by part (c) of the preceding proposition, $\|A-B\| \leq \|A-B\|_2 < \epsilon$, so A can be approximated by finite rank operators. To prove the first statement, let $\mathcal{E}$ be a basis and let $e_1, \dots, e_n \in \mathcal{E}$ such that $\sum_{e \neq e_1,\dots,e_n} \|Ae\|^2 < \epsilon^2$. Define the finite rank operator B by letting $B = A$ on $[e_k : 1 \leq k \leq n]$ and $B = 0$ on $[e_k : 1 \leq k \leq n]^\perp$. So $\|A - B\|_2^2 = \sum_{e\neq e_1,\dots,e_n} \|Ae\|^2 \leq \epsilon^2$. ∎

18.8 Proposition. *If $A \in \mathcal{B}(\mathcal{H})$, the following statements are equivalent.*

(a) $A \in \mathcal{B}_1(\mathcal{H})$.

(b) $|A|^{1/2} \in \mathcal{B}_2$.

(c) *A is the product of two Hilbert–Schmidt operators.*

(d) *$|A|$ is the product of two Hilbert–Schmidt operators.*

Proof. Let $A = W|A|$ be the polar decomposition of A.

(a) *implies* (b). For any vector e, $\| \, |A|^{1/2}e\|^2 = \langle |A|e, e\rangle$.

(b) *implies* (c). Here $A = (W|A|^{1/2})(|A|^{1/2})$ and from (b) and Proposition 18.6, both of these factors are in $\mathcal{B}_2$.

(c) *implies* (d). If $A = BC$ and B and C are in $\mathcal{B}_2$, then $|A| = (W^*B)C$. By Proposition 18.6, $W^*B \in \mathcal{B}_2$.

(d) *implies* (a). Suppose $|A| = BC$, where B and $C \in \mathcal{B}_2$. For any orthonormal basis $\mathcal{E}$ and $e \in \mathcal{E}$, $\langle |A|e, e\rangle = \langle Ce, B^*e\rangle \le \|Ce\| \, \|B^*e\|$. Hence

$$\begin{aligned}\sum_{\mathcal{E}} \langle |A|e, e\rangle &\le \sum_{\mathcal{E}} \|Ce\| \, \|B^*e\| \\ &\le \left[\sum_{\mathcal{E}} \|Ce\|^2\right]^{\frac{1}{2}} \left[\sum_{\mathcal{E}} \|B^*e\|^2\right]^{\frac{1}{2}} \\ &= \|C\|_2 \|B\|_2. \ \blacksquare\end{aligned}$$

18.9 Proposition. *If $A \in \mathcal{B}_1$ and $\mathcal{E}$ is a basis, then $\sum_{\mathcal{E}} |\langle Ae, e\rangle| < \infty$ and $\sum_{\mathcal{E}} \langle Ae, e\rangle$ is independent of the choice of basis.*

Proof. Since $A \in \mathcal{B}_1$, it is possible to write $A = C^*B$, where B and C are Hilbert–Schmidt operators. Since $\|(B - \lambda C)e\|^2 \ge 0$ for any choice of scalar λ,

$$2\text{Re}\, \bar{\lambda}\langle Be, Ce\rangle \le \|Be\|^2 + |\lambda|^2 \, \|Ce\|^2.$$

Choosing λ with $|\lambda| = 1$ and $\bar{\lambda}\langle Be, Ce\rangle = |\langle Be, Ce\rangle|$ shows that $|\langle Ae, e\rangle| = |\langle Be, Ce\rangle| \le \frac{1}{2}[\, \|Be\|^2 + \|Ce\|^2\,]$ for any unit vector e. So for an orthonormal basis $\mathcal{E}$, $\sum_{\mathcal{E}} |\langle Ae, e\rangle| \le \frac{1}{2}[\, \|B\|_2^2 + \|C\|_2^2\,] < \infty$.

To see that $\sum_{\mathcal{E}} \langle Ae, e\rangle$ is independent of the choice of a basis, observe that $\text{Re}\,\langle Ae, e\rangle = \frac{1}{4}[\, \|(B + C)e\|^2 - \|(B - C)e\|^2\,]$. Hence $\text{Re} \sum_{\mathcal{E}} \langle Ae, e\rangle = \frac{1}{4}[\, \|B + C\|_2^2 - \|B - C\|_2^2\,]$. Replacing A by iA gives $\text{Im} \sum_{\mathcal{E}} \langle Ae, e\rangle = \frac{1}{4}[\, \|iB + C\|_2^2 - \|iB - C\|_2^2\,]$. The independence of $\sum_{\mathcal{E}} \langle Ae, e\rangle$ from the choice of a basis is now immediate. $\blacksquare$

In light of the preceding proposition it is possible to define the trace of an operator in $\mathcal{B}_1$.

18.10 Definition. If $A \in \mathcal{B}_1(\mathcal{H})$ and $\mathcal{E}$ is an orthonormal basis, define

$$\operatorname{tr} A \equiv \sum_{\mathcal{E}} \langle Ae, e \rangle.$$

This number, $\operatorname{tr} A$, is called the *trace* of A.

18.11 Theorem. (a) *$\mathcal{B}_1(\mathcal{H})$ is an ideal of $\mathcal{B}(\mathcal{H})$ and $\|\cdot\|_1$ is a norm on $\mathcal{B}_1(\mathcal{H})$.*

(b) *Every trace class operator is compact. If A is a compact operator and $\alpha_1, \alpha_2, \ldots$ are the eigenvalues of $|A|$, each repeated as often as its multiplicity, then $A \in \mathcal{B}_1$ if and only if $\{\alpha_n\} \in \ell^1$, and in this case $\|A\|_1 = \sum \alpha_n$.*

(c) $\operatorname{tr} : \mathcal{B}_1 \to \mathbb{C}$ *is a positive definite linear functional. That is, if $A \in \mathcal{B}_1, A \geq 0$, and $A \neq 0$, then $\operatorname{tr} A > 0$.*

(d) *The normed space $(\mathcal{B}_1, \|\cdot\|_1)$ contains the finite rank operators, $\mathcal{B}_{00}$, as a dense linear manifold. Hence $\mathcal{B}_1$ is a separable normed space if $\mathcal{H}$ is separable.*

(e) *If $A \in \mathcal{B}_1$, then $\operatorname{tr}(AT) = \operatorname{tr}(TA)$ and $|\operatorname{tr}(TA)| \leq \|T\| \, \|A\|_1$ for every operator T in $\mathcal{B}(\mathcal{H})$.*

(f) $\|A\|_1 = \|A^*\|_1$ *for every A in $\mathcal{B}_1(\mathcal{H})$.*

(g) *If $A \in \mathcal{B}_1$ and $T \in \mathcal{B}(\mathcal{H})$, then $\|AT\|_1$ and $\|TA\|_1 \leq \|T\| \, \|A\|_1$.*

Proof. (a). Let A and $B \in \mathcal{B}_1$ and let $A = W|A|, B = V|B|$, and $A + B = U|A+B|$ be the respective polar decompositions. Since each operator in $\mathcal{B}_1$ is the product of two Hilbert–Schmidt operators, $\mathcal{B}_1 \subseteq \mathcal{B}_2 \subseteq \mathcal{B}_0$. Hence $|A+B|$ is a positive compact operator. Thus there is a sequence of orthonormal vectors $\{ e_n \}$ contained in a basis $\mathcal{E}$ such that $|A+B|$ has the diagonalization $|A+B| = \sum \gamma_n e_n \otimes e_n$. Thus (remember to justify the manipulation of the infinite sums that follow since it is not known initially that these sums are convergent)

$$\begin{aligned}
\sum \gamma_n &= \sum \langle |A+B| e_n, e_n \rangle \\
&= \sum \langle (A+B) e_n, U e_n \rangle \\
&= \sum [\, \langle A e_n, U e_n \rangle + \langle B e_n, U e_n \rangle \,] \\
&= \sum [\, \langle |A| e_n, W^* U e_n \rangle + \langle |B| e_n, V^* U e_n \rangle \,] \\
&= \sum \Big[\, \langle |A|^{\frac{1}{2}} e_n, |A|^{\frac{1}{2}} W^* U e_n \rangle + \langle |B|^{\frac{1}{2}} e_n, |B|^{\frac{1}{2}} V^* U e_n \rangle \Big] \\
&\leq \sum \Big[\, \| \, |A|^{\frac{1}{2}} e_n \| \, \| \, |A|^{\frac{1}{2}} W^* U e_n \| + \| \, |B|^{\frac{1}{2}} e_n \| \, \| \, |B|^{\frac{1}{2}} V^* U e_n \| \Big]
\end{aligned}$$

$$\leq \left[\sum \| \, |A|^{\frac{1}{2}} e_n \|^2\right]^{\frac{1}{2}} \left[\sum \| \, |A|^{\frac{1}{2}} W^* U e_n \|^2\right]^{\frac{1}{2}}$$
$$+ \left[\sum \| \, |B|^{\frac{1}{2}} e_n \|^2\right]^{\frac{1}{2}} \left[\sum \| \, |B|^{\frac{1}{2}} V^* U e_n \|^2\right]^{\frac{1}{2}}$$
$$\leq \| \, |A|^{\frac{1}{2}} \|_2^2 + \| |B|^{\frac{1}{2}} \|_2^2$$
$$= \|A\|_1 + \|B\|_1.$$

This shows simultaneously that $A + B \in \mathcal{B}_1$ if A and $B \in \mathcal{B}_1$ and that $\| \cdot \|_1$ satisfies the triangle inequality. The remainder of the proof that $\| \cdot \|_1$ is a norm is left as an exercise.

If $A = XY$ for X and Y in $\mathcal{B}_2$ and $T \in \mathcal{B}(\mathcal{H})$, then $TA = (TX)Y \in \mathcal{B}_1$; similarly, $AT \in \mathcal{B}_1$. This shows that $\mathcal{B}_1$ is an ideal in $\mathcal{B}(\mathcal{H})$ and completes the proof of (a).

(b) Assume that A is compact and let $A = W|A|$ be the polar decomposition of A. Since $|A|$ is a compact positive operator, the Spectral Theorem implies that $|A|$ can be diagonalized. Let $\{e_n\}$ be an orthonormal sequence in a basis $\mathcal{E}$ for $\mathcal{H}$ such that $|A| = \sum \alpha_n e_n \otimes e_n$, $\alpha_n \geq 0$, is the diagonalization of $|A|$. If $A \in \mathcal{B}_1$, then $\infty > \operatorname{tr}(|A|) = \sum_{\mathcal{E}} \langle |A| e, e \rangle = \sum_n \alpha_n$. Conversely, if $\sum_n \alpha_n < \infty$, then it is easily seen that $|A|$ is trace class and thus, since $\mathcal{B}_1$ is an ideal, $A = W|A| \in \mathcal{B}_1$.

(c) It is clear that the trace is a linear functional on $\mathcal{B}_1$ that is positive. Moreover, if A is a positive trace class operator with diagonalization $A = \sum \alpha_n e_n \otimes e_n$, then $\operatorname{tr} A = \sum \alpha_n$. Hence $\operatorname{tr} A = 0$ only when $A = 0$.

(d) The proof of this is like the proof of Corollary 18.7 and is left to the reader.

(e) Let $A \in \mathcal{B}_1$ and write $A = C^* B$ with B and C in $\mathcal{B}_2$. As in the proof of Proposition 18.9, $\operatorname{Re}[\operatorname{tr} A] = \operatorname{Re}[\operatorname{tr}(C^* B)] = \frac{1}{4}[\, \|B + C\|_2^2 - \|B - C\|_2^2] = \frac{1}{4}[\, \|B^* + C^*\|_2^2 - \|B^* - C^*\|_2^2] = \operatorname{Re} \operatorname{tr}(CB^*)$. Replacing B by iB shows that $\operatorname{Im}[\operatorname{tr}(C^* B)] = -\operatorname{Im}[\operatorname{tr}(CB^*)]$. Hence

$$\operatorname{tr}(C^* B) = \overline{\operatorname{tr}(CB^*)}.$$

So for any T in $\mathcal{B}(\mathcal{H})$, $\operatorname{tr}(TA) = \operatorname{tr}((TC^*)B) = \overline{\operatorname{tr}((CT^*)B^*)} = \overline{\operatorname{tr}(C(BT)^*)}$ $= \operatorname{tr}(C^*(BT)) = \operatorname{tr}(AT)$.

For the second part of (e), note that $T|A|^{1/2}$ and $|A|^{1/2}T \in \mathcal{B}_2$. If $A = W|A|$ is the polar decomposition of A, then applying the Cauchy-Schwarz Inequality gives

$$|\operatorname{tr}(TA)| = \sum_{\mathcal{E}} |\langle |A|^{1/2} e, |A|^{1/2} W^* T^* e \rangle|$$

$$\leq \sum_{\mathcal{E}} \| \, |A|^{1/2} e \| \, \| \, |A|^{1/2} W^* T^* e \|$$

$$\leq \left[\sum_{\mathcal{E}} \| \, |A|^{1/2} e \|^2 \right]^{1/2} \left[\sum_{\mathcal{E}} \| \, |A|^{1/2} W^* T^* e \|^2 \right]^{1/2}$$

$$= \| \, |A|^{1/2} \|_2 \, \| \, |A|^{1/2} W^* T^* \|_2 .$$

By Proposition 18.6, $|\text{tr}\,(TA)| \leq \| \, |A|^{1/2} \|_2^2 \, \|W^* T^*\| \leq \| \, |A|^{1/2} \|_2^2 \, \|T\|$. But it is also true that $\|A\|_1 = \| \, |A|^{1/2} \|_2^2$, so the proof of (e) is complete.

(f) If $A \in \mathcal{B}_1$ and $A = W|A|$ is the polar decomposition of A, then $AA^* = W|A|^2 W^*$ and so, by uniqueness of the square root, $|A^*| = W|A|W^*$. Hence $\|A^*\|_1 = \text{tr}\,(|A^*|) = \text{tr}\,(W|A|W^*)$. But by part (e) this last quantity is just $\text{tr}\,(W^*W|A|)$ and $W^*W|A| = |A|$ since W^*W is the projection of $\mathcal{H}$ onto $\text{cl}\,[\text{ran}\,|A|\,]$.

(g) Let $A = W|A|$ and $TA = W_1|TA|$ be the respective polar decompositions; so $|TA| = S|A|$, where $S = W_1^* TW$ and hence $\|S\| \leq \|T\|$. From part (d), $\|TA\|_1 = \text{tr}\,(|TA|) = \text{tr}\,(S|A|) \leq \|S\| \, \|A\|_1 \leq \|T\| \, \|A\|_1$. The second half of (g) can be obtained by combining the first half with part (f). ■

18.12 Corollary. *If $A \in \mathcal{B}_1$ and $\mathcal{E} = \{ e_i \}$ and $\mathcal{F} = \{ f_i \}$ are two bases for $\mathcal{H}$, then*

$$\sum_i |\langle Ae_i, f_i \rangle| \leq \|A\|_1 .$$

Proof. Let α_i be a unimodular scalar such that $\alpha_i \langle Ae_i, f_i \rangle = |\langle Ae_i, f_i \rangle|$ and let U be the unitary operator defined by $Ue_i = \overline{\alpha_i} f_i$. Thus $|\langle Ae_i, f_i \rangle| = \langle Ae_i, Ue_i \rangle = \langle U^* Ae_i, e_i \rangle$. Since $U^*A \in \mathcal{B}_1$, $\sum |\langle Ae_i, f_i \rangle| = \text{tr}\,(U^*A) \leq \|U^*\| \, \|A\|_1 \leq \|A\|_1$. ■

18.13 Theorem. *If $\{g_n\}, \{h_n\}$ are two square-summable sequences from $\mathcal{H}$, then $A = \sum_n g_n \otimes h_n \in \mathcal{B}_1(\mathcal{H})$ and $\|A\|_1 \leq \sum_n \|g_n\| \, \|h_n\|$. Conversely, if A is trace class, then two orthogonal sequences of vectors $\{g_n\}, \{h_n\}$ can be chosen such that $\sum \|g_n\|^2 = \|A\|_1 = \sum \|h_n\|^2$ and $A = \sum_n g_n \otimes h_n$.*

Proof. Suppose $\{g_n\}, \{h_n\}$ are sequences of vectors whose norms are square summable and let $\{e_n\}$ be an orthonormal sequence of vectors in $\mathcal{H}$. Put $G_n = \sum_{k=1}^{n} g_n \otimes e_n$, a finite rank operator. Note that if $n > m$, $\|G_n - G_m\|_2^2 = \sum_{k=m+1}^{n} \|(G_n - G_m)e_k\|^2 = \sum_{k=m+1}^{n} |\langle e_k, g_k \rangle|^2 \leq \sum_{k=m+1}^{n} \|g_k\|^2$, and this can be made arbitrarily small. Thus $G = \sum_n g_n \otimes e_n$ is a well defined Hilbert–Schmidt operator. Since $H = \sum e_n \otimes h_n$ is the adjoint of an operator like G, $H \in \mathcal{B}_2$. It is left to the reader to supply the required details to prove that $A = \sum_n g_n \otimes h_n = GH \in \mathcal{B}_1$.

For the converse, let $A = W|A|$ be the polar decomposition of an operator A in $\mathcal{B}_1$. Write the diagonalization $|A| = \sum_n \alpha_n e_n \otimes e_n$, where each $\alpha_n > 0$ and $\{e_n\}$ is a basis for $(\ker|A|)^\perp = (\ker A)^\perp$. Put $h_n = \alpha_n^{1/2} e_n$ and $g_n = Wh_n$. Clearly $\{h_n\}$ is an orthogonal sequence. Because W is an isometry on $(\ker A)^\perp$, its initial space, $\{g_n\}$ is an orthogonal sequence and $\|g_n\|^2 = \|h_n\|^2 = \alpha_n$. So $\sum \|g_n\|^2 = \sum \|h_n\|^2 = \sum \alpha_n = \operatorname{tr}(|A|) = \|A\|_1$. It is easy to see that $A = \sum_n g_n \otimes h_n$. ■

Note also that if $\{g_n\}, \{h_n\}$ are the sequences chosen as in the converse of the preceding theorem, then $\|A\|_1 = \sum \|g_n\| \, \|h_n\|$. This section concludes with a statement of a result for $\mathcal{B}_2$ analogous to Theorem 18.11. The proof is similar and is left therefore to the interested reader.

18.14 Theorem. (a) *If $\langle A, B\rangle \equiv \operatorname{tr}(B^*A)$ for A and B in $\mathcal{B}_2$, then $\langle\cdot,\cdot\rangle$ is an inner product on $\mathcal{B}_2$, $\|\cdot\|_2$ is the norm defined by this inner product, and $\mathcal{B}_2$ is a Hilbert space with respect to this inner product.*

(b) *Every Hilbert–Schmidt operator is compact. If A is a compact operator and $\alpha_1, \alpha_2, \ldots$ are the eigenvalues of $|A|$, each repeated as often as its multiplicity, then $A \in \mathcal{B}_2$ if and only if $\{\alpha_n\} \in \ell^2$ and in this case $\|A\|_2 = \sum \alpha_n^2$.*

(c) *The Hilbert space $\mathcal{B}_2$ contains $\mathcal{B}_{00}$ as a dense linear manifold and is, therefore, separable when the underlying Hilbert space $\mathcal{H}$ is separable.*

Exercises.

1. Show that $\|g \otimes h\| = \|g \otimes h\|_1 = \|g \otimes h\|_2 = \|g\| \, \|h\|$.

2. If $A \in \mathcal{B}(\mathcal{H})$, show that A is trace class if and only if $\sum_{\mathcal{E}} |\langle Ae, e\rangle| < \infty$ for every orthonormal basis $\mathcal{E}$. Give an example of an orthonormal basis and an operator A that is not trace class but such that this series converges absolutely for this particular basis.

3. If $A \in \mathcal{B}$, show that $A \in \mathcal{B}_1$ if and only if the real and imaginary parts of its Cartesian decomposition are in $\mathcal{B}_1$. If $A = A^*$, then $A \in \mathcal{B}_1$ if and only if $A = A_+ - A_-$, where $A_\pm$ are positive trace class operators. Do analogous statements hold for $\mathcal{B}_2$?

4. Let $A \in \mathcal{B}(\mathcal{H})$ and show that A is compact if and only if there are orthonormal sequences $\{e_n\}$ and $\{f_n\}$ and scalars $\{\alpha_n\}$ such that $\alpha_n \to 0$ and $A = \sum \alpha_n e_n \otimes f_n$, where this infinite sum converges in the operator norm. Compute A^*, $|A|$, and $\|A\|$ in terms of this decomposition. Discuss uniqueness of the sum $\sum \alpha_n e_n \otimes f_n$.

5. Define (or look up) the tensor product of two Hilbert spaces. For a Hilbert space $\mathcal{H}$ define the Hilbert space $\mathcal{H}^\#$ to be the same set $\mathcal{H}$ with the same definition of vector addition but where scalar multiplication and the inner product are defined by: $\alpha \cdot^\# f = \bar{\alpha} f$ and $\langle f, g\rangle_\# = \langle g, f\rangle$. Show that

there is an isomorphism $U : \mathcal{H} \otimes \mathcal{H}^{\#} \to \mathcal{B}_2$ such that $U(g \otimes h) = g \otimes h$. (Sorry for the ambiguous notation, but, in fact, this exercise shows that the notation is not so ambiguous after all.)

6. Let (X, Ω, μ) be a σ-finite measure space. For k in $L^2(\mu \times \mu)$, define $T_k : L^2(\mu) \to L^2(\mu)$ by $(T_k f)(x) = \int k(x,y) f(y)\, d\mu(y)$. Show that T_k is a Hilbert–Schmidt operator on $L^2(\mu)$ and the map $k \to T_k$ is an isomorphism of $L^2(\mu \times \mu)$ onto $\mathcal{B}_2(L^2(\mu))$.

7. For $p > 0$, define the class of operators $\mathcal{B}_p = \{ A : |A|^{p/2} \in \mathcal{B}_2 \}$. This is called the *Schatten p-class.* If $A \in \mathcal{B}_p$, let $\|A\|_p \equiv \| \, |A|^{p/2} \|^{2/p}$. It can be shown that $\mathcal{B}_p$ is a vector space, $\| \cdot \|_p$ is a norm on $\mathcal{B}_p$, and with respect to this norm $\mathcal{B}_p$ is complete. The interested reader can consult Schatten [1960] or Ringrose [1971a].

§19. The dual spaces of the compact operators and the trace class

For A in $\mathcal{B}_1$ define $\Phi_A : \mathcal{B}_0 \to \mathbb{C}$ by

$$\Phi_A(C) = \operatorname{tr}(CA) = \operatorname{tr}(AC)$$

for every compact operator C; Φ_A is a well defined linear functional on $\mathcal{B}_0$.

19.1 Theorem. *The map $A \to \Phi_A$ is an isometric isomorphism of $\mathcal{B}_1$ onto $\mathcal{B}_0^*$.*

Proof. By Theorem 18.11.(d), $\sup\{|\operatorname{tr}(AC)| : C \in \mathcal{B}_0 \text{ and } \|C\| \leq 1\} \leq \|A\|_1$. This says that Φ_A is a bounded linear functional on $\mathcal{B}_0$ and $\|\Phi_A\| \leq \|A\|_1$. Define $\rho : \mathcal{B}_1 \to \mathcal{B}_0^*$ by $\rho(A) = \Phi_A$; ρ is a linear mapping and the preceding inequality shows that $\|\rho(A)\| \leq \|A\|_1$ for all A in $\mathcal{B}_1$.

It must be shown that ρ is surjective and $\|\rho(A)\| \geq \|A\|_1$ for A in $\mathcal{B}_1$. In fact, both these facts will be shown simultaneously. Let $\Phi \in \mathcal{B}_0^*$ and define $[g,h] = \Phi(g \otimes h)$ for all g and h in $\mathcal{H}$. It follows that $|\,[g,h]\,| \leq \|\Phi\| \, \|g\| \, \|h\|$ for all g and h, so that $[\cdot,\cdot]$ is a bounded sesquilinear form. Hence there is a bounded operator A on $\mathcal{H}$ such that $\langle Ag, h\rangle = [g,h] = \Phi(g \otimes h)$ for all g and h in $\mathcal{H}$. It will now be shown that $A \in \mathcal{B}_1$ and $\Phi = \Phi_A$.

If $C \in \mathcal{B}_{00}$, then $C = \sum_{k=1}^n g_k \otimes h_k$ for vectors $g_1, \ldots, g_n, h_1, \ldots, h_n$ in $\mathcal{H}$ (Exercise 16.3). Thus

$$\begin{aligned}\Phi(C) = \Phi\left(\sum_{k=1}^n g_k \otimes h_k\right) &= \sum_{k=1}^n \langle A g_k, h_k\rangle \\ &= \sum_{k=1}^n \operatorname{tr}(A(g_k \otimes h_k)) = \operatorname{tr}(AC).\end{aligned}$$

If it can be shown that $A \in \mathcal{B}_1$; then both Φ and Φ_A are bounded linear functionals on $\mathcal{B}_0$, and the preceding equation demonstrates that they agree on the dense manifold $\mathcal{B}_{00}$. It then follows that $\Phi = \Phi_A$ and ρ is proven to be surjective.

To see that $A \in \mathcal{B}_1$, let $A = W|A|$ be the polar decomposition of A and let $\mathcal{E}$ be an orthonormal basis. If E is a finite subset of $\mathcal{E}$, then $C_E = [\sum_{e\in E} e\otimes e]W^*$ is a contraction in $\mathcal{B}_{00}$ and so $\|\Phi\| \geq |\Phi(C_E)| = |\Phi(\sum_{e\in E} e\otimes (We))| = \sum_{e\in E} |\langle Ae, We\rangle| = \sum_{e\in E}\langle |A|e, e\rangle$. Allowing E to increase shows that $\|\Phi\| \geq \|A\|_1$ and so $A \in \mathcal{B}_1$. Thus $\Phi = \Phi_A$. But this also says that $\|\Phi_A\| \geq \|A\|_1$. This shows that ρ is an isometry, completing the proof of the theorem. ■

What is the dual of $\mathcal{B}_1$? If $\mathcal{B}_0$ and $\mathcal{B}_1$ are the analogues of c_0 and ℓ^1, respectively, then the fact that $(\ell^1)^* = \ell^\infty$ might suggest that the dual of $\mathcal{B}_1$ is $\mathcal{B}$. This is indeed the case.

For B in $\mathcal{B}$, define $\Psi_B : \mathcal{B}_1 \to \mathbb{C}$ by

$$\Psi_B(A) = \operatorname{tr}(BA) = \operatorname{tr}(AB).$$

Clearly Ψ_B is a linear functional on $\mathcal{B}_1$.

19.2 Theorem. *The mapping $B \to \Psi_B$ is an isometric isomorphism of $\mathcal{B}(\mathcal{H})$ onto ${\mathcal{B}_1}^*$.*

Proof. The fact that $\|\Psi_B\| \leq \|B\|$ follows from Theorem 18.11.e; thus $\Psi_B \in {\mathcal{B}_1}^*$. Define $\rho : \mathcal{B} \to {\mathcal{B}_1}^*$ by $\rho(B) = \Psi_B$; clearly ρ is linear. If $\epsilon > 0$, choose a unit vector g such that $\|Bg\| > \|B\| - \epsilon$. Now choose a unit vector h with $\langle Bg, h\rangle = \|Bg\|$. If $C = g \otimes h$, then $C \in \mathcal{B}_1$ and $\|C\|_1 = 1$. Thus $\|\Psi_B\| \geq |\operatorname{tr}(BC)| = \langle Bg, h\rangle = \|Bg\| > \|B\| - \epsilon$. Since ϵ was arbitrary, $\|\Psi_B\| = \|B\|$ and ρ is an isometry. It remains to show that ρ is surjective.

Let $\Psi \in {\mathcal{B}_1}^*$; as in the proof of Theorem 19.1, there is an operator B in $\mathcal{B}$ such that $\langle Bg, h\rangle = \Psi(g \otimes h)$ for all g and h in $\mathcal{H}$. As before, it follows that $\Psi(T) = \Psi_B(T)$ for every finite rank operator T. Since $\mathcal{B}_{00}$ is dense in $\mathcal{B}_1$ and both Ψ and Ψ_B are bounded linear functionals, $\Psi = \Psi_B$. ■

The use of the trace leads to another interpretation of Theorem 8.1.

19.3 Proposition. $(\mathcal{B}, \mathrm{WOT})^* = (\mathcal{B}, \mathrm{SOT})^* = \mathcal{B}_{00}$.

Proof. Note that for $C = \sum_{k=1}^n f_k \otimes g_k$, $\operatorname{tr}(CT) = \sum_{k=1}^n \langle Tf_k, g_k\rangle$. ■

Exercises.

1. Fix an orthonormal basis $\mathcal{E}$ for $\mathcal{H}$ and define $\eta : c_0(\mathcal{E}) \to \mathcal{B}_0$ by $\eta(\{\alpha_e\}) = \sum \alpha_e e \otimes e$. (a) Show that η is an isometry whose range is the set of compact normal operators that have $\mathcal{E}$ as eigenvectors. (b) Compute $\eta^* : \mathcal{B}_1(\mathcal{H}) \to \ell^1$ and $\eta^{**} : \ell^\infty \to \mathcal{B}$. (c) Characterize the range of η^{**}. Now assume that $\mathcal{H}$ is separable so that $\mathcal{E}$ is countable. (d) Show that there is no bounded projection of $\mathcal{B}$ onto $\mathcal{B}_0$ by using the fact that there is no bounded projection of ℓ^∞ onto c_0. (e) Putting $\mathcal{E} = \{e_n\}$ and $T = \sum n^{-1} e_n \otimes e_n$, show that $\operatorname{ran} \eta = \{T\}'' \cap \mathcal{B}_0$.

2. If $\mathcal{X}$ is any Banach space, consider the natural map $\mathcal{X} \to \mathcal{X}^{**}$. Show that the dual of this map, $\mathcal{X}^{***} \to \mathcal{X}^*$, is a projection of norm 1. (Here $\mathcal{X}^*$ is considered, in the natural way, as a subspace of its second dual, $(\mathcal{X}^*)^{**} = \mathcal{X}^{***}$.) Apply this to the space $\mathcal{X} = \mathcal{B}_0$ to show that if $L \in \mathcal{B}^*$, then there is a unique trace class operator A and a unique linear functional L_0 in $\mathcal{B}^*$ such that: (i) $L = L_0 + \Phi_A$; (ii) $L_0(T) = 0$ for every compact operator T; (iii) $\|L\| = \|L_0\| + \|A\|_1$. Show that this operator A satisfies $\langle Ag, h\rangle = L(g \otimes h)$ for all g and h in $\mathcal{H}$.

3. For a measure μ on a compact space X, identify, via the Radon–Nikodym Theorem, the space $L^1(\mu)$ with $\{\nu \in M(X) : |\nu| << \mu\}$. If $N = \int z\, dE(z)$ is a normal operator with μ as its scalar-valued spectral measure, define $\alpha : \mathcal{B}_1(\mathcal{H}) \to L^1(\mu)$ by $\alpha(T)(\Delta) = \operatorname{tr}(TE(\Delta))$. Show that α is a contractive surjection. What is $\alpha^* : L^\infty(\mu) \to \mathcal{B}(\mathcal{H})$?

§20. The weak-star topology

There is another topology on $\mathcal{B}(\mathcal{H})$ that must be discussed; the weak* topology that $\mathcal{B}(\mathcal{H})$ has as the dual of the trace class operators. Here the defining seminorms are given by $T \to |\operatorname{tr}(TA)|$, where A can be any trace class operator. The reader should be alert that in some other books this topology is called the *ultraweak* or *σ-weak topology.* This is traditional terminology and stems from the fact that the topic was introduced before the theory of Banach spaces and weak* topologies had fully established itself in mathematics. There is no inherent advantage in calling this the weak* topology save to be explicit that that is what it is and that the full force of Banach space theory is available.

20.1 Proposition. (a) *If $\mathcal{H}$ is separable, then the closed unit ball of $\mathcal{B}(\mathcal{H})$ with the weak* topology is a compact metric space.*

(b) *The weak* topology and the* WOT *agree on bounded subsets of $\mathcal{B}(\mathcal{H})$.*

(c) *A sequence in $\mathcal{B}(\mathcal{H})$ converges weak* if and only if it converges* WOT.

Proof. (a) In fact, when $\mathcal{H}$ is separable, $\mathcal{B}(\mathcal{H})$ is the dual of the separable Banach space $\mathcal{B}_1(\mathcal{H})$.

(b) It is easy to see that the weak* topology is bigger than the WOT, and so the identity map $i : (\mathcal{B}(\mathcal{H}), \text{weak}^*) \to (\mathcal{B}(\mathcal{H}), \text{WOT})$ is continuous. If $\mathcal{S}$ is a closed bounded subset of $\mathcal{B}(\mathcal{H})$, consider $i : (\mathcal{S}, \text{weak}^*) \to (\mathcal{S}, \text{WOT})$. Since closed subsets of $\mathcal{S}$ are weak* compact and i is continuous, i is also a closed map. Hence i is a homeomorphism.

(c) By the Principle of Uniform Boundedness, WOT convergent sequences must be uniformly bounded. Thus part (c) follows from part (b). ■

Here is another interpretation of the weak* topology. Indeed this is the way that the ultraweak topology is usually defined in older books.

20.2 Proposition. *A net $\{T_i\}$ in $\mathcal{B}(\mathcal{H})$ converges weak* to 0 if and only if for any two sequences $\{f_n\}$ and $\{g_n\}$ of vectors from $\mathcal{H}$ with $\sum \|f_n\|^2 < \infty$ and $\sum \|g_n\|^2 < \infty$, $\sum_n \langle T_i f_n, g_n\rangle \to 0$ with i.*

Proof. According to Theorem 18.13, an operator A is in $\mathcal{B}_1$ if and only if $A = \sum_n f_n \otimes g_n$, where $\{f_n\}, \{g_n\}$ are square summable sequences of vectors. For such an operator A, $\operatorname{tr}(AT) = \sum_n \langle T f_n, g_n\rangle$. ■

As was said above, the entire theory of weak* topologies on Banach spaces can be applied to $\mathcal{B}(\mathcal{H})$. One part of that theory that is important is the Krein–Smulian Theorem, which states that for a Banach space $\mathcal{X}$ and a convex subset S in $\mathcal{X}^*$, S is weak* closed in $\mathcal{X}^*$ if and only if $S \cap r(\text{ball}\,\mathcal{X}^*)$ is weak* closed for all $r > 0$. (See [ACFA], V.12.1.) If $\mathcal{X}$ is separable, the weak* topology is metrizable on bounded sets and so the Krein–Smulian Theorem becomes, in this situation, the statement that a convex subset S in $\mathcal{X}^*$ is weak* closed if and only if it is weak* sequentially closed. Since $\mathcal{B}_1$ is separable when $\mathcal{H}$ is, the following proposition is a consequence. The details are left to the reader.

20.3 Proposition. *Assume $\mathcal{H}$ is a separable Hilbert space.*

(a) *If $\mathcal{S}$ is a convex subset of $\mathcal{B}(\mathcal{H})$, then $\mathcal{S}$ is weak* closed if and only if it is* WOT *sequentially closed.*

(b) *If $\mathcal{X}$ is a separable Banach space and $\rho : \mathcal{X}^* \to \mathcal{B}(\mathcal{H})$ is a linear map, the following statements are equivalent:*

(i) $\rho : (\mathcal{X}^*, \text{weak}^*) \to (\mathcal{B}, \text{weak}^*)$ *is continuous;*

(ii) $\rho : (\mathcal{X}^*, \text{weak}^*) \to (\mathcal{B}, \text{WOT})$ *is sequentially continuous;*

(iii) *there is a bounded linear mapping $\tau : \mathcal{X} \to \mathcal{B}_1$ such that $\rho = \tau^*$.*

(c) *If $\mathcal{X}$ is a separable Banach space and $\rho : (\mathcal{X}^*, \text{weak}^*) \to (\mathcal{B}(\mathcal{H}), \text{WOT})$ is a sequentially continuous isometry, then $\mathcal{W} = \rho(\mathcal{X}^*)$ is weak* closed in $\mathcal{B}(\mathcal{H})$ and $\rho : \mathcal{X}^* \to \mathcal{W}$ is a weak* homeomorphism.*

20.4 Proposition. *If (X, Ω, μ) is a σ-finite measure space and $\{\phi_i\}$ is a net in $L^\infty(\mu)$, then the following statements are equivalent.*

(a) $\phi_i \to 0$ *weak* in* $L^\infty(\mu)$.

(b) $M_{\phi_i} \to 0$ WOT *in* $\mathcal{B}(L^2(\mu))$.

(c) $M_{\phi_i} \to 0$ *weak* in* $\mathcal{B}(L^2(\mu))$.

Proof. The equivalence of (a) and (b) was shown in (10.5). Weak* convergence always implies WOT convergence, so assume (b). If $\{f_n\}$ and $\{g_n\}$ are two square summable sequences in $L^2(\mu)$, then $h = \sum_n f_n \overline{g_n} \in L^1(\mu)$. Factor h as the product $h_1 \overline{h_2}$, with h_1, h_2 in $L^2(\mu)$. Then $\sum_n \langle M_{\phi_i} f_n, g_n \rangle = \int \phi_i h \, d\mu = \langle M_{\phi_i} h_1, h_2 \rangle \to 0$. Thus (c) holds by Proposition 20.2. ■

20.5 Definition. A Banach algebra $\mathcal{A}$ is a *dual algebra* if there is a Banach space $\mathcal{X}$ such that $\mathcal{A}$ is isometrically isomorphic to $\mathcal{X}^*$ and multiplication in $\mathcal{A}$ is separately weak* continuous. That is, if $a \in \mathcal{A}$, then the maps $x \to ax$ and $x \to xa$ are continuous in the weak* topology.

This is an abstract version of the definition of a dual algebra given in Bercovici, Foias, and Pearcy [1985] where a dual algebra is defined as a weak* closed subalgebra of $\mathcal{B}(\mathcal{H})$. In fact, if $\mathcal{A}$ is a weak* closed subalgebra of $\mathcal{B}(\mathcal{H})$, $\mathcal{A} = (\mathcal{B}_1(\mathcal{H})/\mathcal{A}_\perp)^*$. $L^\infty(\mu)$ is another example of a dual algebra. In fact, the next section will show that every von Neumann algebra is a dual algebra.

20.6 Definition. If $\mathcal{A}$ and $\mathcal{B}$ are dual algebras, $\rho : \mathcal{A} \to \mathcal{B}$ is a *dual algebra homomorphism* if ρ is a Banach algebra homomorphism that is weak* continuous. The map ρ is a *dual algebra isomorphism* if it is an isometric isomorphism that is a weak* homeomorphism.

The proof of the next result is immediate from Proposition 20.3.

20.7 Proposition. *Assume that $\mathcal{A}$ and $\mathcal{B}$ are dual algebras that are duals of separable Banach spaces. If $\rho : \mathcal{A} \to \mathcal{B}$ is an isometric isomorphism that is weak* sequentially continuous, then ρ is a dual algebra isomorphism.*

20.8 Definition. If μ is a compactly supported measure on the plane, let $P^\infty(\mu)$ be the weak* closure of the polynomials in $L^\infty(\mu)$.

The next result follows from (10.5) and the preceding results in this section.

20.9 Proposition. *If μ is a compactly supported measure on the plane, $\{ M_\phi : \phi \in P^\infty(\mu) \}$ is a dual algebra and $\phi \to M_\phi$ is a dual algebra isomorphism.*

This section concludes with a result that will be useful as well as helpful in underscoring a distinction between various concepts encountered later in this book. Since $\mathcal{B}_1$ is the Banach space dual of $\mathcal{B}_0$, it too has a weak* topology. To avoid confusion, employ the notation $\sigma(\mathcal{B}_1, \mathcal{B}_0)$ to denote the weak* topology on the trace class when considered as the dual of the compact operators. The weak* topology $\sigma(\mathcal{B}(\mathcal{H}), \mathcal{B}_1)$ on $\mathcal{B}(\mathcal{H})$ as the dual of the trace class has already been discussed. A reference to the weak* topology on $\mathcal{B}(\mathcal{H})$ always means this topology $\sigma(\mathcal{B}(\mathcal{H}), \mathcal{B}_1)$. (In fact, $\mathcal{B}_1$ is the unique predual of $\mathcal{B}(\mathcal{H})$, but that is another matter.)

Now if the space of finite rank operators, $\mathcal{B}_{00}$, is given the trace norm, its dual is $\mathcal{B}(\mathcal{H})$ since $\mathcal{B}_{00}$ is dense in the trace class. However the topology $\sigma(\mathcal{B}(\mathcal{H}), \mathcal{B}_{00})$ is the WOT. Also $\mathcal{B}_1$ is the dual of $\mathcal{B}_{00}$ if the space of finite rank operators is given the operator norm. So it is legitimate to discuss the $\sigma(\mathcal{B}_1, \mathcal{B}_{00})$ topology. After you have taken a moment to sort out your confusion, consider the following.

Recall that if $\mathcal{X}$ is a Banach space and $S \subseteq \mathcal{X}^*$, then the *preannihilator* of S, $S_\perp$, is the set $S_\perp = \{x \in \mathcal{X} : s^*(x) = 0 \text{ for all } s^* \textit{ in } S\}$.

20.10 Proposition. *If $\mathcal{A}$ is a weak* closed linear subspace of $\mathcal{B}(\mathcal{H})$, then $\mathcal{A}_\perp$ is the norm closed linear span of $\mathcal{A}_\perp \cap \mathcal{B}_{00}$ if and only if $\mathcal{A}$ is* WOT *closed.*

Proof. Assume that $\mathcal{A}$ is WOT closed. Applying the Hahn–Banach Theorem to the normed space $(\mathcal{B}_{00}, \|\cdot\|_1)$ shows that $\mathcal{A} = (\mathcal{A}_\perp \cap \mathcal{B}_{00})^\perp$. Thus $\mathcal{A}_\perp \cap \mathcal{B}_{00}$ is dense in $\mathcal{A}_\perp$. Conversely, assume that $\mathcal{A}_\perp \cap \mathcal{B}_{00}$ is dense in $\mathcal{A}_\perp$. Because $\mathcal{A}$ is weak* closed, $\mathcal{A} = (\mathcal{A}_\perp)^\perp = (\mathcal{A}_\perp \cap \mathcal{B}_{00})^\perp$. Thus $\mathcal{A}$ is WOT closed. ∎

Exercises.

1. Let $\mathcal{L}$ be a linear manifold in $\mathcal{B}_1$ that contains $\mathcal{B}_{00}$ and consider the topology $(\mathcal{L})$ defined on $\mathcal{B}(\mathcal{H})$ by the seminorms $\{p_A : A \in \mathcal{L}\}$, where $p_A(T) = |\mathrm{tr}\,(AT)|$. Show that $(\mathcal{B}, (\mathcal{L}))^* = \mathcal{L}$ and that $(\mathcal{L})$ and WOT agree on bounded sets in $\mathcal{B}(\mathcal{H})$.

2. Prove a converse to Proposition 20.1.a. That is, show that if ball $\mathcal{B}(\mathcal{H})$ with the weak* topology is metrizable, then $\mathcal{H}$ is separable.

§21. Inflation and the topologies

This section will examine the relationship between the various topologies on $\mathcal{B}(\mathcal{H})$ and the inflation of the operators (6.3). One such relationship appeared in Proposition 8.6, where inflation and the attachment of a manifold of operators was used to characterize the strong closure of the manifold.

In many respects the first pair of results will function as lemmas. They have, however, some independent interest and will have additional use in this book.

21.1 Proposition. *If $1 \leq d \leq \infty$ and $C \in \mathcal{B}_1(\mathcal{H}^{(d)})$ with the matrix representation $C = (C_{jk})$, then:*

(a) *$K = \sum_k C_{kk}$ converges weak* in $\mathcal{B}_1(\mathcal{H})$ and $\|\sum_k C_{kk}\|_1 \leq \|C\|_1$;*

(b) *if $T \in \mathcal{B}(\mathcal{H})$ and K is as in (a), then*

$$\operatorname{tr}(T^{(d)}C) = \operatorname{tr}(TK) = \sum_{k=1}^{d} \operatorname{tr}(TC_{kk}),$$

where this series is absolutely convergent when $d = \infty$.

Proof. The only difficult part of the proof is when $d = \infty$. So assume this is the case. (The proof for finite d is left to the reader.)

(a) Let P_k be the projection of $\mathcal{H}^{(\infty)}$ onto the k-th coordinate space. Note that $C_{kk} = P_k C P_k$, so clearly each $C_{kk} \in \mathcal{B}_1(\mathcal{H})$. Let $1 \leq n < m < \infty$; so $\sum_{k=n}^{m} C_{kk} \in \mathcal{B}_1(\mathcal{H})$. If $L \in \mathcal{B}_0(\mathcal{H}^{(\infty)})$, then

$$\begin{aligned}
\left|\operatorname{tr}\left(L\sum_{k=n}^{m} C_{kk}\right)\right| &= \left|\sum_{k=n}^{m} \operatorname{tr}(LP_kCP_k)\right| \\
&= \left|\sum_{k=n}^{m} \operatorname{tr}(C(P_kLP_k))\right| \\
&= \left|\operatorname{tr}\left(C\sum_{k=n}^{m} P_kLP_k\right)\right| \\
&\leq \|C\|_1 \left\|\sum_{k=n}^{m} P_kLP_k\right\| \\
&= \|C\|_1 \sup\{\|P_kLP_k\| : n \leq k \leq m\} \\
&\leq \|C\|_1 \, \|L\|,
\end{aligned}$$

the penultimate line being justified because the projections P_k are pairwise orthogonal. This says two things. First, for any $n \geq 1$

$$\left\| \sum_{k=1}^{n} C_{kk} \right\|_1 = \sup \left\{ \left| \operatorname{tr} \left(L \sum_{k=1}^{n} P_k C P_k \right) \right| : L \in \operatorname{ball} \mathcal{B}_0(\mathcal{H}) \right\} \leq \|C\|_1.$$

Also for a fixed compact operator L on $\mathcal{H}^{(\infty)}$, $\|P_k L P_k\| \to 0$ (Why?). Hence for any $\epsilon > 0$ there is an integer n_0 such that $\|P_k L P_k\| < \epsilon$ for all $k \geq n_0$. Letting $m \to \infty$ in the first estimate shows that for $n \geq n_0$,

$$\left| \operatorname{tr} \left(L \sum_{k=n}^{\infty} P_k C P_k \right) \right| \leq \epsilon \|C\|_1.$$

Since L was an arbitrary compact operator, this implies that the sequence of partial sums of $\sum_k C_{kk}$ converges weak* in $\mathcal{B}_1(\mathcal{H})$.

(b) Let K be as in part (a), and let $\mathcal{E} = \{ e_i : i \in I \}$ be a basis for $\mathcal{H}$. Set $e_{ki} = e_i$ for all $k \geq 1$ and i in I. For each $k \geq 1$, $\mathcal{E}_k = \{ e_{ki} : i \in I \}$ is a basis for P_k, and $\{ e_{ki} : k \geq 1, i \in I \}$ is a basis for $\mathcal{H}^{(\infty)}$. Thus for any bounded operator T, $T^{(\infty)} C \in \mathcal{B}_1(\mathcal{H}^{(\infty)})$. Hence $\operatorname{tr}(T^{(\infty)} C) = \sum_{k,i} \langle T^{(\infty)} C e_{ki}, e_{ki} \rangle$, and this series converges absolutely. Rearranging the order of summation yields

21.2 $$\operatorname{tr}(T^{(\infty)} C) = \sum_{i \in I} \sum_{k=1}^{\infty} \langle T^{(\infty)} C e_{ki}, e_{ki} \rangle.$$

Fixing k and i for the moment,

$$\langle T^{(\infty)} C e_{ki}, e_{ki} \rangle = \langle T C_{kk} e_i, e_i \rangle = \operatorname{tr}\left((e_i \otimes e_i) T C_{kk} \right).$$

Now $(e_i \otimes e_i) T = e_i \otimes (T^* e_i)$ is a rank one operator, so the weak* convergence of $K = \sum_k C_{kk}$ implies that for each i in I,

$$\begin{aligned} \langle T K e_i, e_i \rangle &= \operatorname{tr}\left((e_i \otimes (T^* e_i)) K \right) \\ &= \sum_{k=1}^{\infty} \langle T C_{kk} e_{ki}, e_{ki} \rangle \\ &= \sum_{k=1}^{\infty} \langle T^{(\infty)} C e_{ki}, e_{ki} \rangle, \end{aligned}$$

and the series is absolutely convergent. Substituting in Equation 21.2 gives that $\operatorname{tr}(T^{(\infty)} C) = \sum_{i \in I} \langle T K e_i, e_i \rangle = \operatorname{tr}(TK)$.

Now reverse the order of summation in (21.2) to get

$$\operatorname{tr}(T^{(\infty)}C) = \sum_{k=1}^{\infty}\sum_{i\in I}\langle TC_{kk}e_{ki}, e_{ki}\rangle = \sum_{k=1}^{\infty}\operatorname{tr}(TC_{kk}). \quad \blacksquare$$

The next result has some built-in redundancies. If the dimension d of the inflation is infinite, the statement about the rank of the trace class operator is superfluous. On the other hand, if $d < \infty$, the statement that the operator is trace class is extraneous.

21.3 Proposition. (a) *If $C : \mathcal{H} \to \mathcal{H}$ is a trace class operator with rank at most d, $1 \le d \le \infty$, then there are vectors f, g in $\mathcal{H}^{(d)}$ such that $\|f\|^2 = \|g\|^2 \le \|C\|_1$ and*

21.4 $$\langle T^{(d)}f, g\rangle = \operatorname{tr}(TC)$$

for all operators T in $\mathcal{B}(\mathcal{H})$.

(b) *Conversely, if $1 \le d \le \infty$ and $f, g \in \mathcal{H}^{(d)}$, then there is a trace class operator $C : \mathcal{H} \to \mathcal{H}$ having rank at most d such that* (21.4) *holds.*

(c) *If $C \in \mathcal{B}_1(\mathcal{H}^{(\infty)})$, then there are vectors f, g in $\mathcal{H}^{(\infty)}$ such that $\|f\|^2 = \|g\|^2 \le \|C\|_1$ and*

21.5 $$\langle T^{(\infty)}f, g\rangle = \operatorname{tr}(T^{(\infty)}C)$$

for all operators T in $\mathcal{B}(\mathcal{H})$.

(d) *Conversely, if $f, g \in \mathcal{H}^{(\infty)}$, then there is an operator C in $\mathcal{B}_1(\mathcal{H}^{(\infty)})$ such that* (21.5) *holds.*

Proof. Again the proof is only given for the case that $d = \infty$.

(a) By Theorem 18.13 there are vectors $f = (f_k), g = (g_k)$ in $\mathcal{H}^{(\infty)}$ such that $C = \sum_k f_k \otimes g_k$ and $\|f\|^2 = \|g\|^2 = \|C\|_1$. It is elementary to check that $\langle T^{(\infty)}f, g\rangle = \sum_k\langle Tf_k, g_k\rangle = \operatorname{tr}(TC)$.

(b) If the vectors f and g are given, let $C = \sum_k f_k \otimes g_k$; so $C \in \mathcal{B}_1(\mathcal{H})$. Again, it is routine to show that (21.4) holds.

(c) If $C = (C_{jk}) \in \mathcal{B}_1(\mathcal{H}^{(\infty)})$, let $K = \sum_k C_{kk}$. By Proposition 21.1, $K \in \mathcal{B}_1(\mathcal{H})$ and $\|K\|_1 \le \|C\|_1$. So there are vectors $f = (f_k), g = (g_k)$ in $\mathcal{H}^{(\infty)}$ such that $K = \sum_k f_k \otimes g_k$ and $\sum_n \|f_n\|^2 = \sum_n \|g_n\|^2 = \|K\|_1$ (18.13). Put $f = (f_n)$ and $g = (g_n)$; so $\|f\|^2, \|g\|^2 \le \|C\|_1$. By (21.1), it follows that for any operator T in $\mathcal{B}(\mathcal{H})$,

$$\operatorname{tr}(T^{(\infty)}C) = \operatorname{tr}(TK) = \sum_{k=1}^{\infty}\langle Tf_k, g_k\rangle = \langle T^{(\infty)}f, g\rangle.$$

(d) In fact this part was only written here to achieve symmetry. Just take $C = f \otimes g$. ■

21.6 Proposition. *If $\{T_i\}$ is a net in $\mathcal{B}(\mathcal{H})$, then the following statements are equivalent.*

(a) $T_i \to 0$ weak* *in* $\mathcal{B}(\mathcal{H})$.

(b) $T_i^{(\infty)} \to 0$ weak* *in* $\mathcal{B}(\mathcal{H}^{(\infty)})$.

(c) $T_i^{(\infty)} \to 0$ WOT *in* $\mathcal{B}(\mathcal{H}^{(\infty)})$.

Proof. If $C \in \mathcal{B}_1(\mathcal{H}^{(\infty)})$, then Proposition 21.1 implies there is an operator K in $\mathcal{B}_1(\mathcal{H})$ such that for all i, $\operatorname{tr}(CT_i^{(\infty)}) = \operatorname{tr}(KT_i)$. Hence (a) implies (b). It's trivial that (b) implies (c). Now assume that (c) holds and let $K \in \mathcal{B}_1(\mathcal{H})$. By Proposition 21.3.a there are vectors f, g in $\mathcal{H}^{(\infty)}$ such that $\operatorname{tr}(KT_i) = \langle T_i^{(\infty)} f, g\rangle$, which converges to 0 by part (c). ■

21.7 Corollary. *If $\mathcal{S}$ is a subset of $\mathcal{B}(\mathcal{H})$ and $\mathcal{S}_1$ is the* weak* *closure of $\mathcal{S}$ in $\mathcal{B}(\mathcal{H})$, then the weak operator topology closure and the* weak* *closure of $\{ S^{(\infty)} : S \in \mathcal{S} \}$ are the same and equal to $\mathcal{S}_1^{(\infty)}$. Consequently a linear subspace $\mathcal{S}$ in $\mathcal{B}(\mathcal{H})$ is* weak* *closed if and only if $\mathcal{S}^{(\infty)}$ is* WOT *closed.*

The next result follows from the preceding one and constitutes an extension of the Double Commutant Theorem.

21.8 Proposition. *If $\mathcal{A}$ is a C^*-subalgebra of $\mathcal{B}(\mathcal{H})$ that contains the identity, then $\mathcal{A}''$ is the weak* closure of $\mathcal{A}$. Consequently, every von Neumann algebra is a dual algebra.*

Proof. Let $\mathcal{A}_1$ be the weak* closure of $\mathcal{A}$. By Corollary 21.7, the WOT and weak* closure of $\mathcal{A}^{(\infty)}$ are the same and equal to $\mathcal{A}_1^{(\infty)}$. But the Double Commutant Theorem implies that the WOT closure of $\mathcal{A}^{(\infty)}$ is $\{\mathcal{A}^{(\infty)}\}''$. Thus $\mathcal{A}_1^{(\infty)} = \{\mathcal{A}^{(\infty)}\}'' = [\mathcal{A}'']^{(\infty)}$ by Proposition 12.2. Therefore $\mathcal{A}_1$ must equal $\mathcal{A}''$. ■

Exercises.

1. Let $\rho \colon \mathcal{B}_1(\mathcal{H}^{(d)}) \to \mathcal{B}_1(\mathcal{H})$ be the map defined by $\rho((C_{jk})) = \sum_k C_{kk}$. Show that ρ is a linear contraction and its adjoint $\rho^* : \mathcal{B}(\mathcal{H}) \to \mathcal{B}(\mathcal{H}^{(d)})$ is given by $\rho(T) = T^{(d)}$.

2. In Proposition 21.1, does the series $\sum_k C_{kk}$ converge in the trace norm?

3. Define an additional topology, σ-SOT, on $\mathcal{B}(\mathcal{H})$ as follows. If $\bar{h} = \{h_k\}$ is a sequence of vectors with $\sum_k \|h_k\|^2 < \infty$, let $p_{\bar{h}}(T) = [\sum_k \|Th_k\|^2]^{1/2}$.

The σ-SOT topology is that which is defined by the collection of all such seminorms $p_{\bar{h}}$. Some call this the *ultrastrong topology.* For a net $\{T_i\}$ in $\mathcal{B}(\mathcal{H})$, show that the following statements are equivalent. (a) $T_i \to 0$ (σ-SOT) in $\mathcal{B}(\mathcal{H})$. (b) $T_i^{(\infty)} \to 0$ (σ-SOT) in $\mathcal{B}(\mathcal{H}^{(\infty)})$. (c) $T_i^{(\infty)} \to 0$ (SOT) in $\mathcal{B}(\mathcal{H}^{(\infty)})$.

Chapter 4

Some Non-Normal Operators

This chapter considers a number of operators that are not normal and have received the attention of various mathematicians over the years. The treatment here of these operators cannot be called exhaustive. Indeed, most of these operators have monographs devoted to their explication.

§22. Algebras and lattices

The chapter starts with some general theory about algebras and lattices of invariant subspaces. As in [ACFA], the term "subspace" means a closed linear manifold. The notation $\mathcal{M} \leq \mathcal{H}$ is used to denote that $\mathcal{M}$ is a subspace of $\mathcal{H}$. One of the modern developments in operator theory is an emphasis on asymmetric operator algebras; that is, algebras that are not closed under taking adjoints. This first section sets the stage for much that follows.

For two subspaces $\mathcal{M}$ and $\mathcal{N}$, define the *join* and *meet* of $\mathcal{M}$ and $\mathcal{N}$, $\mathcal{M} \vee \mathcal{N}$ and $\mathcal{M} \wedge \mathcal{N}$, respectively, by

$$\begin{aligned} \mathcal{M} \vee \mathcal{N} &= \operatorname{cl}\{f + g : f \in \mathcal{M}, g \in \mathcal{N}\}, \\ \mathcal{M} \wedge \mathcal{N} &= \mathcal{M} \cap \mathcal{N}. \end{aligned}$$

22.1 Proposition. *With the definition of join and meet given above, the collection of all subspaces of a Hilbert space forms a complete lattice.*

The proof of this is an exercise. There is some interest in characterizing the collection of all subspaces of a Hilbert space as a lattice. It's a very difficult problem. A variety of special cases of this problem also have some interest, with many of these special cases having been solved.

22.2 Definition. If $\mathcal{A}$ is a subset of $\mathcal{B}(\mathcal{H})$, define the *lattice of invariant subspaces* of $\mathcal{A}$ by

$$\operatorname{Lat}\mathcal{A} \equiv \{\mathcal{M} \leq \mathcal{H} : A\mathcal{M} \subseteq \mathcal{M} \text{ for all } A \text{ in } \mathcal{A}\}.$$

If $\mathcal{L}$ is a collection of closed subspaces of $\mathcal{H}$, define

$$\operatorname{Alg}\mathcal{L} \equiv \{A \in \mathcal{B}(\mathcal{H}) : A\mathcal{M} \subseteq \mathcal{M} \text{ for all } \mathcal{M} \text{ in } \mathcal{L}\}.$$

There is considerable interest in finding those lattices that can be realized as the lattice of invariant subspaces of an operator or set of operators. Radjavi and Rosenthal [1973] is a good place to start a pursuit of such questions. Recall the definition of $\operatorname{Ref}\mathcal{A}$ from 8.4.

22.3 Proposition. *Let $\mathcal{A}$ be a set of operators on the Hilbert space $\mathcal{H}$ and let $\mathcal{L}$ be a collection of subspaces of $\mathcal{H}$.*

(a) $\operatorname{Lat}\mathcal{A}$ *is a complete lattice under the operations* $\vee$ *and* $\wedge$.

(b) $\operatorname{Alg}\mathcal{L}$ *is a* WOT *closed subalgebra of* $\mathcal{B}(\mathcal{H})$.

(c) $\mathcal{L} \subseteq \operatorname{Lat}\operatorname{Alg}\mathcal{L}$ *and* $\mathcal{A} \subseteq \operatorname{Alg}\operatorname{Lat}\mathcal{A}$.

(d) $\operatorname{Alg}\operatorname{Lat}\operatorname{Alg}\mathcal{L} = \operatorname{Alg}\mathcal{L}$ *and* $\operatorname{Lat}\operatorname{Alg}\operatorname{Lat}\mathcal{A} = \operatorname{Lat}\mathcal{A}$.

(e) *If $\mathcal{A}$ is an algebra that contains the identity, then* $\operatorname{Alg}\operatorname{Lat}\mathcal{A} = \operatorname{Ref}\mathcal{A}$.

Proof. The proofs of parts (a), (b), and (c) are routine. In light of (c), $\operatorname{Alg}\operatorname{Lat}\operatorname{Alg}\mathcal{L} \subseteq \operatorname{Alg}\mathcal{L}$. On the other hand, if $T \in \operatorname{Alg}\mathcal{L}$ and $\mathcal{M} \in \operatorname{Lat}\operatorname{Alg}\mathcal{L}$, then, by the definition of $\operatorname{Lat}\operatorname{Alg}\mathcal{L}$, $T\mathcal{M} \subseteq \mathcal{M}$. Hence $T \in \operatorname{Alg}\operatorname{Lat}\operatorname{Alg}\mathcal{L}$. The proof of the other equality in (d) is similar.

To see (e), let $T \in \operatorname{Ref}\mathcal{A}$ and $\mathcal{M} \in \operatorname{Lat}\mathcal{A}$. For any h in $\mathcal{M}$, $Th \in [\mathcal{A}h] \leq \mathcal{M}$, so $T \in \operatorname{Alg}\operatorname{Lat}\mathcal{A}$. Conversely, assume $T \in \operatorname{Alg}\operatorname{Lat}\mathcal{A}$ and $h \in \mathcal{H}$. Since $\mathcal{A}$ is an algebra containing the identity, $h \in [\mathcal{A}h]$ and $[\mathcal{A}h] \in \operatorname{Lat}\mathcal{A}$. Thus $Th \in [\mathcal{A}h]$ so that, by definition, $T \in \operatorname{Ref}\mathcal{A}$. ■

22.4 Examples. (a) Let J be the $n \times n$ matrix

$$\begin{bmatrix} 0 & 1 & 0 & 0 \cdots & 0 \\ 0 & 0 & 1 & 0 \cdots & 0 \\ & & & \cdot & \\ & & & \cdot & \\ & & & \cdot & \\ & & & \cdot & \\ 0 & 0 & 0 & 0 \cdots & 0 \end{bmatrix}$$

acting on $\mathbb{C}^n$ and let $e_1, \ldots, e_n$ be the usual basis. So $Je_k = e_{k-1}$ for $2 \leq k \leq n$ and $Je_1 = 0$. (J is the usual Jordan n-cell appearing in Jordan canonical forms.) If $\mathcal{M}_k$ is the linear span of $\{e_1, \ldots, e_k\}$ for $1 \leq k \leq n$ and $\mathcal{M}_0 = (0)$, then it is a nice exercise in linear algebra to show that $\operatorname{Lat} J = \{\mathcal{M}_k : 0 \leq k \leq n\}$. Note that $\operatorname{Alg}\operatorname{Lat} J$ is the collection of all upper triangular matrices.

(b) If $\mathcal{A}$ is a von Neumann algebra, $\operatorname{Lat}\mathcal{A} = \{ P : P \text{ is a projection in } \mathcal{A}' \}$. This follows because $A^* \in \mathcal{A}$ whenever $A \in \mathcal{A}$, so every invariant subspace for $\mathcal{A}$ reduces $\mathcal{A}$.

(c) If $\mathcal{A}_\mu = \{M_\phi : \phi \in L^\infty(\mu)\}$, then an interpretation of part (b), using the fact that $\mathcal{A}_\mu' = \mathcal{A}_\mu$, shows that $\operatorname{Lat}\mathcal{A}_\mu = \{ L^2(\mu|\Delta) : \Delta \text{ is measurable} \}$.

(d) If $\mathcal{A}$ is a von Neumann algebra, $\operatorname{Alg}\operatorname{Lat}\mathcal{A} = \mathcal{A}$. If $T \in \operatorname{Alg}\operatorname{Lat}\mathcal{A}$, then for every projection P in $\mathcal{A}'$, $P^\perp TP = 0$. But $P^\perp \in \mathcal{A}'$ whenever P is, so that $PTP^\perp = 0$. So $PT = TP$ for every projection P in the commutant of $\mathcal{A}$. Since the projections generate $\mathcal{A}'$ as a von Neumann algebra (13.3.e), $T \in \mathcal{A}'' = \mathcal{A}$.

A special case of (22.4.d) that is of particular importance occurs when μ is a compactly supported measure on $\mathbb{C}$. Since $L^\infty(\mu)$ is the weak* closed algebra generated by the functions z and $\bar{z}$, the following holds.

22.5 Examples. Let μ be a compactly supported measure on the plane and define the normal operator N on $L^2(\mu)$ by $Nf = zf$ for all f in $L^2(\mu)$.

(a) $\operatorname{Lat}\{N, N^*\} = \{L^2(\mu|\Delta) : \Delta \text{ is a Borel set}\}$.

(b) $\operatorname{Alg}\operatorname{Lat}\{N, N^*\} = \mathcal{A}_\mu$.

The characterization of $\operatorname{Lat} N$ for the operator N defined above is an unsolved problem of considerable difficulty. Even if μ is the restriction of area measure to a bounded subset of $\mathbb{C}$, this problem is unsolved and, at present, seems intractable.

22.6 Definition. If $\mathcal{S}$ is a set of operators, define the following.

(a) $P(\mathcal{S})$ is the smallest norm closed algebra containing $\mathcal{S}$ and the identity.

(b) $P^\infty(\mathcal{S})$ is the smallest weak* closed algebra containing $\mathcal{S}$ and the identity.

(c) $W(\mathcal{S})$ is the smallest WOT closed algebra containing $\mathcal{S}$ and the identity.

Note that $P(\mathcal{S}) \subseteq P^\infty(\mathcal{S}) \subseteq W(\mathcal{S})$. Of particular interest in this book is the case where $\mathcal{S}$ consists of a single operator. Here $P(T)$, $P^\infty(T)$, and $W(T)$ are the closures of the polynomials in T in the appropriate topologies.

Finding an example of an operator T such that $P^\infty(T) \neq W(T)$ is hard. The first such example was exhibited in Wogen [1987]. This will be seen later in §61.

To find an algebra that is weak* closed but not WOT closed is considerably easier.

22.7 Example. Let C be a trace class operator that does not have finite rank and put $\mathcal{C} = \{T \in \mathcal{B}(\mathcal{H}) : \operatorname{tr}(TC) = 0\}$. So $\mathcal{C}$ is a weak* closed linear subspace of $\mathcal{B}(\mathcal{H})$ that is not WOT closed. If $\mathcal{A}$ equals the collection of operators on $\mathcal{H} \oplus \mathcal{H}$ represented by the 2×2 matrices of the form

$$\begin{bmatrix} \alpha & A \\ 0 & \alpha \end{bmatrix}$$

where $A \in \mathcal{C}$ and $\alpha \in \mathbb{C}$, then $\mathcal{A}$ is a weak* closed algebra that is not WOT closed.

The next result is from Conway and Wu [1977].

22.8 Proposition. *Let T_1 and T_2 be two operators and consider the following statements.*

(a) $P(T_1 \oplus T_2) = P(T_1) \oplus P(T_2)$.

(b) $P^\infty(T_1 \oplus T_2) = P^\infty(T_1) \oplus P^\infty(T_2)$.

(c) $W(T_1 \oplus T_2) = W(T_1) \oplus W(T_2)$.

(d) $\operatorname{Lat}(T_1 \oplus T_2) = \operatorname{Lat}(T_1) \oplus \operatorname{Lat}(T_2)$.

(e) $\operatorname{Alg}\operatorname{Lat}(T_1 \oplus T_2) = \operatorname{Alg}\operatorname{Lat}(T_1) \oplus \operatorname{Alg}\operatorname{Lat}(T_2)$.

Each of these conditions implies the next and conditions (d) *and* (e) *are equivalent.*

Proof. It is easy to see that (a) is equivalent to the condition that there is a sequence of polynomials $\{p_n\}$ such that $\|p_n(T_1) - 1\| \to 0$ and $\|p_n(T_2)\| \to 0$. This immediately gives that (a) implies (b). Similarly (b) implies (c).

It is always the case that $\operatorname{Lat}(T_1 \oplus T_2) \supseteq \operatorname{Lat}(T_1) \oplus \operatorname{Lat}(T_2)$. Let P_k be the projection of $\mathcal{H}_1 \oplus \mathcal{H}_2$ onto $\mathcal{H}_k$. If (c) holds, then $P_k \in W(T_1 \oplus T_2)$. Thus $P_k\mathcal{M} \subseteq \mathcal{M}$ whenever $\mathcal{M} \in \operatorname{Lat}(T_1 \oplus T_2)$. It follows that $\mathcal{M} = P_1(\mathcal{M}) \oplus P_2(\mathcal{M}_2)$ and that $P_k(\mathcal{M}) \in \operatorname{Lat} T_k$. This proves (d).

Assume (d). The fact that $\mathcal{H}_k \in \operatorname{Lat}(T_1 \oplus T_2)$ implies that $\operatorname{Alg}\operatorname{Lat}(T_1 \oplus T_2) \subseteq \operatorname{Alg}\operatorname{Lat}(T_1) \oplus \operatorname{Alg}\operatorname{Lat}(T_2)$. That is, if $A \in \operatorname{Alg}\operatorname{Lat}(T_1 \oplus T_2)$, then $A = A_1 \oplus A_2$ with $A_k \in \operatorname{Alg}\operatorname{Lat} T_k$. If $A_k \in \operatorname{Alg}\operatorname{Lat} T_k$ and $\mathcal{M}_k \in \operatorname{Lat} T_k$, then $A_1 \oplus A_2(\mathcal{M}_1 \oplus \mathcal{M}_2) \subseteq \mathcal{M}_1 \oplus \mathcal{M}_2$. So $\operatorname{Alg}\operatorname{Lat}(T_1 \oplus T_2) \supseteq \operatorname{Alg}\operatorname{Lat}(T_1) \oplus \operatorname{Alg}\operatorname{Lat}(T_2)$.

Now assume (e). Clearly $P_1 = 1 \oplus 0 \in \text{Alg Lat}\,(T_1) \oplus \text{Alg Lat}\,(T_2)$, so if (e) is true, $P_1 \in \text{Alg Lat}\,(T_1 \oplus T_2)$. Thus for each $\mathcal{M}$ in $\text{Lat}\,(T_1 \oplus T_2)$, $P_1\mathcal{M} \subseteq \mathcal{M}$. It is now an easy argument to show that $\mathcal{M} = P_1\mathcal{M} \oplus P_2\mathcal{M} \in \text{Lat}\, T_1 \oplus \text{Lat}\, T_2$. ■

Recall that for a compact subset K of $\mathbb{C}$, $\widehat{K}$ denotes the polynomially convex hull of K. That is, $\widehat{K} = \{z : |p(z)| \le \max\{|p(w)| : w \in K\}$ for every polynomial $p\}$. A compact set K is *polynomially convex* if $K = \widehat{K}$. Half the next result is due to Crimmins and Rosenthal [1967], who showed that the lattice of a direct sum of operators splits whenever their spectra have disjoint polynomially convex hulls.

22.9 Proposition. *If T_1 and T_2 are two operators, then $P(T_1 \oplus T_2) = P(T_1) \oplus P(T_2)$ if and only if $\widehat{\sigma(T_1)} \cap \widehat{\sigma(T_2)} = \emptyset$.*

Proof. If $\widehat{\sigma(T_1)} \cap \widehat{\sigma(T_2)} = \emptyset$, there is a sequence of polynomials $\{p_n\}$ such that $|p_n(z) - 1| \to 0$ for z in a neighborhood of $\widehat{\sigma(T_1)}$ and $p_n(z) \to 0$ for z in a neighborhood of $\widehat{\sigma(T_2)}$. This implies that $\|p_n(T_1) - 1\| \to 0$ and $\|p_n(T_2)\| \to 0$. Thus $1 \oplus 0 \in P(T_1 \oplus T_2)$ and it follows that $P(T_1 \oplus T_2) = P(T_1) \oplus P(T_2)$.

Conversely, assume that $P(T_1 \oplus T_2) = P(T_1) \oplus P(T_2)$. Thus there is a sequence of polynomials $\{p_n\}$ such that $\|p_n(T_1) - 1\| \to 0$ and $\|p_n(T_2)\| \to 0$. Since the spectral radius of an operator is less than its norm, $\sup\{|p_n(z) - 1| : z \in \sigma(T_1)\} \le \|p_n(T_1) - 1\| \to 0$ and $\sup\{|p_n(z)| : z \in \sigma(T_2)\} \le \|p_n(T_2)\| \to 0$. This shows that the polynomially convex hulls of the spectra are disjoint. ■

It is natural to ask for examples that demonstrate that, with the exception of (d) and (e), the statements in Proposition 22.8 are not equivalent. Some of these can be found with the technique at hand; others require considerably more.

22.10 Example. There are two operators T_1, T_2 such that $P(T_1 \oplus T_2) \ne P(T_1) \oplus P(T_2)$, but $P^\infty(T_1 \oplus T_2) = P^\infty(T_1) \oplus P^\infty(T_2)$. In fact the operators can be chosen so that the identity map on polynomials extends to an isometric isomorphism between $P(T_1 \oplus T_2)$ and both $P(T_1)$ and $P(T_2)$.

Let μ be Lebesgue measure on $[0,1]$ and let $\nu = \sum_n 2^{-n}\delta_{a_n}$, where $\{a_n\}$ is a dense sequence in $[0,1]$. Let M and N be multiplication by the independent variable on $L^2(\mu)$ and $L^2(\nu)$, respectively. So both M and N are hermitian and $\sigma(M) = \sigma(N) = [0,1]$. Thus for any continuous function f on $[0,1]$, $\|f(M \oplus N)\| = \|f(M)\| = \|f(N)\| = \sup\{|f(t)| : t \in [0,1]\}$. Since the polynomials are dense in the continuous functions on the interval, the map $p(M \oplus N) \to p(M)$, defined for polynomials p, extends to an isometric isomorphism of $P(M \oplus N) = C^*(M \oplus N)$ onto $P(M) = C^*(M)$. Similarly, these algebras are isometrically isomorphic with $P(N) = C^*(N)$. Thus, most definitely, $P(M \oplus N) \neq P(M) \oplus P(N)$.

If $\Delta = \{a_n\}$ and $\Lambda = [0,1] \setminus \Delta$, then the Borel set Δ carries the measure ν and Λ carries μ. Thus a function f in $L^2(\mu + \nu)$ can be written as $f = f|\Lambda + f|\Delta$, the sum of a function in $L^2(\mu)$ and a function in $L^2(\nu)$. This shows that in this natural way, $L^2(\mu+\nu) = L^2(\mu) \oplus L^2(\nu)$. (Verify!) Similarly $L^\infty(\mu + \nu) = L^\infty(\mu) \oplus L^\infty(\nu)$.

But using (20.9) it follows that $P^\infty(M) = \{ M_\phi : \phi \in P^\infty(\mu) \}$. Since μ is carried by the interval, this means that $P^\infty(M) = \mathcal{A}_\mu = \{M_\phi : \phi \in L^\infty(\mu)\}$. Similarly, $P^\infty(N) = \mathcal{A}_\nu$ and $P^\infty(M \oplus N) = \mathcal{A}_{\mu+\nu}$. Since $L^\infty(\mu + \nu) = L^\infty(\mu) \oplus L^\infty(\nu)$, this establishes that $P^\infty(M \oplus N)$ splits.

Here is another example illustrating the same phenomenon. Of course it is not needed for the purposes of this exposition, but it gives a nice connection between operator theory and function theory. The reader will need to know the F and M Riesz Theorem (25.4): if μ is a measure on the circle $\partial\mathbb{D}$ and $\int p\, d\mu = 0$ for all polynomials p, then μ is absolutely continuous with respect to Lebesgue measure.

22.11 Example. Let m be Lebesgue measure on $\partial\mathbb{D}$ and let ν be any positive measure on $\partial\mathbb{D}$ that is singular to m. As in the previous example, it must be shown that $P^\infty(m + \nu) = P^\infty(m) \oplus P^\infty(\nu)$. Now it is always true that $P^\infty(m + \nu) \subseteq P^\infty(m) \oplus P^\infty(\nu)$. So it must be shown that if $f \in L^1(m + \nu)$ and $0 = \int pf\, d(m + \nu)$ for every polynomial p, then $0 = \int pf\, dm = \int qf\, d\nu$ for all polynomials p and q. But applying the F and M Riesz Theorem to the measure $f(m + \nu)$ shows that $fm + f\nu \ll m$. Since $\nu \perp m$, this implies that $f\nu$ must be the zero measure. This proves what was sought.

The reader who knows the F and M Riesz Theorem will also recognize that $P^\infty(m) = H^\infty$, the space of bounded analytic functions. Thus in the preceding example, $P^\infty(m + \nu) = H^\infty \oplus L^\infty(\nu)$.

Necessary and sufficient conditions that $P^\infty(M \oplus N) = P^\infty(M) \oplus P^\infty(N)$ for normal operators M and N are known (Conway [1977]). However the statement of the conditions relies on some deep function theoretic results of Sarason [1972].

Exercises.

1. If T_1 and T_2 are operators on a finite dimensional space, show that all the conditions in Proposition 22.8 are equivalent to the condition that T_1 and T_2 have disjoint spectra.

2. How do the conditions $\{T_1 \oplus T_2\}' = \{T_1\}' \oplus \{T_2\}'$ and $\{T_1 \oplus T_2\}'' = \{T_1\}'' \oplus \{T_2\}''$ fit into the scheme of implications in Proposition 22.8?

3. If $R(T)$ is defined as the norm closure of $\{f(T) : f$ is a rational function with poles off $\sigma(T)\}$, show that $R(T_1 \oplus T_2) = R(T_1) \oplus R(T_2)$ if and only if $\sigma(T_1) \cap \sigma(T_2) = \emptyset$. It is also possible to talk of Lat $R(T)$ and formulate and prove results similar to those in Proposition 22.8.

4. If μ is arc length measure on $\partial\mathbb{D}$, show that for $|\lambda| < 1$, $P^\infty(N_\mu \oplus \lambda) \neq P^\infty(N_\mu) \oplus P^\infty(\lambda) = P^\infty(N_\mu) \oplus \mathbb{C}$. (The dimension of the space that the scalar λ acts on is irrelevant.) However $P^\infty(N_\mu \oplus 1) = P^\infty(N_\mu) \oplus \mathbb{C}$.

5. Generalize Example 22.10 by showing that for hermitian operators A and B, $P^\infty(A \oplus B) = P^\infty(A) \oplus P^\infty(B)$ if and only if the scalar-valued spectral measures for A and B are mutually singular.

§23. Isometries

Every unitary operator is an isometry, but the converse is well known to be false. The standard example is called the unilateral shift. This operator is, undoubtedly, the most widely studied (and understood) non-normal operator.

23.1 Definition. If ℓ^2 is the set of square summable sequences, the *unilateral shift* is the operator $S : \ell^2 \to \ell^2$ defined by

$$S(\alpha_0, \alpha_1, \ldots) = (0, \alpha_0, \alpha_1, \ldots).$$

(There is a very good reason for starting the enumeration of the sequences from 0. The reason for this will become apparent later.)

It is easy to verify that S is an isometry and clearly it fails to be surjective. The article 'the' is used in the title of this operator since there are other unilateral shifts.

23.2 Definition. An operator S acting on a Hilbert space $\mathcal{H}$ is called a *unilateral shift* if there is a sequence of pairwise orthogonal subspaces $\mathcal{H}_0, \mathcal{H}_1, \ldots$ such that $\mathcal{H} = \mathcal{H}_0 \oplus \mathcal{H}_1 \oplus \cdots$ and S maps $\mathcal{H}_n$ isometrically onto $\mathcal{H}_{n+1}$ for all $n \geq 0$.

The unilateral shift is a unilateral shift where each $\mathcal{H}_n$ is the one dimensional space spanned by the n-th basis vector. It is clear that in Definition 23.2 each of the spaces $\mathcal{H}_n$ has the same dimension, though each may be infinite dimensional. Call this common dimension the *multiplicity* of S. In fact, it is easy to see that if S is a unilateral shift (with the notation as above), then $\operatorname{ran} S = \mathcal{H}_0^\perp$ and the multiplicity of S is $\dim \mathcal{H}_0 = \dim (\operatorname{ran} S)^\perp$. The proof of the next result is left as an exercise.

23.3 Proposition. *Two unilateral shifts are unitarily equivalent if and only if they have the same multiplicity.*

23.4 Corollary. *If S is a unilateral shift on $\mathcal{H}$, $\mathcal{L} = \mathcal{H} \ominus (S\mathcal{H})$, and T is defined on $\mathcal{L} \oplus \mathcal{L} \oplus \cdots$ by $T(f_0 \oplus f_1 \oplus \cdots) = 0 \oplus f_0 \oplus f_1 \oplus \cdots$, then S and T are unitarily equivalent.*

Proof. In fact, T is a unilateral shift with the same multiplicity as S. ■

23.5 Proposition. *If T is defined on $\mathcal{L} \oplus \mathcal{L} \oplus \cdots$ by $T(f_0 \oplus f_1 \oplus \cdots) = 0 \oplus f_0 \oplus f_1 \oplus \cdots$, then $T^*(f_0 \oplus f_1 \oplus \cdots) = f_1 \oplus f_2 \oplus \cdots$.*

The proof of the preceding proposition is a straightforward exercise. This operator T is usually called the *backward shift* of multiplicity $\dim \mathcal{L}$.

If S is an isometry and $\mathcal{M}$ is an invariant subspace, then clearly $S|\mathcal{M}$ is also an isometry. If $S\mathcal{M} = \mathcal{M}$, then the restriction of S to $\mathcal{M}$ is a surjective isometry, and hence is unitary. More can be said.

23.6 Proposition. *If S is an isometry and $\mathcal{M}$ is a subspace such that $S\mathcal{M} = \mathcal{M}$, then $\mathcal{M}$ reduces S, $S|\mathcal{M}$ is unitary, and $\mathcal{M} \subseteq \bigcap_n S^n\mathcal{H}$.*

Proof. Recall that an operator S is an isometry if and only if $S^*S = 1$. Thus $S\mathcal{M} = \mathcal{M}$ implies that $S^*\mathcal{M} = S^*S\mathcal{M} = \mathcal{M}$; so $\mathcal{M}$ reduces S. Clearly $S|\mathcal{M}$ is unitary. Also $\mathcal{M} = S^n\mathcal{M} \subseteq S^n\mathcal{H}$ for every $n \geq 1$. ■

If S is an isometry, it fails to be unitary if it is not surjective. So for an isometry S consider the closed subspaces $S\mathcal{H}, S^2\mathcal{H}, S^3\mathcal{H}, \ldots$. Either all these spaces are equal (in which case S is unitary) or no two of them are equal (Why?). Also $S\mathcal{H} \supseteq S^2\mathcal{H} \supseteq \ldots$; so consider $\mathcal{L}_\infty \equiv \bigcap_n S^n\mathcal{H}$. Note that since S is one-to-one and $S^{n+1}\mathcal{H} \subseteq S^n\mathcal{H}$, $S\mathcal{L}_\infty = \bigcap_n S^{n+1}\mathcal{H} = \mathcal{L}_\infty$. So by the preceding proposition, $S|\mathcal{L}_\infty$ is unitary and $\mathcal{L}_\infty$ reduces S. Also every invariant subspace $\mathcal{M}$ with $S\mathcal{M} = \mathcal{M}$ is contained in $\mathcal{L}_\infty$; that is, $\mathcal{L}_\infty$ is the "unitary part" of S. All these matters are collected together in the next result.

23.7 The von Neumann–Wold Decomposition. *If S is an isometry on $\mathcal{H}$ and $\mathcal{L}_\infty = \bigcap_n S^n\mathcal{H}$, then $\mathcal{L}_\infty$ reduces S, $S|\mathcal{L}_\infty$ is unitary, and $S|\mathcal{L}_\infty^\perp$ is a unilateral shift. Thus every isometry on a Hilbert space is the direct sum of a unilateral shift and a unitary.*

Proof. Put $\mathcal{L}_n = S^n\mathcal{H}$ for $n \geq 0$ (so $\mathcal{L}_0 = \mathcal{H}$); clearly $\mathcal{L}_n \supseteq \mathcal{L}_{n+1}$. Put $\mathcal{H}_n = \mathcal{L}_n \ominus (\mathcal{L}_{n+1})$. The fact that S is an isometry implies that each $\mathcal{L}_n$, and hence each $\mathcal{H}_n$, is closed.

Claim 1. $\mathcal{H}_n \perp \mathcal{H}_m$ for $n \neq m$.

In fact, if $n < m$, then $\mathcal{H}_m \subseteq \mathcal{L}_m \subseteq \mathcal{L}_{n+1} \subseteq \mathcal{H}_n^\perp$.

Claim 2. $\mathcal{L}_\infty^\perp = \mathcal{H}_0 \oplus \mathcal{H}_1 \oplus \cdots$.

If $m > n$, then $\mathcal{H}_n \perp \mathcal{L}_m$. Hence $\mathcal{H}_n \perp \bigcap_{m \geq n+1} \mathcal{L}_m = \mathcal{L}_\infty$. Thus $\mathcal{H}_0 \oplus \mathcal{H}_1 \oplus \cdots \subseteq \mathcal{L}_\infty^\perp$. On the other hand, if $f \in (\mathcal{H}_0 \oplus \mathcal{H}_1 \oplus \cdots)^\perp$, then for every $n \geq 1$, $f \in (\mathcal{H}_0 \oplus \cdots \oplus \mathcal{H}_n)^\perp = \mathcal{L}_{n+1}$. Thus $f \in \mathcal{L}_\infty$.

Claim 3. $\mathcal{L}_\infty$ reduces S, $S\mathcal{L}_\infty = \mathcal{L}_\infty$, and $S|\mathcal{L}_\infty$ is unitary.

This is a consequence of the preceding proposition, as discussed prior to the statement of this theorem.

Claim 4. $S\mathcal{H}_n = \mathcal{H}_{n+1}$.

It is easy to see that $S\mathcal{H}_n \subseteq S^{n+1}\mathcal{H}$. Also if $f \in \mathcal{H}_n$ and $h \in \mathcal{H}$, then $\langle Sf, S^{n+2}h \rangle = \langle f, S^{n+1}h \rangle = 0$. Hence $S\mathcal{H}_n \subseteq \mathcal{H}_{n+1}$. Conversely, if $f \in \mathcal{H}_{n+1} = \mathcal{L}_{n+1} \ominus (\mathcal{L}_{n+2})$, then there is a vector g in $\mathcal{L}_n$ with $Sg = f$. If $h \in \mathcal{L}_{n+1}$, then $Sh \in \mathcal{L}_{n+2}$ and so $0 = \langle f, Sh \rangle = \langle Sg, Sh \rangle = \langle g, h \rangle$. Hence $g \in \mathcal{L}_n \ominus (\mathcal{L}_{n+1}) = \mathcal{H}_n$ and $f = Sg \in S\mathcal{H}_n$.

Combining Claims 2 and 4, it follows that $S|\mathcal{L}_\infty^\perp$ is a unilateral shift of multiplicity $\dim \mathcal{H}_0$. ■

An isometry is said to be *pure* if there is no reducing subspace on which it is unitary. In light of Proposition 23.6, this is equivalent to saying that there is no invariant subspace $\mathcal{M}$ such that $S|\mathcal{M}$ is unitary.

23.8 Corollary. *An isometry is pure if and only if it is a unilateral shift.*

Proof. If S is pure, the preceding theorem implies that S must be a unilateral shift. Conversely, if S is a unilateral shift and $\mathcal{M}$ is an invariant subspace for S such that $S\mathcal{M} = \mathcal{M}$, then Proposition 23.6 implies $\mathcal{M} \subseteq \bigcap_n S^n\mathcal{H} = (0)$. ■

23.9 Definition. An operator U on a Hilbert space $\mathcal{K}$ is a *bilateral shift* if there are pairwise orthogonal subspaces $\{\mathcal{K}_n : n \in \mathbb{Z}\}$ such that $\mathcal{K} = \bigoplus_n \mathcal{K}_n$ and U maps $\mathcal{K}_n$ isometrically onto $\mathcal{K}_{n+1}$ for all n in $\mathbb{Z}$. The *multiplicity* of U is the dimension of $\mathcal{K}_0$ ($= \dim \mathcal{K}_n$ for all n).

It is easy to see that a bilateral shift is unitary and its multiplicity is a complete unitary invariant. Define *the bilateral shift* to be the bilateral shift of multiplicity 1.

There is a relation between unilateral and bilateral shifts, as you might expect. For example, if U is a bilateral shift relative to the spaces $\{\mathcal{K}_n\}$ and $\mathcal{H} = \mathcal{K}_0 \oplus \mathcal{K}_1 \oplus \cdots$, then $\mathcal{H}$ is invariant for U and $S = U|\mathcal{H}$ is a unilateral shift. Note that the multiplicity of S is the same as the multiplicity of U. A converse of this is also true and the astute reader may see it now. Indeed a more general result is true, but some preliminary groundwork needs to be done.

Suppose that S is a unilateral shift on $\mathcal{H} = \mathcal{H}_0 \oplus \mathcal{H}_1 \oplus \cdots$, $S\mathcal{H}_n = \mathcal{H}_{n+1}$. If e_0 is a unit vector in $\mathcal{H}_0$ and $e_n = S^n e_0$ for $n \geq 1$, then $\{e_n\}$ is an orthonormal sequence. Using this technique produces the following result, the details of the proof being left as an exercise.

23.10 Proposition. *If S is a unilateral shift on $\mathcal{H}$ of multiplicity α and $\{e_i : 1 \leq i \leq \alpha\}$ is a basis for $\mathcal{H} \ominus (S\mathcal{H})$, then $\{S^n e_i : n \geq 0$ and $1 \leq i \leq \alpha\}$ is a basis for $\mathcal{H}$.*

23.11 Corollary. *If S is a unilateral shift of multiplicity α and $1 \leq \beta < \alpha$, then there is an invariant subspace $\mathcal{M}$ for S such that $S|\mathcal{M}$ is a unilateral shift of multiplicity β.*

23.12 Theorem. *If S is an isometry on $\mathcal{H}$, then there is a Hilbert space $\mathcal{K}$ containing $\mathcal{H}$ and a unitary $U\colon \mathcal{K} \to \mathcal{K}$ such that $U\mathcal{H} \subseteq \mathcal{H}$ and $U|\mathcal{H} = S$. Moreover $\mathcal{K}$ can be chosen to be the smallest reducing subspace for U that contains $\mathcal{H}$; in this case, U and $\mathcal{K}$ are unique up to unitary equivalence.*

Proof. First assume that S is a unilateral shift. By Corollary 23.3 it may be assumed that $\mathcal{H} = \mathcal{L} \oplus \mathcal{L} \oplus \cdots$ and $S(f_0 \oplus f_1 \oplus \cdots) = 0 \oplus f_0 \oplus f_1 \oplus \cdots$. For convenience write this as $\mathcal{H} = \mathcal{L}_0 \oplus \mathcal{L}_1 \oplus \cdots$, where $\mathcal{L}_n = \mathcal{L}$ for all $n \geq 0$. Let $\mathcal{K} = \bigoplus_{n=-\infty}^{\infty} \mathcal{L}_n$, where $\mathcal{L}_n = \mathcal{L}$ for all n. Define U on $\mathcal{K}$ by $U(\cdots \oplus f_{-2} \oplus f_{-1} \oplus \widehat{f_0} \oplus f_1 \oplus f_2 \oplus \cdots) = \cdots \oplus f_{-2} \oplus \widehat{f_{-1}} \oplus f_0 \oplus f_1 \oplus f_2 \oplus \cdots$, where the $\widehat{\ }$ denotes the location of the zero-th coordinate. Clearly U is a unitary; in fact, U is a bilateral shift of the same multiplicity as S.

It is also clear that $U\mathcal{H} \subseteq \mathcal{H}$ and $U|\mathcal{H} = S$. Suppose $\mathcal{K}'$ is a reducing subspace for U and $\mathcal{H} \subseteq \mathcal{K}'$. So $\mathcal{K}' \supseteq U^{*n}\mathcal{L}_0$ for every $n \geq 1$. But $\mathcal{L}_{-n} = U^{*n}\mathcal{L}_0$ and so it follows that $\mathcal{K}' = \mathcal{K}$.

If $\mathcal{R}$ is any Hilbert space containing $\mathcal{H}$ and $V : \mathcal{R} \to \mathcal{R}$ is a unitary such that $V\mathcal{H} \subseteq \mathcal{H}$ and $V|\mathcal{H} = S$, then it can be checked that the subspaces $V^n\mathcal{L}_0$, $n \in \mathbb{Z}$, are closed and pairwise orthogonal and that V maps $V^n\mathcal{L}_0$ isometrically onto $V^{n+1}\mathcal{L}_0$. Thus $\mathcal{M} = [V^n\mathcal{L}_0 : n \in \mathbb{Z}]$ reduces V and $V|\mathcal{M}$ is a bilateral shift that extends S and has the same multiplicity as S. Save for a few details to be supplied by the reader, this establishes uniqueness in the case that S is a shift.

Now for the arbitrary isometry. Using the von Neumann–Wold decomposition, $S = S_1 \oplus S_0$ on $\mathcal{H} = \mathcal{H}_1 \oplus \mathcal{H}_0$, where S_1 is a unilateral shift and S_0 is unitary. Let U_1 be a bilateral shift on $\mathcal{K}_1 \supseteq \mathcal{H}_1$ such that $U_1|\mathcal{H}_1 = S_1$. Put $\mathcal{K} = \mathcal{K}_1 \oplus \mathcal{H}_0$ and $U = U_1 \oplus S_0$. The verification that U has the appropriate properties is left to the reader. The uniqueness follows from the uniqueness statement when the isometry is a shift. ∎

If U is the unitary extension of the isometry S for which $\mathcal{K}$ is the smallest reducing subspace of U that contains $\mathcal{H}$, then U is called the *minimal unitary extension* of S. This terminology is justified by the uniqueness part of the theorem.

23.13 Corollary. *The minimal unitary extension of a unilateral shift of multiplicity α is a bilateral shift of multiplicity α.*

23.14 Corollary. *Let S be a unilateral shift of multiplicity α and let U be its minimal unitary extension acting on $\mathcal{K}$. If $\{e_i : 1 \leq i \leq \alpha\}$ is a basis for $\mathcal{H} \ominus (S\mathcal{H})$, then $\{U^n e_i : n \in \mathbb{Z}, 1 \leq i \leq \alpha\}$ is a basis for $\mathcal{K}$.*

The next result will be of importance later. Actually if the multiplicity theory of normal operators ([ACFA], §IX.10) were available, the proof of the next proposition would be immediate. Part of that theory says that if N is a normal operator and $\mathcal{R}$ is a reducing subspace, $N|\mathcal{R}$ has a multiplicity function dominated by the multiplicity function of N. This has not introduced any inconsistencies in the language, and the definition of multiplicity for a bilateral shift meshes with the definition of multiplicity of normal operators. Thus if U is a bilateral shift of multiplicity α and $\mathcal{R}$ is a reducing subspace such that $U|\mathcal{R}$ is also a bilateral shift, then the multiplicity of $U|\mathcal{R}$ is at most α. This, in conjunction with Corollary 23.13, could be used to prove the next proposition. Multiplicity theory for normal operators has not been developed in this book at this point, so an independent proof of this result is needed. Fortunately it is straightforward.

23.15 Proposition. *If S is a unilateral shift of multiplicity α acting on $\mathcal{H}$ and $\mathcal{M}$ is an invariant subspace for S, then $S|\mathcal{M}$ is a unilateral shift of multiplicity $\beta \leq \alpha$.*

Proof. It is clear that $S|\mathcal{M}$ is an isometry. Since S is pure it must be that $S|\mathcal{M}$ is pure. By Corollary 23.8, $S|\mathcal{M}$ is a unilateral shift; let β be the multiplicity of $S|\mathcal{M}$. It remains to show that $\beta \leq \alpha$. To do this it suffices to assume that $\alpha < \infty$.

Let U, acting on $\mathcal{K}$, be the minimal unitary extension of S and let $\{e_i : 1 \leq i \leq \alpha\}$ be an orthonormal basis for $\mathcal{H} \ominus (S\mathcal{H})$. By the preceding corollary, $\{U^n e_i : n \in \mathbb{Z}, 1 \leq i \leq \alpha\}$ is a basis for $\mathcal{K}$. Let $f_1, \ldots, f_\beta$ be a basis for $\mathcal{M} \ominus (S\mathcal{M})$. Hence

$$\begin{aligned}
\beta &= \sum_{j=1}^{\beta} \|f_j\|^2 \\
&= \sum_{j=1}^{\beta} \sum_{i=1}^{\alpha} \sum_{n=0}^{\infty} |\langle f_j, U^n e_i \rangle|^2 \\
&= \sum_{j=1}^{\beta} \sum_{i=1}^{\alpha} \sum_{n=0}^{\infty} |\langle U^{*n} f_j, e_i \rangle|^2 \\
&= \sum_{i=1}^{\alpha} \sum_{j=1}^{\beta} \sum_{n=0}^{\infty} |\langle U^{*n} f_j, e_i \rangle|^2 \\
&\leq \sum_{i=1}^{\alpha} \|e_i\|^2 \qquad \text{(by Bessel's Inequality)} \\
&= \alpha \quad \blacksquare
\end{aligned}$$

The preceding result is from Halmos [1961], though the proof is due to Halperin; see Halmos [1982], Solution 15. Also see Robertson [1965].

Finally consider the spectral properties of unilateral shifts. Recall (see [ACFA]) that for any operator T in $\mathcal{B}(\mathcal{H})$, the *approximate point spectrum* of T, $\sigma_{ap}(T)$, is defined to be the set of complex numbers λ satisfying $\inf\{\|(T-\lambda)f\| : \|f\| = 1\} = 0$. Equivalently, $\lambda \in \sigma_{ap}(T)$ if and only if either $\ker(\lambda - T) = (0)$ or $\operatorname{ran}(\lambda - T)$ is not closed ([ACFA], Proposition VII.6.4). Thus $\sigma_{ap}(T) = \sigma_\ell(T)$, the *left spectrum* of T ([ACFA], Example VII.3.4). Also $\sigma_{\ell e}(T)$ is the *left essential spectrum* of T; that is, the left spectrum of the coset of T in the Calkin algebra $\mathcal{B}(\mathcal{H})/\mathcal{B}_0(\mathcal{H})$. The *right essential spectrum* and the *essential spectrum* of T are denoted by $\sigma_{re}(T)$ and $\sigma_e(T)$, respectively. The reader should read §XI.2 of [ACFA] for the properties of these subsets of $\sigma(T)$ as well as §XI.5, where some additional information can be found. This will be reviewed later in this book in §37.

23.16 Proposition. *If S is the unilateral shift of multiplicity* 1, *then the following statements are valid.*

(a) $\sigma(S) = \operatorname{cl}\mathbb{D}, \sigma_{ap}(S) = \partial\mathbb{D}, \sigma_p(S) = \emptyset$.

(b) *For* $|\lambda| < 1$, $\dim\ker(S^* - \lambda) = 1$ *and* $(1, \lambda, \lambda^2, \ldots) \in \ker(S^* - \lambda)$.

(c) $\sigma_e(S) = \sigma_{\ell e}(S) = \sigma_{re}(S) = \partial\mathbb{D}$.

Proof. Since $\|S\| = 1$, $\sigma(S) \subseteq \operatorname{cl}\mathbb{D}$. If $\lambda \in \mathbb{C}$, $\lambda \neq 0$, $h = (\alpha_0, \alpha_1, \dots) \in \ell^2$, and $Sh = \lambda h$, then $0 = \lambda\alpha_0, \alpha_0 = \lambda\alpha_1, \dots$. Hence $0 = \alpha_0 = \alpha_1 = \cdots$. Therefore, $\sigma_p(S) = \emptyset$.

(b) If $|\lambda| < 1$, put $k_\lambda = (1, \lambda, \lambda^2, \dots)$. From (23.5) it follows that $S^*(\alpha_0, \alpha_1, \dots) = (\alpha_1, \alpha_2, \dots)$. Therefore $S^*k_\lambda = (\lambda, \lambda^2, \dots) = \lambda k_\lambda$. That is, λ is an eigenvalue of S^* with eigenvector k_λ. To complete the proof of (b) it is necessary to show that $\mathbb{C}k_\lambda$ is the entire eigenspace for S^* corresponding to λ. If $h = (\alpha_0, \alpha_1, \dots) \in \ker(S^* - \lambda)$, then $(\alpha_1, \alpha_2, \dots) = \lambda(\alpha_0, \alpha_1, \dots)$. Thus $\alpha_n = \lambda\alpha_{n-1}$ for all $n \geq 1$. Thus $\alpha_n = \lambda^n\alpha_0$ for all $n \geq 0$ and so $h = \alpha_0 k_\lambda$. This proves (b).

(a) Part of (a) was established at the start of this proof. Since $\sigma(S)^* = \sigma(S^*)$, part (b) implies that $\mathbb{D} \subseteq \sigma(S)$. Thus $\sigma(S) = \operatorname{cl}\mathbb{D}$. Now if $|\lambda| < 1$ and $h \in \ell^2$, then $\|(S - \lambda)h\| \geq |\,\|Sh\| - |\lambda|\,\|h\|\,| = (1 - |\lambda|)\|h\|$. Thus $\lambda \notin \sigma_{ap}(S)$ and so $\sigma_{ap}(S) \subseteq \partial\mathbb{D}$. But in general the boundary of the spectrum of an operator is contained in the approximate point spectrum. Therefore $\sigma(S) = \partial\mathbb{D}$. This completes the proof of (a).

(c) Since $\sigma_{ap}(S) = \partial\mathbb{D}$, $\operatorname{ran}(S - \lambda)$ is closed and $\ker(S - \lambda) = (0)$ whenever $|\lambda| < 1$. Therefore $\sigma_{\ell e}(S)$ and $\sigma_e(S)$ are subsets of $\partial\mathbb{D}$. Since $\dim\ker(S^* - \lambda) = 1$ when $|\lambda| < 1$, it is also true that $\sigma_{re}(S) \subseteq \partial\mathbb{D}$. But since S has no eigenvalues, points λ in $\partial\mathbb{D}$ belong to the approximate spectrum of S because $\operatorname{ran}(S - \lambda)$ is not closed. Therefore $\partial\mathbb{D} \subseteq \sigma_{\ell e}(S) \cap \sigma_{re}(S)$. ∎

Because a shift of multiplicity α is the direct sum of α copies of the shift of multiplicity 1, the proof of the next result can be obtained from the preceding one.

23.17 Proposition. *If S is a unilateral shift of multiplicity α, then the following statements are valid.*

(a) $\sigma(S) = \operatorname{cl}\mathbb{D}, \sigma_{ap}(S) = \partial\mathbb{D}, \sigma_p(S) = \emptyset$.

(b) *For* $|\lambda| < 1$, $\operatorname{ran}(S - \lambda)$ *is closed and* $\dim\ker(S^* - \lambda) = \alpha$.

(c) $\sigma_{\ell e}(S) = \partial\mathbb{D}$.

(d) *If* $\alpha < \infty, \sigma_{re}(S) = \sigma_e(S) = \partial\mathbb{D}$.

(e) *If* $\alpha = \infty, \sigma_{re}(S) = \sigma_e(S) = \operatorname{cl}\mathbb{D}$.

Exercises.

1. Prove Proposition 23.3.
2. Show that a normal isometry is unitary.
3. If S is a unilateral shift of multiplicity α and n is a positive integer, show that S^n is a unilateral shift of multiplicity $n\alpha$.

4. Show that a bilateral shift is unitary and two bilateral shifts are unitarily equivalent if and only if they have the same multiplicity.

5. Define $S\colon L^2(0,\infty) \to L^2(0,\infty)$ by $(Sf)(x) = f(x-1)$ if $x \geq 1$ and $(Sf)(x) = 0$ for $x \leq 1$. Show that S is a unilateral shift and calculate its multiplicity. Obtain a convenient form for its minimal unitary extension. Find S^*.

6. Compute the adjoint of a bilateral shift.

7. Prove that a bilateral shift and its adjoint are unitarily equivalent.

8. Give an example of a unitary operator U that is not a bilateral shift and satisfies $U \cong U^*$.

9. Let S_1, S_2 be unilateral shifts on $\mathcal{H}_1, \mathcal{H}_2$ with minimal unitary extensions U_1, U_2 acting on $\mathcal{K}_1, \mathcal{K}_2$. Show that if $X\colon \mathcal{H}_1 \to \mathcal{H}_2$ is an operator such that $XS_1 = S_2X$, then there is a unique operator $Y\colon \mathcal{K}_1 \to \mathcal{K}_2$ such that: (i) $Y|\mathcal{H}_1 = X$; (ii) $YU_1 = U_2Y$; (iii) $\|Y\| = \|X\|$.

10. Let S be the unilateral shift and put $T = S \oplus S^*$; prove the following statements. (a) $\sigma(T) = \sigma_\ell(T) = \sigma_r(T) = \operatorname{cl}\mathbb{D}$ and $\sigma_p(T) = \mathbb{D}$. (b) $\sigma_e(T) = \partial\mathbb{D}$, and if $|\lambda| < 1$, then $\operatorname{ind}(T-\lambda) = 0$. (See Chapter XI in [ACFA] for the properties of the Fredholm index.) (c) There is a rank 1 operator F such that $T + F$ is the bilateral shift. (d) There is a sequence of rank 1 operators $\{F_n\}$ such that $F_n \to 0$ and $\sigma(T + F_n) = \partial\mathbb{D}$ for all n. What does this say about the continuity of the spectrum?

11. If T is an isometry on a Banach space $\mathcal{X}$, show that either $\sigma(T) = \operatorname{cl}\mathbb{D}$ or $\sigma(T) \subseteq \partial\mathbb{D}$.

§24. Unilateral and bilateral shifts

This section makes a more detailed study of the unilateral shift of multiplicity 1. To this end, a brief foray into function theory is needed.

Let U be the bilateral shift on $\ell^2(\mathbb{Z})$: $Ue_n = e_{n+1}$, where $\{e_n\}$ is the usual basis for $\ell^2(\mathbb{Z})$. U is a star-cyclic unitary operator and is thus unitarily equivalent to multiplication by z, the independent variable, on $L^2(\mu)$ for some measure μ supported on $\partial\mathbb{D}$ (11.2). The introduction of function theory begins by showing that an isomorphism that implements such a unitary equivalence is the Fourier transform. Indeed, the Fourier transform is the unique isomorphism that implements this unitary equivalence and takes e_0 to the constantly 1 function.

Normalized Lebesgue measure on $\partial\mathbb{D}$ will be denoted by m and the Lebesgue spaces of this measure will be denoted by $L^p(\partial\mathbb{D})$ or simply L^p.

24.1 Definition. If $f \in L^1$, then the *Fourier transform* of f is the function $\widehat{f}: \mathbb{Z} \to \mathbb{C}$ defined by

$$\widehat{f}(n) = \int f \bar{z}^n \, dm.$$

Note that since m is a finite measure, $L^p \subseteq L^1$ for $1 \le p \le \infty$. Hence $\widehat{f}$ is defined for f in L^p and all p. A *trigonometric polynomial* is a function p in $C(\partial \mathbb{D})$ of the form $p(z) = \sum_{k=-n}^{n} a_k z^k$.

24.2 Lemma. *The trigonometric polynomials are uniformly dense in the space of continuous functions on the circle and, hence, dense in L^p for $1 \le p < \infty$; they are weak* dense in L^∞. Thus $\{z^n : n \in \mathbb{Z}\}$ is an orthonormal basis for L^2.*

Proof. The first part follows easily from The Stone–Weierstrass Theorem. The last statement only needs the calculation necessary to show that the functions z^n are orthonormal. ∎

24.3 Theorem. *If $f \in L^2$, then $\widehat{f} \in \ell^2(\mathbb{Z})$. If $V: L^2 \to \ell^2(\mathbb{Z})$ is defined by $Vf = \widehat{f}$, then V is an isomorphism and if W denotes multiplication by z on L^2, then VWV^{-1} is the bilateral shift.*

Proof. The first part, that $\widehat{f} \in \ell^2(\mathbb{Z})$, as well as the statement that V is an isometry, is a direct consequence of Parseval's Identity and the fact that $\{z^n\}$ is a basis for L^2. If $f = z^n$, then it is straightforward to check that $\widehat{f}(k) = 0$ if $k \neq n$ and $\widehat{f}(n) = 1$. That is, $\widehat{f}$ is the n-th basis vector in $\ell^2(\mathbb{Z})$. Thus $\operatorname{ran} V$ is dense and so V must be an isomorphism.

If $\{e_n\}$ is the usual basis in $\ell^2(\mathbb{Z})$, then the preceding paragraph shows that $V(z^n) = e_n$. Thus $VW(z^n) = V(z^{n+1}) = e_{n+1} = UV(z^n)$. This completes the proof. ∎

24.4 Corollary. *If U is a bilateral shift of any multiplicity, then $\sigma(U) = \partial \mathbb{D}$.*

Recall that the spectral properties of the unilateral shift S of multiplicity α were determined in Proposition 23.17.

24.5 Definition. For $1 \le p \le \infty$, H^p = all functions f in L^p such that $\widehat{f}(n) = 0$ for $n < 0$.

The letter *H* here stands for "Hardy." G H Hardy was a pioneer in the study of these spaces and they are universally called *Hardy spaces.*

Note that H^p is the annihilator of $\{z^{-n} : n \geq 1\} = \{\bar{z}^n : n \geq 1\}$. Since $\{z^{-n} : n \geq 1\} \subseteq L^q$ for the index q that is dual to p, H^p is a closed subspace of L^p. (H^∞ is a weak* closed subspace of L^∞.)

Now return to the notation in Theorem 24.3. If $\ell^2 \equiv \ell^2(\mathbb{N} \cup \{0\})$, then $V^{-1}\ell^2 = H^2$. Since $V^{-1}e_n = z^n$, it follows that the analytic polynomials are dense in H^2. Also, the unilateral shift, $U|\ell^2$, is unitarily equivalent to $W|H^2$. This information is gathered together in the next theorem.

24.6 Theorem. *The analytic polynomials are dense in H^2 and the unilateral shift is unitarily equivalent to multiplication by z on H^2. Moreover, the isomorphism that implements this unitary equivalence is the restriction of the Fourier transform.*

Thus the unilateral shift will be identified with multiplication by z on H^2 and both will be denoted by S. Since two unilateral shifts are unitarily equivalent if they have the same multiplicity, it would have been possible to observe that multiplication by z on H^2 is a unilateral shift since it shifts the basis $\{z^n : n \geq 0\}$ and conclude that it is unitarily equivalent to the shift on ℓ^2. However, it is important to realize that this unitary equivalence is the classical Fourier transform. This allows the application of function theory to the study of the shift.

In (23.16) it was shown that when $|\lambda| < 1$, the vector $(1, \lambda, \lambda^2, \dots) \in \ell^2$ and is an eigenvector for S^* corresponding to the eigenvalue λ. Equivalently, $\ker(S - \lambda)^*$ is generated by $(1, \bar{\lambda}, \bar{\lambda}^2, \dots)$. Mapping this vector into H^2 by the inverse Fourier transform gives the function

$$\sum_{n=0}^{\infty} \bar{\lambda}^n z^n = (1 - \bar{\lambda}z)^{-1} \equiv k_\lambda(z).$$

Note that k_λ is a bounded function, and hence belongs to $H^2 \cap L^\infty = H^\infty$. The next proposition records this information along with an additional property of the function k_λ.

24.7 Proposition. *If $|\lambda| < 1$ and $k_\lambda(z) = (1 - \bar{\lambda}z)^{-1}$, then $k_\lambda \in H^\infty$, $\|k_\lambda\|_2 = (1 - |\lambda|^2)^{-1/2}$, $\langle p, k_\lambda \rangle = p(\lambda)$ for all polynomials, and $S^* k_\lambda = \bar{\lambda} k_\lambda$.*

Proof. To calculate the norm of k_λ, apply Parseval's Identity to the expansion $k_\lambda = \sum_n \bar{\lambda}^n e_n$. Also if $n \geq 0$, then $\langle z^n, k_\lambda \rangle = \lambda^n$. Hence $\langle p, k_\lambda \rangle = p(\lambda)$ for every polynomial. The remainder of the proof follows from Proposition 23.16. ∎

The reason that the eigenvector k_λ is distinguished from its scalar multiples is the fact that $\langle p, k_\lambda \rangle = p(\lambda)$ for all polynomials p. This "reproducing" property will be seen to be important. In fact the function k_λ is called the *reproducing kernel* for H^2 at λ.

It has been established that H^2 is a closed subspace of L^2 and $\{z^n : n \geq 0\}$ is an orthonormal basis for H^2. So if $f \in H^2$, $f = \sum_{n=0}^{\infty} \widehat{f}(n) z^n$, where $\widehat{f}(n)$ is the n-th Fourier coefficient and $\sum_{n=0}^{\infty} |\widehat{f}(n)|^2 = \|f\|^2 < \infty$. It follows that the radius of convergence of the power series $\sum_n \widehat{f}(n) z^n$ is at least 1. Thus each f in H^2 is associated with a unique analytic function defined on the open unit disk, $\mathbb{D}$. This connection with analytic functions can be exploited and, as a result, deep information can be obtained about H^2. The next section explores this connection in more detail..

24.8 Proposition. *If $\phi \in H^\infty$ and $f \in H^2$, then $\phi f \in H^2$. Conversely, if $\phi \in L^\infty$ and $\phi f \in H^2$ for all f in H^2, then $\phi \in H^\infty$.*

Proof. Let $\phi \in L^\infty$ and $k \in \mathbb{Z}$. For any n in $\mathbb{Z}$ it follows that $\widehat{(\phi z^k)}(n) = \int \phi z^k \bar{z}^n \, dm = \int \phi \bar{z}^{(n-k)} \, dm = \widehat{\phi}(n-k)$. If $\phi \in H^\infty$, $k \geq 0$, and $n \leq -1$, then $\widehat{\phi z^k}(n) = \widehat{\phi}(n-k) = 0$ since $n - k \leq -1$. Thus $\phi z^k \in H^2$ for all $k \geq 0$. Hence $\phi p \in H^2$ for all polynomials p. Since ϕ is bounded, $\phi H^2 \subseteq H^2$ by Proposition 24.6.

Conversely, assume that $\phi \in L^\infty$ and $\phi f \in H^2$ for all f in H^2. In particular, $\phi = \phi 1 \in H^2$. Hence the negative Fourier coefficients of ϕ vanish and so $\phi \in H^\infty$. ■

24.9 Corollary. *H^∞ is a subalgebra of L^∞.*

Proof. If $\phi, \psi \in H^\infty$ and $f \in H^2$, then $(\phi\psi)f = \phi(\psi f)$. But the preceding proposition implies that $\psi f \in H^2$. Applying the proposition again shows that $\phi(\psi f) \in H^2$. The converse of the proposition now shows that $\phi\psi \in H^\infty$. ■

Thus for each ϕ in H^∞ the operator $\phi(S)$ can be defined on H^2 by $\phi(S) f = \phi f$. The more traditional notation is $\phi(S) = T_\phi$, the *analytic Toeplitz operator with symbol* ϕ.

The stage is now set to prove one of the most celebrated theorems in analysis, Beurling's Theorem. This will be derived as a consequence of a somewhat more general result that will be useful in further developments.

24.10 Theorem. *If μ is a measure supported on $\partial\mathbb{D}$ and $U = N_\mu$, then $\mathcal{M}$ is an invariant subspace for U if and only if there is a Borel function ϕ and a Borel subset Δ of $\partial\mathbb{D}$ such that $\phi\chi_\Delta = 0$ a.e. $[\mu]$, $|\phi|^2\mu = m$, and*

$$\mathcal{M} = \phi H^2 \oplus L^2(\mu|\Delta),$$

though it is possible for either of these direct summands to be absent.

Proof. Assume that such a function ϕ and set Δ exist. For h in H^2, $\int |\phi h|^2\, d\mu = \int |h|^2\, dm$. Thus the mapping $h \to \phi h$ is an isometry from H^2 into $L^2(\mu)$. Hence $\mathcal{M}_0 = \phi H^2$ is a closed subspace of $L^2(\mu)$ and clearly $\mathcal{M}_0 \in \operatorname{Lat} U$. It is even clearer that $L^2(\mu|\Delta) \in \operatorname{Lat} U$. Also the fact that $\phi\chi_\Delta = 0$ a.e. $[\mu]$ implies that $\phi H^2 \perp L^2(\mu|\Delta)$ in $L^2(\mu)$. Therefore $\mathcal{M} = \phi H^2 \oplus L^2(\mu|\Delta) \in \operatorname{Lat} U$.

For the converse assume that $\mathcal{M} \in \operatorname{Lat} U$. So $U|\mathcal{M}$ is an isometry and, by Theorem 23.7, decomposes into the direct sum of a unilateral shift and a unitary. It will be shown that if $U|\mathcal{M}$ is either a shift or a unitary, then $\mathcal{M}$ has one of the two forms. Since the condition that $\phi\chi_\Delta = 0$ a.e. $[\mu]$ is equivalent to the condition that $\phi \in L^2(\mu|\Delta)^\perp$, this will complete the proof.

If $U|\mathcal{M}$ is unitary, $\mathcal{M}$ is a reducing subspace for U (23.6). By Corollary 12.7 there is a Borel set Δ such that $\mathcal{M} = L^2(\mu|\Delta)$.

Now suppose $\mathcal{M} \in \operatorname{Lat} U$ and $U|\mathcal{M}$ is a shift. Since $U|\mathcal{M}$ has a star-cyclic unitary extension (*viz.*, U), the minimal unitary extension of $U|\mathcal{M}$ is the restriction of U to some reducing subspace. This reducing subspace has the form $L^2(\mu|\Lambda)$ for some Borel set Λ contained in $\partial\mathbb{D}$. This says two things. First, the minimal unitary extension of $U|\mathcal{M}$, $V \equiv U|L^2(\mu|\Lambda)$, is star cyclic, and so $U|\mathcal{M}$ is a shift of multiplicity 1 (Corollary 23.14). Second, $V \cong N_m$, the bilateral shift. Therefore $\mu|\Lambda$ and m are mutually absolutely continuous (11.4). The rest of the proof consists in giving these two statements more specificity.

Since $U|\mathcal{M}$ is a shift of multiplicity 1, there is an isomorphism $W\colon H^2 \to \mathcal{M}$ such that $WS = UW$, where S is multiplication by z on H^2. Let $\phi = W(1)$; so $\phi \in L^2(\mu)$. Because 1 is a cyclic vector for S, ϕ is a cyclic vector for $U|\mathcal{M}$. Also $\{\, z^n\phi : n \in \mathbb{Z}\}$ is an orthonormal basis for $L^2(\mu|\Lambda)$ that is shifted by V. So W extends to an isomorphism $W\colon L^2 \to L^2(\mu|\Lambda)$ such that $WN_m = VW$. Since $W(1) = \phi$, the proof of Corollary 11.4 shows that $Wh = \phi h$ for all h in L^2 and $m = |\phi|^2\mu$. It also follows that $\mathcal{M} = WH^2 = \phi H^2$. ■

Note that if the measure μ in the preceding theorem is such that there is an invariant subspace $\mathcal{M}$ with $U|\mathcal{M}$ not unitary, then the existence of the function ϕ implies that $m << \mu$. This essentially proves the following.

24.11 Corollary. *If μ is a measure supported on $\partial\mathbb{D}$ and $U = N_\mu$, then U has a non-reducing invariant subspace if and only if $m << \mu$.*

Proof. Half the proof was noted before the statement of the corollary. If $m << \mu$, then there is a function ϕ in $L^2(\mu)$ such that $m = |\phi|^2\mu$. It is left to the reader to check that $\mathcal{M} = \phi H^2$ is an invariant subspace for U that is not reducing. ■

The direct sum decomposition in the preceding theorem is precisely the von Neumann–Wold decomposition of the isometry $N_\mu|\mathcal{M}$. If the invariant subspace $\mathcal{M}$ has the form $\mathcal{M} = \phi H^2$ for a function ϕ such that $|\phi|^2\mu = m$, say that $\mathcal{M}$ is a *pure* invariant subspace. A more traditional terminology is to call such subspaces *simple invariant subspace.*

24.12 Corollary. *If μ is a measure on $\partial\mathbb{D}$ such that Lebesgue measure is not absolutely continuous with respect to μ, then the polynomials are dense in $L^2(\mu)$.*

Proof. Let $P^2(\mu)$ be the closure of the polynomials in $L^2(\mu)$. So $P^2(\mu)$ is an invariant subspace for N_μ. By the preceding corollary $P^2(\mu)$ is reducing. Hence $P^2(\mu) = L^2(\mu|\Delta)$ for some Borel subset Δ of $\partial\mathbb{D}$. But $1 \in P^2(\mu)$, so $\Delta = \partial\mathbb{D}\ [\mu]$. ■

Now consider the special case that the measure μ in Theorem 24.10 is Lebesgue measure. If $\phi \in L^2$ and $|\phi|^2 m = m$, then $|\phi| = 1$ a.e. Thus it is impossible for there to be a Borel set Δ such that $f\chi_\Delta = 0$ unless $m(\Delta) = 0$. Hence in this case each invariant subspace must have one form or the other. That is, every invariant subspace for the bilateral shift is either pure or reducing. This is summarized as follows.

24.13 Corollary. *If U is the bilateral shift on L^2 and $\mathcal{M} \in \operatorname{Lat} U$, then either there is a Borel set Δ such that $\mathcal{M} = L^2(m|\Delta)$ or there is a function ϕ in L^∞ with $|\phi| = 1$ a.e. and $\mathcal{M} = \phi H^2$.*

The next theorem is from Beurling [1949]. As presented here, it is a corollary of Theorem 24.10. It is, however, a forefather of that result and has had an enormous influence on the development of operator theory and function theory.

24.14 Beurling's Theorem. (Beurling [1949]) *If S is the unilateral shift and $\mathcal{M}$ is a non-zero invariant subspace of S, then there is a function ϕ in H^∞ such that $|\phi| = 1$ a.e. and $\mathcal{M} = \phi H^2$.*

Proof. Let U be the bilateral shift on L^2. If $\mathcal{M}$ is as described in the statement of the theorem, then $\mathcal{M} \in \operatorname{Lat} U$. By Corollary 24.13, either there is a Borel set Δ such that $\mathcal{M} = L^2(m|\Delta)$ or there is a function ϕ in L^∞ with $|\phi| = 1$ a.e. and $\mathcal{M} = \phi H^2$. Since S is a shift, $(0) = \bigcap_n S^n H^2 \supseteq \bigcap_n S^n \mathcal{M}$, so the first alternative is impossible. Hence $\mathcal{M} = \phi H^2 \subseteq H^2$. Since $\phi = \phi \cdot 1 \in \mathcal{M}$, $\phi \in L^\infty \cap H^2 = H^\infty$. ■

24.15 Definition. A function ϕ in H^∞ is *inner* if $|\phi| = 1$ a.e.

There is a catch here in stating that this is Beurling's Theorem. The preceding result is really not the full Beurling Theorem. The full strength of the theorem will only come to the fore with a full understanding of the nature of inner functions. For example, suppose ψ is any function in H^∞ and consider ψH^2. If S is the unilateral shift, it is clear that $S\psi H^2 \subseteq \psi H^2$, so $\text{cl}\,[\psi H^2] \in \text{Lat}\, S$. According to Beurling's Theorem, for any such ψ there is an inner function ϕ with $\phi H^2 = \text{cl}\,[\psi H^2]$. How does one get ϕ from ψ? If ϕ_1 and ϕ_2 are inner functions, then $\phi_1 H^2 \cap \phi_2 H^2$ and $\phi_1 H^2 \vee \phi_2 H^2 \in \text{Lat}\, S$. Thus there are inner functions ψ and χ such that $\phi_1 H^2 \cap \phi_2 H^2 = \psi H^2$, and $\phi_1 H^2 \vee \phi_2 H^2 = \chi H^2$. What is the relation of ψ and χ to ϕ_1 and ϕ_2? Can it ever be that $\phi_1 H^2 \cap \phi_2 H^2$ is (0)?

Before stating with full justification that the lattice of invariant subspaces for the shift is known, such questions should be answerable using the description of the lattice. All these are questions that can be answered, but only after devoting some effort to the study of the function theory associated with the Hardy spaces. This is done in the next section; see Exercise 25.6. Additional material can be found in Chapter III of Conway [1991] and Chapter 20 of Conway [1995] as well as Duren [1970], Hoffman [1962], and Koosis [1980].

The easiest examples of inner functions are the *Blaschke products*,

$$B(z) = \prod_{n=1}^{\infty} \frac{|a_n|}{a_n}\left(\frac{a_n - z}{1 - \bar{a}_n z}\right),$$

where $\{a_n\}$ is a sequence in $\mathbb{D}$ such that $\sum_n (1 - |a_n|) < \infty$. The convergence of this sum is necessary and sufficient that the infinite product converge uniformly on compact subsets of $\mathbb{D}$. This follows from basic results on the convergence of infinite products. See Exercise VII.5.4 in Conway [1978] or §20.2 in Conway [1995]. At this point of the development of this subject, however, it is difficult to see that a Blaschke product is an inner function, though it is not hard to show that a finite Blaschke product is. What is needed is the identification of H^∞ with the algebra of bounded analytic functions on the disk. This will not be done here, though this discussion will be revisited.

Another good example of a non-constant inner function is

$$\phi(z) = \exp\left[\alpha \frac{1+z}{1-z}\right],$$

with $\alpha < 0$. This function plays an important role in §28.

Remarks and suggestions for further reading. 1. Douglas [1972] is a good source for "soft" proofs of a number of results on Hardy spaces. Some of this is presented in the next section. More classical approaches can be found in the references cited above.

2. The material of this section can be generalized to the unilateral shift of any multiplicity. The generalization to shifts of finite multiplicity is due to Lax [1959] and [1961] and the generalization to shifts of infinite multiplicity is due to Halmos [1961]. Halmos's article is a very clean presentation. The book by Helson [1964] is a self-contained treatment. It is Helson and Lowdenslager [1958] that is the source of most of the modern approaches to this material.

3. There are connections between this material and probability that can be found in the references already cited. Another source of the interest in Beurling's Theorem and its generalizations is the result due to Sz.-Nagy [1953] that for every contraction T on a Hilbert space $\mathcal{H}$ there is a unitary operator U acting on a Hilbert space $\mathcal{K}$ that contains $\mathcal{H}$ such that for all $n \geq 0$, $T^n = PU^n$, where P is the orthogonal projection of $\mathcal{K}$ onto $\mathcal{H}$. (This will be proved in §35.) Later it was discovered that under the assumption that T is *completely non-unitary*, that is, there is no reducing subspace for T on which it is unitary, the unitary U can be chosen to be a bilateral shift. This led to a program of trying to study general operator theory with the help of Fourier analysis. This approach to operator theory has had a great deal of success, though one hope, the proof of the Invariant Subspace Theorem, has not been realized. One approach to this is in Sz.-Nagy and Foias [1970]. (See §35 for the beginning of an exposition.) Another is in deBranges and Rovnyak [1966a] and [1966b]. Also see Rosenblum and Rovnyak [1985].

4. Let $\phi \in L^\infty$ and define the operator $T_\phi \colon H^2 \to H^2$ by $T_\phi f = P(\phi f)$, where P is the orthogonal projection from L^2 onto H^2. This is called the *Toeplitz operator* with symbol ϕ. When $\phi \in H^\infty$, this is the operator $\phi(S)$ defined above and, in this case, there is no need to use the projection P in the definition. There is a vast literature on Toeplitz operators. A good starting point is Douglas [1972].

Exercises.

1. For $f = \sum_{n\geq 0} c_n z^n$ in H^2, extend f to an analytic function on $\mathbb{D}$ by letting $f(z) = \sum_{n\geq 0} c_n z^n$ for $|z| < 1$. Show that $\langle f, k_\lambda \rangle = f(\lambda)$.

2. If $\phi \in H^\infty$ and $|\lambda| < 1$, show that $\phi(S)^* k_\lambda = \phi(\lambda)^* k_\lambda$. ($H^\infty \subseteq H^2$ and so $\phi(\lambda)$ is defined for λ in $\mathbb{D}$ as in Exercise 1.)

3. Show that for every ϕ in H^∞, $\sigma(\phi(S)) = \mathrm{cl}\,[\phi(\mathbb{D})]$.

4. Show that if U is the bilateral shift on L^2 and $\mathcal{M} \in \operatorname{Lat} U$ such that $\mathcal{M}$ does not reduce U, then there is a ϕ in L^∞ with $|\phi| = 1$ such that $\mathcal{M} = \phi H^2$. Discuss the uniqueness of ϕ.

5. Let p be a polynomial that does not vanish on $\partial\mathbb{D}$ and show that $\ker p(S) = (0)$, $\operatorname{ran} p(S)$ is closed, and

$$\dim\,[\operatorname{ran} p(S)]^{\perp} = -\frac{1}{2\pi}\int_{|z|=1}\frac{p'(z)}{p(z)}\,dz.$$

§25. Some results on Hardy spaces

This section explores some of the properties of the Hardy spaces needed to establish certain properties of the unilateral shift S, defined as multiplication by z on H^2. Here the identification of the functions in H^p with analytic functions on the disk will be avoided to some extent. In so doing the development in Douglas [1972] will be used. Nevertheless, strong use will be made of the fact that each function in H^2 gives rise to an analytic function defined on the unit disk as was mentioned in (Exercise 24.1).

Consider the following. If $f \in H^2$ and $f(z) = \sum_{n=0}^{\infty} a_n z^n$ is its Fourier expansion, this same series converges uniformly on compact subsets of $\mathbb{D}$. Indeed, if $|z| \le r < 1$, then

$$\begin{aligned}\sum_{n=m}^{\infty}|a_n z^n| &\le \left[\sum_{n=m}^{\infty}|a_n|^2\right]^{1/2}\left[\sum_{n=m}^{\infty}|z|^{2n}\right]^{1/2}\\ &\le \|f\|_2\left[\sum_{n=m}^{\infty} r^{2n}\right]^{1/2}\end{aligned}$$

which can be made as small as desired. Therefore it is possible to identify H^2 with the space of analytic functions on the unit disk whose Taylor coefficients are square summable.

25.1 Proposition. *If f is a real-valued function in H^1, then f is a constant.*

Proof. Let $\alpha = \int f\,dm$; so $\alpha \in \mathbb{R}$. By virtue of belonging to H^1, $\int f z^n\,dm = 0$ for all $n \ge 1$, so $\int (f-\alpha) z^n\,dm = 0$ for all $n \ge 0$. Taking complex conjugates shows that this equation is valid for every integer n. Thus $f - \alpha$ annihilates all the trigonometric polynomials. By Lemma 24.2, $f - \alpha = 0$ in L^1. ■

25.2 Corollary. *If ϕ is an inner function such that $\bar{\phi} = 1/\phi \in H^2$, then ϕ is constant.*

Proof. By hypothesis, $\phi + \bar{\phi}$ and $(\phi - \bar{\phi})/i$ are two real-valued functions in H^2, which are constant by the preceding proposition. ∎

The proof of the next important result uses Beurling's Theorem. This is sometimes called the F and M Riesz Theorem after its discoverers. That name will be reserved, however, for the result that follows the next one.

25.3 Theorem. *If f is a non-zero function in H^2, then $m(\{z \in \partial\mathbb{D} : f(z) = 0\}) = 0$.*

Proof. Let Δ be a Borel subset of the circle and put $\mathcal{M} = \{h \in H^2 : h(z) = 0 \text{ a.e. on } \Delta\}$. It is easy to check that $\mathcal{M}$ is an invariant subspace for the shift. By Beurling's Theorem, if $\mathcal{M} \neq (0)$, there is an inner function ϕ such that $\mathcal{M} = \phi H^2$. But $\phi = \phi \cdot 1 \in \mathcal{M}$ and, so, must vanish on Δ. But $|\phi| = 1$ a.e. It must be that $\mathcal{M} = (0)$. ∎

25.4 The F and M Riesz Theorem. *If μ is a complex-valued, Borel measure on $\partial\mathbb{D}$ such that $\int z^n \, d\mu = 0$ for all $n \geq 1$, then there is a function f in H^1 such that $\mu = fm$.*

Proof. Let v be a Borel function such that $|v| = 1$ and $\mu = v|\mu|$. So

25.5
$$\int_{\partial\mathbb{D}} z^n v \, d|\mu| = \int z^n \, d\mu = 0$$

for $n \geq 1$. That is, if $\mathcal{M} = [z^n : n \geq 1]$ in $L^2(|\mu|)$, $\bar{v} \perp \mathcal{M}$ in $L^2(|\mu|)$. Let $W = N_{|\mu|}$. Clearly $\mathcal{M} \in \operatorname{Lat} W$. Also note that $\mathcal{M} \neq (0)$, since z never vanishes. By Theorem 24.10 there is a function g in $\mathcal{M}$ and a Borel set Δ such that $|g|^2|\mu| = m$ and $\mathcal{M} = gH^2 \oplus L^2(|\mu| \,|\Delta)$, though one of these direct summands may be absent.

Now $\bar{v}\chi_\Delta \in L^2(|\mu| \,|\Delta)$, so (25.5) implies that $\bar{v}\chi_\Delta = 0$. Hence $|\mu|(\Delta) = \int_\Delta |v|^2 \, d|\mu| = \int \bar{v}\chi_\Delta \, d\mu = 0$. Thus $\mathcal{M} = gH^2$ with $|g|^2 \,|\mu| = m$; in particular, $m << |\mu|$. Also $z \in gH^2$, so there is a function h in H^2 such that $z = gh$ in $L^2(|\mu|)$. Because z is never 0, $g(z) \neq 0$ a.e. $[\mu]$. Therefore $|\mu|$ and m are mutually absolutely continuous. Let $f = d\mu/dm$; so $f \in L^1$. But then for $n \geq 1$, $\int f z^n \, dm = \int z^n \, d\mu = 0$, so that $f \in H^1$. ∎

25.6 Definition. A function f in H^2 is said to be an *outer function* if $H^2 = [z^n f : n \geq 0]$.

So in fact, a function in H^2 is outer if and only if it is a cyclic vector for the shift.

25.7 Theorem. *If f is a non-zero function in H^2, then there is an inner function ϕ and an outer function g in H^2 such that $f = \phi g$. If $f \in H^\infty$, then $g \in H^\infty$.*

Proof. Observe that $\mathcal{M} \equiv [z^n f : n \geq 0] \in \operatorname{Lat} S$. By Beurling's Theorem there is an inner function ϕ such that $\mathcal{M} = \phi H^2$. Let $g \in H^2$ such that $f = \phi g$ and put $\mathcal{N} = [z^n g : n \geq 0]$. Again there is an inner function ψ such that $\mathcal{N} = \psi H^2$. Note that $\phi H^2 = [z^n f : n \geq 0] = [z^n \phi g : n \geq 0] = \phi\psi H^2$. Therefore there is a function $h \in H^2$ such that $\phi = \phi\psi h$. Multiplying both sides of the equation by $\bar{\phi}$ shows that $1 = \psi h$, or $\bar{\psi} = h \in H^2$. Hence ψ is a constant by Corollary 25.2. So $\mathcal{N} = H^2$ and g is outer.

Now assume that $f \in H^\infty$ with the inner-outer factorization $f = \phi g$. Thus $|g| = |f|$ a.e. on $\partial\mathbb{D}$, and g must also be bounded. ■

25.8 Proposition. *If $g, h \in H^2$ and g is outer, then $|h| \leq |g|$ if and only if there is a function ϕ in* ball H^∞ *such that $h = \phi g$.*

Proof. One of the two implications is clear. So suppose that $|h| \leq |g|$. Because g is outer, there is a sequence of polynomials $\{p_n\}$ such that $\|1 - gp_n\|_2 \to 0$. By hypothesis, $\int |p_n h - p_k h|^2\, dm = \int |p_n - p_k|^2 |h|^2\, dm \leq \int |p_n - p_k|^2 |g|^2\, dm = \|p_n g - p_k g\|_2^2$ for all k, n. That is, $\{p_n g\}$ is a Cauchy sequence. So there is a function ϕ in H^2 such that $\|p_n h - \phi\|_2 \to 0$. For any $n \geq 1$,

$$\begin{aligned}\int |\phi g - h|\, dm &\leq \int |g(\phi - p_n h)|\, dm + \int |(gp_n - 1)h|\, dm \\ &\leq \|g\|_2\, \|\phi - p_n h\|_2 + \|h\|_2\, \|gp_n - 1\|_2\end{aligned}$$

and this converges to 0. Therefore $h = \phi g$. Since $g \in H^2$, it differs from 0 a.e.; thus $\phi = h/g$ a.e. on the circle. Hence ϕ is bounded by 1. Since $\phi \in H^2$, this shows that $\phi \in \operatorname{ball} H^\infty$. ■

As a consequence of this result it follows that outer functions are determined to within a constant multiple by their absolute values.

25.9 Corollary. *If g_1 and g_2 are outer functions in H^2 such that $|g_1| = |g_2|$* a.e., *then there is a unimodular constant α such that $g_1 = \alpha g_2$.*

Proof. Applying the proposition twice, there are functions ϕ and ψ in ball H^∞ such that $g_1 = \phi g_2$ and $g_2 = \psi g_1$. So $g_1 = \phi\psi g_1$. By cancellation, $1 = \phi\psi$. Thus, $\bar{\phi} = \psi \in H^\infty$. By Corollary 25.2, ϕ is constant. ■

25.10 Corollary. *For $j = 1, 2$, let ϕ_j be an inner function and let g_j be an outer function in H^2. If $\phi_1 g_1 = \phi_2 g_2$, then there is a unimodular constant α such that $g_1 = \alpha g_2$ and $\phi_1 = \bar{\alpha}\phi_2$.*

Proof. The hypothesis implies that $|g_1| = |g_2|$, so the preceding corollary yields the constant α such that $g_1 = \alpha g_2$. Thus $\phi_2 g_2 = \alpha\phi_1 g_2$. Since H^2 functions differ from zero a.e., g_2 can be canceled and so $\phi_2 = \alpha\phi_1$. ∎

In light of this corollary it is correct to talk about *the inner-outer factorization* of functions in H^2. Proposition 25.8 also raises the question of which positive functions in L^2 are the absolute value of an outer function. Realize that for each f in L^2, $[z^n f : n \geq 0] \in \operatorname{Lat} U$, where U is the bilateral shift. According to Corollary 24.13, this invariant subspace is either pure or reducing.

25.11 Theorem. *If $f \in L^2$, then there is an outer function g such that $|f| = |g|$ if and only if $[z^n f : n \geq 0]$ is a pure invariant subspace for the bilateral shift.*

Proof. Fix f and put $\mathcal{M} = [z^n f : n \geq 0]$. Suppose there is an outer function g such that $|f| = |g|$. Then there is a function ϕ in L^∞ with $|\phi| = 1$ a.e. and $f = \phi g$. Thus $fz^n = \phi(z^n g)$. Because g is outer it follows that $\mathcal{M} = \phi H^2$, so that $\mathcal{M}$ is pure.

Now assume that $\mathcal{M}$ is pure. By Corollary 24.13 there is a function ϕ in L^∞ with $|\phi| = 1$ a.e. and $\mathcal{M} = \phi H^2$. But $f \in \mathcal{M}$, so there is a function g in H^2 such that $f = \phi g$. Note that $|f| = |g|$. But the mapping $h \to \phi h$ is an isometry of H^2 onto $\mathcal{M}$, and so this mapping takes $\{ z^n g \}$ onto the total set $\{ z^n f \}$ in $\mathcal{M}$. It must be that $\{ z^n g \}$ is total in H^2. Hence, g is outer. ∎

25.12 Corollary. *If $f \in L^2$ and* $\inf |f| > 0$, *then there is an outer function g in H^2 such that $|f| = |g|$.*

Proof. Since $\inf |f| > 0$, there cannot be a non-trivial Borel set Δ such that $[z^n f : n \geq 0] = L^2(m|\Delta)$. ∎

25.13 Theorem. *If $f \in H^1$ and $f \neq 0$, then there is an outer function g in H^2 such that $|f| = |g|^2$.*

Proof. Let $h = |f|^{1/2} \in L^2$ and consider $\mathcal{M} = [z^n h : n \geq 0]$. Clearly $\mathcal{M}$ is an invariant subspace for the bilateral shift. The theorem follows from Theorem 25.11 if it can be shown that $\mathcal{M}$ is pure. (If it were known at this point that functions in H^1 vanish almost nowhere, as is the case for functions in H^2, the proof would be finished as the space $\mathcal{M}$ could not be reducing. Though this is true, indeed is a consequence of this theorem, it is not available now.)

Suppose $\mathcal{M}$ is reducing. So if $k > 0$, $\bar{z}^k h \in \mathcal{M}$. Therefore there is a sequence of polynomials $\{p_n\}$ such that $\|p_n h - \bar{z}^k h\|_2 \to 0$. Thus

$$\begin{aligned} 0 = \lim_{n\to\infty} \|\bar{z}^k h - p_n h\|_2^2 &= \int |h^2 \bar{z}^{2k} - 2h^2 p_n \bar{z}^k + h^2 p_n^2|\, dm \\ &= \int |h^2 \bar{z}^k - h^2(2p_n - p_n^2 z^k)|\, dm \\ &= \|h^2 \bar{z}^k - h^2(2p_n - p_n^2 z^k)\|_1. \end{aligned}$$

This says that $h^2\bar{z}^k \in \mathcal{L}$, where $\mathcal{L}$ is the closure in L^1 of $\{\, z^n h^2 : n \geq 0 \,\}$. Let $\phi \in L^\infty$ such that $|\phi| = 1$ a.e. and $f = \phi h^2$. So $\bar{z}^k f = \phi(\bar{z}^k h^2) \in \phi\mathcal{L} = [\, z^n f : n \geq 0]_{L^1} \subseteq H^1$. Thus $f \in z^k H^1$ for all $k \geq 1$. Computing Fourier coefficients, this implies $f = 0$, a contradiction. Thus $\mathcal{M}$ is pure. ■

25.14 Corollary. *If $f \in H^1$ and $f(z) = 0$ on a subset of $\partial\mathbb{D}$ having positive measure, then $f = 0$.*

This leads to a very important corollary.

25.15 Corollary. *If $f \in H^1$, then there are functions g_1, g_2 in H^2 such that $f = g_1 g_2$ and $|g_1|^2 = |g_2|^2 = |f|$.*

Proof. Without loss of generality it can be assumed that f is not the zero function. By the preceding theorem there is an outer function g in H^2 such that $|g|^2 = |f|$. Let $\{p_n\}$ be a sequence of polynomials such that $p_n g \to 1$ in H^2. Thus $\|fp_n^2 - fp_m^2\|_1 \leq \|fp_n(p_n - p_m)\|_1 + \|fp_m(p_n - p_m)\|_1 \leq \|gp_n\|_2\, \|g(p_n - p_m)\|_2 + \|gp_m\|_2\, \|g(p_n - p_m)\|_2$. Since $\sup_n \|gp_n\|_2 < \infty$, this shows that $\{fp_n^2\}$ is a Cauchy sequence in H^1; let h be its limit. By passing to a subsequence if necessary, it can also be assumed that $f(z)p_n^2(z) \to h(z)$ and $g(z)p_n(z) \to 1$ a.e. on $\mathbb{D}$. So for almost all z, $h(z)g(z)^2 = \lim_n f(z)[p_n(z)g(z)]^2 = f(z)$. Thus $|f(z)| = |h(z)|\, |g(z)|^2 = |h(z)|\, |f(z)|$ a.e. Since f vanishes almost nowhere, $|h(z)| = 1$ a.e. Thus $h \in H^1 \cap L^\infty = H^\infty$; that is h is an inner function. Letting $g_1 = g$ and $g_2 = hg$ completes the proof. ■

If $g, h \in H^2$, there are sequences of polynomials $\{p_n\}$ and $\{q_n\}$ such that $\|g - p_n\|_2 \to 0$ and $\|h - q_n\|_2 \to 0$. It is a routine manipulation with the triangle inequality and the Cauchy–Schwarz Inequality to show that $\|gh - p_n q_n\|_1 \to 0$. In light of the preceding corollary, the next one is immediate.

25.16 Corollary. *The polynomials are dense in* H^1.

Exercises.

1. Show that Proposition 25.8 characterizes outer functions by proving the following proposition. If a function g in H^2 has the property that whenever $f \in H^2$ and $|f| \leq |g|$ a.e., there is a function ϕ in ball H^∞ such that $f = \phi g$, then g is outer.

2. Let $f \in L^2$ such that the invariant subspace $[z^n f : n \geq 0]$ for the bilateral shift is reducing. What is the Borel set Δ such that $[z^n f : n \geq 0] = L^2(\mu|\Delta)$?

3. If $\phi \in H^\infty$, show that ϕ is invertible in the algebra H^∞ if and only if ϕ is invertible in L^∞ and ϕ is an outer function.

4. Since the annihilator of the Banach space H^1 in L^∞ is the space H_0^∞, show that there is a natural isometric isomorphism between $(H^1)^*$ and L^∞/H_0^∞.

5. Let $A = \{ f \in C(\partial\mathbb{D}) : \int f z^n \, dm = 0 \; for \; n \geq 1 \}$. So $A = C(\partial\mathbb{D}) \cap H^\infty$. (a) Show that A is a Banach algebra. (b) Use the F and M Riesz Theorem to show that there is a natural isometric isomorphism between $(C(\partial\mathbb{D})/A)^*$ and H_0^1. (c) What is the dual space of A? (d) Show that the double dual of $C(\partial\mathbb{D})/A$ is L^∞/H^∞. (e) Take the image of $C(\partial\mathbb{D})/A$ in its double dual, L^∞/H^∞, and then take the inverse image of that in L^∞ under the quotient map from L^∞ onto L^∞/H^∞. Show that this equals $H^\infty + C(\partial\mathbb{D})$. Show that $H^\infty + C(\partial\mathbb{D})$ is a norm closed subalgebra of L^∞.

6. Define $\mathbb{I}$ to be the set of all inner functions whose first non-zero term in its Fourier expansion is positive. If $\phi, \psi \in \mathbb{I}$, show that $\phi H^2 = \psi H^2$ if and only if $\phi = \psi$. Observe that $\mathbb{I}$ is a semigroup under multiplication and it has an identity. If $\phi, \psi \in \mathbb{I}$, say that ϕ *divides* ψ if there is a function θ in $\mathbb{I}$ such that $\psi = \theta\phi$. In symbols this is denoted by $\phi|\psi$. For ϕ_1 and ϕ_2 in $\mathbb{I}$, say that an inner function ϕ is the *least common multiple* of ϕ_1 and ϕ_2 if $\phi_1|\phi$, $\phi_2|\phi$, and if $\psi \in \mathbb{I}$ with $\phi_1|\psi$ and $\phi_2|\psi$, then $\phi|\psi$. Symbolically, write $\phi = \operatorname{lcm}(\phi_1, \phi_2)$. Similarly, define ϕ to be the *greatest common divisor* of ϕ_1 and ϕ_2 if $\phi|\phi_1$, $\phi|\phi_2$, and if ψ is any other common divisor of ϕ_1 and ϕ_2, then $\psi|\phi$. Denote this by $\phi = \gcd(\phi_1, \phi_2)$. (At this point it is uncertain that the greatest common divisor or least common multiple of any two inner functions exists.) Order $\mathbb{I}$ by declaring that $\phi \leq \psi$ if $\phi|\psi$. Show that the map $\phi \to \phi H^2$ defines an order anti-isomorphism between $\mathbb{I}$ with the order just defined and $\operatorname{Lat} S$ with its usual order. Consequently, for any two functions ϕ_1 and ϕ_2 in $\mathbb{I}$, $\phi_1 H^2 \vee \phi_2 H^2 = \gcd(\phi_1, \phi_2) H^2$ and $\phi_1 H^2 \wedge \phi_2 H^2 = \operatorname{lcm}(\phi_1, \phi_2) H^2$. In particular, least common multiples and greatest common divisors always exist. Show that the intersection of any two non-zero invariant subspaces for the shift cannot be the space (0).

§26. The functional calculus for the unilateral shift

Now to apply the information on Hardy spaces to the shift. Recall that for ϕ in H^∞, $\phi(S)$ is the operator defined on H^2 as multiplication by ϕ. This produces a functional calculus for the shift: $\phi \to \phi(S)$. In this section this functional calculus is explored. The next theorem gives the basic properties of the functional calculus by identifying its range and continuity properties.

26.1 Theorem. *If S is the unilateral shift on H^2, then*

$$\{S\}' = \{S\}'' = P^\infty(S) = \{\phi(S) : \phi \in H^\infty\}.$$

Moreover the map $\phi \to \phi(S)$ of H^∞ onto $P^\infty(S)$ is an isometric algebraic isomorphism and a weak homeomorphism.*

Proof. If p is a polynomial, then it is easy to see that $p(S) \in \{S\}''$. So by general operator theory, $P^\infty(S) \subseteq \{S\}'' \subseteq \{S\}'$. Define ρ: $H^\infty \to \mathcal{B}(H^2)$ by $\rho(\phi) = \phi(S)$. The reader can check that ρ is an algebraic homomorphism. Also for ϕ in H^∞ and f in H^2, $\|\phi f\|^2 = \int |\phi f|^2\, dm \leq \|\phi\|_\infty^2 \|f\|^2$ and so $\|\phi(S)\| \leq \|\phi\|_\infty$.

Suppose $\{\phi_i\}$ is a net in H^∞ and $\phi_i \to 0$ weak*. If $f_n, g_n \in H^2$ such that $\sum_n \|f_n\|^2 < \infty, \sum_n \|g_n\|^2 < \infty$, then $\sum_n \langle \phi_i(S) f_n, g_n\rangle = \int \phi_i h\, dm$, where $h = \sum f_n \bar{g}_n \in L^1$. So $\phi_i(S) \to 0$ (weak*). That is, the map ρ : $(H^\infty, \text{weak}^*) \to (\mathcal{B}(H^2), \text{weak}^*)$ is continuous. Since the polynomials are weak* dense in H^∞, this shows that $\rho(H^\infty) \subseteq P^\infty(S) \subseteq \{S\}'' \subseteq \{S\}'$. It must be shown that $\{S\}' \subseteq \rho(H^\infty)$.

Let $T \in \{S\}'$ and put $\phi = T(1)$; so $\phi \in H^2$. It follows that for any polynomial p, $T(p) = \phi p$. If $f \in H^2$, let $\{p_n\}$ be a sequence of polynomials such that $p_n \to f$ in H^2. By passing to a subsequence if necessary, it may be assumed that $p_n(z) \to f(z)$ a.e. $[m]$. Thus $\phi p_n = T(p_n) \to T(f)$ in H^2 and $\phi p_n \to \phi f$ *a.e.* $[m]$. Thus $Tf = \phi f$ for all f in H^2. It is necessary to show that ϕ is bounded.

Without loss of generality it can be assumed that $\|T\| = 1$. Algebraic manipulation shows that $T^k f = \phi^k f$ for f in H^2 and $k \geq 1$. Hence $\|\phi^k f\|_2 \leq \|f\|_2$ for all $k \geq 1$. Taking $f = 1$ shows that $\int |\phi|^{2k}\, dm \leq 1$ for all $k \geq 1$. If $\Delta = \{z \in \partial\mathbb{D} : |\phi(z)| > 1\}$, then $1 \geq \int_\Delta |\phi|^{2k}\, dm$ for all $k \geq 1$. If $m(\Delta) \neq 0$, then this sequence of integrals converges monotonically to ∞, giving a contradiction. Thus $m(\Delta) = 0$, and so $\phi \in L^\infty$ and $\|\phi\|_\infty \leq 1$. This shows that $\phi H^2 = TH^2 \subseteq H^2$; so, in fact, $\phi \in H^\infty$ and $T = \phi(S)$. This completes the proof that $\{S\}' = \{S\}'' = P^\infty(S) = \{\phi(S) : \phi \in H^\infty\}$.

But this argument also shows that for any ϕ in H^∞, $\|\phi\|_\infty \leq \|\phi(S)\|$ and hence ρ is an isometry of H^∞ onto $P^\infty(S)$. Since ρ is weak* continuous, Proposition 20.3.c implies that $\rho(H^\infty) = P^\infty(S)$ is a weak* closed algebra in $\mathcal{B}(H^2)$ and ρ is a weak* homeomorphism. ■

What's a functional calculus without a spectral mapping theorem? Establishing this is equivalent to determining when the operator $\phi(S)$ is invertible.

26.2 Proposition. *If $\phi \in H^\infty$, then $\phi(S)$ is an invertible operator if and only if ϕ is invertible as an element of the Banach algebra H^∞.*

Proof. If $\phi(S)$ is invertible, then it is easy to check that $\phi(S)^{-1} \in \{S\}'$. By the preceding theorem, $\phi(S)^{-1} = \psi(S)$ for some ψ in H^∞. Thus $(\phi\psi)(S) = 1$ and so ψ is the inverse of ϕ in H^∞. The converse is clear. ∎

It follows from Proposition 24.7 that the reproducing kernel, $k_\lambda(z) = (1 - \bar{\lambda}z)^{-1} = \sum_{n=0}^\infty \bar{\lambda}^n z^n$, defined for $|\lambda| < 1$ and z on $\partial\mathbb{D}$, belongs to H^∞. Note that if $r < 1$, this series expansion for k_λ converges uniformly for $|\lambda| \le r$ and z in $\partial\mathbb{D}$. Thus for f in H^1 it is possible to define

26.3
$$\begin{aligned}\widehat{f}(\lambda) &= \int f\bar{k}_\lambda \, dm \\ &= \sum_{n=0}^\infty \left(\int f\bar{z}^n \, dm\right) \lambda^n \\ &= \sum_{n=0}^\infty \widehat{f}(n)\lambda^n,\end{aligned}$$

where $\widehat{f}(n)$ is the n-th Fourier coefficient of f. (The reader's pardon is solicited for using the hat notation, $\widehat{\ }$, both for the Fourier transform and the definition of the associated function on the disk. This is traditional.) Also the uniform convergence of the series expansion of k_λ for $|\lambda| \le r < 1$ shows that the convergence of the series $\sum_n \widehat{f}(n)\lambda^n$ is uniform for $|\lambda| \le r < 1$. The next proposition summarizes this discussion.

26.4 Proposition. *If for each f in H^1, the function $\widehat{f}\colon \mathbb{D} \to \mathbb{C}$ is defined as in (26.3), then $\widehat{f}$ is an analytic function with the following properties.*

(a) *If $\widehat{f}(\lambda) = 0$ for all λ in $\mathbb{D}$, then $f = 0$.*

(b) *If f is a function that is analytic in a neighborhood of the closed disk, then $f|\partial\mathbb{D} \in H^\infty$ and $\widehat{f}(\lambda) = f(\lambda)$ for $|\lambda| < 1$.*

Proof. If $\widehat{f}(\lambda) = 0$ on $\mathbb{D}$, then (26.3) shows that $\widehat{f}(n) = 0$ for all $n \ge 0$. Since $f \in H^1$ it follows that $\widehat{f}(n) = 0$ for all n in $\mathbb{Z}$. Thus f is an L^1 function that annihilates all powers of z, so it annihilates all the trigonometric polynomials. But the trigonometric polynomials are dense in the space of continuous functions by Lemma 24.2, so $f = 0$.

(b) Let f be analytic in a neighborhood of the closed disk and let $f(\lambda) = \sum_{n=0}^{\infty} a_n \lambda^n$ be its power series expansion with radius of convergence R. By hypothesis, $R > 1$. An easy calculation shows that $\widehat{f}(n) = a_n$. So the power series is also the Fourier series and clearly $f|\partial\mathbb{D} \in H^\infty$. ∎

Since $H^p \subseteq H^1$ for $1 \le p \le \infty$, $\widehat{f}$ is defined for f in any Hardy space.

26.5 Lemma. *If $f, g \in H^2$, then $(\widehat{fg})(\lambda) = \widehat{f}(\lambda)\widehat{g}(\lambda)$ for all λ in $\mathbb{D}$.*

Proof. Let $\{p_n\}$ and $\{q_n\}$ be two sequences of polynomials that converge in H^2 to f and g, respectively. An easy application of the Cauchy–Schwarz Inequality shows that $p_n q_n \to fg$ in H^1. Since $k_\lambda \in H^\infty$, $(\widehat{p_n q_n})(\lambda) \to (\widehat{fg})(\lambda)$. But $(\widehat{p_n q_n}) = p_n(\lambda)q_n(\lambda) = \widehat{p}_n(\lambda)\widehat{q}_n(\lambda)$ by the preceding lemma. Since it is also true that $p_n(\lambda) \to \widehat{f}(\lambda)$ and $q_n(\lambda) \to \widehat{g}(\lambda)$, this concludes the proof. ∎

26.6 Proposition. *If $\lambda \in \mathbb{D}$ and $\rho_\lambda \colon H^\infty \to \mathbb{C}$ is defined by $\rho_\lambda(\phi) = \widehat{\phi}(\lambda)$, then ρ_λ is a multiplicative linear functional on the Banach algebra H^∞. Also the mapping $\lambda \to \rho_\lambda$ is a homeomorphism of $\mathbb{D}$ onto a subset of the maximal ideal space of H^∞.*

Proof. Since $H^\infty \subseteq H^2$, the fact that ρ_λ is multiplicative follows from the preceding lemma. Linearity is clear because k_λ is an L^1 function, each ρ_λ is bounded. For $\lambda, \lambda_1, \lambda_2, \ldots$ in $\mathbb{D}$, the statement that $\rho_{\lambda_n} \to \rho_\lambda$ in $\mathcal{M}_\infty$, the maximal ideal space of H^∞, is equivalent to the statement that $\widehat{\phi}(\lambda_n) \to \widehat{\phi}(\lambda)$ for every ϕ in H^∞. Since $\widehat{\phi}$ is analytic, $\rho_{\lambda_n} \to \rho_\lambda$ in $\mathcal{M}_\infty$ whenever $\lambda_n \to \lambda$ in the disk. Since $z \in H^\infty$, the converse is also true. ∎

Note that $\|\rho_\lambda\| = 1$. In fact, if $\phi \in H^\infty$ and $\|\phi\|_\infty < 1$, $\|\phi^n\|_\infty \to 0$. Since ρ_λ is bounded, $0 = \lim_n \rho_\lambda(\phi) = \rho_\lambda(\phi)^n$. Hence $|\rho_\lambda(\phi)| < 1$. This implies that $\|\rho_\lambda\| \le 1$ and evaluating ρ_λ at the identity establishes equality. Thus the following corollary holds.

26.7 Corollary. *If $\phi \in H^\infty$, then $|\widehat{\phi}(\lambda)| \le \|\phi\|_\infty$ for all λ in the disk.*

This corollary says that the map $\phi \to \widehat{\phi}$ is a contraction from H^∞ into $H^\infty(\mathbb{D})$, the algebra of bounded analytic functions on the disk. In fact this mapping is an isometric bijection. The proof of this is not terribly difficult, though it requires more classical analysis than seems appropriate here. The interested reader can see Theorem 20.1.5 in Conway [1995] or Theorem III.6.5 in Conway [1991].

With the background information established, the sought after theorem can be proved.

26.8 Theorem. *A function ϕ in H^∞ is invertible if and only if $|\widehat{\phi}(\lambda)|$ is bounded below on $\mathbb{D}$.*

Proof. Assume that ϕ is invertible in H^∞ and let $\mathcal{M}_\infty$ be the maximal ideal space of H^∞. So the Gelfand transform of ϕ is bounded below on $\mathcal{M}_\infty$. Thus $|\widehat{\phi}|$ is bounded below on $\mathbb{D}$ by the preceding proposition.

Conversely, assume there is an $\epsilon > 0$ such that $|\widehat{\phi}(\lambda)| \geq \epsilon$ for all λ in $\mathbb{D}$. Define the analytic function $\psi \colon \mathbb{D} \to \mathbb{C}$ by $\psi(\lambda) = 1/\widehat{\phi}(\lambda)$. So ψ is bounded on $\mathbb{D}$ with $|\psi(\lambda)| \leq \epsilon^{-1}$. (If it were known that H^∞ and $H^\infty(\mathbb{D})$ could be identified, the proof would now be finished.) Let $\psi(\lambda) = \sum_n a_n\lambda^n$ be the power series expansion of ψ on the disk.

For $0 < r < 1$, let $\psi_r(\lambda) = \psi(r\lambda)$. Since ψ_r is analytic in a neighborhood of the closed disk, $\psi_r \in H^\infty \subseteq H^2$. Also the power series expansion of ψ_r is given by

$$\psi_r(\lambda) = \sum_{n=0}^{\infty} a_n r^n \lambda^n,$$

and this is also its Fourier series. That is $\widehat{\psi_r}(n) = a_n r^n$. Therefore

$$\sum_{n=0}^{\infty} |a_n|^2 r^{2n} = \|\psi_r\|_2^2 \leq \|\psi_r\|_\infty^2 \leq \|\psi\|_\infty^2 \leq \frac{1}{\epsilon^2}.$$

Letting $r \to 1-$ shows that $\sum_n |a_n|^2 < \infty$. Thus $f(z) = \sum_{n=0}^{\infty} a_n z^n \in H^2$ and $\widehat{f}(\lambda) = \psi(\lambda)$ for $|\lambda| < 1$.

By Lemma 26.5 $\widehat{\phi f}(\lambda) = \widehat{\phi}(\lambda)\widehat{f}(\lambda) = \widehat{\phi}(\lambda)\psi(\lambda) = 1$. But then Proposition 26.4.b implies $\phi f = 1$ in H^1.

Define $f_r(z) = f(rz)$. So $f_r \in H^2$ and

$$\lim_{r\to 1-} \int |f - f_r|^2\, dm = \lim_{r\to 1-} \sum_{n=0}^{\infty} |a_n|^2(1 - r^{2n}) = 0.$$

Thus there is a sequence $\{r_j\}$ converging upward to 1 such that $f(r_j z) = f_{r_j}(z) \to f(z)$ a.e. on the circle. But $|f_{r_j}(z)| = |\psi(r_j z)| \leq \epsilon^{-1}$, so $|f(z)| \leq \epsilon^{-1}$ a.e. on $\partial\mathbb{D}$. Thus $f \in H^2 \cap L^\infty = H^\infty$. Since $\phi f = 1$ in H^1, ϕ is invertible in H^∞. ∎

Now for the spectral mapping theorem for this functional calculus for the shift.

26.9 Theorem. *If $\phi \in H^\infty$, then*

$$\sigma(\phi(S)) = \mathrm{cl}\,[\widehat{\phi}(\mathbb{D})].$$

Proof. Note that for any scalar α, $\phi(S) - \alpha = (\phi - \alpha)(S)$. By Proposition 26.2, $\phi(S) - \alpha$ is an invertible operator if and only if $\phi - \alpha$ is invertible in H^∞. Thus $\sigma(\phi(S))$ is precisely the spectrum of ϕ in H^∞. By the preceding theorem, this equals $\text{cl}\,[\widehat{\phi}(\mathbb{D})]$. ∎

This section closes with an application of this correspondence between functions in the Hardy spaces and analytic functions on the disk.

26.10 Proposition. *The polynomials are weak* sequentially dense in H^∞.*

Proof. If $\phi \in H^\infty$ and $r < 1$, let $\phi_r(z) = \widehat{\phi}(rz)$ for all z in the circle. By Corollary 26.7, $\|\phi_r\|_\infty \le \|\phi\|_\infty$, so $\phi_r \in L^\infty$. Computing the Fourier coefficients gives that $\widehat{\phi_r}(n) = r^n\widehat{\phi}(n)$. Hence $\phi_r \in H^\infty$. Also it is easy to see that the Fourier series of ϕ_r converges absolutely and uniformly on $\partial\mathbb{D}$. Since this series converges in H^2 norm to ϕ_r, $\phi_r(z) = \sum_n r^n\widehat{\phi}(n)z^n$, with the convergence being uniform and absolute.

It is left to the reader to check that $\int \phi_r z^n\, dm(z) \to \int \phi z^n\, dm$ for all n in $\mathbb{Z}$. Since the trigonometric polynomials are dense in L^1 and $\|\phi_r\|_\infty \le \|\phi\|_\infty$, $\phi_r \to \phi$ weak* in H^∞. Taking a sequence of r converging to 1 and approximating each ϕ_r by a truncation of its Fourier series produces a sequence of polynomials that converges to ϕ weak* in H^∞. ∎

Exercises.

1. Show that for $|\lambda| < 1$, the homomorphism $\rho_\lambda \colon H^\infty \to \mathbb{C}$ defined in Proposition 26.6 is weak* continuous. Conversely, show that if $\rho \colon H^\infty \to \mathbb{C}$ is any weak* continuous homomorphism, there is a λ in the disk with $\rho = \rho_\lambda$.

2. Show that I is a weak* closed ideal of H^∞ if and only if there is an inner function ϕ such that $I = \phi H^\infty$.

§27. Weighted shifts

If $A \in \mathcal{B}(\mathcal{H})$ and $A = S|A|$ is the polar decomposition of A, then S is an isometry if and only if $\ker A = (0)$. It may be that S is the unilateral shift. If this is the case and $|A|$ is a diagonalizable positive operator, then A is called a weighted shift.

27.1 Definition. An operator A is a *unilateral weighted shift* if there is an orthonormal basis $\{e_n : n \ge 0\}$ and a sequence of scalars $\{\alpha_n\}$ such that $Ae_n = \alpha_n\, e_{n+1}$ for all $n \ge 0$. Say that A is a weighted shift with weight sequence $\{\alpha_n\}$.

It is not hard to see that if A is as in the definition, then $A = SD$, where S is the unilateral shift and D is the diagonal operator with $De_n = \alpha_n e_n$

for all n. Thus $|A| = |D|$ and $|A|e_n = |\alpha_n|\, e_n$ for all n. So $\{e_n\}$ is precisely the basis of eigenvectors for $|A|$. Also note that a necessary and sufficient condition for A to be bounded is that the weight sequence be bounded.

Bilateral weighted shifts are defined analogously. An operator A is a *bilateral weighted shift* with weight sequence $\{\alpha_n : n \in \mathbb{Z}\}$ if there is an orthonormal basis $\{e_n : n \in \mathbb{Z}\}$ such that $Ae_n = \alpha_n\, e_{n+1}$ for all n.

Throughout this section and beyond, results occur in pairs; one for unilateral shifts and one for bilateral shifts. If there is no dramatic difference in the proofs, only one of the proofs will be given and the other will be left without comment to the reader.

The fundamental reference for weighted shifts is Shields [1974]. Nikolskii [1985] is a complete study of the subject, illustrating deep connections between these operators and classical analysis. In addition, weighted shifts constitute a testing ground for operator theory. Whenever an operator theorist thinks of a new concept, the first reaction, after seeing what it says in the finite dimensional case, is to examine this for weighted shifts. Are there examples of weighted shifts having this property? (Often, it seems, the next reaction is to have a graduate student characterize all the weighted shifts with the property.) The result of such an inquiry will often indicate whether the topic has sufficient merit to warrant further investigation.

In this section only the surface of the subject will be skimmed. The first result establishes that it will always suffice to assume that the weight sequence consists of non-negative numbers.

27.2 Proposition. *If A is a unilateral (bilateral) weighted shift with weight sequence $\{\alpha_n\}$, then A is unitarily equivalent to the unilateral (bilateral) weighted shift with weight sequence $\{|\alpha_n|\}$.*

Proof. If $\{e_n\}$ is the basis for $\mathcal{H}$ that is shifted by A, define $U : \mathcal{H} \to \mathcal{H}$ by $Ue_n = \lambda_n e_n$, where $|\lambda_n| = 1$ for all n. It follows that $UAU^* e_n = \bar{\lambda}_n \lambda_{n+1} \alpha_n\, e_{n+1}$. If $\lambda_0 = 1$, then UAU^* is a weighted shift with weight sequence $\{\lambda_1\alpha_0, \bar{\lambda}_1\lambda_2\alpha_1, \bar{\lambda}_2\lambda_3\alpha_2, \dots\}$. It is easy to verify that $\{\lambda_n\}$ can be chosen so that this sequence becomes $\{|\alpha_0|, |\alpha_1|, |\alpha_2|, \dots\}$. ∎

What happens if one of the weights is 0? If $\alpha_n = 0$, then it is easy to see that $\mathcal{L} = [e_k : 0 \le k \le n]$ and $\mathcal{M} = [e_k : k \ge n+1]$ are both invariant subspaces for the weighted shift A. Since $\mathcal{L} \perp \mathcal{M}$, these spaces reduce A. Thus $A = B \oplus C$, where $B = A|\mathcal{L}$ and $C = A|\mathcal{M}$. Moreover C is a weighted shift with weight sequence $\{\alpha_{n+1}, \alpha_{n+2}, \dots\}$ and B is a nilpotent operator on a finite dimensional space. In fact, if α_n is the first non-zero weight, B is a cyclic nilpotent operator. This leads to two possibilities.

If the weight sequence contains an infinite number of 0's, A is unitarily equivalent to the direct sum of cyclic nilpotent operators on finite dimensional spaces. Such operators as this are not readily understood; indeed, they can be quite complicated. For example, their spectrum can be any size disk of radius at most $\|A\|$, though it must be a disk since such an operator is unitarily equivalent to $e^{i\theta}$ times itself for any choice of θ. See Halmos [1982], Problem 104.

If there are a finite number of 0's in the weight sequence, then A is unitarily equivalent to the direct sum of a finite number of cyclic nilpotent operators on finite dimensional spaces and a weighted shift with non-zero weights. The direct sum of a finite number of cyclic nilpotent operators on finite dimensional spaces is just the arbitrary nilpotent on a finite dimensional space. This is well understood. This leaves the weighted shift with no zero weights.

In light of these comments and the preceding proposition, the following binding agreement is appropriate.

Henceforth, all weight sequences will be strictly positive.

Also, to avoid repetitive statements about notation, in this section A will always denote a weighted shift with weight sequence $\{\alpha_n\}$, $\alpha_n > 0$ for $n \geq 0$, and A will shift the basis $\{e_n\}$. Similar notational conventions apply to bilateral shifts. The proof of the next proposition is an exercise.

27.3 Proposition. (a) *If A is a unilateral shift, then $A^*e_0 = 0$ and $A^*e_n = \alpha_{n-1}e_{n-1}$ for $n \geq 1$. Consequently, $(A^*A - AA^*)e_0 = \alpha_0^2 e_0$ and $(A^*A - AA^*)e_n = (\alpha_n^2 - \alpha_{n-1}^2)e_n$ for $n \geq 1$.*

(b) *If A is a bilateral shift, then $A^*e_n = \alpha_{n-1}e_{n-1}$ for all n. Consequently, $(A^*A - AA^*)e_n = (\alpha_n^2 - \alpha_{n-1}^2)e_n$ for all n.*

27.4 Corollary. *If A is a unilateral (bilateral) weighted shift, S (respectively, U) is the corresponding unilateral (bilateral) unweighted shift, and $D =$ diag $\{\alpha_n\}$, then $A = SD$ $(A = UD)$ is the polar decomposition of A.*

27.5 Corollary. *A weighted shift A is compact if and only if $\alpha_n \to 0$. It belongs to $\mathcal{B}_p$ if and only if $\{\alpha_n\} \in \ell^p$.*

27.6 Proposition. *If A is a weighted shift and $\omega \in \partial\mathbb{D}$, then $A \cong \omega A$.*

Proof. If $Ue_n = \omega^n e_n$ for all n, then a computation shows that $UAU^* = \omega A$. ∎

As a consequence of the preceding proposition, $\sigma(A) = \sigma(\omega A) = \omega\sigma(A)$; that is, the spectrum of a weighted shift must be circularly symmetric.

The spectral properties of certain weighted shifts will be presented shortly. The determination of the spectrum of an arbitrary unilateral or bilateral weighted shift is quite entailed. For a unilateral shift, the spectrum is the disk centered at the origin whose radius is the spectral radius of the operator. Even stating the result that characterizes such a thing as the approximate point spectrum of an arbitrary shift requires considerable foundation. This is all done in Shields [1974]. Here the specialization is made to certain weighted shifts that will be encountered later.

Use of the essential spectrum of an operator and the Fredholm theory will be made here in a small way. The reader who so chooses can jump ahead to §37 or read Chapter XI in [ACFA].

27.7 Proposition. (a) *If A is a unilateral weighted shift, then $\sigma_p(A) = \emptyset$.*

(b) *If A is a unilateral weighted shift such that $\alpha_n \to \alpha_+$, then $\sigma_e(A) = \{\lambda : |\lambda| = \alpha_+\}$, $\sigma(A) = \{\lambda : |\lambda| \le \alpha_+\}$, and for $|\lambda| < \alpha_+$, $A - \lambda$ is Fredholm and* $\operatorname{ind}(\lambda - A) = -1$.

(c) *If A is a bilateral weighted shift and $\alpha_n \to \alpha_\pm$ as $n \to \pm\infty$, then $\sigma_e(A) = \{\lambda : |\lambda| = \alpha_+ \text{ or } \alpha_-\}$. Also $\sigma_p(A) \subseteq \{\lambda : \alpha_+ \le |\lambda| \le \alpha_-\}$. In particular, if $\alpha_- < \alpha_+$, $\sigma_p(A) = \emptyset$, $\sigma(A) = \{\lambda : \alpha_- \le |\lambda| \le \alpha_+\}$, and for $\alpha_- < |\lambda| < \alpha_+$, $A - \lambda$ is Fredholm with* $\operatorname{ind}(A - \lambda) = -1$.

Proof. (a) The fact that $\sigma_p(A) = \emptyset$ should be left as an exercise, but the computation needed to prove it is also needed for the bilateral case. If $f = \sum_n \widehat{f}(n)e_n \in \mathcal{H}$, then $(A - \lambda)f = \sum_n \alpha_n \widehat{f}(n)e_{n+1} - \sum_n \lambda\widehat{f}(n)e_n$. So if $(A - \lambda)f = 0$, $\lambda\widehat{f}(0) = 0$ and $\lambda\widehat{f}(n) = \alpha_{n-1}\widehat{f}(n-1)$ for $n \ge 1$. It is easy to see that these equations are valid only if $f = 0$.

(b) Now assume that $\lim \alpha_n = \alpha_+$. If $\alpha_+ = 0$, then A is compact since $|A|$ is. If $\alpha_+ > 0$ and S is the unilateral shift relative to the basis $\{e_n\}$, then it follows that $A - \alpha_+ S$ is a unilateral weighted shift whose weight sequence converges to 0. Hence $A - \alpha_+ S$ is compact and so $\sigma_e(A) = \sigma_e(\alpha_+ S) = \{\lambda : |\lambda| = \alpha_+\}$. Also for $|\lambda| < \alpha_+$, $A - \lambda \in \mathcal{F}$ and $\operatorname{ind}(A - \lambda) = \operatorname{ind}(\alpha_+ S - \lambda) = -1$. In particular, this implies that $\{\lambda : |\lambda| \le \alpha_+\} \subseteq \sigma(A)$.

It remains to show that if $|\lambda| > \alpha_+$, then $A - \lambda$ is invertible. But such a λ is not in $\sigma_e(A)$ so that $A - \lambda \in \mathcal{F}$ and so $\operatorname{ind}(A - \lambda) = \operatorname{ind}(\alpha_+ S - \lambda) = 0$. Since $\ker(A - \lambda) = (0)$, it follows that $\operatorname{ran}(A - \lambda)^\perp = (0)$ and so $A - \lambda$ is invertible.

(c) Let $\mathcal{H}_+ = [e_n : n \ge 0]$ and $\mathcal{H}_- = [e_{-n} : n \ge 1]$; so $\mathcal{H} = \mathcal{H}_- \oplus \mathcal{H}_+$. Define unilateral weighted shifts $A_\pm$ on $\mathcal{H}_\pm$ by letting $A_+ = A|\mathcal{H}_+$ and $A_- e_{-n} = \alpha_{-n} e_{-n-1}$ for $n \ge 1$. Also define the rank one operator F on $\mathcal{H}$ by letting $Fe_{-1} = \alpha_{-1}e_0$ and $Fe_n = 0$ for $n \neq -1$. It is left to the reader to check that $A = (A_-^* \oplus A_+) + F$. Since F has finite rank, part (a) implies that $\sigma_e(A) = \sigma_e(A_-^* \oplus A_+) = \{\lambda : |\lambda| = \alpha_+ \text{ or } \alpha_-\}$.

To prove the statement about $\sigma_p(A)$, note that if $(A-\lambda)f = 0$, then, as in part (a), $\lambda\widehat{f}(n) = \alpha_{n-1}\widehat{f}(n-1)$ for all n in $\mathbb{Z}$. Without loss of generality it can be assumed that $\widehat{f}(0) = 1$. Hence it is possible to solve these equations to obtain $\widehat{f}(n) = (\alpha_0 \cdots \alpha_{n-1})\lambda^{-n}$ and $\widehat{f}(-n) = \lambda^n(\alpha_{-1}\cdots\alpha_{-n})^{-1}$ for $n \geq 1$. Thus

$$\|f\|^2 = \sum_{n=1}^{\infty} |\lambda|^{2n}(\alpha_{-1}\cdots\alpha_{-n})^{-2} + 1 + \sum_{n=1}^{\infty} |\lambda|^{-2n}(\alpha_0\cdots\alpha_{n-1})^2.$$

So if $|\lambda| < \alpha_+$, let $|\lambda| < r < \alpha_+$ and choose m such that $r < \alpha_n$ for $n \geq m$. Hence

$$\begin{aligned}
&|\lambda|^{-2n}(\alpha_0\cdots\alpha_{n-1})^2 \\
&\quad = \left[|\lambda|^{-2m}(\alpha_0\cdots\alpha_{m-1})^2\right]\left[|\lambda|^{-2n+2m}(\alpha_m\cdots\alpha_{n-1})^2\right] \\
&\quad > \left[|\lambda|^{-2m}(\alpha_0\cdots\alpha_{m-1})^2\right] \\
&\quad > 0,
\end{aligned}$$

contradicting the fact that the above series converges. Thus it must be that $|\lambda| \geq \alpha_+$. In a similar fashion, $|\lambda| \leq \alpha_-$.

Now assume that $\alpha_- < \alpha_+$. Clearly $\sigma_p(A) = \emptyset$. Also let $A_\pm$ and F be defined as above. If $\alpha_- < |\lambda| < \alpha_+$, then $A - \lambda$ is Fredholm and $\operatorname{ind}(A-\lambda) = \operatorname{ind}(A_-^* \oplus A_+ - \lambda) = \operatorname{ind}(A_-^* - \lambda) + \operatorname{ind}(A_+ - \lambda)$. But from part (a), $A_-^* - \lambda$ is invertible and $\operatorname{ind}(A_+ - \lambda) = -1$. Thus $\operatorname{ind}(A-\lambda) = -1$.

If $|\lambda| < \alpha_-$, then part (a) implies that $\operatorname{ind}(A-\lambda) = \operatorname{ind}(A_-^* - \lambda) + \operatorname{ind}(A_+ - \lambda) = 1 + (-1) = 0$. But $\ker(A-\lambda) = (0)$, so $A - \lambda$ is invertible. Similarly, $A-\lambda$ is invertible for $|\lambda| > \alpha_+$ and so $\sigma(A) = \{\lambda : \alpha_- \leq |\lambda| \leq \alpha_+\}$. ∎

Here is a collection of weighted shifts. A weighted shift A is called a *Donoghue shift* if it is a unilateral shift whose weight sequence is monotone decreasing and belongs to ℓ^2. So a Donoghue shift is compact (27.5) and has spectrum equal to $\{0\}$ (27.7.b). These shifts are so named because Donoghue [1957] was the first to note their invariant subspace structure.

For any unilateral weighted shift A, certain invariant subspaces are clear. Namely $[e_k : k \geq n]$ clearly belongs to $\operatorname{Lat} A$ for every $n \geq 0$. Are there others? It was shown that for the unilateral shift S there are many others. Donoghue [1957] showed that for the weight sequence $\{2^{-n}\}$ there are no others. Later Nikolskii [1965] established the following result.

27.8 Theorem. *If A is a Donoghue shift and $\mathcal{M}_n = [e_k : k \geq n]$, then*

$$\operatorname{Lat} A = \{\, \mathcal{M}_n : n \geq 0 \,\} \cup (0).$$

Proof. It turns out to be somewhat easier to find $\operatorname{Lat} A^*$. So put $B = A^*$ and let $\mathcal{M} \in \operatorname{Lat} B$. It is required to show that there is an integer $n \geq 0$ such that $\mathcal{M} = \mathcal{N}_n \equiv [\, e_k : 0 \leq k \leq n] = \mathcal{M}_{n+1}^{\perp}$. For convenience let $\mathcal{N}_\infty = \mathcal{H}$. Note that if E is any subset of the extended non-negative integers, $[\{ \mathcal{N}_k : k \in E \}] = \mathcal{N}_n$, where n is the largest member of E. Therefore it suffices to show that if h is a non-zero vector in $\mathcal{H}$, then there is an $n \geq 0$ such that $\left[B^k h : k \geq 0 \right] = \mathcal{N}_n$.

Start with a computation. It is left to the reader to show that for $n \geq 1$,

$$B^n e_k = \begin{cases} 0 & \text{if } 0 \leq k \leq n-1 \\ (\alpha_{k-1} \cdots \alpha_{k-n}) e_{k-n} & \text{if } k \geq n. \end{cases}$$

Fix the vector $h = \sum_{n=0}^{\infty} t_n e_n$ and put $\mathcal{M} = \left[B^k h : k \geq 0 \right]$. Consider two cases. In the first case only a finite number of the coefficients t_k are not 0. So assume $h = \sum_{k=0}^{n} t_k e_k$, where $t_n \neq 0$. Compute.

$$\begin{aligned} Bh &= t_1 \alpha_0\, e_0 + \cdots + t_n \alpha_{n-1}\, e_{n-1} \\ B^2 h &= t_2 \alpha_1 \alpha_0\, e_0 + \cdots + t_n \alpha_{n-1} \alpha_{n-2}\, e_{n-2} \\ &\ \ \vdots \\ B^{n-1} h &= t_{n-1} \alpha_{n-2} \cdots \alpha_0\, e_0 + t_n \alpha_{n-1} \cdots \alpha_1\, e_1 \\ B^n h &= t_n \alpha_{n-1} \cdots \alpha_0\, e_0. \end{aligned}$$

Thus $e_0 \in \mathcal{M}$. Using this and the formula for $B^{n-1}h$ it follows that $e_1 \in \mathcal{M}$. Continue and get that $\mathcal{N}_n \subseteq \mathcal{M}$. Since it is clear that $\mathcal{M} \subseteq \mathcal{N}_n$, this completes the proof of the first case.

For the second case assume that $t_n \neq 0$ for infinitely many n. Again compute to see that

$$B^n h = \sum_{k=n}^{\infty} t_k (\alpha_{k-1} \cdots \alpha_{k-n}) e_{k-n}.$$

Suppose $t_n \neq 0$. So

$$\frac{1}{t_n \alpha_{n-1} \cdots \alpha_0} B^n h = e_0 + \sum_{k=n+1}^{\infty} \frac{t_k}{t_n} \left(\frac{\alpha_{k-1} \cdots \alpha_{k-n}}{\alpha_{n-1} \cdots \alpha_0} \right) e_{k-n}.$$

Now use the fact that $\{\alpha_k\}$ is a decreasing sequence. If $k \geq n+1$, $\alpha_{k-1} \leq \alpha_{n-2}$. Similarly $\alpha_{k-j} \leq \alpha_{n-(j-1)}$ for $1 \leq j \leq n$. Thus

$$\frac{\alpha_{k-1}\cdots\alpha_{k-n}}{\alpha_{n-1}\cdots\alpha_0} \leq \frac{\alpha_{k-1}}{\alpha_0}$$

for all $k \geq n+1$. Therefore

$$\left\| \sum_{k=n+1}^{\infty} \frac{t_k}{t_n}\left(\frac{\alpha_{k-1}\cdots\alpha_{k-n}}{\alpha_{n-1}\cdots\alpha_0}\right)e_{k-n}\right\|^2 \leq \sum_{k=n+1}^{\infty}\left|\frac{t_k}{t_n}\right|^2\left(\frac{\alpha_{k-1}}{\alpha_0}\right)^2.$$

Let $\epsilon > 0$. Because the weight sequence is square summable, it is possible to choose N such that

$$\sum_{k=N+1}^{\infty}\left(\frac{\alpha_{k-1}}{\alpha_0}\right)^2 < \epsilon^2.$$

But since $\{t_k\}$ converges to 0, there is an $n \geq N$ such that $|t_n| \geq \max\{\,|t_k| : k \geq n\,\}$. Therefore, with this choice of n,

$$\left\|\frac{1}{t_n\alpha_{n-1}\cdots\alpha_0}B^n h - e_0\right\|^2 \leq \sum_{k=n+1}^{\infty}\left|\frac{t_k}{t_n}\right|^2\left(\frac{\alpha_{k-1}}{\alpha_0}\right)^2 < \epsilon^2.$$

This shows that $e_0 \in \mathcal{M}$.

Thus for every $n \geq 1$,

$$B^n h - t_n(\alpha_{n-1}\cdots\alpha_0)e_0 = \sum_{k=n+1}^{\infty} t_k(\alpha_{k-1}\cdots\alpha_{k-n})e_{k-n} \in \mathcal{M}.$$

Applying the same reasoning as above shows that $e_1 \in \mathcal{M}$. Continue. This shows that $e_n \in \mathcal{M}$ for all $n \geq 0$ and so $\mathcal{M} = \mathcal{H}$. ∎

Exercises.

1. Give necessary and sufficient conditions that two weighted shifts be unitarily equivalent.

2. Give necessary and sufficient conditions that two weighted shifts be similar.

3. Show that a weighted shift with non-zero weights is irreducible.

4. Show that if A is a unilateral weighted shift with $\|A\| \leq 1$, then $A^{*n} \to 0$ (SOT). Is the condition that A be a contraction needed?

5. Which unilateral and bilateral weighted shifts are normal?

6. If T is an arbitrary operator in $\mathcal{B}(\mathcal{H})$, define $m(T) \equiv \inf\{\|Tf\| : \|f\| = 1\}$. (a) Show that if T is invertible, then $m(T) = \|T^{-1}\|^{-1}$. (b) If $T \in \mathcal{B}(\mathcal{H})$ and $\lambda \in \sigma_{ap}(T)$, then $|\lambda| \geq m(T)$. (c) If A is a weighted shift with increasing weights, show that $\|A\| = \lim_n m(A^n)^{1/n}$. (Hint: Find a formula for $m(A^n)$ in terms of the weights.)

§28. The Volterra operator

Renew acquaintance with the Volterra operator, defined in Example II.1.7 of [ACFA]. For convenience the definition and some of its basic properties will be recalled here.

28.1 Definition. The operator $V\colon L^2[0,1] \to L^2[0,1]$ defined by

$$Vf(x) = \int_0^x f(t)\,dt$$

for all f in $L^2[0,1]$ is called the *Volterra operator* .

The next result appeared in Example VII.6.14 of [ACFA].

28.2 Proposition. *The Volterra operator is compact,* $\sigma_p(V) = \emptyset$*, and* $\sigma(V) = \{0\}$.

It can be shown that $\|V\| = 2/\pi$. See Halmos [1982], Solution 188. One of the more interesting facts about the Volterra operator is the determination of its invariant subspace lattice.

28.3 Theorem. $\mathcal{M} \in \operatorname{Lat} V$ *if and only if there is an* α *in* $[0,1]$ *such that*

$$\mathcal{M} = L^2[\alpha, 1].$$

Note that Theorem 28.3 says that $\operatorname{Lat} V$, like the lattice of invariant subspaces of a Donoghue shift, is totally ordered. Such an operator is said to be *unicellular*. (Personally, I don't like the term, but we seem to be stuck with it.)

The proof of this result was independently discovered by Brodskii [1957], Donoghue [1957], and Kalisch [1957]. Earlier it was proved for the case of the real-valued space $L^2[0,1]$ in Dixmier [1949]. A complete proof of Theorem 28.3 will not be given. Instead the invariant subspace lattice will be determined for an operator that Sarason [1965b] shows is intimately related to the Volterra operator. Indeed, the argument could be made that this section is misnamed as the discussion of the properties of the next operator is far more complete than the discussion of the properties of the Volterra operator.

For $0 < \alpha \le 1$, let ϕ_α be the inner function

$$\phi_\alpha(z) = \exp\left[\alpha \frac{z+1}{z-1}\right].$$

For convenience define $\phi_0 \equiv 1$. Define the operator T on $\mathcal{H} \equiv H^2 \ominus \phi_1 H^2$ by

28.4 $$Tf = P(zf),$$

where P is the projection of H^2 onto $\mathcal{H}$. So T is the compression of the shift to $(\phi_1 H^2)^\perp$. If $u \in H^\infty$, define the operator $u(T)\colon \mathcal{H} \to \mathcal{H}$ by

$$u(T)f \equiv P(uf)$$

for all f in $\mathcal{H}$. It is easy to see that $u(T)$ is a bounded operator with $\|u(T)\| \leq \|u\|_\infty$.

28.5 Proposition. *The operator T defined in (28.4) is a contraction, $\sigma(T) = \{1\}$, and the map $u \to u(T)$ is a contractive, multiplicative functional calculus for T.*

Proof. That T is a contraction is clear. It was observed that $\|u(T)\| \leq \|u\|_\infty$ and it is easy to show it is linear. It will be shown that this functional calculus is multiplicative. (Before starting, the reader is asked to observe that this part of the proof is valid if ϕ_1 is any inner function.) If $u, v \in H^\infty$ and $f \in \mathcal{H}$, then $uvf = u\,P(vf) + u\,P^\perp(vf)$. Now $P^\perp(vf) \in \phi_1 H^2$ and so $u\,P^\perp(vf) \in \phi_1 H^2$. Therefore, $(uv)(T)f = P(uvf) = P[u\,P(vf)] = P[u\,v(T)f] = u(T)v(T)f$. This shows multiplicativity.

Because T is a contraction, $\sigma(T) \subseteq \operatorname{cl}\mathbb{D}$. Let $\zeta \in \operatorname{cl}\mathbb{D}$, $\zeta \neq 1$. Thus ϕ_1 is analytic in a neighborhood of ζ and does not vanish there. If $u(z) = (\phi_1(z) - \phi_1(\zeta))/(z - \zeta)$, then $u \in H^\infty$; thus $u(T)$ is a bounded operator. From the properties of the functional calculus, $u(T)(T - \zeta) = \phi_1(T) - \omega$, where $\omega = \phi_1(\zeta) \neq 0$. Since $\phi_1(T) = 0$, $-\omega^{-1}u(T)$ is a bounded inverse for $T - \zeta$. Thus $\sigma(T) \subseteq \{1\}$. Since the spectrum cannot be empty, equality holds. ■

To prove the desired result, a standard result from classical analysis is needed that may be familiar to many readers. This could be avoided if the Factorization Theorem for bounded analytic functions on the disk was known (Conway [1995], 20.3.7). If this were available, it would be possible to determine all the inner factors of the function ϕ_1 and this would give all the invariant subspaces. On the other hand, the proof of the Factorization Theorem usually involves an application of the next result, so there is less loss than might have been suspected.

28.6 Herglotz's Theorem. *If u is a non-negative harmonic function on $\mathbb{D}$, there is unique positive measure μ on $\partial\,\mathbb{D}$ such that*

$$u(z) = \int_{\partial\mathbb{D}} P_z(w)\, d\mu(w),$$

where $P_z\colon \partial\,\mathbb{D} \to [0, \infty)$ is the Poison kernel.

A proof can be found in several places, including Conway [1995], Theorem 19.1.11. Also recall that $P_z(w) = \text{Re}\,[(1+z\bar{w})/(1-z\bar{w})]$, for z in $\mathbb{D}$ and w in $\partial\mathbb{D}$. In other words, if, for w in $\partial\mathbb{D}$, the Möbius transformation

$$M_w(z) = \frac{z+w}{z-w},$$

is defined, then

$$P_z(w) = -\text{Re}\,M_1(z\bar{w}) = -\text{Re}\,M_w(z).$$

Let $\mathbb{H}$ denote the left half plane, $\{\,z : \text{Re}\,z < 0\,\}$.

28.7 Proposition. *If $\mathcal{C}$ is the set of analytic functions $f\colon \mathbb{D} \to \mathbb{H}$ such that $f(0) = -1$, then $\mathcal{C}$ is a convex set and the extreme points of $\mathcal{C}$ are $\{\,M_w : w \in \partial\mathbb{D}\,\}$.*

Proof. If $f \in \mathcal{C}$, then $-\text{Re}\,f$ is a non-negative harmonic function on the disc; thus, Herglotz's Theorem applies. Let μ be a positive measure on the circle such that $\text{Re}\,f(z) = \int \text{Re}\,M_w(z)\,d\mu(w)$ for all z in $\mathbb{D}$. Since $f(0) = -1$, μ is a probability measure on $\partial\mathbb{D}$. Thus each function f in $\mathcal{C}$ has the representation

$$f(z) = f_\mu(z) \equiv \int M_w(z)\,d\mu(w).$$

If $\mathcal{P}$ denotes the collection of probability measures on $\partial\mathbb{D}$, it has just been shown that the map $\mu \to f_\mu$ is an affine bijection of $\mathcal{P}$ onto $\mathcal{C}$. Thus the extreme points of $\mathcal{C}$ are precisely the images of the extreme points of $\mathcal{P}$ under this map. Since the extreme points of $\mathcal{P}$ are the unit point masses ([ACFA], V.8.4), this proves the proposition. ■

28.8 Theorem. *If T is the operator defined in* (28.4), *then*

$$\text{Lat}\,T = \{\,\phi_\alpha H^2 \ominus \phi_1 H^2 : 0 \le \alpha \le 1\,\}.$$

Proof. It is left to the reader to check that each space $\phi_\alpha H^2 \ominus \phi_1 H^2$ with $0 \le \alpha \le 1$ belongs to $\text{Lat}\,T$. On the other hand, if $\mathcal{M} \in \text{Lat}\,T$, then $\mathcal{M} + \phi_1 H^2$ is an invariant subspace for the shift. Therefore there is an inner function ϕ such that $\mathcal{M} + \phi_1 H^2 = \phi H^2$. Since $\phi_1 \in \phi_1 H^2 \subseteq \phi H^2$, there is a function ψ in H^2 such that $\phi_1 = \phi\psi$. Because $|\phi| = |\phi_1| = 1$ a.e., ψ is also an inner function.

Now think of ϕ_1, ϕ, ψ as analytic functions on $\mathbb{D}$. Since ϕ_1 never vanishes, neither do ϕ and ψ. Also $e^{-1} = \phi_1(0)$, so, by replacing ϕ and ψ by suitable unimodular multiples if necessary, it can be assumed that both $\phi(0)$ and

$\psi(0)$ are positive real numbers. Thus there are analytic functions f and g on $\mathbb{D}$ such that $\phi = \exp f, \psi = \exp g$, $\alpha = -f(0) > 0$, and $\beta = -g(0) > 0$. Hence for $|z| < 1$, $\phi_1(z) = \exp[M_1(z)] = \exp[f(z) + g(z)]$. By a standard connectedness argument this implies that for all z in $\mathbb{D}$

$$\begin{aligned} M_1(z) &= f(z) + g(z) \\ &= \alpha\frac{f(z)}{\alpha} + \beta\frac{g(z)}{\beta}. \end{aligned}$$

So $-1 = M_1(0) = -\alpha - \beta$, and therefore $\alpha, \beta \in [0,1]$ with $\alpha + \beta = 1$. Also, both f/α and g/β belong to the convex set $\mathcal{C}$ defined in Proposition 28.7. By that proposition, $f = \alpha M_1$ and $g = \beta M_1$, so that $\phi = \phi_\alpha$ and $\psi = \phi_\beta$.

In particular, $\mathcal{M} + \phi_1 H^2 = \phi_\alpha H^2$, so that $\mathcal{M} = [\mathcal{M} + \phi_1 H^2] \ominus \phi_1 H^2 = \phi_\alpha H^2 \ominus \phi_1 H^2$. ■

Sarason [1965b] shows that the operator T defined in (28.4) is unitarily equivalent to $W = (1-V)(1+V)^{-1}$, with V the Volterra operator. The proof depends on a version of the Paley-Wiener Theorem for the Hardy space of a half-plane, an exposition of which ventures too far from the subject matter of this book. Under the Hilbert space isomorphism that implements this unitary equivalence, the invariant subspaces of T are mapped onto the spaces $L^2[\alpha, 1]$. Hence $\operatorname{Lat} W = \{\, L^2[\alpha,1] : 0 \le \alpha \le 1 \,\}$. It remains to show that $\operatorname{Lat} W = \operatorname{Lat} V$. Since, on a neighborhood of $\{0\}$, $h(z) = (1-z)(1+z)^{-1}$ is the uniform limit of polynomials, $\operatorname{Lat} W \supseteq \operatorname{Lat} V$. Similarly, h^{-1} is the uniform limit of a sequence of polynomials in a neighborhood of $\{1\}$, so $\operatorname{Lat} V = \operatorname{Lat} h^{-1}(W) \supseteq \operatorname{Lat} W$.

Given Theorem 28.3, it is possible to prove the Titchmarsh Convolution Theorem. See Radjavi and Rosenthal [1973], page 71.

The operator T defined in (28.4) has its generalizations. The first such is obtained by replacing the inner function ϕ_1 in the definition by an arbitrary inner function. This is done in Exercise 2 below where some results are presented. For any inner function ϕ, Ahern and Clark [1970] construct an explicit isomorphism from the direct sum of three L^2 spaces onto the space $H^2 \ominus \phi H^2$.

Another generalization of the operator in (28.4) is obtained by a use of the Sz.-Nagy–Foias functional calculus for completely non-unitary contractions (see §35.). For such contractions T it is possible to define $\phi(T)$ for any bounded analytic function ϕ. An operator T is said to be of class C_0 if it is a non-unitary contraction and there is a function ϕ in H^∞ such that $\phi(T) = 0$. In this case the set of such functions in H^∞ that annihilate T is a weak* closed ideal, and so there is an inner function ϕ that generates this ideal. The operators in Exercise 2 are examples of operators of class C_0. Bercovici [1988] is an exposition of this subject.

Exercises.

1. Show that the Volterra operator is Hilbert–Schmidt.

2. Let ϕ be any inner function and put $\mathcal{H}_\phi = H^2 \ominus \phi H^2$. As in (28.4), define S_ϕ in $\mathcal{B}(\mathcal{H}_\phi)$ by $S_\phi f = P_\phi(zf)$ for all f in $\mathcal{H}_\phi$, where P_ϕ is the projection of H^2 onto $\mathcal{H}_\phi$. Prove the following facts. (Some of these may be difficult to do without using the structure of inner functions.) (a) Show that $\sigma(S_\phi) = \{\lambda \in \mathbb{D} : \phi(\lambda) = 0\} \cup \{\lambda \in \partial\mathbb{D} : \phi$ has no analytic continuation to a neighborhood of $\lambda\}$. (b) If $u \in H^\infty$, define $u(S_\phi)$ on $\mathcal{H}_\phi$ by $u(S_\phi)f = P_\phi(uf)$ for all f in $\mathcal{H}_\phi$. Show that the map $u \to u(S_\phi)$ is a multiplicative, linear contraction with kernel ϕH^∞. (c) (Sarason [1967]) If $H^\infty/\phi H^\infty$ has the quotient norm, show that $u + \phi H^\infty \to \phi(S_\phi)$ is an isometry onto $\{S_\phi\}'$.

3. If T is the operator defined in (28.4), what is $\operatorname{Lat} T^2$?

§29. Bergman operators

For a bounded open subset G of $\mathbb{C}$, let $L^p(G)$ denote the Lebesgue space for area measure restricted to G. Area measure will be denoted by Area or $d\mathcal{A}$. Let $L^p_a(G)$ denote the Bergman space of those functions in $L^p(G)$ that are analytic; technically, those elements in $L^p(G)$ that are equal almost everywhere to a function that is analytic on G. This was defined in [ACFA], §I.1, where it was also proved that $L^p_a(G)$ is a Banach space and $L^2_a(G)$ is a Hilbert space.

29.1 Definition. The *Bergman operator* for G is the operator S defined on $L^p_a(G)$ by $(Sf)(z) = zf(z)$.

Note that since G is bounded, S is a bounded operator. Also note that multiplication by any bounded analytic function defines a bounded operator on $L^p_a(G)$.

The treatment begins with a discussion of the spectral properties. Here only the Hilbert space case, $p = 2$, is considered. The Bergman operator is a subnormal operator (§30, below) and is an important example of this class. There will be a discussion of the Fredholm properties of the operator, but the reader who chooses may ignore this. In the long run, however, this is not a good policy.

29.2 Theorem. *If S is the Bergman operator for the bounded open set G, then S is a bounded operator and the following statements hold.*

(a) *$\sigma(S) = \operatorname{cl} G$ and $\sigma_p(S) = \emptyset$.*

(b) *For λ in G there is a k_λ in $L^2_a(G)$ such that $\langle f, k_\lambda\rangle = f(\lambda)$ for all f in $L^2_a(G)$.*

(c) *For* λ *in* G, $\operatorname{ran}(S-\lambda) = \{f \in L^2_a(G) : f(\lambda) = 0\}$, $\operatorname{ran}(S-\lambda)$ *is closed,* $S-\lambda$ *is Fredholm, and* $\operatorname{ind}(S-\lambda) = -1$.

Proof. (a) If $\lambda \notin \operatorname{cl} G$, then $(z-\lambda)^{-1}$ is a bounded analytic function on G. Multiplication by this function defines the inverse of $S-\lambda$. Thus $\sigma(S) \subseteq \operatorname{cl} G$. On the other hand, if $\lambda \in G$, then $1 \notin (z-\lambda)L^2_a(G) = (S-\lambda)L^2_a(G)$, and so $G \subseteq \sigma(S)$. Since the spectrum is closed, $\operatorname{cl} G \subseteq \sigma(S)$. Note that if $0 = (S-\lambda)f = (z-\lambda)f$, then $f(z) = 0$ for $z \neq \lambda$; hence $\ker(S-\lambda) = (0)$ for λ in G.

(b) Corollary I.12 in [ACFA] implies that for λ in G, the linear functional $f \to f(\lambda)$ is bounded on $L^2_a(G)$. Hence there is a unique k_λ in $L^2_a(G)$ such that $f(\lambda) = \langle f, k_\lambda \rangle$ if $f \in L^2_a(G)$. This proves (b).

(c) Note that $\{f \in L^2_a(G) : f(\lambda) = 0\}$ is the kernel of the linear functional defined by k_λ, and so this linear manifold is closed. Clearly $\operatorname{ran}(S-\lambda) \subseteq \{f \in L^2_a(G) : f(\lambda) = 0\}$. Conversely, if $f \in L^2_a(G)$ and $f(\lambda) = 0$, then there is an analytic function $g\colon G \to \mathbb{C}$ such that $f = (z-\lambda)g$. It will now be shown that $g \in L^2_a(G)$. From here it follows that $f = (S-\lambda)g$, thus establishing that $\operatorname{ran}(S-\lambda) = \{f \in L^2_a(G) : f(\lambda) = 0\}$. To show that $g \in L^2_a(G)$, let $\delta > 0$ such that $B = \{z : |z-\lambda| \le \delta\} \subseteq G$. By Corollary I.12 of [ACFA] there is a constant $C > 0$ such that $|f(z)| \le C\|f\|$ for every f in $L^2_a(G)$ and z in B. If $|z-\lambda| = \delta$, then $|g(z)| = |z-\lambda|^{-1}|f(z)| \le \delta^{-1}C\|f\|$. By the Maximum Principle, this inequality holds in B. Also $\int_{G\setminus B} |g|^2\, dA = \int_{G\setminus B} |z-\lambda|^{-2}|f|^2\, dA \le \delta^{-2}\|f\|^2$. Combining these facts shows that $g \in L^2_a(G)$.

Finally, if $f \in L^2_a(G)$, then $f = (f - f(\lambda)) + f(\lambda)$, so that $\operatorname{ran}(S-\lambda) + \mathbb{C} = L^2_a(G)$ and $\dim[\operatorname{ran}(S-\lambda)]^\perp = 1$. Since $\ker(S-\lambda) = (0)$, this shows that $\operatorname{ind}(S-\lambda) = -1$. ∎

The exact determination of the essential spectrum of the Bergman operator is not trivial. This is done in Axler, Conway, and McDonald [1982]; the result will be stated here without proof. If $\lambda \in \partial G$, say that λ is a *removable boundary point* of G if for every f in $L^2_a(G)$ there is a neighborhood U of λ such that f has an analytic continuation to $G \cup U$. Call the points in ∂G that are not removable the *essential boundary points.*

29.3 Theorem. (Axler, Conway, and McDonald [1982]) *If G is an open bounded subset of the plane and S is the Bergman operator for G, then:*

(a) $\sigma_e(S)$ *is the essential boundary points of* G*;*

(b) *if* $\lambda \in \operatorname{cl} G$ *and* $\lambda \notin \sigma_e(S)$, *then* $\operatorname{ind}(S-\lambda) = -1$.

It is not difficult to see that $\partial[\text{cl}\, G] \subseteq \sigma_e(S)$ and it is rather easy to show that isolated points of ∂G are removable boundary points. Also Area $(\partial G) =$ Area $(\sigma_e(S))$. (See Exercises 3 and 5.)

29.4 Example. Let K be a compact subset of $\mathbb{D}$ such that Area $K > 0$ and int $K = \emptyset$. Also assume that the support of the restriction of area measure to K is the set K itself; that is, if U is an open set and $U \cap K \neq \emptyset$, then Area $(U \cap K) \neq 0$. (Give an example of such a set.) If $G = \mathbb{D} \setminus K$, then the essential spectrum of the Bergman operator for G is $\partial\mathbb{D} \cup K = \partial G$. In fact if $f(z) = \int_K (z - \zeta)^{-1}\, dA(\zeta)$, then f is a bounded analytic function on G (see §31 below or verify directly) that does not analytically continue to any point of K. Hence the essential spectrum of the Bergman operator on G is $\partial\mathbb{D} \cup K$.

It is shown in Axler, Conway, and McDonald [1982] that $\lambda \in \sigma_e(S)$ if and only if for every neighborhood U of λ, the set $U \cap \partial G$ has positive logarithmic capacity.

29.5 Definition. For a bounded open set G, let $P^2(G)$ be the closure of the polynomials in $L^2(G)$. $R^2(G)$ is the closure of the set of rational functions with poles off cl G.

It follows that $P^2(G) \subseteq R^2(G) \subseteq L^2_a(G)$ and $P^2(G)$ and $R^2(G)$ are invariant subspaces for the Bergman operator. It is not difficult to construct an example of a set G for which $P^2(G) \neq R^2(G)$. (Like what?) The region G defined in Example 29.4 has the property that $R^2(G) \neq L^2_a(G)$.

If K is a compact subset of $\mathbb{C}$, then the open set $\mathbb{C} \setminus K$ has at most a countable number of components, exactly one of which is unbounded. Call the boundary of this unique unbounded component of $\mathbb{C} \setminus K$ the *outer boundary* of K. Note that the outer boundary of K is a subset of ∂K. In fact, the outer boundary of K is precisely $\partial\widehat{K}$, the boundary of the polynomially convex hull of K.

29.6 Definition. A *Carathéodory region* is a bounded open connected subset of $\mathbb{C}$ whose boundary equals the outer boundary of its closure.

29.7 Proposition. *A Carathéodory region is a component of the interior of the polynomially convex hull of its closure and, hence, is simply connected.*

Proof. Let $K = \widehat{G}$ and let H be the component of int K that contains G; it must be shown that $H = G$. Suppose there is a point z_1 in $H \setminus G$ and fix a point z_0 in G. Let $\gamma\colon [0,1] \to H$ be a path such that $\gamma(0) = z_0$ and $\gamma(1) = z_1$. Put $\alpha = \inf\{\, t : \gamma(t) \in H \setminus G \}$. Thus $0 < \alpha \leq 1$ and $\gamma(t) \in G$ for $0 \leq t < \alpha$. Since $H \setminus G$ is relatively closed in H, $w = \gamma(\alpha) \notin G$. Thus

$w \in \partial G$. But since G is a Carathéodory region, $\partial G = \partial K$. Hence $w \in \partial K$. But $w \in H \subseteq \operatorname{int} K$, a contradiction.

It is left as an exercise for the reader to show that the components of the interior of any polynomially convex subset of $\mathbb{C}$ are simply connected. ■

There are simply connected regions that are not Carathéodory regions. For example, let G be the slit disk, $\mathbb{D} \setminus (-1, 0]$. Carathéodory regions tend to be well behaved simply connected regions, however there can be some rather bizarre ones.

29.8 Example. A *cornucopia* is an open ribbon G that lies outside the closed unit disk and winds about the unit circle, approaching the circle asymptotically. So each point of $\partial\mathbb{D}$ belongs to ∂G.

If G is the cornucopia, then $\operatorname{cl} G$ consists of the closed ribbon together with $\partial\mathbb{D}$. Hence $\mathbb{C} \setminus \operatorname{cl} G$ has 2 components: the unbounded component and $\mathbb{D}$. Nevertheless G is a Carathéodory region. It is worth observing that the image of the cornucopia under the Möbius transformation z^{-1} is not a Carathéodory region.

29.9 Proposition. *For a Carathéodory region G, $G = \operatorname{int}[\operatorname{cl} G]$. If G is a simply connected region such that $G = \operatorname{int}[\operatorname{cl} G]$ and $\mathbb{C} \setminus \operatorname{cl} G$ is connected, then G is a Carathéodory region.*

Proof. Exercise. ■

29.10 Lemma. *If $\{f_n\}$ is a sequence in $L^2_a(G)$, then $\{f_n\}$ converges weakly to f if and only if $\sup_n \|f_n\| < \infty$ and $f_n(z) \to f(z)$ for all z in G.*

Proof. If $f_n \to f$ weakly, then $\sup_n \|f_n\| < \infty$ by the Principle of Uniform Boundedness. Also $f_n(z) = \langle f_n, k_z\rangle \to \langle f, k_z\rangle = f(z)$. If τ is the topology of pointwise convergence on $L^2_a(G)$, then this shows that the identity map $i\colon (L^2_a(G), \text{weak}) \to (L^2_a(G), \tau)$ is continuous. Since $(\operatorname{ball} L^2_a(G), \text{weak})$ is compact and τ is a Hausdorff topology, i must be a homeomorphism on bounded sets. ■

29.11 Proposition. *Let G be a bounded Carathéodory region and suppose $\{G_n\}$ is a sequence of bounded simply connected regions such that $\operatorname{cl} G_{n+1} \subseteq G_n$ and $\bigcap\{G_n : n \geq 1\} = \widehat{G}$. If a is a point in G and, for each n, $\phi_n\colon G \to G_n$ is the Riemann map such that $\phi_n(a) = a$ and $\phi'(a) > 0$, then $\phi_n(z) \to z$ uniformly on compact subsets of G.*

Proof. Let $K = \widehat{G}$ and let $\psi_n\colon G \to G$ be the restriction of ϕ_n^{-1} to G. Now $\{\phi_n\}$ and $\{\psi_n\}$ are uniformly bounded sequences of analytic functions on G.

Let ϕ and ψ be cluster points of $\{\phi_n\}$ and $\{\psi_n\}$ in the topology of uniform convergence on compact sets. It will be shown that $\phi(z) = z$ and so z is the only cluster point of $\{\phi_n\}$; this will complete the proof of the proposition.

By passing to a subsequence if necessary, it can be assumed that $\phi_n(z) \to \phi(z)$ and $\psi_n(z) \to \psi(z)$ uniformly on compact subsets of G. Hurwitz's Theorem implies that either ϕ (respectively, ψ) is one-to-one or it is constant. It will be shown that neither function is constant. To do this, let $\bar{B}(a;r) \subseteq G$ and let $M \geq |\phi_n'(z)|$ for all $n \geq 1$ and all z in $\bar{B}(a;r)$. Thus if $|w-a| \leq r/M$, then $|\phi_n(w) - a| = |\phi_n(w) - \phi_n(a)| \leq M|w-a| \leq r$. So $\phi_n(B(a;r/M)) \subseteq \bar{B}(a;r)$. Fix w with $|w-a| \leq r/M$ and put $z_n = \phi_n(w)$. There is a point z in $B(a;r)$ and a subsequence $\{z_{n_k}\}$ such that $z_{n_k} \to z$. Hence $w = \psi_{n_k}(z_{n_k}) \to \psi(z)$. That is, $\psi(z) = w$. Therefore $\bar{B}(a;r/M) \subseteq \psi(G)$ and ψ is not constant. Hence ψ is one-to-one. Also $z_{n_k} = \phi_{n_k}(w) \to \phi(w)$, so that $\phi(w) = z$. Therefore

$$\psi(\phi(w)) = w$$

for $|w-a| \leq r/M$. This implies that ϕ is not constant.

Since ϕ is not constant, $\phi(G)$ is a connected open set. From the conditions imposed on $\{G_n\}, \phi(G) \subseteq K$. But $a = \phi(a) \in \phi(G)$. Because G is a Carathéodory region, Proposition 29.9 implies that $\phi(G) \subseteq G$. But $\psi(\phi(w)) = w$ on $\bar{B}(a;r/M)$ and hence $\psi(\phi(w)) = w$ for all w in G. This implies that $\psi\colon G \to G$ is a Riemann map. Clearly $\psi(a) = a$ since $\psi_n(a) = a$ for each $n \geq 1$; by similar reasoning $\psi'(a) > 0$. By the uniqueness of Riemann maps, $\psi(z) = z$ for all z in G. Thus $\phi(z) = z$ on G. ∎

The next result was found independently by Farrell [1934] and Markusevic [1934].

29.12 Theorem. *If G is a bounded Carathéodory region, then $P^2(G) = L^2_a(G)$*

Proof. Let $K = \widehat{G}$ and let $\tau\colon \mathbb{D} \to \mathbb{C}_\infty \setminus K$ be a Riemann map with $\tau(0) = \infty$. Put $G_n = \mathbb{C} \setminus \tau(\{z : |z| \leq 1 - 1/n\})$. The sequence $\{G_n\}$ satisfies the hypothesis of the preceding proposition. So fix a in G and let ϕ_n be the Riemann map of G onto G_n with $\phi_n(a) = a$ and $\phi_n'(a) > 0$. By Proposition 29.11, $\phi_n(z) \to z$ uniformly on compact subsets of G. Let $\psi_n = \phi_n^{-1}\colon G_n \to G$. Fix f in $L^2_a(G)$ and put $f_n = (f \circ \psi_n)\psi_n'$. Thus f_n is analytic in a neighborhood of K and so, by Runge's Theorem, f_n can be approximated uniformly on K by polynomials. Thus $f_n|G \in P^2(G)$.

Also $\|f_n\|^2 \leq \int_{G_n} |f_n|^2\, dA = \int_{G_n} |f \circ \psi_n|^2 |\psi_n'|^2\, dA = \int_G |f|^2\, dA = \|f\|^2$ by the change of variables formula for area integrals. If $z \in G$, $\psi_n(z) \to z$

and $\psi_n'(z) \to 1$. Therefore $f_n(z) \to f(z)$ as $n \to \infty$. By Lemma 29.10, $f_n \to f$ weakly and so $f \in P^2(G)$. ■

Rubel and Shields [1964] prove that if G is a bounded open set whose boundary coincides with the boundary of its polynomially convex hull and if $f \in H^\infty(G)$, then there exists a sequence of polynomials $\{p_n\}$ such that $\|p_n\|_G \le \|f\|_G$ and $p_n(z) \to f(z)$ uniformly on compact subsets of G. (Note that this condition on G is the same as the condition for a region to be a Carathéodory region, but G is not assumed to be connected here. Exercise 12 asks the reader to supply a proof of this result for the case that G is a Carathéodory region.) A corollary of this result is that the algebra $\{ M_p : p \text{ is a polynomial} \}$ is SOT dense in $\{ M_\phi : \phi \in H^\infty(G) \}$ acting on the Bergman space $L^2_a(G)$. In particular, one can approximate in this sense with polynomials the H^∞ function that is 1 on the open unit disk and 0 on the cornucopia, a fact that is not a consequence of Runge's Theorem.

Some hypothesis is needed in Theorem 29.12 besides the simple connectedness of G. For example, if $G = \mathbb{D} \setminus (-1, 0]$, then $z^{1/2} \in L^2_a(G)$ but $z^{1/2} \notin P^2(G)$. In fact it is not difficult to see that the functions in $P^2(G)$ are precisely those functions in $L^2_a(G)$ that have an analytic continuation to $\mathbb{D}$.

An exact description of the functions in $P^2(G)$ is difficult, though many properties of these functions can be given. Exercise 13 shows that if G is an annulus, then every f in $P^2(G)$ has an analytic extension to the open disk. In general, if U is a bounded component of $\mathbb{C} \setminus [\operatorname{cl} G]$ such that ∂U is disjoint from the outer boundary of G, then every function in $P^2(G)$ has an analytic extension to $G \cup [\operatorname{cl} U]$ that belongs to $P^2(G \cup [\operatorname{cl} U])$, though the norm of the extension is larger.

What happens if U is a bounded component of $\mathbb{C}\setminus[\operatorname{cl} G]$ and ∂U meets the outer boundary of G? Only partial answers are available for this question. The subject constitutes a continuing subject of research. See Mergeljan [1953], Cima and Matheson [1985], and Brennan [1977].

The next theorem can be proved by reasoning similar to that used to prove Theorem 29.12, or Theorem 29.12 can be used to prove it. See Mergeljan [1953] for details.

29.13 Theorem. *Let G be a bounded region in $\mathbb{C}$ such that $G = \operatorname{int}[\operatorname{cl} G]$ and $\mathbb{C} \setminus [\operatorname{cl} G]$ has components $U_1, \dots, U_m$. Let $K_j = \partial U_j$ and let K_0 be the outer boundary of G; assume $K_i \cap K_j = \emptyset$ for $i \ne j$ and fix a point z_j in U_j, $1 \le j \le m$. If $f \in L^2_a(G)$, then there is a sequence $\{f_n\}$ of rational functions with poles in $\{\infty, z_1, ..., z_m\}$ such that $f_n \to f$ in $L^2_a(G)$. In particular, $R^2(G) = L^2_a(G)$.*

Remarks. 1. The space $L^p_a(G)$ is called the Bergman space because of the work of Bergman ([1947], [1950]).

2. To the best of my knowledge, the first place the operator multiplication by z on $L^2_a(G)$ is studied is Halmos, Lumer, and Schäffer [1953].

3. The question of when $P^2(G)$ or $R^2(G)$ equals $L^2_a(G)$ has attracted considerable attention. There are several related questions. For example, if ω is a positive function in $L^1(G)$, let $L^2_a(\omega dA)$ denote the space of analytic functions in $L^2(\omega dA)$; this is a weighted Bergman space. When are the polynomials or rational functions dense in $L^2_a(\omega dA)$? For that matter, when is the weighted Bergman space complete? For some results and further references, see Adams and Hedberg [1996], Brennan [1977] and [1979], Hedberg [1972a] and [1972b], and Mergeljan [1953].

4. The Bergman operator on a weighted Bergman space was studied in Elias [1988] where, amongst other things, the spectral properties are established for a large class of weights.

5. The question of the density of the polynomials in $L^2_a(G)$ is the same as the question of when the Bergman operator for G has the function 1 as a cyclic vector. The question arises as to whether the Bergman operator can be cyclic even if 1 is not a cyclic vector. The answer is yes. See Akeroyd, Khavinson, and Shapiro [1991].

6. Kyung Hee Jin [1989] studies unbounded Bergman operators.

If S is the Bergman operator for G and $\phi \in H^\infty(G)$, the bounded analytic functions on G, then $\phi(S)$ can be defined as multiplication by ϕ on $L^2_a(G)$. It is easy to see that $\phi(S)$ is a bounded operator and $\|\phi(S)\| \le \|\phi\|_\infty$. This defines the functional calculus for the Bergman operator.

29.14 Theorem. *If S is the Bergman operator on any bounded region G, then the following hold.*

(a) *$\phi \to \phi(S)$ is an isometric isomorphism of $H^\infty(G)$ onto $\{S\}'$.*

(b) *If $\{\phi_i\}$ is a uniformly bounded net in $H^\infty(G)$ such that $\phi_i(z) \to 0$ for all z in G, then $\phi_i(S) \to 0$ (WOT).*

(c) *For each non-constant ϕ in $H^\infty(G)$, $\sigma_p(\phi(S)) = \emptyset$, $\overline{\phi(G)} \subseteq \sigma_p(\phi(S)^*)$, and $\sigma(\phi(S)) = \text{cl}\,[\phi(G)]$.*

Proof. (c) If $\lambda \in \sigma_p(\phi(S))$, then there is a non-zero function f in $L^2_a(G)$ such that for all z in G, $0 = [\phi(S) - \lambda]f(z) = (\phi(z) - \lambda)f(z)$. This can only happen if $\phi \equiv \lambda$. Now let $a \in G$. For each f in L^2_a, $\langle \phi(S)^* k_a, f\rangle = \langle k_a, \phi f\rangle = \overline{\phi(a)}\overline{f(a)} = \overline{\phi(a)}\langle k_a, f\rangle$. Therefore, $\phi(S)^* k_a = \overline{\phi(a)} k_a$ and so $\overline{\phi(G)} \subseteq \sigma_p(\phi(S)^*)$. This also implies that $\text{cl}\,[\phi(G)] \subseteq \sigma(\phi(S))$. Conversely, if $\lambda \notin \operatorname{cl}[\phi(G)]$, $(\phi - \lambda)^{-1} \in H^\infty(G)$ and multiplication by this function is $(\phi(S) - \lambda)^{-1}$.

(a) It has already been observed that $\|\phi(S)\| \le \|\phi\|_\infty$. Since the norm of an operator is at least as big as its spectral radius, part (c) implies the reverse inequality. It is left to the reader to check that the map $\phi \to \phi(S)$ is a homomorphism. It remains to verify that the image is the commutant of S.

Clearly $\phi(S) \in \{S\}'$. If $AS = SA$, then $A^*S^* = S^*A^*$. So for each z in G, $\ker(S^* - \bar{z}) = \mathbb{C}k_z$ is invariant for A^*. Thus for each z in G there is a complex number $\phi(z)$ such that $A^*k_z = \overline{\phi(z)}k_z$. Since $\phi(z) \in \sigma(A)$, $|\phi(z)| \le \|A\|$; that is, $\phi\colon G \to \mathbb{C}$ is a bounded function. On the other hand, $\phi(z) = \langle 1, A^*k_z\rangle = \langle A(1), k_z\rangle$ for each z in G, so that $\phi = A(1)$ and hence ϕ is an analytic function. That is, $\phi \in H^\infty(G)$ and $\|\phi\|_\infty \le \|A\|$. Moreover, for every f in $L^2_a(G)$, $(Af)(z) = \langle Af, k_z\rangle = \langle f, A^*k_z\rangle = \langle f, \overline{\phi(z)}k_z\rangle = \phi(z)f(z)$; so $A = \phi(S)$.

(b) If $\{\phi_i\}$ is a uniformly bounded net in $H^\infty(G)$ such that $\phi_i(z) \to 0$ for all z in G, then for each f in $L^2_a(G)$, $\phi_i f \to 0$ weakly in $L^2_a(G)$ by an argument similar to the proof of Lemma 29.10. Therefore $\langle \phi_i f, g\rangle \to 0$ for all f, g in $L^2_a(G)$ and so $\phi_i(S) \to 0$ (WOT). ∎

The proof of part (a) of the preceding theorem is from Shields and Wallen [1971].

Even though the commutant of the Bergman operator was found with minimal pain, finding $P^\infty(S)$ is far more complicated. If the open set G is rather nice, then $P^\infty(S)$ is not complicated and can be easily found. (See Exercises 15, 16, and 17.) But consider the following example.

Let Γ be the usual Cantor ternary set in $[0, 1]$ and put $G = B(1/2; 1)\backslash\Gamma$. If S is the Bergman operator for G, then it follows that $\sigma_e(S) = \Gamma \cup \partial B(1/2; 1)$. However, $P^\infty(S) = H^\infty(B(1/2; 1))$ because any bounded analytic function on G has an analytic continuation to $B(1/2; 1)$. This is due to the fact that though Γ has positive logarithmic capacity, it has zero analytic capacity. See §V.12 in Conway [1991].

Many problems concerning Bergman operators remain unsolved. For example, what are the invariant subspaces? It is known in the case where $G = \mathbb{D}$, that the structure of Lat S is fantastically complicated. In fact, there are spaces $\mathcal{M}$ and $\mathcal{N}$ in Lat S with $\mathcal{N} \subseteq \mathcal{M}$, such that $\dim(\mathcal{M} \ominus \mathcal{N}) = \infty$ and $\mathcal{L} \in$ Lat S whenever $\mathcal{N} \subseteq \mathcal{L} \subseteq \mathcal{M}$. (Apostol, Bercovici, Foias, and Pearcy [1985]. Also see [Bercovici, Foias, and Pearcy [1985], page 101.) That is, Lat S contains copies of the lattice of all subspaces of an infinite dimensional Hilbert space. In fact the invariant subspaces of the Bergman operator for the disk is the subject of concerted research. A sampling of papers is:

Aleman, Richter, and Ross [1996], Aleman, Richter, and Sundberg [1996], Duren, Khavinson, Shapiro, and Sundberg [1993] and [1994], Hedenmalm [1991] and [1993], Hedenmalm, Korenblum, and Zhu [1996], Hedenmalm, Richter, and Seip [1996]. Also see Zhu [1990].

This section closes with the following.

29.15 Proposition. *The Bergman operator for the unit disk is a weighted shift.*

Proof. Let normalized area measure of $\mathbb{D}$ define the norm on $L^2_a(\mathbb{D})$: $\|f\|^2 = \pi^{-1}\int_{\mathbb{D}} |f(z)|^2\, dA(z)$. Thus

$$\|z^n\|^2 = \frac{1}{\pi}\int_0^1\int_0^{2\pi} r^{2n+1}\, d\theta\, dr = \frac{1}{n+1}.$$

Using polar coordinates shows that if $e_n = \sqrt{n+1}\, z^n$, then $\{ e_n : n \geq 0 \}$ is an orthonormal sequence in $L^2_a(\mathbb{D})$. If $f \in L^2_a(\mathbb{D})$ has the power series expansion $f(z) = \sum_{n=0}^{\infty} a_n z^n$, then $\|f\|^2 = \sum_{n=0}^{\infty}(n+1)|a_n|^2$. Thus an analytic function on the disk belongs to $L^2_a(\mathbb{D})$ if and only if its Taylor coefficients satisfy $\sum_{n=0}^{\infty}(n+1)^{-1}|a_n|^2 < \infty$. This implies that $\{ e_n : n \geq 0 \}$ is a basis for $L^2_a(\mathbb{D})$. If $f(z) = \sum_n a_n z^n \in L^2_a(\mathbb{D})$, then

$$f = \sum_{n=0}^{\infty} \frac{a_n}{\sqrt{n+1}}\, e_n$$

is its Fourier series in $L^2_a(\mathbb{D})$. It is easy to see that the Bergman operator S for the disk is a weighted shift relative to this basis with weight sequence

$$\left\{\sqrt{\frac{n+1}{n+2}}\right\}. \quad \blacksquare$$

Exercises.

For these exercises, G will always be a bounded open subset of $\mathbb{C}$ and S is the Bergman operator for G.

1. If f is analytic in the punctured disk $G = \{ z : 0 < |z| < 1 \}$, for which values of p does the condition $\int_G |f|^p\, dA < \infty$ imply that f has a removable singularity at 0?

2. Let G be a bounded open set and let $G_1, G_2, \ldots$ be its components. If S_n is the Bergman operator for G_n, show that S is unitarily equivalent to $\bigoplus_n S_n$.

3. Show that $\partial[\operatorname{cl} G] \subseteq \sigma_e(S)$.

4. If $|z| < 1$, find a formula for the evaluation function k_z in $L^2_a(\mathbb{D})$.

5. Verify the statements in Example 29.4. Use the same technique to show that Area $(\partial G) =$ Area $(\sigma_e(S))$ for any open set G.

6. If $G = \mathbb{D} \setminus \{0, 1/2, 1/3, ...\}$, find $\sigma_e(S)$.

7. If $G = \mathbb{D} \setminus [-1/2, 1/2]$, find $\sigma_e(S)$.

8. Find an example of a set G such that $0 \in \operatorname{int}\{\operatorname{cl} G\}$ and $(z - n^{-1})^{-1} \in L^2_a(G)$ for all $n \geq 1$.

9. Show that if K is a polynomially convex subset of $\mathbb{C}$, then the components of int K are simply connected.

10. If G is a Carathéodory region, then $\partial G = \partial[\operatorname{cl} G]$.

11. Give an example of a simply connected region G with $G = \operatorname{int}[\operatorname{cl} G]$ that is not a Carathéodory region.

12. Use the technique of the proof of Theorem 29.12 to show that if G is a bounded Carathéodory region and $\phi \in H^\infty(G)$, there is a sequence of polynomials $\{p_n\}$ such that $\|p_n\|_\infty \leq \|\phi\|_\infty$ and $p_n(z) \to \phi(z)$ for all z in $\mathbb{D}$.

13. If G is a bounded open set in $\mathbb{C}$ and K is a compact subset of G, then every function f in $P^2(G \setminus K)$ has an analytic continuation to G that belongs to $P^2(G)$. Show that if G is connected, then the restriction map $f \to f|(G \setminus K)$ is a bijection of $P^2(G)$ onto $P^2(G \setminus K)$.

14. Let $\{a_n\}$ be an increasing sequence of positive numbers such that $1 = \lim_n a_n$. Choose $r_1, r_2, ...$ such that the closed balls $B_n = \bar{B}(a_n; r_n)$ are pairwise disjoint and contained in $\mathbb{D}$; put $G = \mathbb{D} \setminus \bigcup_n B_n$. Show that each f in $P^2(G)$ has an analytic continuation to $\mathbb{D}$. Must this continuation belong to $L^2_a(D)$?

15. Prove that the Bergman operator for G is irreducible if and only if G is connected.

16. Show that if S is the Bergman operator for $\mathbb{D}$, then $P^\infty(S) = \{S\}' = H^\infty(\mathbb{D})$.

17. If S is the Bergman operator for $G = \mathbb{D} \setminus [-1/2, 1/2]$, then $P^\infty(S) = H^\infty(\mathbb{D})$ and $\{S\}' = H^\infty(G)$. So $P^\infty(S) \neq \{S\}'$.

18. If S is the Bergman operator for $G = \{z : 1/2 < |z| < 1\}$, then $\{S\}' = H^\infty(G)$ and $P^\infty(S) = H^\infty(\mathbb{D})$. However, $H^\infty(G) = R^\infty(S)$, the weak* closure of the rational functions in S.

19. Let G_1 and G_2 be bounded open sets with $\operatorname{cl} G_2 \subsetneq G_1$ and define $R\colon L^2_a(G_1) \to L^2_a(G_2)$ by $Rf = f|G_2$. Prove that R is a compact operator. In fact, R^*R is a trace class operator on $L^2_a(G_1)$.

20. Let $L^\infty(G)$ denote $L^\infty(\text{Area}\,|G)$ and consider $H^\infty(G)$ as a linear manifold contained in $L^\infty(G)$. Apply the Krein–Smulian Theorem to show

that $H^\infty(G)$ is weak* closed in $L^\infty(G)$. Also show that a sequence $\{\phi_n\}$ in $H^\infty(G)$ converges to 0 weak* if and only if it is uniformly bounded and $\phi_n(z) \to 0$ for all z in G.

21. Show that the Bergman operator for an annulus centered at the origin is a bilateral weighted shift.

§30. Subnormal operators

This section introduces a class of operators that includes the normal operators, but is far more extensive.

30.1 Definition. An operator S on a Hilbert space $\mathcal{H}$ is *subnormal* if there is a Hilbert space $\mathcal{K}$ containing $\mathcal{H}$ and a normal operator N on $\mathcal{K}$ such that $N\mathcal{H} \subseteq \mathcal{H}$ and $N|\mathcal{H} = S$.

When discussing a subnormal operator, the notation just introduced will be assumed. Loosely speaking, a subnormal operator is one that has a normal extension. Needless to say, each normal operator is subnormal. Here are some more examples. The Bergman operators are subnormal. If G is a bounded open set in the plane and S is the Bergman operator on $L^2_a(G)$, then the operator defined as multiplication by z on $L^2(G)$ is a normal extension of S.

30.2 Example. Let μ be a compactly supported measure on $\mathbb{C}$ and let $P^2(\mu)$ denote the closure in $L^2(\mu)$ of the analytic polynomials. Define $S_\mu\colon P^2(\mu) \to P^2(\mu)$ by $(S_\mu f)(z) = zf(z)$ for all f in $P^2(\mu)$. If N_μ is defined as in Proposition 2.5, then N_μ is normal. In fact, N_μ is the general star-cyclic normal operator (11.2). If p is an analytic polynomial, then $N_\mu p = zp$ is also a polynomial. By passing to limits it follows that N_μ leaves $P^2(\mu)$ invariant, and, thus, S_μ is a subnormal operator with N_μ as a normal extension.

Here is a generalization of the preceding example. For a compact subset K in the plane, let Rat K denote the algebra of all rational functions with poles off K.

30.3 Example. Let K be a compact subset of $\mathbb{C}$ and define $R^2(K,\mu)$ to be the closure in $L^2(\mu)$ of Rat (K). For f in $R^2(K,\mu)$, let $Sf = zf$. Then S is subnormal and N_μ is a normal extension.

By Runge's Theorem Example 30.2 is a special case of the preceding example where K is the polynomially convex hull of the support of μ. It is easy to see that the operator S_μ defined in (30.2) is cyclic. In fact, every cyclic subnormal operator is unitarily equivalent to S_μ for some appropriate choice of μ. See Theorem 30.15 below.

There is an extensive literature on subnormal operators. Conway [1991] is a good place to start. In this section only the basics will be touched on; an operator theoretic appetizer.

Maintain the notation in the definition and let P be the projection of $\mathcal{K}$ onto $\mathcal{H}$. It is easy to see that for f in $\mathcal{H}$, $S^{*n}f = PN^{*n}f$. Thus $\|S^{*n}f\| \le \|N^{*n}f\| = \|N^n f\|$. This fact will be used in the next proof as well as in other proofs in this section. In particular, $\langle SS^*f, f\rangle = \|S^*f\|^2 \le \|Nf\|^2 = \langle S^*Sf, f\rangle$. Hence $S^*S - SS^* \ge 0$. This says that a subnormal operator is *hyponormal*, a class of operators that will not be explored in this book but which also has a life of its own. See Martin and Putinar [1989].

30.4 Theorem. *If S is an operator on $\mathcal{H}$, then the following statements are equivalent.*

(a) *S is subnormal.*

(b) (Halmos [1950]) *If $f_0, \dots, f_n \in \mathcal{H}$, then*

30.5
$$\sum_{j,k} \langle S^j f_k, S^k f_j\rangle \ge 0;$$

and there is a constant $c > 0$ such that for any $f_0, \dots, f_n$ in $\mathcal{H}$,

30.6
$$\sum_{j,k} \langle S^{j+1} f_k, S^{k+1} f_j\rangle \le c \sum_{j,k} \langle S^j f_k, S^k f_j\rangle.$$

(c) (Szymanski [1990]) *There is a constant $c > 0$ such that* (30.6) *holds for any $f_0, \dots, f_n$ in $\mathcal{H}$.*

(d) (Bram [1955]) *Condition* (30.5) *holds for any $f_0, \dots, f_n$ in $\mathcal{H}$.*

(e) (Bunce and Deddens [1977]) *If $B_0, \dots, B_n$ belong to the C^*-algebra generated by S, then*

$$\sum_{j,k} B_j^* S^{*k} S^j B_k \ge 0.$$

Proof. (a) *implies* (b). Let N be a normal operator on $\mathcal{K}$ with $\mathcal{H} \subseteq \mathcal{K}$ and $N|\mathcal{H} = S$. Let P be the orthogonal projection from $\mathcal{K}$ onto $\mathcal{H}$. If $f_0, \dots, f_n$

are vectors in $\mathcal{H}$, then

$$\begin{aligned}\sum_{j,k}\langle S^j f_k, S^k f_j\rangle &= \sum_{j,k}\langle N^j f_k, N^k f_j\rangle = \sum_{j,k}\langle N^{*k}N^j f_k, f_j\rangle \\ &= \sum_{j,k}\langle N^j N^{*k} f_k, f_j\rangle = \sum_{j,k}\langle N^{*k} f_k, N^{*j} f_j\rangle \\ &= \left\|\sum_k N^{*k} f_k\right\|^2.\end{aligned}$$

Thus (30.5) holds. If $g_k = Sf_k$ and each f_k in the preceding equations is replaced by g_k, then it follows that

$$\begin{aligned}\sum_{j,k}\langle S^{j+1} f_k, S^{k+1} f_j\rangle &= \sum_{j,k}\langle S^j g_k, S^k g_j\rangle \\ &= \left\|\sum_k N^{*k} N f_k\right\|^2 \\ &= \left\|N\sum_k N^{*k} f_k\right\|^2 \\ &\le \|N\|^2 \sum_{j,k}\langle S^j f_k, S^k f_j\rangle.\end{aligned}$$

Therefore (30.6) is valid with $c = \|N\|^2$.

(b) *implies* (c). Clear.

(c) *implies* (d). First observe that it suffices to assume that $\|S\| < 1$. In fact for an arbitrary constant $\rho > 0$, if $S_1 = \rho^{-1}S$ and $h_k = \rho^{-k} f_k$, then

$$\sum_{j,k}\langle S_1^j f_k, S_1^k f_j\rangle = \sum_{j,k}\langle S^j h_k, S^k h_j\rangle$$

and

$$\sum_{j,k}\langle S_1^{j+1} f_k, S_1^{k+1} f_j\rangle = \rho^{-2}\sum_{j,k}\langle S^{j+1} h_k, S^{k+1} h_j\rangle.$$

Thus (30.5) and (30.6) hold if and only if the same conditions hold for S_1. Thus it can be assumed that $\|S\| < 1$.

Next take just two vectors in (30.6), $f_0 = 0$ and f_1 arbitrary, and see that the constant c in question must satisfy $c \ge \|S\|^2$. Since $\|S\| < 1$ and (30.6) remains valid for any larger value of c, it is clear that it may be assumed that $c \ge 1$.

Let $\mathcal{K} = \mathcal{H}^{(\infty)}$ and let $\mathcal{K}_0$ be the set of all finitely non-zero sequences in $\mathcal{K}$. Let M be the infinite operator matrix whose (j,k) entry is $S^{*k}S^j$ and let M act on $\mathcal{K}_0$. If $f = (f_0, \ldots, f_n, 0, \ldots) \in \mathcal{K}_0$, then

$$\begin{aligned}\sum_j \|(Mf)_j\|^2 &= \sum_j \left\| \sum_k S^{*k}S^j f_k \right\|^2 \\ &\le \sum_j \left[\sum_k \|S\|^{k+j}\|f_k\| \right]^2 \\ &\le \sum_j \left[\sum_k \|S\|^{2k+2j} \right] \left[\sum_k \|f_k\|^2 \right] \\ &\le (1 - \|S\|^2)^{-2}\|f\|^2.\end{aligned}$$

Since $\|S\| < 1$, $Mf \in \mathcal{K}$ and M extends to a bounded operator on $\mathcal{K}$.

Note that $\langle Mf, f\rangle = \sum_{j,k}\langle S^j f_k, S^k f_j\rangle$. This shows that M is hermitian. To prove that (30.5) holds, it must be shown that M is positive. Now

$$\langle S^{(\infty)*}MS^{(\infty)}f, f\rangle = \sum_{j,k}\langle S^{j+1}f_k, S^{k+1}f_j\rangle,$$

so that $S^{(\infty)*}MS^{(\infty)} \le cM$ on $\mathcal{K}$ by (30.6). Iterating this inequality shows that

$$\begin{aligned}\langle S^{(\infty)*2}MS^{(\infty)2}f, f\rangle &= \langle (S^{(\infty)*}MS^{(\infty)})S^{(\infty)}f, S^{(\infty)}f\rangle \\ &\le c\langle MS^{(\infty)}f, S^{(\infty)}f\rangle \\ &= c\langle S^{(\infty)*}MS^{(\infty)}f, f\rangle \\ &\le c^2\langle Mf, f\rangle.\end{aligned}$$

Continuing shows that $\langle Mf, f\rangle \ge c^{-n}\langle S^{(\infty)*n}MS^{(\infty)n}f, f\rangle$ for all f in $\mathcal{K}$. But $c \ge 1$ implies that $\{c^{-n}\}$ is a bounded sequence. Also $\|S^{(\infty)*n}MS^{(\infty)n}\| \le \|S\|^{2n}\|M\|$ and this converges to 0 since $\|S\| < 1$. Thus

$$0 = \lim_n c^{-n}\langle S^{(\infty)*n}MS^{(\infty)n}f, f\rangle \le \langle Mf, f\rangle$$

and (30.5) is established.

(d) *implies* (e). If $B_0, \ldots, B_n \in C^*(S)$ and $f \in \mathcal{H}$, let $f_k = B^k f$ and write out (30.5) to obtain that (e) is valid.

(e) *implies* (d). A routine application of Zorn's Lemma shows that any operator is the direct sum of operators that have a star-cyclic vector. Thus it suffices to prove that (30.5) holds when S is assumed to have a star-cyclic vector e_0; that is, assume that $\mathcal{H} = \text{cl}\,[C^*(S)e_0]$. If $B_0, \dots, B_n \in C^*(S)$, then (e) implies that (30.5) holds for $f_k = B_k e_0$. But since (30.5) holds for a dense set of vectors, it holds for all vectors.

(d) *implies* (a). Without loss of generality it can be assumed that $\|S\| < 1$. Start by defining $\mathcal{K}_0$, $\mathcal{K}$, and M as in the proof that (c) implies (d). As in that proof, M is a bounded hermitian operator. Condition (30.5) is equivalent to the condition that M is a positive operator. Define $A = S^{(\infty)}$ on $\mathcal{K}$. For f in $\mathcal{K}$, $(A^*MAf)_j = S^*(MAf)_j = S^* \sum_k S^{*k} S^j (Af)_k = S^* \sum_k S^{*k} S^{j+1} f_k$. Thus

$$\begin{aligned}
\|A^*MAf\|^2 &= \sum_{j=0}^{\infty} \left\| S^* \sum_{k=0}^{\infty} S^{*k} S^{j+1} f_k \right\|^2 \\
&= \sum_{j=1}^{\infty} \left\| S^* \sum_{k=0}^{\infty} S^{*k} S^{j} f_k \right\|^2 \\
&\le \sum_{j=1}^{\infty} \left\| \sum_{k=0}^{\infty} S^{*k} S^{j} f_k \right\|^2 \\
&= \|Mf\|^2.
\end{aligned}$$

Thus $(A^*MA)^2 \le M^2$. Since both M and A^*MA are positive operators, this implies $A^*MA \le M$ ([ACFA], Exercise VIII.3.12). That is, $\langle MAf, Af \rangle \le \langle Mf, f \rangle$ for all f in $\mathcal{K}$. Writing this out shows that

$$\sum_{j,k} \langle S^{j+1} f_k, S^{k+1} f_j \rangle \le \sum_{j,k} \langle S^j f_k, S^k f_j \rangle,$$

which is condition (30.6) with $c = 1$.

Let $\widetilde{\mathcal{L}}$ be the set of all finitely non-zero sequences of elements of $\mathcal{H}$ and define a sesquilinear form on $\widetilde{\mathcal{L}}$ by

$$[f, g] = \sum_{j,k} \langle S^k f_j, S^j g_k \rangle,$$

for all $f = (f_0, f_1, \dots)$ and $g = (g_0, g_1, \dots)$ in $\widetilde{\mathcal{L}}$. By (30.5), $[f, f] \ge 0$. Let $\mathcal{L}_0 \equiv \{ f \in \widetilde{\mathcal{L}} : [f, f] = 0 \}$. It is left as an exercise for the reader to show that $\mathcal{L}_0$ is a linear manifold in $\widetilde{\mathcal{L}}$ and $[f + \mathcal{L}_0, g + \mathcal{L}_0] \equiv [f, g]$ defines an inner product on $\widetilde{\mathcal{L}}/\mathcal{L}_0$. Let $\mathcal{L}$ be the completion of $\widetilde{\mathcal{L}}/\mathcal{L}_0$ with respect to the norm defined by this inner product.

Define a function N_0 on $\widetilde{\mathcal{L}}$ by $(N_0f)_k = Sf_k$. It is easy to see that N_0 is linear and $[N_0f, N_0f] = \sum_{j,k}\langle S^{k+1}f_j, S^{j+1}f_k\rangle \leq [f, f]$ by (30.6). It follows that $N_0\mathcal{L}_0 \subseteq \mathcal{L}_0$ and N_0 induces a contraction N on $\mathcal{L}$ defined by $N(f + \mathcal{L}_0) = N_0f + \mathcal{L}_0$.

Now define B_0 on $\widetilde{\mathcal{L}}$ by $(B_0f)_0 = 0$ and $(B_0f)_k = f_{k-1}$ for $k \geq 1$. It is left to the reader to show that (30.6) implies that $[B_0f, B_0f] \leq [f, f]$ for all f in $\widetilde{\mathcal{L}}$. Thus B_0 induces a contraction B on $\mathcal{L}$. For f, g in $\widetilde{\mathcal{L}}$,

$$\begin{aligned}
[f, B_0g] &= \sum_{j,k}\langle S^kf_j, S^j(B_0g)_k\rangle \\
&= \sum_{j\geq 0}\sum_{k\geq 1}\langle S^kf_j, S^jg_{k-1}\rangle \\
&= \sum_{j\geq 0}\sum_{k\geq 1}\langle S^{k-1}(N_0f)_j, S^jg_{k-1}\rangle \\
&= [N_0f, g].
\end{aligned}$$

It follows that $B = N^*$. But also for any f in $\widetilde{\mathcal{L}}$, $(N^*Nf)_0 = 0 = (NN^*f)_0$. While for $k \geq 1$, $(N^*Nf)_k = (Nf)_{k-1} = Sf_{k-1}$ and $(NN^*f)_k = S(N^*f)_k = Sf_{k-1}$. Thus N is a normal operator.

If for each vector f in $\mathcal{H}$ $\widehat{f}$ is defined as the sequence in $\widetilde{\mathcal{L}}$ that equals f in the 0-coordinate and 0 in the remaining ones, then $f \to \widehat{f}$ is an isometric embedding of $\mathcal{H}$ into $\widetilde{\mathcal{L}}$. This leads to an isometric embedding of $\mathcal{H}$ into $\mathcal{L}$. Clearly this image of $\mathcal{H}$ is invariant for N and $N|\mathcal{H} = S$. ■

One virtue of condition (e) in the preceding theorem is that it permits the definition of subnormality in a C*-algebra context.

30.7 Definition. If $\mathcal{A}$ is a C*-algebra, define s in $\mathcal{A}$ to be *subnormal* if $\sum_{j,k} a_j^* s^{*k} s^j a_k \geq 0$ for any choice of $a_0, \ldots, a_n$ in $C^*(s)$.

From here it is easy to see that if $\mathcal{A}$ and $\mathcal{B}$ are C*-algebras and $\rho\colon \mathcal{A} \to \mathcal{B}$ is a *-homomorphism, then ρ maps subnormal elements of $\mathcal{A}$ onto subnormal elements of $\mathcal{B}$. In particular, if $(\rho, \mathcal{H})$ is a representation of $\mathcal{A}$, $\rho(s)$ is a subnormal operator on $\mathcal{H}$ whenever s is a subnormal element of $\mathcal{A}$.

The normal extension of a subnormal operator is never unique. If N is a normal extension of S and M is any normal operator, then $N \oplus M$ is also a normal extension of S. Note that if N is a normal extension of S, then $\mathcal{K}$ contains

30.8 $$\left[\{ N^{*k}h : h \in \mathcal{H}, k \geq 1 \}\right].$$

If $\mathcal{L}$ denotes this space, then it is easy to see that $\mathcal{L}$ is a reducing subspace for N that contains $\mathcal{H}$. Thus $N|\mathcal{L}$ is also a normal extension of S. Moreover, if $\mathcal{R}$ is any reducing subspace for N that contains $\mathcal{H}$, then $\mathcal{R}$ must contain $\mathcal{L}$. This leads to the following definition.

30.9 Definition. If S is a subnormal operator on $\mathcal{H}$ and N is a normal extension of S to $\mathcal{K}$, then N is a *minimal normal extension* of S if $\mathcal{K}$ equals the space defined in (30.8).

If S is the subnormal operator defined in Example 30.3, then N_μ is a minimal normal extension.

30.10 Proposition. *If S is a subnormal operator, any two minimal normal extensions are unitarily equivalent.*

Proof. For $p = 1, 2$ let N_p be a minimal normal extension of S acting on $\mathcal{K}_p \supseteq \mathcal{H}$. Define U on $\mathcal{K}_1$ by the formula $U(N_1^{*n}h) = N_2^{*n}h$ for h in $\mathcal{H}$ and $n \geq 0$. It must be shown that this indeed defines an operator that is an isomorphism.

If $h_1, \ldots, h_m \in \mathcal{H}$ and $n_1, \ldots, n_m \geq 0$, then

$$\begin{aligned}
\left\| \sum_k N_2^{*n_k} h_k \right\|^2 &= \left\langle \sum_k N_2^{*n_k} h_k, \sum_j N_2^{*n_j} h_j \right\rangle \\
&= \sum_{j,k} \langle N_2^{n_j} h_k, N_2^{n_k} h_j \rangle \\
&= \sum_{j,k} \langle S^{n_j} h_k, S^{n_k} h_j \rangle \\
&= \sum_{j,k} \langle N_1^{n_j} h_k, N_1^{n_k} h_j \rangle \\
&= \left\| \sum_k N_1^{*n_k} h_k \right\|^2 .
\end{aligned}$$

This shows simultaneously that

$$U\left[\sum_k N_1^{*n_k} h_k\right] = \sum_k N_2^{*n_k} h_k$$

is a well defined linear operator from a dense linear manifold in $\mathcal{K}_1$ onto a dense linear manifold in $\mathcal{K}_2$ and U is an isometry. Therefore U extends to an isomorphism of $\mathcal{K}_1$ onto $\mathcal{K}_2$. Also for all h in $\mathcal{H}$, $Uh = h$. Thus for h

in $\mathcal{H}$ and $n \geq 0$, $UN_1N_1^{*n}h = UN_1^{*n}Sh = N_2^{*n}Sh = N_2N_2^{*n}h = N_2UN_1^{*n}h$. That is, $UN_1 = N_2U$, so that N_1 and N_2 are unitarily equivalent. ■

Now it is legitimate to speak of *the* minimal normal extension of a subnormal operator. It is, therefore, unambiguous to define the *normal spectrum* of a subnormal operator S, $\sigma_n(S)$, as the spectrum of its minimal normal extension.

30.11 Proposition. *If S is a subnormal operator, the following hold.*

(a) (Halmos [1952]) $\sigma_n(S) \subseteq \sigma(S)$.

(b) $\sigma_{ap}(S) \subseteq \sigma_n(S)$ *and* $\partial\sigma(S) \subseteq \partial\sigma_n(S)$.

(c) (Bram [1955]) *If U is a bounded component of $\mathbb{C} \setminus \sigma_n(S)$, then either $U \cap \sigma(S) = \emptyset$ or $U \subseteq \sigma(S)$.*

Proof. (a) It suffices to show that if S is invertible, then N is invertible. (Why?) If $N = \int z dE(z)$ is the spectral decomposition of N, $\epsilon > 0$, and $\mathcal{M} = E(B(0;\epsilon))\mathcal{K}$, then for f in $\mathcal{M}$, $\|N^k f\| \leq \epsilon^k \|f\|$ for all $k \geq 1$. So if $f \in \mathcal{M}$ and $h \in \mathcal{H}$,

$$\begin{aligned} |\langle f, h\rangle| &= |\langle f, S^k S^{-k} h\rangle| = |\langle f, N^k S^{-k} h\rangle| \\ &= |\langle N^{*k} f, S^{-k} h\rangle| \leq \|N^{*k} f\| \, \|S^{-k} h\| \\ &\leq \epsilon^k \|f\| \, \|S^{-k} h\| \leq \epsilon^k \|S^{-1}\|^k \, \|f\| \, \|h\|. \end{aligned}$$

Letting $k \to \infty$ shows that if $\epsilon < \|S^{-1}\|^{-1}$, then $\langle f, h\rangle = 0$. Thus $\mathcal{M} \perp \mathcal{H}$ if $\epsilon < \|S^{-1}\|^{-1}$ and so $\mathcal{H} \subseteq \mathcal{M}^\perp$. But this implies that $N|\mathcal{M}^\perp$ is a normal extension of S. By the minimality of N, $\mathcal{M} = (0)$ and so N is invertible.

(b) If $\lambda \in \sigma_{ap}(S)$, there is a sequence of unit vectors $\{h_n\}$ in $\mathcal{H}$ such that $\|(\lambda - S)h_n\| \to 0$. But $(\lambda - S)h_n = (\lambda - N)h_n$. Hence $\sigma_{ap}(S) \subseteq \sigma_{ap}(N) = \sigma_n(S)$. Also if $\lambda \in \partial\sigma(S)$, then $\lambda \in \sigma_{ap}(S)$ and so $\lambda \in \sigma_n(S)$. But part (a) implies that $\lambda \notin \operatorname{int} \sigma_n(S)$ and so $\lambda \in \partial\sigma_n(S)$.

(c) (Due to S K Parrott) Let U be a bounded component of the complement of $\sigma(S)$ and put $U_+ = U \setminus \sigma(S)$ and $U_- = U \cap \sigma(S)$. So $U = U_- \cup U_+$, $U_+ \cap U_- = \emptyset$, and U_+ is open. But part (b) implies $U_- = U \cap [\operatorname{int} \sigma(S)]$, so that U_- is also open. By the connectedness of U, either U_+ or U_- is empty, completing the proof. ■

If $r(S)$ denotes the spectral radius of S, then $r(S) \leq \|S\| \leq \|N\| = r(N)$. But since $\sigma_n(S) \subseteq \sigma(S)$, this proves the following.

30.12 Corollary. *The norm of a subnormal operator equals its spectral radius and the norm of its minimal normal extension.*

Here is the promised characterization of the operators appearing in Examples 30.2 and 30.3.

30.13 Definition. If $A \in \mathcal{B}(\mathcal{H})$, $e_0 \in \mathcal{H}$, and K is a compact subset of $\mathbb{C}$ containing $\sigma(A)$, then e_0 is a Rat(K) *cyclic vector* for A if $\{u(A)e_0 : u \in \text{Rat}(K)\}$ is dense in $\mathcal{H}$. An operator is Rat(K) *cyclic* if it has a Rat(K) cyclic vector. In the case that $K = \sigma(S)$, A is a *rationally cyclic operator.*

Recall that e_0 is a cyclic vector for A (A is a cyclic operator) if $\{p(A)e_0 : p \text{ is a polynomial}\}$ is dense in $\mathcal{H}$. By Runge's Theorem, this is equivalent to the statement that e_0 is a Rat$(\widehat{\sigma(A)})$ cyclic vector for A.

Note that if S is subnormal and $N = \text{mne}\, S$, then the fact that $\sigma(N) \subseteq \sigma(S)$ implies that if K contains $\sigma(S)$, then $u(N)$ is well defined for any u in Rat(K)

30.14 Theorem. *If S is subnormal and has a* Rat(K) *cyclic vector e_0, then there is a unique compactly supported measure μ on K and an isomorphism $U: \mathcal{K} \to L^2(\mu)$ such that:*

(a) $U\mathcal{H} = R^2(K, \mu)$;

(b) $Ue_0 = 1$;

(c) $UNU^{-1} = N_\mu$;

(d) *if $V = U|\mathcal{H}$, V is an isomorphism of $\mathcal{H}$ onto $R^2(K,\mu)$ and $VSV^{-1} = N_\mu|R^2(K,\mu)$.*

Proof. First observe that since N is the minimal normal extension of S, $\mathcal{K} = [\{N^{*n}u(N)e_0 : n \geq 0 \text{ and } u \in \text{Rat}(K)\}]$. It is claimed that e_0 is a $*$-cyclic vector for N. Indeed, let $\mathcal{L} = [\{N^{*n}N^k e_0 : n, k \geq 0\}]$; clearly $\mathcal{L}$ is a reducing subspace for N. On the other hand, the Stone–Weierstrass Theorem implies that $C(K)$ is the uniformly closed linear span of $\{\bar{z}^n z^k : n, k \geq 0\}$. Since Rat$(K) \subseteq C(K)$, $u(N)e_0 \in \mathcal{L}$ for every u in Rat(K). Thus $\mathcal{H} \subseteq \mathcal{L}$, and so $\mathcal{L} = \mathcal{K}$ by the minimality of N. Hence e_0 is a $*$-cyclic vector for N.

Therefore there is a compactly supported measure μ and an isomorphism $U: \mathcal{K} \to L^2(\mu)$ such that $Ue_0 = 1$ and $UNU^{-1} = N_\mu$. That is, (b) and (c) hold. By a standard argument, $U\phi(N) = \phi(N_\mu)U$ for every bounded Borel function ϕ. In particular for u in Rat K, $Uu(S)e_0 = Uu(N)e_0 = u$; by taking limits, (a) follows. Part (d) is an immediate consequence. The proof that the measure μ is unique follows easily from the Stone-Weierstrass Theorem. ∎

The next corollary was obtained independently by Bram [1955] and I M Singer (see Wermer [1955]).

30.15 Corollary. *S is a cyclic subnormal operator if and only if S is unitarily equivalent to S_μ for some compactly supported measure μ on the plane.*

There is an extensive functional calculus for subnormal operators. For any compact subset K in $\mathbb{C}$, define $R(K)$ to be uniform closure in $C(K)$ of $\operatorname{Rat}(K)$, the rational functions with poles off K. Denote by $\|f\|_K$ the supremum norm of the function over K. For a subnormal operator S, it is possible to define $f(S)$ for all the functions f in $R(\sigma(S))$. In fact, if $\lambda \notin \sigma(S)$, then $\lambda \notin \sigma(N)$ by the Proposition 30.11. Also $(N-\lambda)\mathcal{H} = (S-\lambda)\mathcal{H} = \mathcal{H}$. Thus $(S-\lambda)^{-1} = (N-\lambda)^{-1}|\mathcal{H}$. Hence for any function f in $\operatorname{Rat}(\sigma(S))$, $f(S)$, as defined by the Riesz functional calculus, equals $f(N)|\mathcal{H}$ and $f(S)$ is subnormal.

Now from the Riesz functional calculus, $\sigma(f(S)) = f(\sigma(S))$. Since $f(S)$ is subnormal, Corollary 30.12 implies that $\|f\|_{\sigma(S)} = \|f(S)\| \le \|f(N)\| = \|f\|_{\sigma(N)} \le \|f\|_{\sigma(S)}$. That is, the map $f \to f(S)$ is an isometry from $\operatorname{Rat}(\sigma(S))$ into $\mathcal{B}(\mathcal{H})$. This map extends to an isometry from $R(\sigma(S))$ into $\mathcal{B}(\mathcal{H})$. Define $f(S)$ for f in $R(\sigma(S))$ to be the image of f under this isometry. It is not difficulty to see that for any f in $R(\sigma(S))$, $f(N)\mathcal{H} \subseteq \mathcal{H}$ and $f(N)|\mathcal{H} = f(S)$.

30.16 Theorem. *If S is subnormal with minimal normal extension N and for each f in $R(\sigma(S))$, $f(S) = f(N)|\mathcal{H}$, then the map $f \to f(S)$ is a multiplicative linear isometry from $R(\sigma(S))$ into $\mathcal{B}(\mathcal{H})$ that extends the Riesz functional calculus for S. Moreover,*

$$\sigma(f(S)) = f(\sigma(S))$$

for each f in $R(\sigma(S))$.

Proof. All of this proposition, except for the part about $\sigma(f(S))$, was proved prior to the statement. Put $K = \sigma(S)$ and for f in $R(K)$ let $\{f_n\}$ be a sequence in $\operatorname{Rat}(K)$ such that $\|f_n - f\|_K \to 0$. From the Riesz functional calculus it is already known that $\sigma(f_n(S)) = f_n(K)$ for all $n \ge 1$.

If $\lambda \notin f(K)$, then there is an integer N such that $\lambda \notin f_n(K)$ for all $n \ge N$ and, moreover, $(f_n - \lambda)^{-1} \to (f-\lambda)^{-1}$ uniformly on K. But for $n \ge N$, $(f_n - \lambda)^{-1} \in \operatorname{Rat}(K)$, so $g = (f-\lambda)^{-1} \in R(K)$. It is easy to see that $g(S)(f(S) - \lambda) = 1$, so that $\lambda \notin \sigma(f(S))$.

For the converse, let ϵ be an arbitrary positive number and put $\Lambda(\epsilon) = \{z : \operatorname{dist}(z, \sigma(f(S)) < \epsilon\}$. As a consequence of the fact that the invertible operators form an open set, there is an integer N such that $f_n(K) = \sigma(f_n(S)) \subseteq \Lambda(\epsilon)$ for all $n \ge N$. (See Exercise 8.) Choose N sufficiently large that $\|f_n - f\|_K < \epsilon$ for all $n \ge N$. It follows that $f(\sigma(S)) \subseteq \Lambda(2\epsilon)$. Since ϵ was arbitrary, $f(K) \subseteq \sigma(f(S))$. ∎

A knowledge of $R(\sigma(S))$ is important in the study of the subnormal operator S. One example of this is the next proposition. More examples will be seen in the next section. First a lemma.

30.17 Lemma. *If S is subnormal and $\sigma(S) \subseteq \mathbb{R}$, then S is hermitian.*

Proof. Let $N = \operatorname{mne} S$, acting on $\mathcal{K}$. By Proposition 30.11, $\sigma(N) \subseteq \sigma(S) \subseteq \mathbb{R}$. Hence, $N = N^*$. But then every invariant subspace for N reduces N. In particular, $\mathcal{H}$ reduces N. Therefore $\mathcal{K} = \mathcal{H}$ by the minimality of N, so that $S = N$ and S is hermitian. ∎

30.18 Proposition. *If S is subnormal and $R(\sigma(S)) = C(\sigma(S))$, then S is normal.*

Proof. Let $\phi(z) = \operatorname{Re} z$ and $\psi(z) = \operatorname{Im} z$. By the hypothesis both ϕ and ψ belong to $R(\sigma(S))$. By Theorem 30.16, $\phi(S)$ is a subnormal operator and $\sigma(\phi(S)) = \phi(\sigma(S)) \subseteq \mathbb{R}$. Therefore $\phi(S)$ is hermitian. Similarly $\psi(S)$ is hermitian. Since $\phi + i\psi = z$, $S = \phi(S) + i\psi(S)$ is the Cartesian decomposition of S. Since the real and imaginary parts of S commute, S is normal. ∎

The astute reader will realize that only the fact that $\operatorname{Re} z$ and $\operatorname{Im} z$ belong to $R(\sigma(S))$ was needed to achieve the conclusion. But this condition is equivalent to the requirement that $R(\sigma(S)) = C(\sigma(S))$. Indeed, it is clear that $\operatorname{Re} z$ and $\operatorname{Im} z$ belong to $R(K)$ if and only if $\bar{z} \in R(K)$. Since $R(K)$ is an algebra, $\bar{z} \in R(K)$ if and only if $R(K)$ contains all polynomials in z and $\bar{z}$. By the Stone–Weierstrass Theorem this is equivalent to the condition that $R(K) = C(K)$.

The functional calculus for a subnormal operator can be developed even further. To set the stage, let μ be the scalar-valued spectral measure for N, the minimal normal extension of S. If $\phi \in L^\infty(\mu)$, define

30.19 $$\phi(S)f = P(\phi(N)f),$$

where P is the projection onto $\mathcal{H}$. If S is the unilateral shift, $\phi(S)$ is the classical Toeplitz operator with symbol ϕ. See Remark 4 at the end of §24. This gives another functional calculus for subnormal operators that extends the previous one. A difficulty, however, is that the map $\phi \to \phi(S)$ from $L^\infty(\mu)$ into $\mathcal{B}(\mathcal{H})$ is not multiplicative. It does have other properties.

30.20 Proposition. *With this notation and $\phi(S)$ defined as in* (30.19), *then the map $\phi \to \phi(S)$ is a positive, injective, contraction from the C^*-algebra $L^\infty(\mu)$ into $\mathcal{B}(\mathcal{H})$.*

Proof. The fact that the map is a positive contraction is left to the reader. To prove that it is injective, assume $\phi \in L^\infty(\mu)$ and $\phi(S) = 0$. If $\mathcal{K}$ is the space on which N acts, it follows that this is equivalent to the assertion that $\phi(N)\mathcal{H} \subseteq \mathcal{H}^\perp$. Thus for all $n \geq 1$ and h in $\mathcal{H}$, $\phi(N)N^{*n}h = N^{*n}\phi(N)h \in N^{*n}\mathcal{H}^\perp \subseteq \mathcal{H}^\perp$. Since N is the minimal normal extension of S, this implies $\phi(N)\mathcal{K} \subseteq \mathcal{H}^\perp$.

This says that if $\phi(N)$ is represented as a 2×2 matrix operating on $\mathcal{K} = \mathcal{H} \oplus \mathcal{H}^\perp$, then

$$\phi(N) = \begin{bmatrix} 0 & 0 \\ A & B \end{bmatrix}$$

for some operators A, B. Using this matrix to compute both sides of the equation $\phi(N)^*\phi(N) = \phi(N)\phi(N)^*$ and comparing entries, it follows that $A^*A = 0$, so that $A = 0$. Equivalently, $\phi(N)\mathcal{H} = 0$. But this implies that $\phi(N)N^{*n}\mathcal{H} = (0)$ for all $n \geq 1$. Again by the minimality of N, this implies $\phi(N) = 0$. By the functional calculus for normal operators, this implies $\phi = 0$ in $L^\infty(\mu)$. ■

This section concludes with a result on the C*-algebra generated by a subnormal operator. This will be used in the next section.

30.21 Proposition. *If S is a subnormal operator, T is a bounded operator, and $\rho\colon C^*(S) \to C^*(T)$ is a $*$-homomorphism with $\rho(S) = T$, then T is subnormal and the following hold:*

(a) *$\sigma_n(T) \subseteq \sigma_n(S)$;*

(b) *if $u \in C(\sigma_n(S))$, then $u(T)$ is well defined, $u(S) \in C^*(S)$, $u(T) \in C^*(T)$, and $\rho(u(S)) = u(T)$.*

Proof. The fact that T is subnormal follows from the definition of a subnormal element of a C*-algebra and the fact that a $*$-homomorphism takes positive elements to positive elements.

(a) Suppose there is a λ in $\sigma_n(T)\backslash\sigma_n(S)$. Let u be a continuous function on $\sigma_n(S) \cup \sigma_n(T)$ such that $u|\sigma_n(S) = 0$ and $u(\lambda) \neq 0$. Let $\{p_n\}$ be a sequence of polynomials in z and $\bar{z}$ that converges to u uniformly on $\sigma_n(S) \cup \sigma_n(T)$. When defining $p_n(S)$ (and $p_n(T)$), always put the adjoints on the left (see Exercise 9). Because ρ is a $*$-homomorphism, it follows that $\rho(p_n(S)) = p_n(T)$. Thus $\rho(u(S)) = u(T)$. But $u(S) = 0$ since u vanishes on $\sigma_n(S)$. Therefore $u(T) = 0$. By the preceding proposition, $u = 0$ in $L^\infty(\nu)$, where ν is a scalar-valued spectral measure for the minimal normal extension of T. But since the continuous function u does not vanish at a point of $\sigma_n(T) = \operatorname{supp}\nu$, this is impossible.

(b) Let $u \in C(\sigma_n(S))$. By part (a), $u(T)$ is well defined. The remainder of (b) follows by approximating u by polynomials in z and $\bar{z}$ and arguing as in the proof of (a). ∎

The last proposition appears in Feldman [1999]. One might ask if the map $\phi \to \phi(S)$ in Proposition 30.20 is an isometry. This is not always the case, but the subnormals for which this map is isometric can be characterized. See Keough [1981]. Also see Conway [1991], §II.12 for additional information.

Exercises.

1. Define an operator S to be *quasinormal* if it commutes with S^*S. Show that S is quasinormal if and only if the two factors in its polar decomposition commute. If A is a positive operator on a Hilbert space $\mathcal{L}$, define S on $\mathcal{H} = \mathcal{L}^{(\infty)}$ by the matrix

$$S = \begin{bmatrix} 0 & 0 & 0 & \dots \\ A & 0 & 0 & \dots \\ 0 & A & 0 & \dots \\ 0 & 0 & A & \dots \end{bmatrix}.$$

Show that S is quasinormal.

2. Show that a quasinormal operator is subnormal. (Hint: First assume that $\ker S = (0)$. If $S = UA$ is the polar decomposition of S, put $E = UU^*$ and show that

$$N = \begin{bmatrix} S & E^{\perp}A \\ 0 & S^* \end{bmatrix}$$

is normal. To handle the remaining case, show that $\ker S$ reduces S.)

3. Let ν be positive measure on $[0,1]$ and define a measure μ on the closed unit disk by $d\mu(re^{i\theta}) = (2\pi)^{-1}d\theta\, d\nu(r)$. For this measure consider the cyclic subnormal operator S_μ as defined in Example 30.2. Show that S_μ is a weighted shift.

4. Show that a weighted shift is hyponormal if and only if its weight sequence is increasing.

5. Let μ be the measure constructed in Exercise 3 and let $\mathcal{H}$ be the closed linear span of $\{\, |z|^k z^n : k, n \geq 0 \,\}$. If S is defined on $\mathcal{H}$ as multiplication by z, show that S is quasinormal but not normal.

6. If S is subnormal and λ is an eigenvalue for S, show that $\ker(S - \lambda)$ reduces S. Show that the same is true for hyponormal operators.

7. If N is a normal operator on $\mathcal{K}$ and $\mathcal{H} \in \operatorname{Lat} N$, show that $\mathcal{H}$ reduces N if and only if $N|\mathcal{H}$ is normal.

8. If $\{T_n\}$ is a sequence of operators in $\mathcal{B}(\mathcal{H})$ and $\|T_n - T\| \to 0$, show that for every open set U that contains $\sigma(T)$, there is an integer N such that $\sigma(T_n) \subseteq U$ for all $n \geq N$.

9. With the definition in (30.19), show that for all integers $n, k \geq 0$, $(\bar{z}^n z^k)(S) = S^{*n} S^k$.

§31. Essentially normal operators

An operator T is *essentially normal* if its self-commutator, $[T^*, T] \equiv T^*T - TT^*$, is compact. Equivalently, an operator is essentially normal if its image in the Calkin algebra, $\mathcal{B}(\mathcal{H})/\mathcal{B}_0(\mathcal{H})$, is normal. If N is normal and K is a compact operator, then $T = N + K$ is easily seen to be essentially normal. There are, however, essentially normal operators that are not of this form. Indeed, the unilateral shift is one such example.

If S is the unilateral shift, $[S^*, S] = P$, a projection of rank 1. To see that S is not a compact perturbation of a normal operator, it is necessary to use the Fredholm index ([ACFA], Chapter XI). Here S is Fredholm and $\operatorname{ind} S = -1$. For a normal operator N that is Fredholm and any compact operator K, $\operatorname{ind}(N + K) = 0$. For the remainder of this section the reader will be assumed to be familiar with Chapter XI in [ACFA]. This material is reviewed in §37, with reference to [ACFA] for most of the proofs.

Why the interest in essentially normal operators? One reason is success. In Brown, Douglas, and Fillmore [1973] and [1977] the essentially normal operators are characterized up to unitary equivalence modulo the compact operators. Actually they do more than this and establish a link between operator theory and topological K-theory. This result, part of the theory of C*-algebras, is an important development in mathematics so that it is worth studying essentially normal operators. See Davidson [1996] for a treatment of the "BDF" theory.

The purpose of this section is to generate a large collection of examples of essentially normal operators. To do this a few topics from classical analysis need to be explored.

31.1 Lemma. *If K is a compact subset of $\mathbb{C}$, then for every z*

$$\int_K |z - \zeta|^{-1} \, d\mathcal{A}(\zeta) < \infty.$$

Proof. If z is fixed, then $\int_K |z - \zeta|^{-1} \, d\mathcal{A}(\zeta) = \int \chi_K(\zeta)|z - \zeta|^{-1} \, d\mathcal{A}(\zeta) = \int \chi_K(z - \zeta)|z|^{-1} \, d\mathcal{A}(\zeta) = \int_{z-K} |\zeta|^{-1} \, d\mathcal{A}(\zeta)$. If R is so large that $z - K \subseteq B(0; R)$, then a computation using polar coordinates shows that $\int_K |z - \zeta|^{-1} \, d\mathcal{A}(\zeta) \leq \int_{B(0;R)} |\zeta|^{-1} \, d\mathcal{A}(\zeta) = 2\pi R$. ∎

With this established, the properties can be established for a construction that is basic for much of what will be done.

31.2 Proposition. *If K is a compact set having positive area and*

$$f(z) \equiv \int_K (\zeta - z)^{-1}\, d\mathcal{A}(\zeta)$$

for all z in $\mathbb{C}$ and $f(\infty) = 0$, then $f\colon \mathbb{C}_\infty \to \mathbb{C}$ is a continuous function that is analytic on $\mathbb{C}_\infty \setminus K$ with $f'(\infty) = -\text{Area}\, K$. In addition,

$$|f(z)| \le [\pi \text{Area}\,(K)]^{1/2}$$

for all z in $\mathbb{C}$.

Proof. The fact that f is a continuous function on $\mathbb{C}_\infty$ is left as an exercise for the reader. It is also easy to see that f is analytic on $\mathbb{C}\setminus K$ and, since f is continuous at ∞, ∞ is a removable singularity. Because $f(\infty) = 0$, $f'(\infty)$ is the limit of $zf(z)$ as $z \to \infty$. But $zf(z) = \int_K (\zeta/z - 1)^{-1}\, d\mathcal{A}(\zeta) \to -\text{Area}\, K$ as $z \to \infty$ since $\zeta/z \to 0$ uniformly for ζ in K.

It remains to prove the inequality for $|f(z)|$. This inequality is due to Ahlfors and Beurling [1950], though the proof here is from Gamelin and Khavinson [1989]. From the properties already established and the Maximum Modulus Theorem, f attains its maximum value at some point of K. By translating the set K, it can be assumed that $0 \in K$ and f attains its maximum at 0. In addition, if K is replaced by a suitable unimodular multiple of itself, it can be assumed that $f(0) > 0$. Thus

$$|f(z)| \le f(0) = \int_K \text{Re}\, \frac{1}{\zeta}\, d\mathcal{A}(\zeta).$$

Let $c = [\pi/\text{Area}\, K]^{1/2}$ and $a = 1/2c$. It is elementary to see that the closed disk $D = \bar{B}(a; a)$ is $\{z : \text{Re}\,(1/z) \ge c\}$ and that D and K have the same area. Thus $\text{Area}\,(D \cap K) + \text{Area}\,(D \setminus K) = \text{Area}\,(D) = \text{Area}\,(K) = \text{Area}\,(D \cap K) + \text{Area}\,(K \setminus D)$. Hence $\text{Area}\,(D \setminus K) = \text{Area}\,(K \setminus D)$. On the other hand, $\text{Re}\,(1/\zeta) < c$ for ζ in $K \setminus D$ and $\text{Re}\,(1/\zeta) \ge c$ for ζ in $D \setminus K$. Therefore

$$\begin{aligned}
f(0) &\le \int_{K\cap D} \text{Re}\, \frac{1}{\zeta}\, d\mathcal{A} + c\mathcal{A}(K \setminus D) \\
&= \int_{K\cap D} \text{Re}\, \frac{1}{\zeta}\, d\mathcal{A} + c\mathcal{A}(D \setminus K) \\
&\le \int_{K\cap D} \text{Re}\, \frac{1}{\zeta}\, d\mathcal{A} + \int_{D\setminus K} \text{Re}\, \frac{1}{\zeta}\, d\mathcal{A} \\
&= \int_D \text{Re}\, \frac{1}{\zeta}\, d\mathcal{A}.
\end{aligned}$$

It is left to the reader to show that for $0 < r < a$,

$$\int_0^{2\pi} \frac{1}{a + re^{i\theta}}\, d\theta = \frac{2\pi}{a},$$

by converting this to an integral around the circle $z = a + re^{i\theta}$. Converting to polar coordinates shows that

$$\begin{aligned}
\int_D \operatorname{Re} \frac{1}{\zeta}\, d\mathcal{A} &= \int_0^a \int_0^{2\pi} \operatorname{Re} \frac{1}{a + re^{i\theta}} dq r dr \\
&= \int_0^a \operatorname{Re} \int_0^{2\pi} \frac{1}{a + re^{i\theta}} d\theta r dr \\
&= \int_0^a \frac{2\pi}{a} r\, dr \\
&= \pi\, a \\
&= \pi \left[\frac{\operatorname{Area} K}{\pi}\right]^{1/2} \\
&= [\pi \operatorname{Area} K]^{1/2}. \quad \blacksquare
\end{aligned}$$

For an open subset G of $\mathbb{C}$ and $n \geq 1$, $C^n(G)$ is the space of those complex-valued functions on G whose partial derivatives up to and including the n-th order exist and are continuous. $C_c^n(G)$ is the subalgebra of $C^n(G)$ consisting of those functions with compact support in G. (The *support* of a function ϕ is defined as the closure of $\{z : \phi(z) \neq 0\}$ and is abbreviated by $\operatorname{supp} \phi$.) Also $C^n \equiv C^n(\mathbb{C})$ and $C_c^n \equiv C_c^n(\mathbb{C})$.

Functions on $\mathbb{C}$ can, of course, be thought of as functions defined on $\mathbb{R}^2$. Because of the interest in making connections with analytic function theory, it is more convenient to not think of such functions as functions of two real variables. But thinking of a function u defined on an open subset of $\mathbb{C}$ as a function of a single complex variable isn't possible since the differentiability of such functions must be discussed even when the functions are not analytic. Hence it is necessary to introduce the two complex variables z and $\bar{z}$ that are related to the two real variables in the usual way: $z = x + iy$ and $\bar{z} = x - iy$. With this change of variables and the corresponding change in the basis of the space of differentials, the following formulas relating the corresponding derivatives hold:

$$\partial u \equiv \partial_z u \equiv \frac{\partial u}{\partial z} \equiv \frac{1}{2}\left(\frac{\partial u}{\partial x} - i \frac{\partial u}{\partial y}\right),$$

$$\bar{\partial} u \equiv \partial_{\bar{z}} u \equiv \frac{\partial u}{\partial \bar{z}} \equiv \frac{1}{2}\left(\frac{\partial u}{\partial x} + i\frac{\partial u}{\partial y}\right).$$

The usual rules for differentiating algebraic combinations of functions as well as the chain rule remain in force for these new derivatives. This can be verified by direct computation or appeal to the theory of differential forms.

If μ is any compactly supported measure on $\mathbb{C}$, let

$$\widetilde{\mu}(w) = \int \frac{1}{|w-z|}\, d|\mu|(z)$$

when the integral converges, and let $\widetilde{\mu}(w) = \infty$ when the integral diverges. It happens that $\widetilde{\mu}$ is finite a.e. [Area]. Indeed if K is the support of μ and $R > 0$, then

$$\begin{aligned}\int_{B(0;R)} \widetilde{\mu}(w)\, d\mathcal{A}(w) &= \int_{B(0;R)} \int_K \frac{1}{|w-z|}\, d|\mu|(z)\, d\mathcal{A}(w) \\ &= \int_K \int_{B(0;R)} \frac{1}{|w-z|}\, d\mathcal{A}(w)\, d|\mu|(z) \\ &\le C_R \|\mu\|\end{aligned}$$

where the constant C_R exists by Lemma 31.1. Thus $\widetilde{\mu}$ is locally integrable and hence finite a.e. [Area]. This discussion legitimizes the following definition.

31.3 Definition. If μ is a compactly supported measure on the plane, the *Cauchy transform* of μ is the function $\widehat{\mu}$ defined a.e. [Area] by the equation

$$\widehat{\mu}(w) = \int \frac{1}{z-w}\, d\mu(z).$$

So the function f defined in Proposition 31.2 is the Cauchy transform of the characteristic function of the set K. If g is any bounded Borel function on $\mathbb{C}$ that has compact support, define the Cauchy transform of g, $\widehat{g}$, as the Cauchy transform of the measure $g \cdot \text{Area}$. It is routine to show that in this case $\widehat{g}$ is a continuous function on the plane.

31.4 Proposition. *If μ is a compactly supported measure, the following statements hold.*

(a) *$\widehat{\mu}$ is locally integrable.*

(b) *$\widehat{\mu}$ is analytic on $\mathbb{C}_\infty \setminus \operatorname{supp}(\mu)$.*

(c) *For w in $\mathbb{C} \setminus \operatorname{supp}(\mu)$ and $n \geq 0$,*

$$\partial^n \widehat{\mu}(w) = (-1)^n n! \int \frac{1}{(z-w)^{n+1}} \, d\mu(z).$$

(d) *$\widehat{\mu}(\infty) = 0$ and the power series of $\widehat{\mu}$ near ∞ is given by*

$$\widehat{\mu}(w) = \sum_{n=0}^{\infty} \left(\int z^n \, d\mu(z) \right) \frac{1}{w^{n+1}}.$$

Proof. The proof of (a) follows the lines of the discussion preceding the definition. The proofs of (b) and (c) follow by differentiating under the integral sign. Note that as $w \to \infty$, $(z-w)^{-1} = w^{-1}(z/w - 1)^{-1} \to 0$ uniformly for z in any compact set. Hence $\widehat{\mu}$ has a removable singularity at ∞ and $\widehat{\mu}(\infty) = 0$.

It remains to establish (d). This is done by choosing R so that $\operatorname{supp}(\mu) \subseteq B(0; R)$, expanding $(z-w)^{-1} = -w^{-1}(1 - z/w)^{-1}$ in a geometric series for $|w| > R$, and integrating term-by-term. ■

Here, without proof, is a standard consequence of Green's Theorem. The reader can find a proof in Conway [1991], Corollary V.2.6, or Conway [1995], Corollary 13.2.9.

31.5 Proposition. *If $u \in C_c^1$, then for all w in $\mathbb{C}$*

$$u(w) = -\frac{1}{\pi} \int \frac{1}{z-w} \bar{\partial} u(z) \, dA(z).$$

The importance of the Cauchy transform for rational approximation can be glimpsed in the next theorem.

31.6 Theorem. *If K is a compact subset of $\mathbb{C}$ and μ is a measure on K, then $\mu \perp R(K)$ if and only if $\widehat{\mu}(w) = 0$ a.e. [Area] on $\mathbb{C}_\infty \setminus K$.*

Proof. If $\mu \perp R(K)$, then for $w \notin K$, $(z-w)^{-1} \in R(K)$. Hence $\widehat{\mu}(w) = 0$. Conversely, assume that $\widehat{\mu} = 0$ a.e. [Area] off K; since $\widehat{\mu}$ is analytic off K, $\widehat{\mu}$ is identically 0 off K. This implies that all the derivatives of $\widehat{\mu}$ vanish on $\mathbb{C}_\infty \setminus K$. From (31.4.c) and (31.4.d) it follows that μ annihilates all polynomials and all rational functions with poles off K . ■

31.7 Corollary. *If g is a bounded Borel function on the plane having compact support and $g = 0$ a.e. on K, then $\widehat{g} \in R(K)$.*

Proof. If $\mu \perp R(K)$, then

$$\int \widehat{g}\, d\mu = \int \left[\int_{\mathbb{C}\setminus K} \frac{g(\zeta)}{\zeta - z}\, d\mathcal{A}(\zeta) \right] d\mu(z)$$

$$= \int_{\mathbb{C}\setminus K} g(\zeta) \left[\int_K \frac{1}{\zeta - z}\, d\mu(z) \right] d\mathcal{A}(\zeta).$$

But for ζ in $\mathbb{C} \setminus K$, $(\zeta - z)^{-1} \in R(K)$. So this inside integral is 0. ∎

It can be shown that the set of all such Cauchy transforms that appear in this corollary constitutes a dense subalgebra of $R(K)$. See Proposition V.3.12 in Conway [1991].

As was pointed out in the preceding section, for a compact subset K of $\mathbb{C}$, $R(K) = C(K)$ if and only if $\bar{z} \in R(K)$. So the next result has more than a passing interest.

31.8 Theorem. (Alexander [1973]) *If K is a compact subset of $\mathbb{C}$, then*

$$\operatorname{dist}(\bar{z}, R(K)) \le \left[\frac{\operatorname{Area} K}{\pi} \right]^{1/2}.$$

Proof. Choose a function u in C_c^∞ such that $u(z) = \bar{z}$ in a neighborhood of K. Now $u(z) = -\pi^{-1} \int (w - z)^{-1} u(w)\, d\mathcal{A}(w)$ by Proposition 31.5. But $\bar{\partial} u \equiv 1$ near K so that for z in K,

$$\bar{z} = -\frac{1}{\pi} \int_K \frac{1}{w - z}\, d\mathcal{A}(w) - \frac{1}{\pi} \int_{\mathbb{C}\setminus K} \frac{1}{w - z}\, u(w)\, d\mathcal{A}(w).$$

Define $f = \widehat{\chi_{\mathbb{C}\setminus K} u}$. By Corollary 31.7, $f \in R(K)$. Therefore using Proposition 31.2 it follows that

$$\operatorname{dist}(\bar{z}, R(K)) \le \|\bar{z} - f\|_K$$

$$= \left\| \frac{1}{\pi} \int_K \frac{1}{w - z}\, d\mathcal{A}(w) \right\|_K$$

$$\le \left[\frac{\operatorname{Area} K}{\pi} \right]^{1/2}. \quad ∎$$

The next result is called the Hartogs–Rosenthal Theorem. There is another proof that avoids Theorem 31.8 but which still relies on the Cauchy transform. See Conway [1991].

31.9 Corollary. *If K is a compact subset of the plane having zero area, then $R(K) = C(K)$.*

31.10 Corollary. *If S is a subnormal operator whose spectrum has zero area, then S is normal.*

The next result is the reason for doing the analysis performed thus far in this section. Actually this theorem is valid for hyponormal operators. But a simplified proof, due to Axler and Shapiro [1985], is presented that is only valid for the case of a subnormal operator. The proof of the full theorem can be found in Conway [1991], IV.3.1.

31.11 Putnam's Inequality. (Putnam [1970]) *If S is a subnormal operator, then*

$$\|\,[S^*, S]\,\| \leq \frac{1}{\pi}\text{Area}\,(\sigma(S)).$$

Proof. Actually, it will be shown that

31.12 $$\|S^*S - SS^*\| \leq \left\{\text{dist}_{\,C(\sigma(S))}[\bar{z}, R(\sigma(S))]\right\}^2$$

which, by virtue of the preceding theorem, captures Putnam's Inequality.

If P is the projection of $\mathcal{K}$ onto $\mathcal{H}$ and $g \in \mathcal{H}$ with $\|g\| = 1$, then

$$\begin{aligned}
\langle (S^*S - SS^*)g, g\rangle &= \|Sg\|^2 - \|S^*g\|^2 \\
&= \|Ng\|^2 - \|P(N^*g)\|^2 \\
&= \|N^*g\|^2 - \|P(N^*g)\|^2 \\
&= \text{dist}\,[N^*g, \mathcal{H}]^2 \\
&= \inf\{\|N^*g - h\|^2 : h \in \mathcal{H}\} \\
&\leq \inf\{\|N^*g - f(N)g\|^2 : f \in R(\sigma(S))\} \\
&\leq \inf\{\|N^* - f(N)\|^2 : f \in R(\sigma(S))\} \\
&= \left\{\text{dist}_{\,C(\sigma(S))}[\bar{z}, R(\sigma(S))]\right\}^2.
\end{aligned}$$

Since g was arbitrary, this proves (31.12). ■

This section concludes with a characterization of the essentially normal subnormal operators and some consequences.

31.13 Theorem. (Feldman [1999]) *If S is a subnormal operator and ϕ is the Cauchy transform of the characteristic function of $\sigma_e(S)$, then S is essentially normal if and only if $[\phi(S), S] = \phi(S)S - S\phi(S)$ is compact.*

Proof. First set some notation that will remain in effect for the duration of the proof. Let $\pi\colon \mathcal{B}(\mathcal{H}) \to \mathcal{B}/\mathcal{B}_0$ be the natural map and let $\rho\colon C^*(\pi(S)) \to \mathcal{B}(\mathcal{K})$ be a faithful representation. Put $T = \rho(\pi(S))$; so T is a subnormal operator and $\eta = \rho \circ \pi\colon C^*(S) \to C^*(T)$ is a $*$-homomorphism such that $\eta(S) = T$. Note that $\ker \eta = C^*(S) \cap \mathcal{B}_0$.

Assume that S is essentially normal; thus T is normal. By Proposition 31.21, $\eta([\phi(S), S]) = [\phi(T), T] = 0$. Therefore $[\phi(S), S] \in \mathcal{B}_0$.

Now assume that $[\phi(S), S] \in \mathcal{B}_0$. Let u be a continuously differentiable function on $\mathbb{C}$ that has compact support such that $u(z) = \bar{z}$ in a neighborhood of $\sigma(S)$. By Proposition 31.5, $u(z)$ can be represented as the Cauchy transform of $\bar{\partial} u$. Hence

$$\begin{aligned} u(z) &= -\frac{1}{\pi} \int \frac{\bar{\partial} u(\zeta)}{\zeta - z}\, d\mathcal{A}(\zeta) \\ &= -\frac{1}{\pi} \left[\int_{\mathbb{C} \setminus \sigma(S)} \frac{\bar{\partial} u(\zeta)}{\zeta - z}\, d\mathcal{A}(\zeta) + \int_{\sigma(S) \setminus \sigma_e(S)} \frac{1}{\zeta - z}\, d\mathcal{A}(\zeta) \right. \\ &\qquad \left. + \int_{\sigma_e(S)} \frac{1}{\zeta - z}\, d\mathcal{A}(\zeta) \right] \\ &= -\frac{1}{\pi} \left[u_1(z) + u_2(z) + u_3(z) \right]. \end{aligned}$$

Since $u(z) = \bar{z}$ near $\sigma(S)$, $S^* = u(S) = -\frac{1}{\pi} \left[u_1(S) + u_2(S) + u_3(S) \right]$. So S is essentially normal if it can be shown that $[u_j(S), S] \in \mathcal{B}_0$ for $j = 1, 2, 3$.

Now the hypothesis is precisely the statement that $[u_3(S), S]$ is compact. Since u_1 is the Cauchy transform of a bounded Borel function with compact support that vanishes on $\sigma(S)$, Corollary 7 implies that $u_1 \in R(\sigma(S))$ and so $[u_1(S), S] = 0$. The same corollary implies $u_2 \in R(\sigma_e(S))$. Using the notation from the start of the proof, $\sigma(T) = \sigma(\pi(S)) = \sigma_e(S)$. Thus $\eta([u_2(S), S]) = [u_2(T), T] = 0$, so $[u_2(S), S] \in \ker \eta \subseteq \mathcal{B}_0$. ∎

31.14 Corollary. *If S is a subnormal operator and $R(\sigma_e(S)) = C(\sigma_e(S))$, then S is essentially normal.*

Proof. Since $\phi = \widehat{\chi}_{\sigma_e(S)}$ is a continuous function, the hypothesis says that $\phi \in R(\sigma_e(S))$. Again using the notation from the beginning of the proof of the preceding theorem, this implies $[\phi(T), T] = 0$, so that $[\phi(S), S] \in \ker \eta \subseteq \mathcal{B}_0$. ∎

31.15 Corollary. (Aleman [1996]) *If S is a subnormal operator, then S is essentially normal whenever* Area $(\sigma_e(S)) = 0$.

Proof. This is a consequence of the Hartogs–Rosenthal Theorem and the preceding corollary. ■

Also see Conway and Feldman [1997].

31.16 Corollary. *If G is a bounded open set in the plane, the Bergman operator for G is essentially normal.*

Proof. Denote the Bergman operator for G by S. If ϕ is as in Theorem 31.13, then ϕ is analytic off $\sigma_e(S)$, a subset of ∂G (29.2.c). Since ϕ is continuous on the entire plane (Exercise 3) and vanishes at ∞, ϕ is a bounded analytic function on G. Thus $\phi(S) = M_\phi$ commutes with S. ■

Remarks. 1. A pivotal result on essentially normal operators is the Berger–Shaw Theorem (Berger and Shaw [1973a], [1973b], and [preprint]), which was not examined here. This result says that if T is a hyponormal operator and there are a finite number of vectors $h_1, \ldots, h_m$ such that $\mathcal{H} = [\{ r(T)h_j : r \in \text{Rat}(\sigma(T)), 1 \leq j \leq m \}]$, then $[T^*, T]$ is trace class and

$$\text{tr}\,[T^*, T] \leq (m/\pi)\text{Area}\,(\sigma(T)).$$

The reader can see Conway [1991], §IV.2 for an exposition and additional references.

2. Corollary 31.15 is immediate from Theorem 31.13 without appeal to Corollary 31.14. If Area $(\sigma_e(S)) = 0$, then the function ϕ in Theorem 31.14 is the zero function.

3. It is shown in Berger–Shaw [preprint]† that the Bergman operator S for any bounded open set has a trace class self-commutator and $\text{tr}\,[S^*, S] \leq \pi^{-1}\text{Area}\,(\sigma(S))$. Also see Conway [1991], Theorem IV.4.2.

Exercises.

1. If S and T are irreducible operators, and X is a bounded operator such that $SX = XT$ and $S^*X = XT^*$, show that either $X = 0$ or $S \cong T$.

2. (Morrel [1973]) Let S be a pure subnormal operator with rank 1 self-commutator. (a) Show that there are constants α and β such that if $T = \alpha S + \beta$, then there is a unit vector e_0 such that $[T^*, T] = e_0 \otimes e_0$ and $Te_0 \in \ker\,[T^*, T]$. (b) Show that for any subnormal operator, the kernel of the self-commutator is an invariant subspace for the operator. (c) Show that $\text{ran}\, T \subseteq \ker\,[T^*, T]$. (d) Prove that T is quasinormal. (e) Show that T is the unilateral shift.

---

†Through some unfortunate quirk of fate, this paper was never published. The results have been generalized and extended, but the original approach is largely lost. That's unfortunate as this is worth study.

3. If g is a bounded Borel function on $\mathbb{C}$ with compact support, prove that its Cauchy transform is a continuous function on the plane that is analytic off $\operatorname{supp} g$ and vanishes at ∞.

4. This exercise leads the reader through the proof of a decomposition theorem for essentially normal operators. The proof can be found in Theorem IV.5.4 of Conway [1991], though that proof has an error that is corrected here. Let S be an essential normal operator on $\mathcal{H}$, put $\mathcal{A} = C^*(S)$, and $\mathcal{I} = \mathcal{A} \cap \mathcal{B}_0$. (a) Show that $\mathcal{H}_0' = [\mathcal{L}\mathcal{H}]^\perp$ is a reducing subspace for S and $S|\mathcal{H}_0'$ is normal. Observe that $\mathcal{I}|(\mathcal{H} \ominus \mathcal{H}_0')$ is a non-degenerate C*-algebra of compact operators. Let $\mathcal{H}_1, \mathcal{H}_2, \ldots$ be the minimal reducing subspaces for $\mathcal{I}|(\mathcal{H} \ominus \mathcal{H}_0')$, so that $\mathcal{H} \ominus \mathcal{H}_0' = \mathcal{H}_1 \oplus \mathcal{H}_2 \oplus \cdots$ and $\mathcal{I}|\mathcal{H}_k$ is isomorphic to the algebra of compact operators on a Hilbert space. (See §16.) (b) Show that for $k \geq 1$, $\mathcal{H}_k$ reduces S and $S|\mathcal{H}_k$ is irreducible and essentially normal. (c) Show that each of the spaces $\mathcal{H}_k$, $k \geq 1$, is either infinite dimensional or one dimensional. Define $\mathcal{H}_0$ to be the direct sum of $\mathcal{H}_0'$ and all the spaces $\mathcal{H}_k$ that are one dimensional. Put $S_0 = S|\mathcal{H}_0$ and relabel the remaining spaces so that $\mathcal{H}_1, \mathcal{H}_2, \ldots$ are all infinite dimensional. (d) Show that $S = S_0 \oplus S_1 \oplus \cdots$, where S_0 is normal and each S_n, $n \geq 1$, is an irreducible essentially normal operator. (e) Show that this decomposition is unique except for the order of the summands.

Chapter 5

More on C*-Algebras

This chapter develops some ideas for C*-algebras that will be useful in the rest of this book. The treatment here is certainly not exhaustive. The theory of C*-algebras is a vast subject with several treatises on the subject as well as monographs on special topics. The interested reader can examine Davidson [1996], Dixmier [1964], Kadison and Ringrose [1983] and [1986], and Pedersen [1979]. A nice introduction, which was helpful in writing this chapter, is Arveson [1976]

For the remainder of this book it will be an underlying assumption that all C*-algebras have an identity.

§32. Irreducible representations

Recall (16.10) that a representation $\pi : \mathcal{A} \to \mathcal{B}(\mathcal{H})$ is irreducible provided the C*-algebra $\pi(\mathcal{A})$ is irreducible in $\mathcal{B}(\mathcal{H})$. That is, if there are no subspaces of $\mathcal{H}$ that are reducing for $\pi(\mathcal{A})$ except for the trivial ones, (0) and $\mathcal{H}$. Equivalently, π is irreducible if and only if the only projections on $\mathcal{H}$ that commute with the range of π are the trivial projections, 0 and 1.

This section will characterize those states on a C*-algebra such that the corresponding cyclic representations are irreducible. As a consequence it is proved that every C*-algebra has an abundance of irreducible representations, enough to separate points. Some additional results that relate the properties of cyclic representations with those of the states that determine them will also be presented.

If $\mathcal{A}$ is a C*-subalgebra of $\mathcal{B}(\mathcal{H})$, there is a ready way to define states on it. If $h \in \mathcal{H}$, let $\omega_h : \mathcal{B}(\mathcal{H}) \to \mathbb{C}$ by $\omega_h(T) = \langle Th, h\rangle$ for all operators T. Note that ω_h is a positive linear functional on $\mathcal{B}(\mathcal{H})$ and it is a state if h is a unit vector. Thus $\omega_h|\mathcal{A}$ is a state on $\mathcal{A}$ when $\|h\| = 1$. If $\mathcal{A}$ is an abstract C*-algebra and $(\pi, \mathcal{H})$ is a representation, then for any vector h in $\mathcal{H}$, $a \to \omega_h(\pi(a))$ is a positive linear functional on $\mathcal{A}$.

In this section the notation ϕ, π, e will be reserved to mean that ϕ is a state on the C*-algebra, π is the cyclic representation defined by ϕ, and e is the cyclic vector. (See Theorem 7.7.) If ψ is a positive linear functional, the cyclic representation defined by ψ is discussed even though ψ is not a state. Of course there is no problem here as $\|\psi\|^{-1}\psi$ is a state. If $\pi_\psi : \mathcal{A} \to \mathcal{H}_\psi$ is the corresponding representation, then there is a cyclic vector e_ψ for $\pi_\psi(\mathcal{A})$ and $\psi(a) = \langle \pi_\psi(a)e_\psi, e_\psi \rangle$ for all a in $\mathcal{A}$. The only wrinkle here is that e_ψ will not be a unit vector if ψ is not a state.

The next result is a Radon–Nikodym Theorem for positive linear functionals on C*-algebras.

32.1 Proposition. *If $\mathcal{A}$ is a C^*-algebra and ψ is a positive linear functional on $\mathcal{A}$, then $\psi \leq \phi$ if and only if there is a unique operator T in $\pi(\mathcal{A})'$ with $0 \leq T \leq 1$ such that $\psi(a) = \langle \pi(a)Te, e\rangle = \omega_{T^{1/2}e}(\pi(a))$ for all a in $\mathcal{A}$.*

Proof. First assume that $T \in \pi(\mathcal{A})'$ and $0 \leq T \leq 1$. If $a \in \mathcal{A}_+$, then $0 \leq T^{1/2}\pi(a)T^{1/2} = \pi(a)T = \pi(a)^{1/2}T\pi(a)^{1/2} \leq \pi(a)$. Thus $\psi(a) = \langle \pi(a)Te, e\rangle \leq \langle \pi(a)e, e\rangle = \phi(a)$.

Now assume that ψ is a positive linear functional on $\mathcal{A}$ and $\psi \leq \phi$. The Cauchy–Schwarz Inequality implies

$$\begin{aligned} |\psi(b^*a)|^2 &\leq \psi(a^*a)\,\psi(b^*b) \\ &\leq \phi(a^*a)\,\phi(b^*b) = \|\pi(a)e\|^2\|\pi(b)e\|^2. \end{aligned}$$

Thus $[\pi(a)e, \pi(b)e] \equiv \psi(b^*a)$ is a well defined positive semi-definite form on the linear manifold $\pi(\mathcal{A})e$ and it is bounded by 1. Extend this form to $[\pi(\mathcal{A})e] = \mathcal{H}$. Thus there is an operator T on $\mathcal{H}$ with $0 \leq T \leq 1$ such that

$$\psi(b^*a) = \langle T\pi(a)e, \pi(b)e\rangle$$

for all a, b in $\mathcal{A}$. In particular, $\psi(a) = \omega_e(T\pi(a))$.

If $a, b, c \in \mathcal{A}$, then $\langle T\pi(a)\pi(b)e, \pi(c)e\rangle = \langle T\pi(ab)e, \pi(c)e\rangle = \psi(c^*(ab)) = \psi((a^*c)^*b) = \langle T\pi(b), \pi(a^*c)e\rangle = \langle \pi(a)T\pi(b)e, \pi(c)e\rangle$. Since b and c are arbitrary and e is a cyclic vector, $T\pi(a) = \pi(a)T$.

To see that such an operator is unique, assume there is another positive contraction S in $\pi(\mathcal{A})'$ such that $\langle \pi(a)Te, e\rangle = \langle \pi(a)Se, e\rangle$ for all a in $\mathcal{A}$. But then for all a, b in $\mathcal{A}$, $\langle T\pi(a)e, \pi(b)e\rangle = \langle T\pi(b^*a)e, e\rangle = \langle S\pi(a)e, \pi(b)e\rangle$. Since e is a cyclic vector for the representation π this implies $S = T$. ■

32.2 Proposition. *If $(\pi_j, \mathcal{H}_j)$, $j = 1, 2$ are two cyclic representations of the C^*-algebra $\mathcal{A}$ with unit cyclic vectors e_j, then $\pi_1 \cong \pi_2$ via an isomorphism $U: \mathcal{H}_1 \to \mathcal{H}_2$ that satisfies $Ue_1 = e_2$ if and only if*

$$\langle \pi_1(a)e_1, e_1 \rangle = \langle \pi_2(a)e_2, e_2 \rangle$$

for all a in $\mathcal{A}$.

Proof. If $\pi_1 \cong \pi_2$ via an isomorphism U as described in the statement, then $\langle \pi_1(a)e_1, e_2 \rangle = \langle U^*\pi_2(a)Ue_1, e_1 \rangle = \langle \pi_2(a)e_2, e_2 \rangle$.

Now put $\phi_j(a) = \langle \pi_j(a)e_j, e_j \rangle$ and suppose that $\phi_1 = \phi_2$. If $a \in \mathcal{A}$, $\|\pi_2(a)e_2\|^2 = \phi_2(a^*a) = \phi_1(a^*a) = \|\pi_1(a)e_1\|^2$. Thus $U(\pi_1(a)e_1) = \pi_2(a)e_2$ gives a well defined isometry from the manifold $\pi_1(\mathcal{A})e_1$ onto $\pi_2(\mathcal{A})e_2$. Extend U to an isomorphism $U : \mathcal{H}_1 \to \mathcal{H}_2$. Clearly $Ue_1 = e_2$. Moreover, if $a, b \in \mathcal{A}$, then

$$U(\pi_1(a)\pi_1(b)e_1) = \pi_2(ab)e_2 = \pi_2(a)\pi_2(b)e_2 = \pi_2(a)U(\pi_1(b)e_1).$$

It follows that $U\pi_1(a) = \pi_2(a)U$, so that $\pi_1 \cong \pi_2$. ■

32.3 Corollary. *If ψ is a positive linear functional on $\mathcal{A}$ and $\psi \leq \phi$, then π_ψ is equivalent to a subrepresentation of π.*

Proof. Proposition 32.1 implies there is a positive contraction T in $\pi(\mathcal{A})'$ such that $\psi(a) = \langle \pi(a)Te, e \rangle = \omega_{T^{1/2}e}(\pi(a))$ for all a in $\mathcal{A}$. Put $\mathcal{L} = [\pi(\mathcal{A})T^{1/2}e]$. Clearly $\mathcal{L}$ reduces $\pi(\mathcal{A})$. Since

$$\langle \pi_\psi(a)e_\psi, e_\psi \rangle = \psi(a) = \langle \pi(a)T^{1/2}e, T^{1/2}e \rangle$$

for all a, the preceding proposition implies there is an isomorphism $U : \mathcal{H}_\psi \to \mathcal{L}$ such that $Ue_\psi = T^{1/2}e$ and $U\pi_\psi(a)U^* = \pi(a)|\mathcal{L}$ for all a in $\mathcal{A}$. ■

The results above lead naturally to the next definition.

32.4 Definition. A state ϕ on $\mathcal{A}$ is *pure* if for any positive linear functional ψ on $\mathcal{A}$ such that $\psi \leq \phi$, there is a scalar λ with $0 \leq \lambda \leq 1$ such that $\psi = \lambda\phi$.

The reader can check the veracity of the statements in the following.

32.5 Examples. (a) If $\mathcal{A}$ is any C*-algebra and $\phi : \mathcal{A} \to \mathbb{C}$ is a non-zero multiplicative linear functional, then ϕ is a pure state.

(b) If X is a compact space, the pure states of $C(X)$ are precisely the multiplicative linear functionals.

(c) If h is a unit vector, ω_h is a pure state on $\mathcal{B}(\mathcal{H})$.

The next theorem gives several characterizations of irreducible representations. In particular it shows that they are precisely the representations defined by a pure state. After a moment's thought, at least half of this equivalence should not be too surprising. Pure states are those that have no subfunctionals; irreducible representations are those that have no subrepresentations.

32.6 Theorem. *If $\mathcal{A}$ is a C^*-algebra and $(\rho, \mathcal{H})$ is a representation, the following statements are equivalent.*

(a) *ρ is an irreducible representation.*

(b) $\rho(\mathcal{A})' = \mathbb{C}$.

(c) *$\rho(\mathcal{A})$ is strongly dense in $\mathcal{B}(\mathcal{H})$.*

(d) *If h is any non-zero vector in $\mathcal{H}$, then $\mathcal{H} = [\rho(\mathcal{A})h]$.*

(e) *ρ is equivalent to a cyclic representation defined by a pure state.*

Proof. (a) *is equivalent to* (b). $\rho(\mathcal{A})'$ is a von Neumann algebra. Hence $\rho(\mathcal{A})'$ is the norm closed linear span of its projections (13.3.e). Since reducing subspaces for $\rho(\mathcal{A})$ are projections in its commutant, (a) and (b) are equivalent.

(b) *implies* (c). This a direct consequence of the Double Commutant Theorem.

(c) *implies* (d). Condition (c) implies that for a non-zero vector h, $[\rho(\mathcal{A})h] = [\mathcal{B}(\mathcal{H})h] = \mathcal{H}$.

(d) *implies* (a). If E is a non-zero projection in $\pi(\mathcal{A})'$ and $e \in E$, $[\pi(\mathcal{A})e] \subseteq E$. By (d), $E = 1$.

(a) *implies* (e). Let h be any unit vector; so h is a cyclic vector for the representation ρ by (d). Put $\phi = \omega_h \circ \rho$. Let $(\pi, \mathcal{K})$ be the cyclic representation defined by the state ϕ and let e be its cyclic vector. So $\langle \pi(a)e, e\rangle = \phi(a) = \langle \rho(a)h, h\rangle$. By Proposition 32.2 there is an isomorphism $U : \mathcal{H} \to \mathcal{K}$ such that $Uh = e$ and $U\rho(a)U^{-1} = \pi(a)$. Since ρ is irreducible, so is π.

Now to show that ϕ is a pure state on $\mathcal{A}$. If ψ is a positive linear functional on $\mathcal{A}$ such that $\psi \leq \phi$, Proposition 32.1 implies there is an operator T in $\pi(\mathcal{A})'$ with $0 \leq T \leq 1$ and $\psi(a) = \langle \pi(a)Te, e\rangle$ for all a in $\mathcal{A}$. Since π is irreducible, $\pi(\mathcal{A})' = \mathbb{C}$, so there is a scalar λ such that $T = \lambda$. Thus $\psi = \lambda\phi$ and ϕ is pure.

(e) *implies* (a). By (e) it can be assumed that $(\rho, \mathcal{H}) = (\pi_\phi, \mathcal{H}_\phi)$ for some pure state ϕ with cyclic vector $e = e_\phi$. So $\phi(a) = \langle \rho(a)e, e\rangle$. If P is a projection in $\rho(\mathcal{A})'$, $\omega_{Pe} \circ \rho \leq \phi = \omega_e \circ \rho$. Since ϕ is pure, there is a scalar λ in $[0,1]$ such that $\omega_{Pe} \circ \rho = \lambda\omega_e \circ \rho$. Thus for every a in $\mathcal{A}$, $\langle Pe, \rho(a)e\rangle = \omega_{Pe}(\rho(a^*)) = \lambda\omega_e(\rho(a^*)) = \lambda\langle e, \rho(a)e\rangle$. Since $\mathcal{H} = [\rho(\mathcal{A})e]$, $Pe = \lambda e$. Hence, $P\rho(a)e = \rho(a)Pe = \lambda\rho(a)e$. That is, $P = \lambda$. So λ is either 0 or 1 and ρ must be irreducible. ∎

Observe that $\Sigma = \Sigma(\mathcal{A})$, the state space of $\mathcal{A}$, is a weak* compact, convex subset of $\mathcal{A}^*$, the Banach space dual of $\mathcal{A}$. Thus Σ has extreme points. What are they?

32.7 Theorem. *If ϕ is a state on $\mathcal{A}$, then ϕ is pure if and only if ϕ is an extreme point of $\Sigma(\mathcal{A})$.*

Proof. Assume ϕ is pure so that $\pi = \pi_\phi$ is irreducible, and let e be the cyclic vector for this representation. Suppose there are ψ, κ in $\Sigma = \Sigma(\mathcal{A})$ such that $\phi = \alpha\psi + \beta\kappa$, with $\alpha, \beta \geq 0$ and $\alpha + \beta = 1$. Thus $\alpha\psi \leq \phi$. Since ϕ is pure, there is a scalar λ such that $\alpha\psi = \lambda\phi$. But $\alpha = \alpha\psi(1) = \lambda\phi(1) = \lambda$, so $\psi = \phi$. Therefore ϕ is an extreme point of Σ.

For the converse, assume that ϕ is an extreme point and show that ϕ is pure. If ψ is a positive linear functional such that $\psi \leq \phi$, put $\alpha = \psi(1)$. So $\phi = \alpha[\alpha^{-1}\psi] + (1-\alpha)[(1-\alpha)^{-1}(\phi - \psi)]$ and $\alpha^{-1}\psi$ and $(1-\alpha)^{-1}(\phi-\psi) \in \Sigma$. Thus $\phi = \alpha^{-1}\psi$, or $\psi = \alpha\phi$. Hence ϕ is pure. ∎

32.8 Corollary. *If $\mathcal{B}$ is a C^*-subalgebra of $\mathcal{A}$ that contains the identity, then every pure state on $\mathcal{B}$ can be extended to a pure state on $\mathcal{A}$.*

Proof. Recall (7.6) that every state on $\mathcal{B}$ can be extended to a state on $\mathcal{A}$. The trick is to extend a pure state to a pure state. Here a basic result from Banach space theory (Exercise 4) is used. Define $T\colon \Sigma(\mathcal{A}) \to \Sigma(\mathcal{B})$ by $T(\phi) = \phi|\mathcal{B}$. By the first remark this map is surjective. It is easy to check that T is weak* continuous and affine. The corollary now follows by Exercise 3 and the theorem. ∎

32.9 Corollary. *If $a \in \mathcal{A}$, then*

$$\|a\|^2 = \sup\{\, \phi(a^*a) : \phi \text{ is a pure state on } \mathcal{A}\,\}.$$

Proof. If X is compact, the fact that the pure states of $C(X)$ are the point evaluations shows that the corollary is true when $\mathcal{A} = C(X)$. That is, the corollary is true whenever $\mathcal{A}$ is abelian. Now if $a \in \mathcal{A}$, $C^*(a^*a)$, the C*-algebra generated by a^*a and the identity, is abelian. Therefore, applying the preceding corollary gives

$$\begin{aligned}\|a\|^2 = \|a^*a\| &= \sup\{\, \phi(a^*a) : \phi \text{ is a pure state on } C^*(a^*a)\,\}\\ &\leq \sup\{\, \phi(a^*a) : \phi \text{ is a pure state on } \mathcal{A}\,\}. \quad ∎\end{aligned}$$

32.10 Proposition. *If $\mathcal{A}$ is a separable C^*-algebra, there is a sequence of irreducible representations $\{\pi_n\}$ such that $\bigoplus_n \pi_n$ is a faithful representation of $\mathcal{A}$.*

Proof. Since $\mathcal{A}$ is separable, Σ with its weak* topology is a compact metric space and, hence, separable. Thus there is a sequence $\{\phi_n\}$ of pure states that is weak* dense in the set of all pure states. Let $\{\pi_n\}$ be the corresponding irreducible representations.

If $a \in \mathcal{A}$ and $\pi_n(a) = 0$ for all n, then $\phi_n(a^*a) = 0$ for all n, and so $\phi(a^*a) = 0$ for every pure state ϕ. By Corollary 32.9, $\|a\| = 0$. ■

There is more that can be said about irreducible representations. For example, each irreducible representation π is algebraically irreducible. That is, if h is a non-zero vector in $\mathcal{H}$, then $\pi(\mathcal{A})h = \mathcal{H}$. In other words, condition (d) in Theorem 32.6 is true without taking closures (Kadison [1957]). The proof of this can be found in Kadison and Ringrose [1983]. A number of curiosities follow from this fact. For example, in the GNS construction of the representation π from a state ϕ, the left ideal $\mathcal{L}_\phi = \{x \in \mathcal{A} : \phi(x^*x) = 0\}$ is considered and the Hilbert space $\mathcal{H}_\phi$ is defined as the completion of the quotient space $\mathcal{A}/\mathcal{L}_\phi$ with respect to a norm defined by the state ϕ. In the case that ϕ is pure, the quotient is already complete. In fact, this is immediate from the algebraic irreducibility of π as the manifold $\mathcal{A}/\mathcal{L}_\phi$ is invariant under $\pi_\phi(\mathcal{A})$.

Exercises.

1. Show that a subrepresentation of a cyclic representation is cyclic.

2. Formulate and prove a converse to Corollary 32.3.

3. Let $\mathcal{A}$ be a C*-algebra and $\mathcal{I}$ an ideal of $\mathcal{A}$. If $\rho : \mathcal{A} \to \mathcal{B}(\mathcal{H})$ is an irreducible representation, show that $\rho|\mathcal{I}$ is irreducible. Conversely, if $\rho : \mathcal{I} \to \mathcal{B}(\mathcal{H})$ is an irreducible representation, show that ρ has a unique extension to an irreducible representation of $\mathcal{A}$. (See Proposition 6.10.)

4. If $\mathcal{X}$ and $\mathcal{Y}$ are locally convex spaces, K and L are compact convex subsets of $\mathcal{X}$ and $\mathcal{Y}$, respectively, and $T : K \to L$ is a surjective, continuous affine map, then for each extreme point y of L there is an extreme point x of K such that $T(x) = y$.

5. Show that $(\rho, \mathcal{H})$ is an irreducible representation of $\mathcal{A}$ if and only if for every unit vector e in $\mathcal{H}$, $a \to \omega_e(\rho(a))$ is a pure state.

6. If $\mathcal{A}$ is a unital C*-algebra, Σ its state space, and $a \in \operatorname{Re}\mathcal{A}$, define $\widehat{a}\colon \Sigma \to \mathbb{C}$ by $\widehat{a}(\phi) = \phi(a)$. Show that $a \to \widehat{a}$ is an isometric real linear embedding of $\operatorname{Re}\mathcal{A}$ into $C(\Sigma)$.

§33. Positive maps

Positive linear functionals on a C*-algebra have already received considerable attention in this book. Given this and the general sophistication of anyone capable of getting this far, the definition of a positive map between C*-algebras is predictable. But a bit more is desired. If $\mathcal{A}$ is a C*-algebra and $\mathcal{S}$ is a subset of $\mathcal{A}$, define $\mathcal{S}^* = \{a^* : a \in \mathcal{S}\}$.

33.1 Definition. An *operator system* is a linear manifold $\mathcal{S}$ in a C*-algebra such that $1 \in \mathcal{S}$ and $\mathcal{S} = \mathcal{S}^*$.

The terminology is a bit misleading in that all C*-algebras need not consist of operators. This is not a big deal, since every C*-algebra can be embedded into $\mathcal{B}(\mathcal{H})$. Moreover the terminology seems to be standard. At least it appears in Paulsen [1986], which forms the basis for the presentation here as well as in the next two sections.

It's easy to manufacture examples of operator systems. For example if $\mathcal{L}$ is any linear manifold in $\mathcal{A}$ that contains the identity, $\mathcal{S} = \mathcal{L} + \mathcal{L}^*$ is an operator system. Of course the C*-algebra itself is an operator system. It is also helpful to notice that an operator system has an abundance of positive elements. For example, if a is a hermitian element of $\mathcal{S}$, then $\|a\| \pm a \in \mathcal{S}_+$. In fact, $a = \frac{1}{2}(\|a\| + a) - \frac{1}{2}(\|a\| - a)$ so that every hermitian element of an operator system is the difference of two of its positive elements. Since the real and imaginary parts of each element in $\mathcal{S}$ also belongs to $\mathcal{S}$, it follows that an operator system is the linear span of its positive elements.

Many of the results for positive linear functionals on a C*-algebra carry over to the setting of an operator system. Some of these are recorded here. The proofs are identical to the corresponding results in §7.

33.2 Proposition. *Let $\mathcal{S}$ be an operator system and let $f : \mathcal{S} \to \mathbb{C}$ be a linear functional.*

(a) *If f is positive, then f is bounded and $\|f\| = f(1)$.*

(b) *If f is bounded and $\|f\| = f(1)$, then f is positive.*

(c) *If $\mathcal{S}$ is contained in the C*-algebra $\mathcal{A}$ and f is positive, then f can be extended to a positive linear functional on $\mathcal{A}$.*

33.3 Definition. If $\mathcal{A}$ and $\mathcal{B}$ are C*-algebras and $\mathcal{S}$ is an operator system in $\mathcal{A}$, a *positive map* between $\mathcal{S}$ and $\mathcal{B}$ is a linear map $\phi : \mathcal{S} \to \mathcal{B}$ such that $\phi(a) \geq 0$ whenever $a \in \mathcal{S}_+$.

Observe that if $\mathcal{S}$ is an operator system and $\phi\colon \mathcal{S} \to \mathcal{B}$ is a positive map, then $\phi(a) = \phi(a)^*$ whenever $a \in \mathcal{S}$ and $a = a^*$. Indeed this is immediate from the fact seen above that a hermitian element of $\mathcal{S}$ can be written as the difference of two positive elements. Thus $\phi(x^*) = \phi(x)^*$ for all x in $\mathcal{S}$.

It has been established that a positive linear functional ϕ on an operator system is bounded and its norm is $\|\phi\| = \phi(1)$. The same is not completely true of a positive map.

33.4 Proposition. *If $\mathcal{S}$ is an operator system and $\phi : \mathcal{S} \to \mathcal{B}$ is a positive map, then ϕ is bounded and $\|\phi\| \leq 2\|\phi(1)\|$.*

Proof. If a is a hermitian element of $\mathcal{A}$, $-\|a\| \leq a \leq \|a\|$ and so $-\|a\|\phi(1) \leq \phi(a) \leq \|a\|\phi(1)$. Since every element of the operator system can be written in terms of its real and imaginary parts, $\|\phi\| \leq 2\,\|\phi(1)\|$. ∎

33.5 Example. Consider the manifold $\mathcal{S}$ in $C(\partial\mathbb{D})$ spanned by the functions $1, z, \bar{z}$. That is, $\mathcal{S} = \{\,\alpha + \beta z + \gamma\bar{z} : \alpha, \beta, \gamma \in \mathbb{C}\,\}$. Define $\phi : \mathcal{S} \to M_2$ by

$$\phi(\alpha + \beta z + \gamma\bar{z}) = \begin{bmatrix} \alpha & 2\beta \\ 2\gamma & \alpha \end{bmatrix}.$$

It is left to the reader to verify that $f = \alpha + \beta z + \gamma\bar{z}$ is a positive function on the circle if and only if $\gamma = \bar{\beta}$ and $\alpha \geq 2|\beta|$. Thus when f is positive, the matrix $\phi(f)$ is hermitian, has positive diagonal entries, and a positive determinant. Hence $\phi(f)$ is a positive matrix so that ϕ is a positive map. Now $\phi(1) = 1$ and $\|\phi(z)\| = 2$, while $\|z\| = 1$. Thus the constant 2 in the preceding proposition is sharp.

When the range of a positive map is abelian, the estimate can be improved.

33.6 Proposition. *If $\mathcal{S}$ is an operator system and $\phi : \mathcal{S} \to C(X)$ is a positive map, then $\|\phi\| = \|\phi(1)\|$.*

Proof. For x in X, let δ_x be the evaluation functional on $C(X)$. Note that for x in X, $\delta_x \circ \phi$ is a positive linear functional on $\mathcal{S}$. Thus for a in $\mathcal{S}$,

$$\begin{aligned} \|\phi(a)\| &= \sup\{\,|\phi(a)(x)| : x \in X\,\} \\ &= \sup\{\,|\delta_x \circ \phi(a)| : x \in X\,\} \\ &\leq \sup\{\,\|a\|\,(\delta_x \circ \phi)(1) : x \in X\,\} \\ &= \|a\|\,\|\phi(1)\|. \quad \blacksquare \end{aligned}$$

Note that being defined on an operator system contained in an abelian C*-algebra does not suffice as Example 33.5 shows. It turns out that if the positive map is defined on the entirety of an abelian C*-algebra, however, its norm is also $\|\phi(1)\|$. The proof of this takes a bit more effort and requires one of those lemmas whose proof is easy enough to write down, but which is, undoubtedly, no one's first proof.

33.7 Lemma. *If $\mathcal{A}$ is a C^*-algebra and $a_1, \dots, a_n$ are positive elements in $\mathcal{A}$ such that $\sum_i a_i \leq 1$, then for any scalars $\lambda_1, \dots, \lambda_n$ with $|\lambda_i| \leq 1$ for $1 \leq i \leq n$,*

$$\left\| \sum_{i=1}^{n} \lambda_i a_i \right\| \leq 1.$$

Proof. Let A be the matrix in $M_n(\mathcal{A})$ with $\sum_{i=1}^{n} \lambda_i a_i$ in the upper left-hand corner and zeros elsewhere, and let

$$X = \begin{bmatrix} \sqrt{a_1} & 0 & \cdots & 0 \\ \sqrt{a_2} & 0 & \cdots & 0 \\ \cdot & \cdot & \cdots & \\ \sqrt{a_n} & 0 & \cdots & 0 \end{bmatrix}.$$

The reader can check that $A = X^*[\operatorname{diag}\{\lambda_1, \dots, \lambda_n\}]X$. Using the defining equality for a C*-algebra, $\|a^*a\| = \|a\|^2$, each of the three matrices on the right of the preceding equation is seen to have norm at most 1. Thus $\|A\| \leq 1$, and this is what was desired. ∎

33.8 Theorem. *If X is a compact Hausdorff space and $\phi : C(X) \to \mathcal{B}$ is a positive map, then $\|\phi\| = \|\phi(1)\|$.*

Proof. $\phi(1)$ is a positive element. So by replacing ϕ with a positive multiple of itself, it can be assumed that $\|\phi(1)\| = 1$. It must be shown that $\|\phi\| \leq 1$.

Fix $\epsilon > 0$ and let $f \in \operatorname{ball} C(X)$. Cover X by open sets $U_1, \dots, U_n$ such that such that the oscillation of f over each open set is less than ϵ. Pick a point x_i in U_i, so that $|f(x) - f(x_i)| < \epsilon$ for all x in U_i. Let $\{u_1, \dots, u_n\}$ be a partition of unity on X subordinate to $\{U_1, \dots, U_n\}$. Put $\lambda_i = f(x_i)$. It follows that $\|f - \sum_i \lambda_i u_i\| < \epsilon$.

If $p_i = \phi(u_i)$, then $p_i \in \mathcal{B}_+$ and $\sum_i p_i = \phi(\sum_i u_i) = \phi(1) \leq 1$. According to the preceding lemma, $\|\sum_i \lambda_i p_i\| \leq 1$. Therefore $\|\phi(f)\| \leq \|\phi(f) - \phi(\sum_i \lambda_i u_i)\| + 1 \leq \epsilon\|\phi\| + 1$. Hence $\|\phi(f)\| \leq 1$. ∎

This section concludes with a few facts about positive maps, including a proposition about extending them. The first generalizes Proposition 33.2.b.

33.9 Proposition. *If $\mathcal{S}$ is an operator system and $\phi : \mathcal{S} \to \mathcal{B}$ is a bounded linear map such that $\phi(1) = 1$ and $\|\phi\| = 1$, then ϕ is positive.*

Proof. If $\tau : \mathcal{B} \to \mathbb{C}$ is a state, then $\tau \circ \phi$ is a bounded linear functional on $\mathcal{S}$ with $\|\tau \circ \phi\| \leq 1$ and $\tau \circ \phi(1) = 1$. By Proposition 33.2.b, $\tau \circ \phi \geq 0$. Since τ was arbitrary, ϕ is positive. ■

33.10 Proposition. *If $\mathcal{M}$ is a linear manifold in the C^*-algebra $\mathcal{A}$ that contains the identity, $\mathcal{B}$ is a C^*-algebra, and $\phi\colon \mathcal{M} \to \mathcal{B}$ is a contraction such that $\phi(1) = 1$, then ϕ has a unique extension to a positive map $\widetilde{\phi}\colon \mathcal{M} + \mathcal{M}^* \to \mathcal{B}$ given by*

$$\widetilde{\phi}(a + b^*) = \phi(a) + \phi(b)^*$$

for all a, b in $\mathcal{M}$.

Proof. Put $\mathcal{S} = \mathcal{M} + \mathcal{M}^*$. First prove uniqueness. If $\psi : \mathcal{S} \to \mathcal{B}$ is a positive extension of ϕ, then for all a, b in $\mathcal{M}$, $\psi(a + b^*) = \psi(a) + \psi(b^*) = \phi(a) + \psi(b)^* = \phi(a) + \phi(b)^*$.

Now to show that $\widetilde{\phi}$ is well defined. To do this it suffices to show that if $a = a^* \in \mathcal{M}$, then $\phi(a) = \phi(a)^*$. But $\mathcal{S}_1 \equiv \{ a \in \mathcal{M} : a = a^* \}$ is an operator system, and on $\mathcal{S}_1$ it holds that $\|\phi|\mathcal{S}_1\| \leq 1$ and $\phi(1) = 1$. By Proposition 33.9, $\phi|\mathcal{S}_1 \geq 0$, so that ϕ is hermitian on $\mathcal{S}_1$. Thus $\widetilde{\phi}$ is well defined. It is left for the the reader to check that $\widetilde{\phi}$ is linear.

Now to show that $\widetilde{\phi}$ is positive. Let f be a state on $\mathcal{B}$ and consider the linear functional $f \circ \phi$. This functional has norm at most 1 and $f \circ \phi(1) = 1$. Hence $\|f \circ \phi\| = 1$. Let $\widetilde{f}$ be a norm preserving extension of $f \circ \phi$ to $\mathcal{S}$. By Proposition 33.2, $\widetilde{f} \geq 0$. From the first part of this proof (with $\mathcal{B} = \mathbb{C}$), $\widetilde{f}$ is the unique extension with $\widetilde{f}(a+b^*) = (f \circ \phi)(a) + (f \circ \phi)(b)^* = (f \circ \widetilde{\phi})(a+b^*)$. That is, $f \circ \widetilde{\phi} \geq 0$ for every state f on $\mathcal{B}$. Hence $\widetilde{\phi} \geq 0$. ■

A positive map on an operator system need not have a positive extension to the enveloping C*-algebra. Indeed, if $\mathcal{S}$ is the operator system from Example 33.5 and $\phi : \mathcal{S} \to M_2$ the positive map, then ϕ has no positive extension to $C(\partial\,\mathbb{D})$. (Why?)

Exercise.

1. If $\mathcal{S}$ is an operator system, $\mathcal{B}$ is a C*-algebra, and $\phi\colon \mathcal{S} \to \mathcal{B}$ is a linear map, then ϕ is positive if and only if for every state f on $\mathcal{B}$, $f \circ \phi$ is a positive linear functional on $\mathcal{S}$.

§34. Completely positive maps

If $\mathcal{A}$ is a C*-algebra, form a new C*-algebra $M_n(\mathcal{A})$, the algebra of all $n \times n$ matrices with entries in $\mathcal{A}$. How is $M_n(\mathcal{A})$ made into a C*-algebra? Making it an algebra is no problem. Just define the algebraic operations as is always done when matrices are involved: addition and scalar multiplication are defined entry-wise and multiplication is matrix multiplication. Define the adjoint of an element of $M_n(\mathcal{A})$ as the transpose of the matrix whose entries are the adjoints of the original entries. But how to define the norm?

If $\mathcal{A}$ is a subalgebra of $\mathcal{B}(\mathcal{H})$, defining the norm is easy. Since $\mathcal{A}$ acts on $\mathcal{H}$, $M_n(\mathcal{A})$ acts on the inflation $\mathcal{H}^{(n)}$ in the usual way. Thus give $M_n(\mathcal{A})$ the norm it has as a subalgebra of $\mathcal{B}(\mathcal{H}^{(n)})$. But what if $\mathcal{A}$ is not a subalgebra of $\mathcal{B}(\mathcal{H})$? Take a faithful representation $\rho : \mathcal{A} \to \mathcal{B}(\mathcal{H})$ and define $\rho_n : M_n(\mathcal{A}) \to \mathcal{B}(\mathcal{H}^{(n)})$ by $\rho_n([a_{ij}]) = [\rho(a_{ij})]$. It is routine algebra to check that ρ_n is a $*$-isomorphism from the $*$-algebra $M_n(\mathcal{A})$ into $\mathcal{B}(\mathcal{H}^{(n)})$ and that $\rho_n(M_n(\mathcal{A})) = M_n(\rho(\mathcal{A}))$. Now just define the norm on $M_n(\mathcal{A})$ by $\|[a_{ij}]\| \equiv \|\rho_n([a_{ij}])\|$. To see that the definition of the norm does not depend on the choice of representation, suppose $(\rho_1, \mathcal{H}_1)$ is a second faithful representation. Here $\rho_1 \circ \rho^{-1} : M_n(\rho(\mathcal{A})) \to M_n(\rho_1(\mathcal{A}))$ is a $*$-isomorphism. Because a $*$-isomorphism is an isometry (5.7), this shows that the norm on $M_n(\mathcal{A})$ is well defined. This same argument shows that the norm on $M_n(\mathcal{A})$ is the unique norm that makes $M_n(\mathcal{A})$ into a C*-algebra.

34.1 Examples. (a) The C*-algebra $M_n(C(X))$ is naturally identified with $C(X, M_n)$, the algebra of all continuous functions from X into M_n. If $F \in C(X, M_n)$ and $f_{ij}(x)$ is the (i, j) entry of $F(x)$, then $f_{ij} \in C(X)$ and F corresponds to the element $[f_{ij}]$ in $M_n(C(X))$. For a function F in $C(X, M_n)$, $\|F\| \equiv \sup\{\, \|F(x)\| : x \in X\}$ and $F^*(x) = F(x)^*$. The function F is a positive element of $C(X, M_n)$ precisely when $F(x)$ is a positive matrix for every x in X. The details are left to the reader. Checking the equality of the norms follows because of the uniqueness of the norm on $M_n(C(X))$.

(b) The C*-algebra $M_n(\mathcal{B}(\mathcal{H}))$ is naturally identified with $\mathcal{B}(\mathcal{H}^{(n)})$.

If $\mathcal{S}$ is an operator system contained in the C*-algebra $\mathcal{A}$, then consider $M_n(\mathcal{S})$ as a subset of $M_n(\mathcal{A})$. A moment's reflection will convince the reader that for each $n \geq 1$, $M_n(\mathcal{S})$ is an operator system. Thus it is possible to talk about positive maps defined on $M_n(\mathcal{S})$.

34.2 Definition. If $\mathcal{S}$ is an operator system and $\mathcal{B}$ a C*-algebra, a map $\phi : \mathcal{S} \to \mathcal{B}$ is *completely positive* if its natural extension $\phi_n : M_n(\mathcal{S}) \to M_n(\mathcal{B})$, defined by

$$\phi_n([a_{ij}]) = [\phi(a_{ij})],$$

is positive for every $n \geq 1$.

The notation here is a bit unfortunate. There is always the possibility of confusing the maps ϕ_n with members of a sequence. Context is, of course, everything, and vigilance is a self-rewarding virtue.

From Proposition 33.4 it is known that if ϕ is a completely positive map, each of the positive maps ϕ_n is bounded. In fact, $\|\phi_n\| \leq 2\|\phi_n(1)\| = 2\|\phi(1)\|$, so the norms of the maps ϕ_n are uniformly bounded when ϕ is completely positive. Linear maps such that the sequence of maps $\{\phi_n\}$ are uniformly bounded are called *completely bounded*. So it was just established that a completely positive map is completely bounded and $\sup_n \|\phi_n\| \leq 2\|\phi(1)\|$.

34.3 Examples. (a) If $\rho : \mathcal{A} \to \mathcal{B}$ is a $*$-homomorphism, ρ is completely positive. In fact, it can be checked that ρ_n is also a $*$-homomorphism and, hence, positive.

(b) Suppose $\mathcal{K}$ is a Hilbert space and $\mathcal{H}$ is a closed subspace with P the projection of $\mathcal{K}$ onto $\mathcal{H}$. If $\rho : \mathcal{A} \to \mathcal{B}(\mathcal{K})$ is a $*$-homomorphism, and $\phi\colon \mathcal{A} \to \mathcal{B}(\mathcal{H})$ is defined by $\phi(A) = P\rho(A)|\mathcal{H}$, then ϕ is positive. Now observe that for every $n \geq 1$, $\phi_n(A_{ij}) = P^{(n)}\rho_n(A_{ij})|\mathcal{H}^{(n)}$. Thus ϕ is completely positive. Later a converse to this example will be proved.

Are there positive maps that are not completely positive? The answer is yes. Consider the algebra M_n and define $\phi : M_n \to M_n$ by $\phi(A) = A^t$, the transpose of A. Clearly ϕ is positive. However it will now be shown that ϕ_n is not positive, so that ϕ is not completely positive. Let $\{E_{ij}\}$ be the usual matrix units in M_n. That is, E_{ij} is the matrix with a 1 in the (i, j) entry and zeros elsewhere. As a linear transformation, E_{ij} acts on the usual basis $\{e_i\}$ by $E_{ij}e_j = e_i$ and $E_{ij}e_k = 0$ for $k \neq j$. Thus $E_{ij}E_{km} = E_{im}$ when $j = k$ and 0 otherwise. Also $E_{ij}^* = E_{ji}$. Put $E = [E_{ij}] \in M_n(M_n)$. Matrix multiplication shows that $E^2 = nE$. Since E is hermitian, $E \geq 0$. On the other hand, another calculation shows that $\phi_n(E)^2 = 1$, but $\phi_n(E) \neq 1$. Hence $\phi_n(E)$ is a hermitian unitary and so $\sigma(\phi_n(E)) = \{\pm 1\}$. In particular, $\phi_n(E)$ is not positive.

Now to focus on obtaining conditions on a positive map that imply it is completely positive. For an element a of any C*-algebra $\mathcal{A}$ and a matrix $T = [t_{ij}]$ in M_n, define $a \otimes T$ to be the element of $M_n(\mathcal{A})$ whose (i, j) entry is $t_{ij}a$. That is, $a \otimes T = [t_{ij}a]$. This is just formal notation here, though this formal notation can be turned into reality. The tensor product of two C*-algebras can be defined and the algebra $M_n(\mathcal{A})$ is the tensor product $M_n \otimes \mathcal{A}$. This will not be discussed here. (The interested reader can look at the references on C*-algebras.)

The next two results are from Stinespring [1955], who was the first to isolate the concept of complete positivity and recognize its importance and centrality in the development of many parts of operator theory.

34.4 Proposition. *If X is a compact space, then every positive map on $C(X)$ is completely positive.*

Proof. Suppose $\phi : C(X) \to \mathcal{A}$ is a positive map. If $u \in C(X)$ and $T \in M_n$, then, when $M_n(C(X))$ is identified with $C(X, M_n)$, the element $u \otimes T$ corresponds to the function in $C(X, M_n)$ whose value at x is $u(x)T$. So $u \otimes T \geq 0$ if $u \geq 0$ and $T \geq 0$. If $T = [t_{ij}]$, then $\phi_n(u \otimes T) = \phi_n[t_{ij}u] = [t_{ij}\phi(u)] = \phi(u) \otimes T$. Since ϕ is positive, $\phi_n(u \otimes T) \geq 0$ whenever $u \geq 0$ and $T \geq 0$.

Now let F be any positive element of $C(X, M_n)$; so $F(x) \geq 0$ in M_n for all x in X. Let $\epsilon > 0$ and let $\{U_1, \dots, U_n\}$ be an open cover of X with points x_k in U_k such that $\|F(x) - F(x_k)\| < \epsilon$ for x in U_k. Put $T_k = F(x_k)$, a positive matrix. Let $\{u_k\}$ be a partition of unity subordinate to this cover. It is straightforward to check that $\|F(x) - \sum_k u_k(x)T_k\| < \epsilon$ for every x in X. That is, $\|F - \sum_k u_k \otimes T_k\| < \epsilon$. Thus

$$\begin{aligned} \epsilon\|\phi_n\| &> \left\| \phi_n(F) - \phi_n\left[\sum_k u_k \otimes T_k\right] \right\| \\ &= \left\| \phi_n(F) - \sum_k \phi_n(u_k \otimes T_k) \right\|. \end{aligned}$$

But, from the discussion above, $\phi_n(u_k \otimes T_k) \geq 0$ for each k. So $\phi_n(F)$ can be approximated by positive elements and therefore must be positive. ■

34.5 Lemma. *Every positive linear functional on an operator system is a completely positive map.*

Proof. If $\phi : \mathcal{S} \to \mathbb{C}$ is positive, to show that $\phi_n : M_n(\mathcal{S}) \to M_n$ is positive, it must be shown that for any $h = (\lambda_1, \dots, \lambda_n)$ in $\mathbb{C}^n$ and $A = [a_{ij}]$ a positive matrix in $M_n(\mathcal{S})$,

$$\begin{aligned} \langle \phi_n(A)h, h\rangle &= \sum_{i,j} \phi(a_{ij})\lambda_j\bar{\lambda}_i \\ &= \phi\left(\sum_{i,j} \lambda_j\bar{\lambda}_i a_{ij}\right) \\ &\geq 0. \end{aligned}$$

To see this, show that $\sum_{i,j} \lambda_j\bar{\lambda}_i a_{ij} \geq 0$ in $\mathcal{S}$. But if L is the $n \times n$ matrix with the numbers $\lambda_1, \dots, \lambda_n$ in the first column and zeros elsewhere, then a calculation shows that the sum in question is the $(1, 1)$ entry of the matrix product L^*AL. Since this product is a positive matrix, that entry must be positive. ■

34.6 Proposition. *If $\mathcal{S}$ is an operator system and X is a compact space, every positive map $\phi : \mathcal{S} \to C(X)$ is completely positive.*

Proof. Let $A = [a_{ij}]$ be a positive element in $M_n(\mathcal{S})$. To show that $\phi_n(A) \geq 0$ it must be shown that $\phi_n(A)(x)$ is a positive matrix in M_n for every point x in X. Fix an arbitrary point x in X and let δ: $C(X) \to \mathbb{C}$ be the corresponding evaluation functional. But $\phi_n(A)(x) = \delta_n \circ \phi_n(A) = (\delta \circ \phi)_n(A)$. Since $\delta \circ \phi : \mathcal{A} \to \mathbb{C}$ is a positive linear functional, the preceding lemma implies that $(\delta \circ \phi)_n$ is positive and concludes the proof. ■

The reader can consult Paulsen for additional criteria for a positive map to be completely positive. Before going further, it is timely to settle a point concerning phraseology and notation. Suppose $\mathcal{H}$ and $\mathcal{K}$ are Hilbert spaces and $V : \mathcal{H} \to \mathcal{K}$ is an isometry. Hence $V^*V = 1$ and VV^* is the projection P of $\mathcal{K}$ onto $V(\mathcal{H})$. So in the category of Hilbert spaces, $\mathcal{H}$ and $\mathcal{L} \equiv V(\mathcal{H})$ can be identified. If $T \in \mathcal{B}(\mathcal{K})$, $V^*TV \in \mathcal{B}(\mathcal{H})$. Moreover, $V^*TV \cong PT|\mathcal{L}$. For this reason, it will often be said in this situation that $\mathcal{H}$ is contained in $\mathcal{K}$ and discussion will center on $PT|\mathcal{H}$ rather than V^*TV. This will surface in the next theorem.

The following result is the converse to Example 34.3.b.

34.7 Stinespring's Theorem. (Stinespring [1955]) *If $\mathcal{A}$ is a C^*-algebra and $\phi : \mathcal{A} \to \mathcal{B}(\mathcal{H})$ is a completely positive map with $\phi(1) = 1$, then there is a Hilbert space $\mathcal{K}$ containing $\mathcal{H}$ and a $*$-homomorphism ρ: $\mathcal{A} \to \mathcal{B}(\mathcal{K})$ such that*

$$\phi(a) = P_{\mathcal{H}}\, \rho(a)|\mathcal{H}$$

for all a in $\mathcal{A}$.

Proof. Consider the algebraic tensor product $\mathcal{A} \otimes \mathcal{H}$ consisting of all formal sums $\sum_j a_j \otimes h_j$. Define a bilinear form on $\mathcal{A} \otimes \mathcal{H}$ by

$$\left\langle \sum_j a_j \otimes h_j, \sum_i b_i \otimes g_i \right\rangle = \sum_{i,j} \langle \phi(b_i^* a_j) h_j, g_i \rangle.$$

The reader can check that this is a sesquilinear form on $\mathcal{A} \otimes \mathcal{H}$. It is also positive semi-definite because

$$\begin{aligned} \left\langle \sum_j a_j \otimes h_j, \sum_j a_j \otimes h_j \right\rangle &= \sum_{i,j} \langle \phi(a_i^* a_j) h_j, h_i \rangle \\ &= \langle \phi_n([a_i^* a_j]) \bar{h}, \bar{h} \rangle, \end{aligned}$$

where $\bar{h} = (h_1, \ldots, h_n)$ in $\mathcal{H}^{(n)}$ and $[a_i^* a_j]$ is the matrix in $M_n(\mathcal{A})$. But this matrix is positive since

$$[a_i^* a_j] = \begin{bmatrix} a_1 & \ldots & a_n \\ 0 & \ldots & 0 \\ \vdots & \vdots & \vdots \\ 0 & \ldots & 0 \end{bmatrix}^* \begin{bmatrix} a_1 & \ldots & a_n \\ 0 & \ldots & 0 \\ \vdots & \vdots & \vdots \\ 0 & \ldots & 0 \end{bmatrix}.$$

Thus the positivity of the sesquilinear form is a consequence of the complete positivity of ϕ.

Now proceed as in the proof of the Gelfand–Naimark–Segal construction (§7). Let $\mathcal{N} = \{v \in \mathcal{A} \otimes \mathcal{H} : \langle v, v \rangle = 0\}$; it follows that $\mathcal{N}$ is a linear manifold in $\mathcal{A} \otimes \mathcal{H}$. Consider the quotient space $\mathcal{K}_0 \equiv (\mathcal{A} \otimes \mathcal{H})/\mathcal{N}$. As usual, $\langle u + \mathcal{N}, v + \mathcal{N} \rangle = \langle u, v \rangle$ is a well defined inner product on $\mathcal{K}_0$. Let $\mathcal{K}$ be the completion of $\mathcal{K}_0$ with respect to the norm defined by this inner product. Now define $\rho : \mathcal{A} \to \mathcal{B}(\mathcal{K})$ by

$$\rho(a)\left[\sum_j a_j \otimes h_j + \mathcal{N}\right] = \sum_j a a_j \otimes h_j + \mathcal{N}.$$

It must be checked that $\rho(a)$ is a well defined bounded operator on $\mathcal{K}_0$ and hence extends to a bounded operator on $\mathcal{K}$. After these are done, it must be shown that $\rho\colon \mathcal{A} \to \mathcal{B}(\mathcal{K})$ is a $*$-homomorphism. These details are left to the reader.

Now define an operator $V : \mathcal{H} \to \mathcal{K}$ by $Vh = 1 \otimes h + \mathcal{N}$. Note that $\|Vh\|^2 = \langle 1 \otimes h, 1 \otimes h \rangle = \langle \phi(1)h, h \rangle = \|h\|^2$. Thus V is an isometry. If a is any element of $\mathcal{A}$ and $h, g \in \mathcal{H}$, $\langle V^* \rho(a) V h, g \rangle = \langle \rho(a)(1 \otimes h + \mathcal{N}), 1 \otimes g + \mathcal{N} \rangle = \langle a \otimes h + \mathcal{N}, 1 \otimes g + \mathcal{N} \rangle = \langle \phi(a)h, g \rangle$. Thus $V^* \rho(a) V = \phi(a)$. As discussed in the paragraph leading up to this theorem, this means that $\mathcal{H}$ is contained in $\mathcal{K}$ and $\phi(a) = P_{\mathcal{H}} \rho(a)|\mathcal{H}$. ∎

34.8 Corollary. *If $\phi\colon C(X) \to \mathcal{B}(\mathcal{H})$ is a positive map with $\phi(1) = 1$, then there is a Hilbert space $\mathcal{K}$ containing $\mathcal{H}$ and a $*$-homomorphism $\rho : C(X) \to \mathcal{B}(\mathcal{K})$ such that $\phi(f) = P_{\mathcal{H}} \rho(f)|\mathcal{H}$ for all f in $C(X)$.*

Proof. Proposition 34.4 implies that the positive map ϕ is completely positive. So this is immediate from the theorem. ∎

The preceding corollary is due to Naimark [1943], though it is phrased in terms of positive operator-valued measures as follows.

34.9 Corollary. *If P is a positive operator-valued measure defined on the compact space X whose values are operators on the Hilbert space $\mathcal{H}$ and* *such that* $P(X) = 1$, *then there is a Hilbert space* $\mathcal{K}$ *containing* $\mathcal{H}$ *and a spectral measure* E *whose values are projections in* $\mathcal{B}(\mathcal{K})$ *such that* $P(\Delta) = P_{\mathcal{H}}E(\Delta)|\mathcal{H}$ *for every Borel set* Δ *contained in* X.

Proof. Recall that a positive operator-valued measure P associates with every Borel set Δ contained in X a positive operator $P(\Delta)$ in $\mathcal{B}(\mathcal{H})$ in such a way that when $h, g \in \mathcal{H}$, $\Delta \to \langle P(\Delta)h, g\rangle$ is a regular Borel measure. In this way for each f in $C(X)$, there is an operator $\phi(f)$ in $\mathcal{B}(\mathcal{H})$ defined by

$$\langle \phi(f)h, g\rangle = \int f(x)\, d\langle P(x)h, g\rangle$$

for $h, g \in \mathcal{H}$. It is not difficult to see that $\phi : C(X) \to \mathcal{B}(\mathcal{H})$ is a positive map with $\phi(1) = P(X) = 1$. Thus $\mathcal{K}$ and ρ exist as in the preceding corollary. But for the $*$-homomorphism $\rho\colon C(X) \to \mathcal{B}(\mathcal{K})$ there is a spectral measure E such that $\rho(f) = \int f\, dE$ (Theorem 9.8). Approximating characteristic functions by continuous functions completes the proof. ■

34.10 Definition. If $\phi : \mathcal{A} \to \mathcal{B}(\mathcal{H})$ is a completely positive map, $(\rho, \mathcal{K})$ is a representation of $\mathcal{A}$ with $\mathcal{H} \leq \mathcal{K}$, P is the projection of $\mathcal{K}$ onto $\mathcal{H}$, and $\phi(a) = P\rho(a)|\mathcal{H}$ for every a in $\mathcal{A}$, then $(\rho, \mathcal{K})$ is called a *dilation* of ϕ.

So it is possible to rephrase Stinespring's Theorem by saying that every completely positive map has a dilation. How about uniqueness in Stinespring's Theorem? If $(\rho, \mathcal{K})$ is a dilation of $\phi : \mathcal{A} \to \mathcal{B}(\mathcal{H})$, let $\mathcal{K}_1 = [\rho(\mathcal{A})\mathcal{H}]$. So $\mathcal{K}_1$ is a reducing subspace for $\rho(\mathcal{A})$ that contains $\mathcal{H}$. If $\rho_1\colon \mathcal{A} \to \mathcal{B}(\mathcal{K}_1)$ is defined as the subrepresentation $\rho_1(a) = \rho(a)|\mathcal{K}_1$, then it is easy to see that $(\rho_1, \mathcal{K}_1)$ is also a dilation of ϕ. Say that $(\rho, \mathcal{K})$ is a *minimal dilation* provided $\mathcal{K} = \mathcal{K}_1$ (and hence $\rho = \rho_1$). The proof of the next proposition is left to the reader.

34.11 Proposition. *If* $(\rho_i, \mathcal{K}_i)$ *are minimal dilations for the completely positive map* $\phi\colon \mathcal{A} \to \mathcal{B}(\mathcal{H})$ *with* $\phi(1) = 1$, *then there is a Hilbert space isomorphism* $W\colon \mathcal{K}_1 \to \mathcal{K}_2$ *such that* $W\rho_1 W^* = \rho_2$ *and* $Wh = h$ *for all* h *in* $\mathcal{H}$.

Complete positivity and its sister concepts form an interesting and useful part of the theory of operator algebras. They are the natural extension of positive linear functionals. For example it is a result of Arveson [1969] that a completely positive map $\phi : \mathcal{S} \to \mathcal{B}(\mathcal{H})$ always has an extension to a completely positive map defined on the enveloping C*-algebra. The reader can also consult Paulsen [1986] for this as well as other topics in completely positive maps. The next section presents an application of complete positivity, but this section closes with a characterization of subnormality.

34.12 Theorem. *If S is a bounded operator on $\mathcal{H}$, then S is subnormal if and only if there is a compact set X in $\mathbb{C}$ that contains $\sigma(S)$ and a positive linear map $\phi\colon C(X) \to \mathcal{B}(\mathcal{H})$ such that $\phi(1) = 1$, $\phi(z) = S$, and $\phi(\bar{z}z) = S^*S$. If such a positive map exists, then $\sigma(S)_n \subseteq X$ and the range of ϕ is contained in $C^*(S)$.*

Proof. If S is subnormal with minimal normal extension N acting on $\mathcal{K}$, let $X = \sigma(N)$ and define $\phi(\bar{z}^k z^j) = N^{*k}N^j$ for all $k, j \geq 0$. Extend ϕ by linearity to the operator system $\mathcal{S}$ consisting of all polynomials in z and $\bar{z}$. If $p \in \mathcal{S}$ and $p \geq 0$ on X, then $p \geq 0$ on $\sigma(N)$. Thus $p(N^*, N) \geq 0$ by the Spectral Theorem . So ϕ is positive and, hence, extends to a positive map from $C(X)$ into $\mathcal{B}(\mathcal{H})$ that has the desired properties.

Now assume that such a positive map ϕ exists. By Corollary 34.8 there is a Hilbert space $\mathcal{K}$ containing $\mathcal{H}$ and a $*$-homomorphism $\rho : C(X) \to \mathcal{B}(\mathcal{K})$ such that $\phi(f) = P\rho(f)|\mathcal{H}$ for every continuous function f on X. If $N = \rho(z)$, N is a normal operator with spectrum contained in X. Write N as

$$N = \begin{bmatrix} S & B \\ A & C \end{bmatrix}$$

with respect to the decomposition $\mathcal{K} = \mathcal{H} \oplus \mathcal{H}^{\perp}$. A matrix computation shows that

$$N^*N = \begin{bmatrix} S^*S + A^*A & * \\ * & * \end{bmatrix}.$$

But $S^*S = \phi(\bar{z}z) = PN^*N|\mathcal{H}$, so that A^*A, and hence A, must be 0. This says that N leaves $\mathcal{H}$ invariant and $N|\mathcal{H} = S$. Therefore S is subnormal.

Note that $\sigma(S)_n \subseteq \sigma(N) \subseteq X$. Also since N has the above representation as a 2×2 matrix with $A = 0$, for any polynomial p in z and $\bar{z}$ it follows that

$$p(N, N^*) = \begin{bmatrix} p(S, S^*) & * \\ * & * \end{bmatrix}.$$

Thus $\phi(p) = P\rho(p)|\mathcal{H} = p(N, N^*)|\mathcal{H} = p(S, S^*)$. By taking limits it follows that the range of ϕ is contained in $C^*(S)$. ∎

This idea of always putting the adjoints on the left was used as the basis for the hereditary functional calculus of Agler [1988]. The preceding result and proof were shown to me by Nathan Feldman, though the credit for the preceding theorem goes back to Bram [1955] who showed that an operator S is subnormal if and only if there is a positive operator-valued measure Q defined on a compact subset of the plane such that $S^{*n}S^m = \int \bar{z}^n z^m \, dQ(z)$ for all $n, m \geq 0$.

The next corollary extends Proposition 30.21.

34.13 Corollary. *If S is a subnormal operator and $\phi: C^*(S) \to \mathcal{B}(\mathcal{K})$ is a positive map such that $\phi(1) = 1$, $\phi(S) = T$, and $\phi(S^*S) = T^*T$, then T is subnormal and $\sigma(T)_n \subseteq \sigma(S)_n$.*

Proof. From the preceding theorem it is known that if $X = \sigma(S)_n$ there is a positive map $\psi: C(X) \to C_m(S)$ such that $\psi(1) = 1$, $\psi(z) = S$, and $\psi(\bar{z}z) = S^*S$. Now apply the theorem to the positive map $\phi \circ \psi$ to conclude that T is subnormal and $\sigma(T)_n \subseteq X = \sigma(S)_n$. ■

Exercises.

1. Let $\mathcal{A}$ be a C*-algebra without an identity and let $\mathcal{A}_1 = \mathcal{A} + \mathbb{C}$ be the algebra obtained by adjoining the identity. If $a \in \mathcal{A}$ and $\alpha \in \mathbb{C}$, show that $a + \alpha \geq 0$ if and only if $\alpha \geq 0$ and $a \geq 0$.

2. Give the details needed to prove the statements in Example 34.3.

3. If $\mathcal{A}$ is a C*-algebra contained in $\mathcal{B}(\mathcal{H})$ and $T: \mathcal{K} \to \mathcal{H}$, show that $\phi : \mathcal{A} \to \mathcal{B}(\mathcal{K})$ defined by $\phi(A) = T^*AT$ is completely positive. Does this remain true if the C*-algebra $\mathcal{A}$ is replaced by an operator system?

4. Let $\{\mathcal{A}_k\}$ be a sequence of C*-algebras and for each $k \geq 1$ let $\phi_k : \mathcal{A}_k \to \mathcal{B}_k$ be a completely positive map. Show that if $\sup_k \|\phi_k\| < \infty$, then $\phi : \bigoplus_k \mathcal{A}_k \to \bigoplus_k \mathcal{B}_k$ defined by $\phi(A_1 \oplus A_2 \oplus \cdots) = \phi_1(A_1) \oplus \phi_2(A_2) \oplus \cdots$ is completely positive.

5. Show that a sufficient condition for $a \otimes T$ to be positive is that a is a positive element of the C*-algebra $\mathcal{A}$ and T is a positive matrix in M_n. Give a necessary condition that $a \otimes T \geq 0$ in M_n.

6. (Bram [1955]) Prove that an operator S is subnormal if and only if there is a positive operator-valued measure Q supported on a compact subset of $\mathbb{C}$ such that $S^{*k}S^j = \int \bar{z}^k z^j \, dQ(z)$ whenever $k, j \geq 0$.

7. (Nathan Feldman) If B is the Bergman operator on $\mathbb{D}$ and S is the unilateral shift, show that there is a positive map $\phi: C^*(B) \to C^*(S)$ such that $\phi(1) = 1$, $\phi(B) = S$, and $\phi(B^*B) = S^*S$, but it is impossible to find a $*$-homomorphism ϕ with these properties.

§35. An application: Spectral sets and the Sz.-Nagy Dilation Theorem

This section demonstrates how the results obtained on positive and completely positive maps shed light on an unsolved problem about single operators. Along the way some additional results that are classical in operator theory, such as the Sz.-Nagy Dilation Theorem for contractions, will be obtained. The treatment here is taken from Paulsen [1988]. Also see Paulsen [1986].

Before defining a spectral set, it is necessary to establish von Neumann's Inequality. If $\|T\| < 1$ and $\lambda \in \partial\mathbb{D}$, define the operator theoretic Poisson kernel

35.1 $$P(\lambda; T) \equiv (1 - \bar{\lambda}T)^{-1} + (1 - \lambda T^*)^{-1} - 1.$$

Because $\|T\| < 1$, $1 - \bar{\lambda}T$ is indeed invertible so that this definition of the Poisson kernel makes sense. Note that if $z \in \mathbb{D}$, then $(1 - \bar{\lambda}z)^{-1} + (1 - \lambda\bar{z})^{-1} - 1$ is the usual Poisson kernel. The next result derives properties of this kernel that it shares with the scalar-valued Poisson kernel.

35.2 Lemma. *If* $\|T\| < 1$ *and* $P(\lambda; T)$ *is defined as in* (35.1), *then:*

(a) $P(\lambda; T) \geq 0$ *for all* λ *in* $\partial\mathbb{D}$;

(b) *if* $p(z)$ *and* $q(z)$ *are polynomials, then*

$$p(T) + q(T)^* = \int_{\partial\mathbb{D}} [p(\lambda) + \overline{q(\lambda)}]P(\lambda; T)\, dm(\lambda),$$

where m *is normalized arc length measure on the circle.*

Proof. First observe that for $|\lambda| = 1$,

$$\begin{aligned} P(\lambda; T) &= (1 - \bar{\lambda}T)^{-1}\left[(1 - \lambda T^*) + (1 - \bar{\lambda}T) \right.\\ &\qquad \left. -(1 - \bar{\lambda}T)(1 - \lambda T^*)\right](1 - \lambda T^*)^{-1} \\ &= (1 - \bar{\lambda}T)^{-1}[1 - TT^*](1 - \lambda T^*)^{-1} \\ &\geq 0. \end{aligned}$$

This proves (a). Also

$$\begin{aligned} P(\lambda; T) &= \sum_{k=0}^{\infty} \bar{\lambda}^k T^k + \sum_{k=0}^{\infty} \lambda^k T^{*k} - 1 \\ &= 1 + \sum_{k=1}^{\infty} (\bar{\lambda}^k T^k + \lambda^k T^{*k}). \end{aligned}$$

Thus for any n in $\mathbb{Z}$,

$$\begin{aligned} \int \lambda^n P(\lambda; T)\, dm(\lambda) &= \int \lambda^n\, dm(\lambda) \\ &\quad + \sum_{k=1}^{\infty} \left\{ \left[\int \lambda^n \bar{\lambda}^k\, dm(\lambda)\right] T^k + \left[\int \lambda^n \lambda^k\, dm(\lambda)\right] T^{*k} \right\}. \end{aligned}$$

A routine calculation now shows that

$$\int \lambda^n P(\lambda; T)\, dm(\lambda) = \begin{cases} T^n, & \text{for } n \geq 0, \\ T^{*n}, & \text{for } n < 0. \end{cases}$$

Part (b) now follows. ∎

35.3 Von Neumann's Inequality. *If T is a contraction and p is a polynomial, then $\|p(T)\| \leq \sup\{\, |p(z)| : |z| < 1\,\}$.*

Proof. Let $\mathcal{T}$ be the space of trigonometric polynomials on the circle. That is, $\mathcal{T} = \{\, p + \bar{q} : p \text{ and } q \text{ are polynomials}\,\}$. Define $\phi : \mathcal{T} \to \mathcal{B}(\mathcal{H})$ by $\phi(p + \bar{q}) = p(T) + q(T)^*$. Now $\mathcal{T}$ is an operator system and so if $\|T\| < 1$, the preceding lemma implies that ϕ is a positive map. If $\|T\| = 1$, then this says that for any $\epsilon > 0$, $p((1+\epsilon)^{-1}T) + q((1+\epsilon)^{-1}T)^* \geq 0$ whenever $p(\lambda) + \overline{q(\lambda)} \geq 0$ on the circle. Letting $\epsilon \to 0$ shows that ϕ is positive for any contraction.

Thus Proposition 33.4 implies that ϕ is bounded. Since the trigonometric polynomials are dense in $C(\partial\,\mathbb{D})$ (24.2), ϕ extends to a positive map $\phi : C(\partial\,\mathbb{D}) \to \mathcal{B}(\mathcal{H})$. But now Theorem 33.8 implies that $\|\phi\| = \|\phi(1)\| = 1$. This completes the proof. ■

35.4 The Sz.-Nagy Dilation Theorem. *If T is a contraction on the Hilbert space $\mathcal{H}$, there is a Hilbert space $\mathcal{K}$ containing $\mathcal{H}$ and a unitary on $\mathcal{K}$ such that for all $n \geq 0$*

$$T^n = PU^n|\mathcal{H},$$

where P is the projection of $\mathcal{K}$ onto $\mathcal{H}$.

Proof. Let $\phi : C(\partial\,\mathbb{D}) \to \mathcal{B}(\mathcal{H})$ be defined as in the proof of von Neumann's Inequality. So ϕ is a positive map. By Proposition 34.4, ϕ is completely positive. By Stinespring's Theorem, there is a Hilbert space $\mathcal{K}$ containing $\mathcal{H}$ and a $*$-homomorphism $\rho \colon C(\partial\,\mathbb{D}) \to \mathcal{B}(\mathcal{K})$ such that $\phi(u) = P\rho(u)|\mathcal{H}$ for every continuous function u on $\partial\,\mathbb{D}$. Put $U = \rho(z)$. Since ρ is a $*$-homomorphism, U is unitary. Note that $T = \phi(z)$. Thus $T^n = \phi(z^n) = P\rho(z^n)|\mathcal{H} = PU^n|\mathcal{H}$. ■

As arranged here, the last result depends on von Neumann's Inequality. It is easy to see that the inequality follows from the dilation result. In addition, the Sz.-Nagy Dilation Theorem can be proved independent of von Neumann's Inequality. See Sz.-Nagy and Foias [1970]. This reference is dedicated to the consequences of the Dilation Theorem. It uses this result to develop an entire theory known as the Sz.-Nagy–Foias model of contractions.

The Dilation Theorem leads to the definition of spectral set. Recall that if K is a set and $f \colon K \to \mathbb{C}$, then $\|f\|_K = \sup\{\, |f(z)| : z \in K\,\}$.

35.5 Definition. If $T \in \mathcal{B}(\mathcal{H})$ and K is a compact subset of $\mathbb{C}$ such that $\sigma(T) \subseteq K$, then K is a *spectral set* for T if

$$\|f(T)\| \leq \|f\|_K$$

for all functions f in Rat (K).

Using Runge's Theorem, von Neumann's Inequality says that the closed unit disk is a spectral set for any contraction. More on spectral sets can be found in Conway [1991]. Here attention is restricted to a few known results and a rephrasing of the main problem for spectral sets. Start with a simple lemma observed in Sarason [1965a]. If $T \in \mathcal{B}(\mathcal{H})$ and $\mathcal{L} \leq \mathcal{H}$, the *compression* of T to $\mathcal{L}$ is the operator $T_{\mathcal{L}} \equiv P_{\mathcal{L}}T|\mathcal{L}$.

35.6 Lemma. *If $\mathcal{L} \leq \mathcal{H}$ and $\mathcal{A}$ is a subalgebra of $\mathcal{B}(\mathcal{H})$ that contains the identity, then the map $T \to T_{\mathcal{L}}$ of $\mathcal{A} \to \mathcal{B}(\mathcal{L})$ is multiplicative if and only if there are invariant subspaces $\mathcal{M}$ and $\mathcal{N}$ for $\mathcal{A}$ such that $\mathcal{N} \leq \mathcal{M}$ and $\mathcal{L} = \mathcal{M} \ominus \mathcal{N}$.*

Proof. Put $P = P_{\mathcal{L}}$. If $h \in \mathcal{L}$, $B \in \mathcal{A}$, and $k = P^{\perp}Bh$, then $Bh = PBh + k = B_{\mathcal{L}}h + k$. Thus

$$(AB)_{\mathcal{L}}h = PABh = PAB_{\mathcal{L}}h + PAk = A_{\mathcal{L}}B_{\mathcal{L}}h + PAk.$$

Thus the map $T \to T_{\mathcal{L}}$ is multiplicative if and only if $0 = PAk = PAP^{\perp}Bh$ for all h in $\mathcal{L}$ and all A, B in $\mathcal{A}$.

Assume that $\mathcal{L} = \mathcal{M} \ominus \mathcal{N}$ for two invariant subspaces $\mathcal{M}$ and $\mathcal{N}$ for $\mathcal{A}$ such that $\mathcal{M} \supseteq \mathcal{N}$. If $h \in \mathcal{L} = \mathcal{M} \ominus \mathcal{N}$ and $A, B \in \mathcal{A}$, then $Bh \in \mathcal{M}$, so that $P^{\perp}Bh \in \mathcal{N}$. Thus $AP^{\perp}Bh \in \mathcal{N}$ so that $PAP^{\perp}Bh = 0$. Therefore $T \to T_{\mathcal{L}}$ is multiplicative.

Now assume that the map is multiplicative. Put $\mathcal{N} = [\{CP^{\perp}Bg : g \in \mathcal{L} \text{ and } B, C \in \mathcal{A}\}]$. Clearly $\mathcal{N} \in \text{Lat}\,\mathcal{A}$ and, by assumption, $\mathcal{N} \subseteq \mathcal{L}^{\perp}$. Put $\mathcal{M} = \mathcal{L} \oplus \mathcal{N}$. If $h, g \in \mathcal{L}$ and $B, C \in \mathcal{A}$, consider $f = h + CP^{\perp}Bg$ in $\mathcal{M}$. For A in $\mathcal{A}$, $Af = Ah + ACP^{\perp}Bg = PAh + [P^{\perp}Ah + ACP^{\perp}Bg]$. But from the definition of $\mathcal{N}$ and the fact that $\mathcal{A}$ is an algebra containing the identity, $[P^{\perp}Ah + ACP^{\perp}Bg] \in \mathcal{N}$. Thus $Af \in \mathcal{M}$ and it follows that $\mathcal{M} \in \text{Lat}\,\mathcal{A}$. Clearly $\mathcal{L} = \mathcal{M} \ominus \mathcal{N}$. ■

A subspace $\mathcal{L}$ of the type described in the preceding lemma is called a *semi-invariant* subspace for $\mathcal{A}$. Using semi-invariant subspaces, it is possible to find examples of spectral sets. Let K be a compact subset of $\mathbb{C}$ and let N be a normal operator on $\mathcal{H}$ with $\sigma(N) \subseteq \partial K$. Let $\mathcal{L}$ be a subspace of $\mathcal{H}$ that is semi-invariant for the algebra $\{f(N) : f \in \text{Rat}\,(K)\}$. Put $T = N_{\mathcal{L}}$. First observe that $\sigma(T) \subseteq K$. In fact, if $\lambda \notin K$, then $f = (z-\lambda)^{-1} \in \text{Rat}\,(K)$ and the semi-invariance of $\mathcal{L}$ implies $(N-\lambda)_{\mathcal{L}}^{-1} = (T-\lambda)^{-1}$. Also if $f \in \text{Rat}\,(K)$, then $\|f(T)\| = \|f(N)_{\mathcal{L}}\| \leq \|f(N)\| = \|f\|_{\partial K}$. Thus K is a spectral set for T. This leads to the following.

35.7 Open Question. *If T is an operator on $\mathcal{H}$ and K is a spectral set for T, is there a Hilbert space $\mathcal{K}$ containing $\mathcal{H}$ and a normal operator N on $\mathcal{K}$ with $\sigma(N) \subseteq \partial K$ such that $\mathcal{H}$ is a semi-invariant for $\{ f(N) : f \in \text{Rat}(K) \}$ and $T = N_{\mathcal{H}}$?*

If T is an operator with spectrum contained in K and N is a normal operator as in (35.7), say that T has a *normal ∂K-dilation N*. So the open question is whether every operator with K as a spectral set has a normal ∂K-dilation? The Sz.-Nagy Dilation Theorem says that the answer to this question is yes when $K = \text{cl}\,\mathbb{D}$. In Agler [1985] it is shown that the answer is also affirmative when K is an annulus.

By the way, there is nothing mysterious about requiring the spectrum of the normal dilation to be in the boundary of K. If there is a normal dilation with spectrum in K, then one can be found with spectrum in the boundary (Conway [1994]).

Here is what the theory of completely positive maps has to say about this problem. Fix the compact set K and let T be an operator on $\mathcal{H}$ having K as a spectral set. Put $\mathcal{R} = \text{Rat}(K)|\partial K$, considered as a subalgebra of $C(\partial K)$, and define $\phi \colon \mathcal{R} \to \mathcal{B}(\mathcal{H})$ by $\phi(f) = f(T)$. Since K is a spectral set, the Maximum Modulus Theorem implies that ϕ is a contraction. Because $\phi(1) = 1$, Proposition 33.10 applies and the extension of ϕ, $\widetilde{\phi} : \mathcal{R} + \mathcal{R}^* \to \mathcal{B}(\mathcal{H})$ defined by $\widetilde{\phi}(f + \bar{g}) = f(T) + g(T)^*$ is a well defined positive map. This leads to the following result, obtained independently by Foias [1959] and Lebow [1963]. The result is also credited to C A Berger, but no publication is cited.

35.8 Theorem. *If K is a compact subset of the plane such that the restriction of $[\text{Rat}(K) + \text{Rat}(K)^*]$ to ∂K is dense in $C(\partial K)$, then every operator with K as a spectral set has a normal ∂K-dilation.*

Proof. Adopt the notation preceding the statement of the theorem. As pointed out there, $\widetilde{\phi}$ is a bounded positive map. As such it extends to a positive map on the closure of $\mathcal{R} + \mathcal{R}^*$. By hypothesis, this closure is all of $C(\partial K)$. That is there is a positive map $\psi : C(\partial K) \to \mathcal{B}(\mathcal{H})$ such that $\psi(f) = f(T)$ for all f in $\text{Rat}(K)$. By Proposition 34.4 this map is completely positive. By Stinespring's Theorem, there is a Hilbert space $\mathcal{K}$ containing $\mathcal{H}$ and a $*$-homomorphism $\rho : C(\partial K) \to \mathcal{B}(\mathcal{K})$ such that $\psi(u) = P\rho(u)|\mathcal{H}$ for all u in $C(\partial K)$. If $N = \rho(z)$, this is the required normal ∂K-dilation. ∎

Let $R(K)$ be the uniform closure of $\text{Rat}(K)$ in $C(K)$. If K satisfies the hypothesis of the preceding theorem, $R(K)$ is said to be a *Dirichlet algebra*. The reader can see Conway [1991] for more on Dirichlet algebras. In particular, it is shown there (V.14.9) that $R(K)$ is a Dirichlet algebra if K

is polynomially convex. A characterization of the compact sets K such that $R(K)$ is a Dirichlet algebra can be found in Gamelin and Garnett [1971].

An example of a compact set K for which $R(K)$ is not a Dirichlet algebra is an annulus. If $K = \{ z : 1 \leq |z| \leq 2 \}$ and $f(z) = \log |z|$, then f is not in the closure of the restriction of $R(K) + R(K)^*$ to ∂K. (See Theorem 15.1.3 in Conway [1995].)

The next result is stated without proof as a fundamental result from the theory of completely positive maps, Arveson's Extension Theorem (Arveson [1969]), is needed for its proof and has not been proved in this book.

35.9 Theorem. (Arveson [1969]) *If the operator T has K as a spectral set, then T has a normal ∂K-dilation if and only if the map $\widetilde{\phi}$: $\text{Rat}\,(K) + \text{Rat}\,(K)^* \to \mathcal{B}(\mathcal{H})$ defined by $\widetilde{\phi}(f+\bar{g}) = f(T) + g(T)^*$ is completely positive.*

Also see Paulsen [1986].

Exercises.

1. In the Sz.-Nagy Dilation Theorem show that the Hilbert space $\mathcal{K}$ can be chosen so that it is the smallest reducing subspace for U that contains $\mathcal{H}$. Call such a unitary dilation *minimal.* Show that two minimal unitary dilations of T are unitarily equivalent via an isomorphism that fixes the vectors in $\mathcal{H}$.

2. Formulate and prove a uniqueness statement for Theorem 35.8.

3. In Theorem 35.9 show that if T has a normal ∂K-dilation, then the map $\widetilde{\phi}$ is completely positive. The converse depends on the Arveson Extension Theorem, which states that a completely positive map into $\mathcal{B}(\mathcal{H})$ has an extension to a completely positive map defined on the enveloping C*-algebra. Given this result, prove the converse of Theorem 35.9.

4. If $\|T\| \leq 1$, show that

$$\begin{bmatrix} T & (1-TT^*)^{\frac{1}{2}} \\ (1-T^*T)^{\frac{1}{2}} & -T^* \end{bmatrix}$$

is a unitary operator on $\mathcal{H} \oplus \mathcal{H}$ and the compression of U to $\mathcal{H} \oplus (0)$ is $T \oplus 0$. What is the compression of U^2 to $\mathcal{H} \oplus (0)$?

5. Let T be any operator and show that there is a compact subset X of $\mathbb{C}$ such that $X \supseteq \sigma(T)$ and there is a positive map ϕ: $C(X) \to \mathcal{B}(\mathcal{H})$ such that $\phi(1) = 1$ and $\phi(z) = T$. (Hint: Let $X = \overline{B}(0; \|T\|)$ and apply the Sz.-Nagy Dilation Theorem to $\|T\|^{-1}T$.) Contrast this with (34.13).

§36. Quasicentral approximate identities

It was stated at the beginning of this chapter that all C*-algebras considered would henceforth be assumed to have an identity. If that is the case, why discuss approximate identities? The answer is that for a variety of applications it is necessary to look at ideals in a C*-algebra and approximate identities for the ideal.

So let $\mathcal{A}$ be a C*-algebra with an identity and let $\mathcal{I}$ be a two-sided ideal of $\mathcal{A}$, which will not be assumed to be closed but which is assumed to be self-adjoint. That is, it will always be assumed that $a^* \in \mathcal{I}$ whenever $a \in \mathcal{I}$. A typical example would be that $\mathcal{A} = \mathcal{B}(\mathcal{H})$ and $\mathcal{I}$ is the ideal of finite rank operators. Another would arise when X is a compact space, $x_0 \in X$, $\mathcal{A} = C(X)$, and $\mathcal{I}$ is the ideal of continuous functions on X with support contained in $X \setminus \{x_0\}$. Still another example is the situation when $\mathcal{I}$ is an arbitrary C*-algebra without identity and $\mathcal{A}$ is the C*-algebra that results when an identity is adjoined to $\mathcal{I}$. With this last example it becomes clear that what is done here applies to all C*-algebras without an identity, though the objective of the section is far more easily obtained in this situation.

36.1 Theorem. *If $\mathcal{A}$ is a C^*-algebra and $\mathcal{I}$ is an ideal, there is an approximate identity $\{e_i\}$ for $\mathcal{I}$ with $0 \leq e_i < 1$ such that for every a in $\mathcal{A}$,*

36.2
$$\lim_i \|e_i a - a e_i\| = 0.$$

Proof. According to Theorem 4.3, $\mathcal{E} = \{e \in \mathcal{I} : 0 \leq e < 1\}$ with its natural order is an approximate identity for $\mathcal{I}$. Observe that $\mathcal{E}$ is a convex set. Consider the C*-algebra $\mathbb{C}^n \otimes \mathcal{A} \equiv \{x_1 \oplus \cdots \oplus x_n : x_k \in \mathcal{A}, 1 \leq k \leq n\}$. Note that $\{e^{(n)} \equiv e \oplus \cdots \oplus e : e \in \mathcal{E}\}$ is an approximate identity for $\mathbb{C}^n \otimes \mathcal{I}$.

Claim. If $\bar{a} = a_1 \oplus \cdots \oplus a_n \in \mathbb{C}^n \otimes \mathcal{A}$ and $e_0 \in \mathcal{E}$, then 0 is in the closure of $\{\bar{a}e^{(n)} - e^{(n)}\bar{a} : e \in \mathcal{E} \text{ and } e \geq e_0\}$.

In fact, the subset of $\mathbb{C}^n \otimes \mathcal{A}$ whose closure is asserted to contain the origin is a convex set. Thus if the claim is false, there is a continuous linear functional $L : \mathbb{C}^n \otimes \mathcal{A} \to \mathbb{C}$ such that $\operatorname{Re} L(\bar{a}e^{(n)} - e^{(n)}\bar{a}) \geq 1$ for all e in $\mathcal{E}$ with $e \geq e_0$. Let $\tau\colon \mathbb{C}^n \otimes \mathcal{A} \to \mathcal{B}(\mathcal{H})$ be the universal representation and let $g, h \in \mathcal{H}$ such that $L(\bar{x}) = \langle \tau(\bar{x})g, h\rangle$ for all $\bar{x} = x_1 \oplus \cdots \oplus x_n$ in $\mathbb{C}^n \otimes \mathcal{A}$ (7.14).

Note that if P is the projection of $\mathcal{H}$ onto $[\tau(\mathbb{C}^n \otimes \mathcal{I})\mathcal{H}]$, then $\lim_e \tau(e^{(n)}) = P$ in the strong operator topology of $\mathcal{B}(\mathcal{H})$. In fact, if $k \in \mathcal{H}$ and $\bar{x} \in \mathbb{C}^n \otimes \mathcal{I}$, then $\lim_e \tau(e^{(n)})\tau(\bar{x})k = \lim_e \tau(e^{(n)}x)k = \tau(\bar{x})k$. Since the net $\{\tau(e^{(n)}) : e \in \mathcal{E}\}$ is bounded, $\lim_e \tau(e^{(n)})k = k$ for all k in $P\mathcal{H}$. On the other hand, if $k \perp P\mathcal{H}$, $\tau(e^{(n)})k = 0$ for all e in $\mathcal{E}$.

Hence

$$\lim_e L(\bar{a}e^{(n)} - e^{(n)}\bar{a}) = \lim_e \langle \tau(\bar{a}e^{(n)} - e^{(n)}\bar{a})g, h\rangle$$
$$= \langle (\tau(\bar{a})P - P\tau(\bar{a}))\, g, h\rangle.$$

But $\mathbb{C}^n \otimes \mathcal{I}$ is an ideal of $\mathbb{C}^n \otimes \mathcal{A}$, and hence $P\mathcal{H}$ is invariant for $\tau(\mathbb{C}^n \otimes \mathcal{A})$. Since this is a C*-algebra, $P\mathcal{H}$ reduces $\tau(\mathbb{C}^n \otimes \mathcal{A})$. Hence the above limit is 0, contradicting the selection of L. This proves the claim.

Now let $\mathcal{F}$ be the collection of all finite subsets of $\mathcal{A}$, where repetitions are allowed. Order $\mathcal{F}$ by inclusion, counting multiplicities. Note that $\mathcal{F}$ is a directed set. Consider the directed set $\mathcal{F} \times \mathcal{E}$, where the ordering is by coordinates. If $i = (F, e_0) \in \mathcal{F} \times \mathcal{E}$ and F has n elements, the claim says there is an e in $\mathcal{E}$ such that $e \geq e_0$ and $\|ea - ae\| < 1/n$ for each a in F. Denote this element e by e_i. This defines a net $\{e_i\}$ in $\mathcal{I}$ and $0 \leq e_i < 1$.

First check that $\{e_i\}$ is an approximate identity for $\mathcal{I}$. Indeed, if $x \in \mathcal{I}$ and $\epsilon > 0$, pick e_0 in $\mathcal{E}$ such that $\|x - xe\| + \|x - ex\| < \epsilon$ for all $e \geq e_0$. Let F be the set in $\mathcal{F}$ with x repeated n times, where $n > \epsilon^{-1}$. If $i \geq (F, e_0)$, then $\|xe_i - x\|, \|e_i x - x\| < \epsilon$. To show that the net has the desired property, let $a \in \mathcal{A}$ and let $\epsilon > 0$. Pick any e_0 in $\mathcal{E}$. Let $n > \epsilon^{-1}$ and let F be the set with a repeated n times. It is left to the reader to check that if $i \geq (F, e_0)$, then $\|e_i a - ae_i\| < \epsilon$. ∎

36.3 Definition. If $\mathcal{I}$ is an ideal of the C*-algebra $\mathcal{A}$, an approximate identity $\{e_i\}$ for $\mathcal{I}$ that satisfies (36.2) for all a in $\mathcal{A}$ is called a *quasicentral approximate identity*.

Of course the reason for the word "quasicentral" here is that such an approximate identity for the ideal acts as though it is commuting with elements of the entire C*-algebra.

36.4 Corollary. *If $\mathcal{I}$ is an ideal in a separable C*-algebra, then $\mathcal{I}$ has a sequential quasicentral approximate identity.*

36.5 Corollary. *There is an increasing net $\{F_i\}$ of positive contractions having finite rank such that $\|F_iT - TF_i\| \to 0$ for all T in $\mathcal{B}(\mathcal{H})$.*

The preceding corollary is not so clear without the application of (36.1).

36.6 Corollary. *If $\mathcal{A}$ is a separable C*-subalgebra of $\mathcal{B}(\mathcal{H})$ and $\mathcal{H}$ is separable, then there is an increasing sequence $\{F_n\}$ of positive contractions having finite rank such that $\|F_nA - AF_n\| \to 0$ for all A in $\mathcal{A}$.*

Proof. Recall that $\mathcal{C} = \mathcal{A} + \mathcal{B}_0$ is a C*-algebra (Exercise 5.2). In deed, if $\pi : \mathcal{B}(\mathcal{H}) \to \mathcal{B}(\mathcal{H})/\mathcal{B}_0$ is the natural map, $\pi(\mathcal{A})$ is a C*-algebra in the Calkin algebra and $\mathcal{A} + \mathcal{B}_0 = \pi^{-1}(\pi(\mathcal{A}))$. Now apply the theorem to the ideal of finite rank operators in the C*-algebra $\mathcal{C}$. ∎

Quasicentral approximate units were introduced in Akemann and Pedersen [1977] and Arveson [1977]. Arveson introduced them as part of his simplification of the proof of a vital piece of the Brown–Douglas–Fillmore theory. He also used quasicentral approximate identities to prove Voiculescu's Theorem, which will be done in §40. To set the stage for that process, this section ends with a small result that will be quite useful in the next chapter.

36.7 Proposition. *If $\epsilon > 0$ and f is a continuous function on $[0, 1]$ such that $f(0) = 0$, then there is a $\delta > 0$ such that whenever $a, e \in \operatorname{ball}\mathcal{A}$ with $e \geq 0$ and $\|ea - ae\| < \delta$, then $\|f(e)a - af(e)\| < \epsilon$. Thus if $\mathcal{I}$ is an ideal of $\mathcal{A}$, $\{e_i\}$ is a quasicentral approximate identity for $\mathcal{I}$, $f(0) = 0$, $f(1) = 1$, and $0 \leq f \leq 1$, then $\{f(e_i)\}$ is a quasicentral approximate identity for $\operatorname{cl}\mathcal{I}$.*

Proof. Fix a in $\operatorname{ball}\mathcal{A}$ and define $D : \mathcal{A} \to \mathcal{A}$ by $D(x) = ax - xa$. Clearly D is a linear operator on $\mathcal{A}$ with $\|D\| \leq 2$. Using the fact that $D(x^{n+1}) = D(x^n)x + x^n D(x)$, it follows that whenever $a, x \in \operatorname{ball}\mathcal{A}$ $\|ax^n - x^n a\| \leq n\|ax - xa\|$. So if f is the function $f(z) = z^n$, the lemma is true. It follows that the first part of the lemma holds when f is any polynomial with $f(0) = 0$.

Now if f is any continuous function with $f(0) = 0$, let p be a polynomial with $p(0) = 0$ and $|f(t) - p(t)| < \epsilon/3$ for $0 \leq t \leq 1$. Choose δ so that $\|p(e)a - ap(e)\| < \epsilon/3$ whenever $a, e \in \operatorname{ball}\mathcal{A}$, $e \geq 0$, and $\|ea - ae\| < \delta$. Thus for all such a, e, $\|f(e)a - af(e)\| \leq \|[f(e) - p(e)]\|\|a\| + \|p(e)a - ap(e)\| + \|a\|\|p(e) - f(e)\| < \epsilon$. This proves the first part of the lemma.

If $\{e_i\}$ is a quasicentral approximate identity for the ideal $\mathcal{I}$, then the first part of the proof shows that $\|f(e_i)a - af(e_i)\| \to 0$ for all a in $\mathcal{A}$. Note that for x in $\mathcal{I}$ and $n \geq 1$, $\|e_i^{n+1}x - x\| \leq \|e_i^n(e_i x - x)\| + \|e_i^n x - x\| \leq \|e_i x - x\| + \|e_i^n x - x\|$ and $\|xe_i^{n+1} - x\| \leq \|xe_i - x\| + \|xe_i^n - x\|$. So an induction argument shows that for any $n \geq 1$, $\{e_i^n\}$ is an approximate identity for $\mathcal{I}$. As in the proof of the first part, it follows that $\{f(e_i)\}$ is an approximate identity for $\mathcal{I}$. (Where is the condition that $f(1) = 1$ used?) ∎

Chapter 6

Compact Perturbations

In mathematics there is more than one way to measure "closeness." The most obvious is in a topological sense. In $\mathcal{B}(\mathcal{H})$ there are several topologies, so this alone furnishes different measures of closeness. In this chapter the point of view will be that two operators are close if their difference is a compact operator. This notion is also explored for representations of a C*-algebra.

§37. Behavior of the spectrum under a compact perturbation

This section reviews a few facts about the behavior of the spectrum of an operator under a compact perturbation. The basis for this section is Chapter XI from [ACFA], where many references to the original sources can be found. Some proofs in this section will be skipped and a reference to [ACFA] made.

In this section, $\pi : \mathcal{B}(\mathcal{H}) \to \mathcal{B}(\mathcal{H})/\mathcal{B}_0(\mathcal{H})$ will denote the natural map. $\mathcal{B}$ and $\mathcal{B}_0$ will denote $\mathcal{B}(\mathcal{H})$ and $\mathcal{B}_0(\mathcal{H})$, respectively, when the underlying Hilbert space is understood. Thus the Calkin algebra is denoted by $\mathcal{B}/\mathcal{B}_0$.

37.1 Definition. An operator T on $\mathcal{H}$ is *left Fredholm* if $\pi(T)$ is left invertible in the Calkin algebra. T is *right Fredholm* if $\pi(T)$ is right invertible. A *semi-Fredholm* operator is one that is either left or right Fredholm and a *Fredholm* operator is one that is both.

Thus an operator T is left Fredholm if and only if there is a bounded operator S and a compact operator K such that $ST = 1 + K$.

37.2 Theorem. *If $T \in \mathcal{B}(\mathcal{H})$, the following statements are equivalent.*

(a) *T is left Fredholm.*

(b) $\operatorname{ran} T$ *is closed and* $\dim \ker T < \infty$.

(c) *There is a bounded operator S and a finite rank operator F such that* $ST = 1 + F$.

(d) *There is no sequence of unit vectors $\{h_n\}$ such that $h_n \to 0$ weakly and* $\lim \|Th_n\| = 0$.

(e) *There is no orthonormal sequence $\{e_n\}$ such that* $\lim \|Te_n\| = 0$.

(f) *There is a $\delta > 0$ such that $\{\, h : \|Th\| \leq \delta \|h\| \,\}$ contains no infinite dimensional manifold.*

Proof. (a) *implies* (b). Let $S \in \mathcal{B}(\mathcal{H})$ and let $K \in \mathcal{B}_0$ such that $ST = 1+K$. Hence $\ker T \subseteq \ker (ST) = \ker (1 + K)$. Since K is compact, its non-zero eigenvalues have finite multiplicity. Hence $\dim \ker T \leq \dim \ker ST < \infty$. The Fredholm Alternative also implies that $\operatorname{ran} ST = \operatorname{ran} (1 + K)$ is closed. Hence there is a constant $c > 0$ such that $c\|h\| \leq \|STh\| \leq \|S\|\,\|Th\|$ whenever $h \perp \ker ST$. This implies that $T([\ker ST]^\perp)$ is closed. But $\operatorname{ran} T = T([\ker ST]^\perp) + T(\ker ST)$ and $T(\ker ST)$ is finite dimensional. Hence $\operatorname{ran} T$ is closed.

(b) *implies* (c). If $T_1 \colon [\ker T]^\perp \to \operatorname{ran} T$ is the restriction of T, then (a) and the Open Mapping Theorem imply that T_1 is invertible. If P is the projection onto $\operatorname{ran} T$, let $S = T_1^{-1}P$. If $h \in \ker T$, then $STh = 0$. If $h \perp \ker T$, then $STh = T_1^{-1}PT_1 h = h$. That is, $ST = 1 - F$, where F is the projection onto the finite dimensional space $\ker T$.

(c) *implies* (d). Let $ST = 1 + F$ for a finite rank operator F. If $\{h_n\}$ is a weakly convergent sequence of unit vectors, then $\|Fh_n\| \to 0$. Hence $\|STh_n\| = \|h_n + Fh_n\| \to 1$. Hence $\liminf_n \|Th_n\| > 0$.

(e) *implies* (f). If (f) is false, then for every $n \geq 1$ there is an infinite dimensional manifold $\mathcal{M}_n$ such that $\|Th\| \leq n^{-1}\|h\|$ for all h in $\mathcal{M}_n$. let e_1 be any unit vector in $\mathcal{M}_1$ and suppose $e_1, \ldots, e_n$ are chosen orthonormal vectors with $e_k \in \mathcal{M}_k$. Let E be the projection onto $[\{\, e_1, \ldots, e_n \,\}]$. Since E has finite rank and $\mathcal{M}_{n+1}$ is infinite dimensional, E cannot be injective on $\mathcal{M}_{n+1}$. That is, there must be a unit vector e_{n+1} in $\mathcal{M}_{n+1}$ that is orthogonal to $\operatorname{ran} E$. This inductively constructs an orthonormal sequence $\{e_n\}$ that contradicts (e).

(f) *implies* (c). Let $T = U|T|$ be the polar decomposition of T with $|T| = \int_0^\infty t\, dE(t)$ the spectral decomposition of $|T|$. If δ is as in (f) and $h \in E[0, \delta]\mathcal{H}$, then

$$\|Th\|^2 = \langle T^*Th, h\rangle$$

$$
\begin{aligned}
&= \langle |T|^2 h, h\rangle \\
&= \int_0^\delta t^2\, dE_{h,h}(t) \\
&\leq \delta^2 E_{h,h}[0,\delta] \\
&= \delta^2 \|h\|^2.
\end{aligned}
$$

According to (f), $E[0,\delta]$ has finite rank. Put $S = \left[\int_\delta^\infty t^{-1}\, dE(t)\right] U^*$. A computation shows that $ST = E(\delta,\infty) = 1 - E[0,\delta]$, proving (c).

Since it is clear that (c) implies (a) and (d) implies (e), this completes the proof. ∎

Theorem XI.2.3 in [ACFA] gives additional conditions that are equivalent to those in the preceding theorem.

37.3 Corollary. *An operator T is Fredholm if and only if* $\operatorname{ran} T$ *is closed and both* $\ker T$ *and* $\ker T^*$ *are finite dimensional.*

It can also be shown that T is Fredholm if and only if both $\dim \ker T$ and $\dim \mathcal{H}/\operatorname{ran} T$ are finite dimensional.

Denote the set of all left Fredholm, right Fredholm, semi-Fredholm, and Fredholm operators, respectively, by $\mathcal{F}_\ell$, $\mathcal{F}_r$, $\mathcal{SF}$, and $\mathcal{F}$. So $\mathcal{SF} = \mathcal{F}_\ell \cup \mathcal{F}_r$ and $\mathcal{F} = \mathcal{F}_\ell \cap \mathcal{F}_r$. The *left essential spectrum* of an operator T, $\sigma_{\ell e}(T)$, is the left spectrum of $\pi(T)$. The *right essential spectrum* and the *essential spectrum* of T are denoted by $\sigma_{re}(T)$ and $\sigma_e(T)$, respectively. So $\sigma_\ell(T) = \{\lambda : T - \lambda \notin \mathcal{F}_\ell\}$.

37.4 Definition. The *Fredholm index* is the function $\operatorname{ind} : \mathcal{SF} \to \mathbb{Z} \cup \pm\infty$ defined by

$$\operatorname{ind} T = \dim \ker T - \dim \ker T^*.$$

37.5 Theorem. (a) *If T is a left (respectively, right) Fredholm operator and K is compact, then $T + K$ is left (respectively, right) Fredholm and* $\operatorname{ind}(T+K) = \operatorname{ind} T$.

(b) *If T is a semi-Fredholm operator, there is an $\epsilon > 0$ such that if A is any operator with $\|A\| < \epsilon$, $T + A$ is semi-Fredholm and* $\operatorname{ind}(T+A) = \operatorname{ind} T$. *That is, the set $\mathcal{SF}$ is open and the Fredholm index is a continuous function from $\mathcal{SF}$ into $\mathbb{Z} \cup \pm\infty$ with the discrete topology.*

Proof. See §XI.3 in [ACFA]. ∎

Since the range of the index has the discrete topology, it follows that the continuity of the index is just the statement that it is constant on the components of $\mathcal{SF}$. It can be shown that the connected components of $\mathcal{SF}$ are precisely the sets of constancy of the index. See §XI.5 in [ACFA].

The reader should review Proposition 23.16, Proposition 27.7, Theorem 29.2, Theorem 29.3, and Example 29.4 for illustrations of the essential spectrum and the Fredholm index.

If n is an extended integer, let $P_n(T) \equiv \{\lambda : T-\lambda$ is semi-Fredholm and $\operatorname{ind}(T-\lambda) = n\}$. When $n \neq 0$, $P_n(T)$ is a subset of the spectrum of T that is invariant under compact perturbations. That is, if K is a compact operator, then $P_n(T+K) = P_n(T)$. This invariance is part of the usefulness of the Fredholm index. The essential spectrum, as well as the left and right essential spectrum, also are invariant under compact perturbations. Additional subsets of the spectrum also enjoy this invariance.

37.6 Definition. The *Weyl spectrum* of an operator T is defined as

$$\sigma_w(T) = \bigcap\{\sigma(T+K) : K \in \mathcal{B}_0\}.$$

Taking $K = 0$ in this intersection shows that the Weyl spectrum is contained in the the spectrum. The Weyl spectrum is, thus, the largest subset of the spectrum of the operator that is invariant under all compact perturbations.

37.7 Proposition. (Schecter [1965]) *If T is any operator,*

$$\sigma_w(T) = \sigma_e(T) \cup \bigcup\{P_n(T) : n \neq 0\}.$$

Proof. See [ACFA], XI.6.12. ∎

As defined the set $P_0(T)$ includes the resolvent set for T. It may also include part of the spectrum. In fact, if K is a compact operator, $P_0(K) = \mathbb{C} \setminus \{0\}$. If S is the unilateral shift and $T = S \oplus S^*$, then $P_0(T) \cap \sigma(T) = \mathbb{D}$. The set $P_0(T)$ may also include some isolated points of the spectrum. The isolated points of the spectrum can be bothersome for certain arguments. Realize that isolated points of a set belong to its boundary. The next two results give essentially complete information about the isolated points. See §XI.6 in [ACFA] for the proofs.

37.8 Proposition. (Putnam [1968]) *If $\lambda \in \partial\sigma(T)$, then either λ is an isolated point or $\lambda \in \sigma_{\ell e}(T) \cap \sigma_{re}(T)$.*

37.9 Proposition. *If λ is an isolated point of $\sigma(T)$, then the following statements are equivalent.*

(a) $\lambda \notin \sigma_{\ell e}(T) \cap \sigma_{re}(T)$.

(b) *The Riesz idempotent corresponding to the singleton $\{\lambda\}$ has finite rank.*

(c) $T - \lambda \in \mathcal{F}$ *and* $\operatorname{ind}(T - \lambda) = 0$.

Exercises.

1. Assume that T is a Fredholm operator with $\ker T$ and $\ker T^*$ non-zero. Show that for every $\epsilon > 0$ there is a Fredholm operator S with $\|S - T\| < \epsilon$, $\dim\ker S < \dim\ker T$, and $\dim\ker S^* < \dim\ker T^*$. Can these inequalities be reversed?

2. If T is a semi-Fredholm operator such that $\ker T \subseteq \ker T^n$ for all $n \geq 1$, show that there is a $\delta > 0$ such that for $0 < |\lambda| < \delta$, $\dim\ker(T - \lambda) = \dim\ker T$ and $\dim\ker(T - \lambda)^* = \dim\ker T^*$.

3. If $n \neq 0$ and $\lambda \in \partial P_n(T)$, show that $\operatorname{ran}(T - \lambda)$ is not closed.

4. (Stampfli [1974]) If $T \in \mathcal{B}(\mathcal{H})$, show that there is a compact operator K such that $\sigma(T + K) = \sigma_w(T)$.

5. (Conway [1985]) If L and R are compact subsets of $\mathbb{C}$, show that there is an operator T in $\mathcal{B}(\mathcal{H})$ such that $\sigma_\ell(T) = L$ and $\sigma_r(T) = R$ if and only if $\partial L \subseteq R$ and $\partial R \subseteq L$. What kind of control can be imposed over the other parts of the spectrum?

6. If $\pi : \mathcal{B}(\mathcal{H}) \to \mathcal{B}(\mathcal{H})/\mathcal{B}_0(\mathcal{H})$ is the natural map, show that if $A \in \mathcal{B}(\mathcal{H})$ and $\pi(A)$ is hermitian, then there is a hermitian operator B such that $A - B \in \mathcal{B}_0(\mathcal{H})$. If $\pi(A)$ is positive, show there is a positive operator B such that $A - B \in \mathcal{B}_0(\mathcal{H})$.

7. (L G Brown) If $\pi : \mathcal{B}(\mathcal{H}) \to \mathcal{B}(\mathcal{H})/\mathcal{B}_0(\mathcal{H})$ is the natural map and A and B are hermitian operators such that $\pi(A) \leq 0 \leq \pi(B)$, show that there is a hermitian compact operator K such that $A \leq K \leq B$.

§38. $\mathcal{B}_p$ perturbations of hermitian operators

The next result is fundamental for much of what will follow in this section.

38.1 The Weyl–von Neumann Theorem. *If A is a hermitian operator on a separable Hilbert space, $\epsilon > 0$, and $1 < p < \infty$, then there is a diagonalizable self-adjoint operator D such that $A - D \in \mathcal{B}_p$ and $\|A - D\|_p < \epsilon$.*

This result actually holds for unbounded operators, but only the demonstration for bounded operators is given. A slight modification of this proof, however, will furnish the proof of the unbounded case. If A is a self-adjoint operator (possibly unbounded) call a point λ in the spectrum of A an *m-limit point* if λ is a topological limit point or if λ is an isolated eigenvalue of A having infinite multiplicity. The letter 'm' here stands for 'multiplicity.' Note that if A is bounded, the set of m-limit points of $\sigma(A)$ is precisely the essential spectrum of A, $\sigma_e(A)$. Weyl [1909] showed that if two self-adjoint operators differ by a compact operator, their spectra have the same m-limit points. Von Neumann [1935] proved what is essentially the converse. This is usually called the Weyl–von Neumann Theorem.

The conclusion of the Weyl–von Neumann Theorem fails if $p = 1$. In fact, if A is a hermitian operator and K is a hermitian trace class operator, then a result of Kato [1957] and Rosenblum [1957] implies that the absolutely continuous parts of A and $A+K$ are unitarily equivalent. Also see Putnam [1967], page 101.

The theorem requires a pair of lemmas, the proof of the first of which is left to the reader.

38.2 Lemma. *If T has finite rank m, then $\|T\|_p \le m^{1/p}\|T\|$.*

38.3 Lemma. *If A is a hermitian operator, $1 < p < \infty$, $f \in \mathcal{H}$, and $\epsilon > 0$, then there is a finite rank projection P and a hermitian operator K in the Schatten p-class, $\mathcal{B}_p$, such that $f \in P$, $\|K\|_p < \epsilon$, and $A + K$ is reduced by P.*

Proof. Let E be the spectral measure for A and assume $\sigma(A) \subseteq [a,b]$. Partition $[a,b]$ into n equal subintervals, $\Delta_1, \ldots, \Delta_n$, having length $(b-a)/n$. Note that $\sum_1^k E(\Delta_k) = 1$. Put $f_k = E(\Delta_k)f$, $g_k = f_k/\|f_k\|$ if $f_k \neq 0$ and $g_k = 0$ otherwise. So $g_k \in E(\Delta_k)$ and, hence, $g_j \perp g_k$ for $j \neq k$. If λ_k is the midpoint of the interval Δ_k, then $\|(A-\lambda_k)g_k\| \le (b-a)/n$.

Define P to be the projection onto $[\{g_1, \ldots, g_n\}] = [\{f_1, \ldots, f_n\}]$. Since $P^\perp g_k = 0$,

$$\|P^\perp A g_k\| = \|P^\perp(A-\lambda_k)g_k\| \le \frac{(b-a)}{n}.$$

Because g_k belongs to the spectral subspace $E(\Delta_k)$, $Ag_k \in E(\Delta_k)$. Thus $Ag_k \perp g_j$ for $k \neq j$. This implies that $P^\perp A g_k = Ag_k - \sum_j \langle Ag_k, g_j\rangle g_j = Ag_k - \langle Ag_k, g_k\rangle g_k \in E(\Delta_k)$. So $P^\perp A g_k \perp P^\perp A g_j$ for $k \neq j$. Therefore

$$\begin{aligned}\|P^\perp APh\|^2 &= \left\| \sum_k \langle h, g_k\rangle P^\perp A g_k \right\|^2 \\ &= \sum_k |\langle h, g_k\rangle|^2 \, \|P^\perp A g_k\|^2 \\ &\le \left[\frac{b-a}{n}\right]^2 \|h\|^2.\end{aligned}$$

Hence $\|P^\perp AP\| \le (b-a)/n$ and the rank of $P^\perp AP$ is at most n. Therefore, by Lemma 38.2,

$$\|P^\perp AP\|_p \le n^{1/p}\left[\frac{b-a}{n}\right] = \frac{b-a}{n^{1/q}},$$

where $1/p + 1/q = 1$.

Put $B = PAP + P^\perp AP^\perp$ and $K = -[P^\perp AP + PAP^\perp]$. Note that both B and K are self-adjoint and $A = B - K$. Also B is reduced by the space P, K has finite rank, and $\|K\|_p \le 2(b-a)/n^{1/q}$. This can be made as small as desired with a sufficiently large choice of n. (Note that this is the point in the proof where it is necessary to have $p > 1$.) ■

Proof of the Weyl–von Neumann Theorem. Let $\{h_j\}$ be a dense sequence in $\mathcal{H}$ and apply the preceding lemma with $f = h_1$ to get a finite rank projection P_1 and a self-adjoint operator K_1 in $\mathcal{B}_p$ with $\|K_1\|_p < \epsilon/2$, such that $A + K_1$ is reduced by P_1 and $h_1 \in P_1$. Now apply the lemma again to the self-adjoint operator $(A + K_1)|P_1^\perp$ and the vector $f = P_1^\perp h_2$ to get a self-adjoint operator K_2 in $\mathcal{B}_p(P_1^\perp)$ and a projection $P_2 \le P_1^\perp$ such that $P_1^\perp h_2 \in P_2$, $\|K_2\|_p < \epsilon/2^2$, and $A + K_1 + K_2$ is reduced by P_2. Extend K_2 to be defined on all of $\mathcal{H}$ by letting $K_2 h = 0$ for all h in P_1. Note that both h_1 and h_2 belong to $P_1 + P_2$.

Induction produces a sequence of finite rank projections $\{P_n\}$ and a sequence of self-adjoint operators $\{K_n\}$ in $\mathcal{B}_p$ such that:

(i) $P_j P_n = 0$ for $j \neq n$;
(ii) $\|K_n\|_p < \epsilon/2^n$;
(iii) $h_n \in P_1 + \cdots + P_n$;
(iv) $A + K_1 + \cdots K_n$ is reduced by $(P_1 + \cdots + P_n)\mathcal{H}$;
(v) $K_n(P_1 + \cdots + P_{n-1}) = 0$.

If $K = \sum_n K_n$, then $K \in \mathcal{B}_p$ with $\|K\|_p < \epsilon$. Let $D = A + K$; so D is self-adjoint. Properties (i) and (iii) imply that $\sum_n P_n = 1$. Properties (iv) and (v) imply that D is reduced by each of the spaces P_n and, if $D_n = D|P_n$, $D = \bigoplus_n D_n$. Since each of the spaces P_n is finite dimensional, there is a basis for P_n that diagonalizes D_n. Thus D is a diagonalizable operator. ■

There is a small technical modification of the Weyl–von Neumann Theorem that will be useful later.

38.4 Corollary. *If A is a hermitian operator and $\epsilon > 0$, then there is a diagonalizable hermitian operator D and a compact operator K such that $A = D + K$, $\sigma(D) = \sigma_e(D)$, and $\|K\| < \epsilon$.*

Proof. By the theorem there is a diagonal operator D and a compact operator K such that $A = D + K$ and $\|K\| < \epsilon$. Let $\mu_1, \mu_2, \ldots$ be the isolated eigenvalues of D having finite multiplicity, repeated as often as their multiplicity. Let $h_1, h_2, \ldots$ be the corresponding eigenvectors and write $D = D_1 + \sum_j \mu_j h_j \otimes h_j$, where D_1 is a diagonalizable hermitian operator with $\sigma_e(D_1) = \sigma(D_1)$. For each $j \geq 1$, let $\lambda_j \in \sigma_e(D) = \sigma_e(D_1)$ such that $|\lambda_j - \mu_j| = \text{dist}\,(\mu_j, \sigma_e(D))$. Now $\sigma(D) \subseteq \{\lambda : \text{dist}\,(\lambda, \sigma(A)) < \epsilon\}$, so $|\lambda_j - \mu_j| < \epsilon$ for all $j \geq 1$. If $D_2 = D_1 - \sum_j \lambda_j h_j \otimes h_j$ and $K_2 = \sum_j (\mu_j - \lambda_j) h_j \otimes h_j$, then $D = D_2 + K_2$, D_2 is diagonalizable with $\sigma(D_2) = \sigma_e(D_2)$, and K_2 is compact with $\|K_2\| < \epsilon$. ■

§39. The Weyl–von Neumann–Berg Theorem

Berg [1971] partially extended the Weyl–von Neumann Theorem to normal operators. The proof here is due to Halmos [1972] who used the fact that a normal operator is a continuous function of a hermitian operator. The starting point is a somewhat more general result that will produce this fact about normal operators as a corollary. First a definition.

39.1 Definition. For a basis $\{e_k\}$, the *diagonal algebra corresponding to* $\{e_k\}$ is the collection $\mathcal{D} = \mathcal{D}(\{e_k\})$ of all operators T that are diagonalized by this basis. That is, $T \in \mathcal{D}$ if and only if there are scalars $\{\tau_k\}$ such that $Te_k = \tau_k e_k$ for all k.

Note that $\mathcal{D}$ is an abelian von Neumann algebra, though it is not separable. Indeed, it is the von Neumann algebra generated by the rank one projections onto $\mathbb{C}e_k$. As such it is isomorphic to ℓ^∞.

39.2 Theorem. *Assume $\mathcal{H}$ is separable. If $\mathcal{A}$ is an abelian C^*-algebra algebra in $\mathcal{B}(\mathcal{H})$, the following statements are equivalent.*

(a) *$\mathcal{A}$ is separable.*

(b) *There is a sequence of commuting projections $\{P_n\}$ such that $\mathcal{A} \subseteq C^*(\{P_n\})$.*

(c) *There is a hermitian operator A such that $\mathcal{A} \subseteq C^*(A)$.*

(d) *There is a basis for $\mathcal{H}$ such that $\mathcal{A} \subseteq \mathcal{D}_1 + \mathcal{B}_0$, where $\mathcal{D}_1$ is a separable subalgebra of the diagonal algebra corresponding to the basis.*

Proof. (a) *implies* (b). Let X be the maximal ideal space of $\mathcal{A}$ and $\rho : C(X) \to \mathcal{A}$ the inverse of the Gelfand map. So ρ is a representation of $C(X)$ and there is a spectral measure E defined on X such that $\rho(f) = \int f\, dE$ for all f in $C(X)$. Since $\mathcal{A}$ is separable, X is a compact metric space. For each integer $n \geq 1$ let $\{\Delta_j^n : 1 \leq j \leq k_n\}$ be a Borel partition of X into sets having diameter at most $1/n$. Thus $\mathcal{E} = \{E(\Delta_j^n) : n \geq 1, 1 \leq j \leq k_n\}$ has the property that $\mathcal{A} \subseteq C^*(\mathcal{E})$.

(b) *implies* (c). Let $\mathcal{C} = C^*(\{P_n\})$; so $\mathcal{C}$ is abelian. Let X be the maximal ideal space and let $\rho : C(X) \to \mathcal{C}$ be the inverse of the Gelfand map. Put $p_n = \rho^{-1}(P_n)$; so each p_n is a continuous characteristic function on X. (And, consequently, X is totally disconnected.) Define the function $f = \sum_n 3^{-n}(2p_n - 1)$. It is claimed that f generates $C(X)$. Indeed, let $\mathcal{F} = C^*(f)$. The fact that $\{p_n\}$ generates $C(X)$ implies that these functions separate the points of X. So if x and y are distinct points in X, there is a smallest integer k such that $p_k(x) \neq p_k(y)$. Note that for every $n \geq 1$, $p_n(x) - p_n(y) = 0, 1$ or -1. Thus

$$
\begin{aligned}
|f(x) - f(y)| &= 2\left|\sum_{n=k}^{\infty} 3^{-n}(p_n(x) - p_n(y))\right| \\
&\geq 2\,3^{-k} - 2\sum_{n=k+1}^{\infty} 3^{-n} \\
&= 3^{-k}.
\end{aligned}
$$

Thus the function f separates the points of X. By the Stone-Weierstrass Theorem, $\mathcal{F} = C(X)$. Putting $A = \rho(f)$ proves (c).

(c) *implies* (d). It suffices to assume that $\mathcal{A} = C^*(A)$ for some hermitian operator A. According to the Weyl–von Neumann Theorem, $A = D + K$, where D is a diagonal operator with respect to some basis for $\mathcal{H}$ and K is compact. Letting $\mathcal{D}_1 = C^*(D)$ implies that $\mathcal{A} \subseteq \mathcal{D}_1 + \mathcal{B}_0$.

Since it is clear that (d) implies (a), this completes the proof. ■

It is worth pointing out that the first three conditions in this theorem are equivalent without the assumption that $\mathcal{H}$ is separable.

39.3 Corollary. *If $N_1, \ldots, N_n$ are commuting normal operators on a separable Hilbert space, then there is a hermitian operator A and continuous functions $f_1, \ldots, f_n$ on $\mathbb{R}$ such that $N_j = f_j(A)$ for $1 \leq j \leq n$.*

Proof. If $\mathcal{A} = C^*(N_1, \ldots, N_n)$, then the Fuglede–Putnam Theorem implies that $\mathcal{A}$ is an abelian C*-algebra. Clearly it is separable. Thus Theorem 39.2 implies there is a hermitian operator A with $\mathcal{A} \subseteq C^*(A)$. In particular, each $N_j \in \mathcal{A}$. Thus each $N_j = f_j(A)$ for some continuous function f_j on the spectrum of A. The functions can be continuously extended to $\mathbb{R}$ if so desired. ■

39.4 The Weyl–von Neumann–Berg Theorem. *If $N_1, \ldots, N_n$ are commuting normal operators on a separable Hilbert space and $\epsilon > 0$, then there are simultaneously diagonalizable normal operators $D_1, \ldots, D_n$ and compact operators $K_1, \ldots, K_n$ such that for $1 \le j \le n$, $\|K_j\| < \epsilon$ and $N_j = D_j + K_j$.*

Neglecting the estimate on the compact operators in the theorem, the proof is straightforward. Indeed, the preceding corollary furnishes a hermitian operator A and continuous functions f_j on the real line such that $N_j = f_j(A)$. Let $A = D + K$ with D diagonalizable and K compact. It follows that for every $m \ge 1$, $A^m = D^m + K_m$ for some compact operator K_m. Thus for each polynomial p, $p(A) = p(D) + K_p$ for some compact operator K_p. If f is a continuous function on the line, let $\{p_k\}$ be a sequence of polynomials that converges uniformly to f on a closed interval containing $\sigma(A) \cup \sigma(D)$. So $\|p_k(A) - f(A)\| \to 0$ and $\|p_k(D) - f(D)\| \to 0$. Therefore $K_{p_k} = p_k(A) - p_k(D) \to f(A) - f(D)$, which must be compact. Applying this to each of the functions f_j gives that $N_j = f_j(A) = f_j(D) + K_j$ for some compact operator K_j.

The compact operator K in this argument could have been chosen with arbitrarily small norm. But even so, it isn't clear how this affects things when the functional calculus is applied to A. To settle this a lemma is needed.

39.5 Lemma. *If f is a continuous function on $\mathbb{R}$, then the function $A \to f(A)$ is uniformly continuous on bounded sets of hermitian operators.*

Proof. Since the uniform limit of uniformly continuous functions is uniformly continuous, it suffices to prove this for polynomials. Thus it suffices to show that each power function is uniformly continuous of bounded sets of operators. Observe that $T^{n+1} - S^{n+1} = T^n(T - S) + (T^n - S^n)S$. Hence an induction argument proves that for every $n \ge 1$, $T \to T^n$ is uniformly continuous on a bounded set of operators. ■

Proof of the Weyl–von Neumann–Berg Theorem. Let A be a hermitian operator and $f_1, \ldots, f_n$ continuous functions on the line such that $N_j = f_j(A)$ for $1 \le j \le n$. If $\epsilon > 0$, Lemma 39.5 implies there is a $\delta > 0$ such that for $1 \le j \le n$, $\|f_j(X) - f_j(Y)\| < \epsilon$ whenever X and Y are hermitian operators satisfying $\|X\|, \|Y\| \le \|A\|$ and $\|X - Y\| < \delta$. By Corollary 38.4 there is a diagonal hermitian operator D and a compact operator K with $A = D + K$, $\|K\| < \delta$, and $\sigma_e(D) = \sigma(D)$. Since $\sigma_e(D) = \sigma_e(A)$, $\|D\| \le \|A\|$. Thus $\|N_j - f_j(D)\| = \|f_j(A) - f_j(D)\| < \epsilon$ for $1 \le j \le n$. If $D_j = f_j(D)$, then the argument preceding the statement of the lemma shows that $K_j = N_j - D_j$ is compact. ■

The proof of the next corollary is like the proof of Corollary 38.4 and is left to the reader.

39.6 Corollary. *If N is a normal operator on a separable Hilbert space and $\epsilon > 0$, then there is a diagonalizable normal operator D such that $N - D \in \mathcal{B}_0$ and $\sigma(D) = \sigma_e(D)$.*

The question arises whether the Weyl–von Neumann–Berg Theorem can be improved by getting the differences $N_j - D_j$ to be in any Schatten p-class with $1 < p < \infty$. When there is only one normal operator involved, the answer is yes for $p = 2$, the Hilbert–Schmidt operators, and no for some other values of p. For more than one normal operator it may not be possible to get the compact operators to be Hilbert–Schmidt, though they can be restricted to lie in a proper (non-closed) ideal of the compact operators. This can be found in Voiculescu [1979]. Also see Voiculescu [1981] and [1990].

Now for the problem of determining when two normal operators are unitarily equivalent modulo a compact operator.

39.7 Lemma. *If N and M are diagonalizable normal operators on a separable Hilbert space with the $\sigma(N) = \sigma_e(N) = \sigma_e(M) = \sigma(M)$ and $\epsilon > 0$, then there is a unitary operator U such that $N - U^*MU$ is in the trace class with $\|N - U^*MU\|_1 < \epsilon$.*

Proof. Before beginning the proof, observe that if ν is an isolated point in $\sigma(N)$, then it must be an eigenvalue of infinite multiplicity; otherwise it would not belong to the essential spectrum. Likewise, it is an eigenvalue for M having infinite multiplicity. If ν is not an isolated point of $\sigma(N)$, then it is a limit point of $\sigma(M)$ and there is a sequence of distinct eigenvalues of M that converge to ν.

The proof is basically an exercise in bookkeeping. Let $\{\nu_n\}$ and $\{\mu_m\}$ be the eigenvalues of N and M, respectively, each repeated as often as its multiplicity. In light of the preceding paragraph, for each ν_j there is a subsequence of $\{\mu_n\}$ that converges to it. This justifies the various selections made in the induction argument needed to prove the following.

Claim. There is a renumbering of the two sets $\{\nu_{n_k}\}$ and $\{\mu_{m_k}\}$ such that $k \in \{n_1, \ldots, n_{2k}\} \cap \{m_1, \ldots, m_{2k}\}$ and $|\nu_{n_k} - \mu_{m_k}| < \epsilon/2^k$.

To start, let $n_1 = 1$ and let m_1 be the smallest integer with $|\nu_1 - \mu_{m_1}| < \epsilon/2$. If $m_1 > 1$, let $m_2 = 1$; if $m_1 = 1$, let $m_2 = 2$. Let n_2 be the smallest integer greater than 1 with $|\nu_{n_2} - \mu_{m_2}| < \epsilon/2^2$. Now assume that n_k and m_k satisfy the claim for $1 \le k \le 2j$. Let $n_{2j+1} = \min\{n : n \neq n_1, \ldots, n_{2j}\}$ and let m_{2j+1} be the smallest integer greater than $m_1, \ldots, m_{2j}$ with $|\nu_{n_{2j+1}} -$

$\mu_{m_{2j+1}}| < \epsilon/2^{2j+1}$. Note that, by the claim, if $n_k \neq j+1$ for $1 \leq k \leq 2j$, then $n_{2j+1} = j+1$. Now let $m_{2j+2} = \min\{m : m \neq m_1, \ldots, m_{2j+1}\}$ and let n_{2j+2} be the smallest integer greater than $n_1, \ldots, n_{2j+1}$ with $|\nu_{n_{2j+2}} - \mu_{m_{2j+2}}| < \epsilon/2^{2j+2}$. It is not hard to see that the claim is true for $k = j+1$. By induction, the claim holds for all k.

Let $\{e_n\}$ and $\{f_m\}$ be orthonormal bases for $\mathcal{H}$ such that $Ne_n = \nu_n e_n$ and $Mf_m = \mu_m f_m$. If U is defined on $\mathcal{H}$ by $Ue_{n_k} = f_{m_k}$ for all k, then U is a unitary operator. Put $K = N - U^*MU$. So $Ke_{n_k} = (\nu_{n_k} - \mu_{m_k})e_{n_k}$. Thus K is trace class and $\|K\|_1 < \epsilon$. ■

The next result is also referred to as the Weyl–von Neumann–Berg Theorem.

39.8 Theorem. *If N and M are normal operators on a separable Hilbert space with the same essential spectrum, then there is a unitary operator U such that $N - U^*MU$ is compact.*

Proof. According to Corollary 39.6, there are diagonalizable operators N_1, M_1 such that $N - N_1$ and $M - M_1$ are compact and $\sigma(N_1) = \sigma_e(N_1)$ and $\sigma(M_1) = \sigma_e(M_1)$. Since $\sigma_e(N) = \sigma_e(N_1)$ and $\sigma_e(M) = \sigma_e(M_1)$, the hypothesis implies that $\sigma(N_1) = \sigma_e(N_1) = \sigma_e(M_1) = \sigma(M_1)$. By Lemma 39.7 there is a unitary operator U such that $N_1 - U^*M_1U$ is compact. So $N - U^*MU = (N - N_1) + (N_1 - U^*M_1U) + U^*(M_1 - M)U \in \mathcal{B}_0$. ■

The requirement in Theorem 39.8 that $N - U^*MU$ be compact cannot be replaced with the stronger condition that it be in a Schatten p-class, even if both operators are hermitian. If $\{\alpha_n\}$ is a sequence of positive scalars that converges to 0 but is not p-summable for any $p \geq 1$, let K be the diagonal operator with eigenvalues $\{\alpha_n\}$. So K is a compact positive operator and $K \notin \mathcal{B}_p$ for any p. Taking $N = K$ and $M = 0$ shows that the conclusion of the theorem cannot be so strengthened.

It can also be asked if $\|N - U^*MU\|$ can be made arbitrarily small in Theorem 39.8. The answer is no, even for hermitian operators. Indeed, let A be the zero operator on an infinite dimensional space and let B be the direct sum of A and a finite dimensional identity operator. No matter how U is chosen, $A - U^*BU = -U^*BU$ has norm 1. There is a natural interest in knowing when this norm can be made as small as desired. This is an equivalence relation for operators that is now explored. §41 will explore a more general concept for representations.

39.9 Definition. Say that two operators A and B are *approximately unitarily equivalent* if there is a sequence of unitary operators $\{U_n\}$ such that $\|A - U_n^*BU_n\| \to 0$. The corresponding notation is $A \cong_a B$. A and B are *strongly approximately unitarily equivalent* if there is a sequence of unitary operators $\{U_n\}$ such that $\|A - U_n^* B U_n\| \to 0$ and $A - U_n^* B U_n$ is compact for all $n \geq 1$. The corresponding notation is $A \cong_{sa} B$.

If $\{U_n\}$ is a sequence of unitaries such that $\|A - U_n^* B U_n\| \to 0$, then $\sigma_e(U_n^* B U) = \sigma_e(B)$ for all $n \geq 1$. If A and B are normal, this implies that $\sigma_e(B) = \sigma_e(A)$. (See Exercise 1.) As was seen prior to the definition, having the same essential spectrum is not sufficient to guarantee approximate equivalence.

39.10 Proposition. *If N and M are normal operators on a separable Hilbert space, then the following statements are equivalent.*

(a) $N \cong_{sa} M$.

(b) $N \cong_a M$.

(c) $\sigma_e(N) = \sigma_e(M)$ *and* $\dim\ker(N - \lambda) = \dim\ker(M - \lambda)$ *for all λ not in the common essential spectrum.*

Proof. Let E_N and E_M be the spectral measures for N and M. Clearly (a) implies (b). Assume (b) and let $\{U_n\}$ be a sequence of unitaries such that $\|N - U_n^* M U_n\| \to 0$. It is left to the reader (Exercise 1) to show that $\sigma(N) = \sigma(M)$. If λ is an isolated point of $\sigma(N) = X$, then $\chi_{\{\lambda\}}$ is a continuous function on X and it follows that $\chi_{\{\lambda\}}(U_n^* M U_n) \to \chi_{\{\lambda\}}(N)$ in norm. But this characteristic function of the operators is precisely the spectral projection of the singleton $\{\lambda\}$. This says that $E_M(\{\lambda\})$ and $E_N(\{\lambda\})$ have the same rank and so $\dim\ker(N - \lambda) = \dim\ker(M - \lambda)$ for all isolated points λ in X. Since the essential spectrum of a normal operator consists of the points in the spectrum that are not isolated eigenvalues of finite multiplicity, this also shows that $\sigma_e(N) = \sigma_e(M)$.

Now assume (c) and prove (a). Using the Spectral Theorem write $N = N_1 \oplus N_2$ and $M = M_1 \oplus M_2$, where N_1 and M_1 are diagonalizable, each eigenvalue of N_1 and M_1 is isolated and has finite multiplicity, and $\sigma(N_2) = \sigma_e(N_2) = \sigma_e(N) = \sigma_e(M) = \sigma_e(M_2) = \sigma(M_2)$. By the second half of condition (c), N_1 and M_1 have the same eigenvalues and $\dim\ker(N_1 - \lambda) = \dim\ker(M_1 - \lambda)$ for each such eigenvalue λ. Thus $N_1 \cong M_1$ since these normal operators are diagonalizable; let V be a unitary such that $VM_1V^* = N_1$. Now apply Lemma 39.7 to obtain a sequence of unitaries $\{W_n\}$ such that $W_n M_2 W_n^* \to N_2$ and $W_n M_2 W_n^* - N_2 \in \mathcal{B}_0$. Putting $U_n = V \oplus W_n$ shows that $N \cong_{sa} M$. ∎

See Theorem 41.12 for an extension of the preceding proposition to representations of a separable C*-algebra.

Exercises.

1. If $\mathcal{A}$ is a C*-algebra and $\{a_n\}$ is a sequence of normal elements of $\mathcal{A}$ such that $a_n \to a$, show that $\sigma(a_n) \to \sigma(a)$. That is, show that if $\alpha \in \sigma(a)$, there are points α_n in $\sigma(a_n)$ with $\alpha = \lim_n \alpha_n$; and, conversely, if $\alpha_n \in \sigma(a_n)$ for all $n \geq 1$ and $\alpha_n \to \alpha$, then $\alpha \in \sigma(a)$.

2. Show that approximate equivalence and strong approximate equivalence are equivalence relations.

3. If N and M are approximately equivalent normal operators and $f \in C(\sigma(N))$, show that $f(N) \cong_a f(M)$.

§40. Voiculescu's Theorem

Throughout this section only separable Hilbert spaces are considered. This section will mostly be concerned with C*-algebras that are contained in $\mathcal{B}(\mathcal{H})$. So $\mathcal{A}$ will always denote a separable C*-algebra contained in $\mathcal{B}(\mathcal{H})$ and containing the identity. To avoid trivialities, $\mathcal{H}$ will always be assumed to be infinite dimensional. The objective is to prove the following.

40.1 Voiculescu's Theorem. *If $\mathcal{H}$ is separable and $\mathcal{A}$ is a separable C^*-algebra contained in $\mathcal{B}(\mathcal{H})$ and $\phi\colon \mathcal{A} \to \mathcal{B}(\mathcal{K})$ is a completely positive map such that $\phi(1) = 1$ and $\phi(\mathcal{A} \cap \mathcal{B}_0) = 0$, then there is a sequence of isometries $\{V_n\}$ from $\mathcal{K}$ into $\mathcal{H}$ such that:*

(a) $\operatorname{ran} V_n \perp \operatorname{ran} V_m$ *for $n \neq m$, and, hence, $V_n \to 0$* (WOT)*;*

(b) *for every A in $\mathcal{A}$ and $n \geq 1$, $\phi(A) - V_n^* A V_n$ is compact;*

(c) *for all A in $\mathcal{A}$, $\|\phi(A) - V_n^* A V_n\| \to 0$ as $k \to \infty$.*

This result, with a representation substituted for the completely positive map, first appeared in Voiculescu [1976]. The extension to completely positive maps comes from Arveson [1977], whose treatment forms the basis for this section. Theorem 40.1 leads to a generalization of the Weyl–von Neumann–Berg Theorem, which will be seen in §42. Before beginning the proof of the theorem, here is the statement of the result for representations.

40.2 Corollary. *If $\mathcal{A}$ is a separable C^*-algebra contained in $\mathcal{B}(\mathcal{H})$ and $\rho :$ $\mathcal{A} \to \mathcal{B}(\mathcal{K})$ is a representation such that $\rho(\mathcal{A} \cap \mathcal{B}_0) = 0$, then there is a sequence of isometries $\{V_n\}$ from $\mathcal{K}$ into $\mathcal{H}$ such that:*

(a) $\operatorname{ran} V_n \perp \operatorname{ran} V_m$ *for $n \neq m$, and, hence, $V_n \to 0$* (WOT)*;*

(b) *$V_n\phi(A) - AV_n$ is compact for every A in $\mathcal{A}$ and for all $n \geq 1$;*

(c) *for all A in $\mathcal{A}$, $\|V_n\phi(A) - AV_n\| \to 0$ as $n \to \infty$.*

Proof. Representations are completely positive, so isometries $\{V_n\}$ can be obtained as in (40.1). Now observe that

$$\begin{aligned}
&[V_n\rho(A) - AV_n]^*[V_n\rho(A) - AV_n] \\
&\qquad = \rho(A)^*\rho(A) - \rho(A)^*V_n^*AV_n - V_n^*A^*V_n\rho(A) + V_n^*A^*AV_n \\
&\qquad = \rho(A^*)[\rho(A) - V_n^*AV_n] + [\rho(A^*) - V_n^*A^*V_n]\rho(A) \\
&\qquad\quad - [\rho(A^*A) - V_n^*A^*AV_n].
\end{aligned}$$

By the preceding theorem, each of these summands is compact and converges to 0 as $n \to \infty$. ∎

To set the stage for the proof of Theorem 40.1 and the required lemmas, recall that if $\mathcal{A}$ is a C*-algebra contained in $\mathcal{B}(\mathcal{H})$, $\mathcal{A} + \mathcal{B}_0(\mathcal{H})$ is also a C*-algebra (Exercise 1.2).

40.3 Lemma. *If $\mathcal{A}$ is a C^*-algebra contained in $\mathcal{B}(\mathcal{H})$, $A \in \mathcal{A}$, and $K \in \mathcal{B}_0(\mathcal{H})$ such that $A + K \geq 0$, then there is a positive operator A_1 in $\mathcal{A}$ and a positive compact operator K_1 such that $A + K = A_1 + K_1$.*

Proof. Assume that $B = A + K$ is positive. Therefore $B^2 = B^*B = A^*A + K^*A + A^*K + K^*K = A^*A + L$, where L is a compact hermitian operator. Choose $a > 0$ and sufficiently large that $\sigma(B^2) \cup \sigma(A^*A) \cup \sigma(L) \subseteq [-a, a]$. Define $f\colon [-a, a] \to [0, \infty)$ by $f(t) = \sqrt{t}$ if $t \geq 0$ and $f(t) = 0$ if $t \leq 0$ and let $\{\, p_n \,\}$ be a sequence of polynomials such that $p_n(t) \to f(t)$ uniformly on $[-a, a]$. As in the paragraph following the statement of (39.4), $p_n(B^2) = p_n(A^*A) + L_n$, where L_n is a positive compact operator. Now $p_n(B^2) \to B$ and $p_n(A^*A) \to |A|$, so $\{\, L_n \,\}$ converges in norm to positive compact operator K_1 and $A + K = B = |A| + K_1$. ∎

Now assume that $\phi : \mathcal{A} \to \mathcal{B}(\mathcal{K})$ is a completely positive map such that $\phi(\mathcal{A} \cap \mathcal{B}_0) = 0$. Define $\tilde{\phi}\colon \mathcal{A} + \mathcal{B}_0 \to \mathcal{B}(\mathcal{K})$ by $\tilde{\phi}(A + K) = \phi(A)$, when $A \in \mathcal{A}$ and $K \in \mathcal{B}_0$. It follows from the hypothesis that $\tilde{\phi}$ is well defined and linear. An application of the preceding lemma shows that $\tilde{\phi}$ is positive. Indeed, such an application combined with the following observation will show the map to be completely positive. Simply note that $M_n(\mathcal{A} + \mathcal{B}_0) = M_n(\mathcal{A}) + \mathcal{B}_0(\mathcal{H}^{(n)})$ and, using analogous notation, $(\tilde{\phi})_n = \widetilde{\phi_n}$. Thus in discussing a completely positive map $\phi : \mathcal{A} \to \mathcal{B}(\mathcal{K})$ such that $\phi(\mathcal{A} \cap \mathcal{B}_0) = 0$, there is no loss in generality in assuming that $\mathcal{A} \supseteq \mathcal{B}_0$.

The next result comes from Glimm [1960].

40.4 Glimm's Lemma. *If $\mathcal{A}$ is separable and $\omega : \mathcal{A} \to \mathbb{C}$ is a state, then $\omega(\mathcal{A} \cap \mathcal{B}_0) = 0$ if and only if there is a sequence of unit vectors $\{h_n\}$ in $\mathcal{H}$ such that $h_n \to 0$ weakly and $\langle Ah_n, h_n\rangle \to \omega(A)$ for every A in $\mathcal{A}$.*

Proof. If there is a sequence of unit vectors $\{h_n\}$ in $\mathcal{H}$ such that $h_n \to 0$ weakly and $\langle Ah_n, h_n\rangle \to \omega(A)$ for every A in $\mathcal{A}$, it is transparent that ω is a state that annihilates the compact operators in $\mathcal{A}$.

Now for the converse. As in the discussion preceding the lemma, it can be assumed that $\mathcal{A} \supseteq \mathcal{B}_0$. Let Σ denote the collection of all states on $\mathcal{A}$ and let Σ_0 be those states on $\mathcal{A}$ that annihilate $\mathcal{B}_0$. Since $\mathcal{A}$ is separable, Σ with its weak* topology is a compact metric convex space. It follows that Σ_0 is a compact subset of Σ. Let Σ_1 be the collection of all states $\psi : \mathcal{A} \to \mathbb{C}$ for which there is a sequence $\{h_n\}$ of unit vectors in $\mathcal{H}$ such that $h_n \to 0$ weakly and $\psi(A) = \lim_n \langle Ah_n, h_n\rangle$ for A in $\mathcal{A}$. Because compact operators map weakly convergent sequences into norm convergent ones, $\Sigma_1 \subseteq \Sigma_0$. It must be shown that the two sets are equal.

First note that $\Sigma_1 \neq \emptyset$. In fact, let $\{e_n\}$ be any orthonormal sequence in $\mathcal{H}$ and define $\omega_n\colon \mathcal{A} \to \mathbb{C}$ by $\omega_n(A) = \langle Ae_n, e_n\rangle$. By passing to a subsequence if necessary, it can be assumed that $\omega_n \to \omega$ weak* in Σ. It is easy to check that $\omega \in \Sigma_1$.

Claim. Σ_1 is convex.

Let $\phi, \psi \in \Sigma_1$ with associated sequences of unit vectors $\{f_n\}$ and $\{h_n\}$, respectively. Let $\{A_n\}$ be a dense sequence in $\mathcal{A}$, put $g_1 = h_1$, and define a sequence $\{g_n\}$ as follows. Let P_2 be the projection onto $[A_n f_2, A_n^* f_2 : n = 1, 2] \vee [g_1, f_2]$. Since P_2 has finite rank, $\|P_2 h_n\| \to 0$. By choosing $g_2 = P_2^{\perp} h_{n_2}/\|P_2^{\perp} h_{n_2}\|$ for sufficiently large n_2, it follows that g_2 is a unit vector, $g_2 \perp P_2$, and $\|g_2 - h_{n_2}\| \leq 2^{-2}$. Continuing produces a sequence of unit vectors $\{g_k\}$ and a subsequence $\{h_{n_k}\}$ such that $\|g_k - h_{n_k}\| \leq 2^{-k}$ and $g_k \perp P_k$, where P_k is the projection onto $[\{ A_n f_k, A_n^* f_k : 1 \leq n \leq k \}] \vee [\{ f_k, g_i : 1 \leq i < k \}]$.

Since $\|g_k - h_{n_k}\| \to 0$, $\psi(A) = \lim_k \langle Ag_k, g_k\rangle$ for all A in $\mathcal{A}$. If $0 \leq s, t \leq 1$, $s^2 + t^2 = 1$, let $e_k = (sf_k + tg_k)$. Since $g_k \perp f_k$, $\|e_k\| = 1$. Moreover, $e_k \to 0$ weakly and, for $k > n \geq 1$,

$$\begin{aligned}\langle A_n e_k, e_k\rangle &= \big[s^2\langle A_n f_k, f_k\rangle + st\langle A_n f_k, g_k\rangle \\ &\qquad + st\langle A_n g_k, f_k\rangle + t^2\langle A_n g_k, g_k\rangle\big] \\ &= \big[s^2\langle A_n f_k, f_k\rangle + t^2\langle A_n g_k, g_k\rangle\big] .\end{aligned}$$

Thus $s^2\phi(A_n) + t^2\psi(A_n) = \lim_{k\to\infty}\langle A_n e_k, e_k\rangle$. Since $\{ A_n \}$ is dense in $\mathcal{A}$, this holds with A_n replaced by an arbitrary element of $\mathcal{A}$. Therefore $s^2\phi + t^2\psi \in \Sigma_1$, whence the claim.

Claim. Σ_1 is weak* closed in Σ.

Let $\{\psi_n\}$ be a sequence in Σ_1 and assume $\psi_n \to \psi$ in Σ. Let $\{h_{nk}\}$ be a sequence of unit vectors such that if $\omega_{nk}(A) = \langle Ah_{nk}, h_{nk}\rangle$, then, as $k \to \infty$,

$\omega_{nk} \to \psi_n$ weak* and $h_{nk} \to 0$ weakly in $\mathcal{H}$. Thus ψ belongs to the weak* closure of $\{\omega_{nk} : n \geq 1, k \geq 1\}$ and hence, since $\mathcal{H}$ is separable, there is a sequence of unit vectors $\{h_k\}$ such that $\psi(A) = \lim_k \langle Ah_k, h_k\rangle$. Now the fact that $\Sigma_1 \subseteq \Sigma_0$ implies that $\psi \in \Sigma_0$. So for every compact operator K, $\langle Kh_k, h_k\rangle \to 0$. This implies $h_k \to 0$ weakly (Exercise 1) so that $\psi \in \Sigma_1$.

Now for a final claim. For A in $\mathcal{A}$, let $\Sigma_1(A) = \{\psi(A) : \psi \in \Sigma_1\}$ and define $\Sigma_0(A)$ similarly.

Claim. If A is a hermitian element of $\mathcal{A}$, $\Sigma_1(A) = \Sigma_0(A)$.

Let $\alpha = \min \sigma_e(A)$, $\beta = \max \sigma_e(A)$. Proposition 7.8 implies that $\Sigma_0(A) = [\alpha, \beta]$. Clearly $[\alpha, \beta] \supseteq \Sigma_1(A)$. Since $\alpha \in \sigma_e(A)$, there is an orthonormal sequence $\{e_n\}$ in $\mathcal{H}$ such that $\langle Ae_n, e_n\rangle \to \alpha$ (37.2.e). Define $\omega_n(T) = \langle Te_n, e_n\rangle$ for T in $\mathcal{A}$. So $\omega_n \in \Sigma(\mathcal{A})$. Since the weak* topology on $\Sigma(\mathcal{A})$ is metrizable, by passing to a subsequence it can be assumed that $\omega_n \to \omega_\alpha$ weak* in $\Sigma(\mathcal{A})$. Thus $\omega_\alpha \in \Sigma_1$ and $\omega_\alpha(A) = \alpha$. Similarly there is a ω_β in Σ_1 such that $\omega_\beta(A) = \beta$. Since Σ_1 is convex, $\Sigma_1(A)$ is a closed interval. But α and β belong to $\Sigma_1(A)$, so $\Sigma_1(A) = [\alpha, \beta] = \Sigma_0(A)$.

Now if $\omega \in \Sigma_0$ but $\omega \notin \Sigma_1$, the Hahn–Banach Theorem implies there is an A in $\mathcal{A}$ such that Re $\omega(A) >$ Re $\psi(A)$ for all ψ in Σ_1. Replacing A by its real part implies that there is a hermitian A in $\mathcal{A}$ such that $\omega(A) > \psi(A)$ for all ψ in Σ_1. By the last claim, $\omega(A) > \max \sigma_e(A)$. This contradicts the last claim. ∎

Voiculescu's Theorem is a generalization of Glimm's Lemma to completely positive maps. Indeed, if $\omega : \mathcal{A} \to \mathbb{C}$ is a state that annihilates the compact operators, ω is completely positive (34.5). Let $\{V_n\}$ be the sequence of isometries given by Voiculescu's Theorem. But V_n maps $\mathbb{C}$ into $\mathcal{H}$. Thus there is a vector h_n in $\mathcal{H}$ such that $V_n(\lambda) = \lambda h_n$ for all complex numbers λ. The fact that V_n is an isometry shows that $\|h_n\| = 1$. The fact that $V_n \to 0$ (WOT) implies that $h_n \to 0$ weakly. A calculation shows that $V_n^* A V_n$ is the complex number $\langle Ah_n, h_n\rangle$. This recaptures Glimm's Lemma and provides the additional fact that the sequence $\{h_n\}$ can be chosen to be orthonormal.

Given that Glimm's lemma gives a necessary and sufficient condition on a state that it annihilate the compact operators, one can ask whether the converse of Voiculescu's Theorem is also true. This is the case, but it will not be needed. A proof is sketched in Exercise 2.

Thus Glimm's Lemma can be viewed as proving Voiculescu's Theorem for the case that the Hilbert space $\mathcal{K}$ is one dimensional. The next step is to prove a version of the theorem for the case that $\mathcal{K}$ has dimension n.

40.5 Lemma. *If $\phi : \mathcal{A} \to M_n$ is a completely positive map such that $\phi(\mathcal{A} \cap \mathcal{B}_0) = 0$ and $\phi(1) = 1$ and $\mathcal{N}$ is a finite dimensional subspace of $\mathcal{H}$, then there is a sequence of isometries $V_k : \mathbb{C}^n \to \mathcal{H}$ such that:*

(a) $\|\phi(A) - V_k^* A V_k\| \to 0$ *for every A in $\mathcal{A}$;*

(b) $\operatorname{ran} V_k \leq \mathcal{N}^\perp$ *for all $k \geq 1$.*

Proof. As above, it can be assumed that $\mathcal{B}_0 \subseteq \mathcal{A}$. Consider $\phi_n : M_n(\mathcal{A}) \to M_{n^2}$ defined by $\phi_n([A_{ij}]) = [\phi(A_{ij})]$. Let $e_1, \dots, e_n$ be the usual basis for $\mathbb{C}^n$ and put $\overline{e} = (e_1, \dots, e_n)/\sqrt{n}$ in $\mathbb{C}^{n^2}$. So $\|\overline{e}\| = 1$. Define $\omega : M_n(\mathcal{A}) \to \mathbb{C}$ by

$$\begin{aligned}\omega([A_{ij}]) &= \langle \phi_n([A_{ij}])\overline{e}, \overline{e}\rangle \\ &= \frac{1}{n} \sum_{i,j=1}^{n} \langle \phi(A_{ij})e_j, e_i\rangle.\end{aligned}$$

Because ϕ is completely positive, ω is positive; since $\phi(1) = 1$, $\omega(1) = 1$. Also, $\mathcal{B}_0(\mathcal{H}^{(n)}) = M_n(\mathcal{B}_0) \subseteq M_n(\mathcal{A})$, so $\omega(\mathcal{B}_0(\mathcal{H}^{(n)})) = (0)$. By Glimm's Lemma there is a sequence of unit vectors $\{\overline{h}^k\}$ in $\mathcal{H}^{(n)}$ such that $\overline{h}^k \to 0$ weakly and $\omega([A_{ij}]) = \lim_{k\to\infty} \langle [A_{ij}]\overline{h}^k, \overline{h}^k\rangle$ for all $[A_{ij}]$ in $M_n(\mathcal{A})$.

Define $T_k : \mathbb{C}^n \to \mathcal{H}$ by $T_k e_i = \sqrt{n}\, h_i^k$ for $1 \leq i \leq n$. If $\alpha_1, \dots, \alpha_n \in \mathbb{C}$ and E_{ij} is the element of $M_n(\mathcal{A})$ with 1 in the (i,j) place and 0 elsewhere, then

$$\begin{aligned}\left\| T_k \sum_{i=1}^{n} \alpha_i e_i \right\|^2 &= \sum_{i,j} \alpha_i \overline{\alpha}_j n \langle h_i^k, h_j^k\rangle \\ &= n \sum_{i,j} \alpha_i \overline{\alpha}_j \langle E_{ji}\overline{h}^k, \overline{h}^k\rangle \\ &= n\langle [\alpha_j \overline{\alpha}_i]\overline{h}^k, \overline{h}^k\rangle.\end{aligned}$$

If $k \to \infty$, this converges to

$$\begin{aligned}n\omega([\alpha_j\overline{\alpha}_i]) &= n\langle \phi_n[\alpha_j\overline{\alpha}_i]\overline{e}, \overline{e}\rangle_{\mathbb{C}^{n^2}} \\ &= n\langle [\alpha_j\overline{\alpha}_i]\overline{e}, \overline{e}\rangle_{\mathbb{C}^{n^2}} \\ &= \sum_{i,j} \alpha_i\overline{\alpha}_j \langle e_j, e_i\rangle_{\mathbb{C}^n} \\ &= \left\| \sum_{i=1}^{n} \alpha_i e_i \right\|^2.\end{aligned}$$

That is, $\|T_k\overline{\alpha}\| \to \|\overline{\alpha}\|$ for every $\overline{\alpha}$ in $\mathbb{C}^n$. Thus $\langle (T_k^*T_k - 1)\overline{\alpha}, \overline{\alpha}\rangle \to 0$ for every $\overline{\alpha}$ in $\mathbb{C}^n$. This implies $T_k^*T_k \to 1$ (WOT) in $\mathcal{B}(\mathbb{C}^n)$; since the space is finite dimensional, $T_k^*T_k \to 1$. Thus for all sufficiently large k, T_k is injective. This implies that if $T_k = W_k|T_k|$ is the polar decomposition of T_k, W_k is an isometry for all large k. Since $|T_k^*| = (T_k^*T_k)^{1/2} \to 1$, $\|T_k - W_k\| \to 0$.

Now fix i and j, $1 \leq i,j \leq n$. If $A \in \mathcal{A}$, let $\widetilde{A}$ be the element of $M_n(\mathcal{A})$ whose (i,j) entry is A and whose other entries are 0. Thus $\langle(\phi(A) - T_k^*AT_k)e_j, e_i\rangle = n[\omega(\widetilde{A}) - \langle Ah_j^k, h_i^k\rangle] \to 0$ as $k \to \infty$. Thus $\|\phi(A) - T_k^*AT_k\| \to 0$ and so $\|\phi(A) - W_k^*AW_k\| \to 0$. So $\{W_k\}$ satisfies (a).

At this point there has been no mention of the finite dimensional space $\mathcal{N}$. It is now necessary to adjust the isometries W_k so that their ranges lie in $\mathcal{N}^\perp$. If $h \in \mathcal{H}$ and E is the rank one projection onto $\mathbb{C}h$, this implies $\|EW_k\|^2 = \|W_k^*EW_k - \phi(E)\| \to 0$. Thus for any $\overline{\alpha}$ in $\mathbb{C}^n$, $\langle W_k\overline{\alpha}, h\rangle = \langle EW_k\overline{\alpha}, h\rangle \to 0$. So $W_k \to 0$ (WOT).

Let P be the projection onto $\mathcal{N}$. Because P has finite rank and $W_k \to 0$ (WOT), $\|P_{\mathcal{N}}W_k\| \to 0$. Therefore $(P^\perp W_k)^*A(P^\perp W_k) \to \phi(A)$ for every A in $\mathcal{A}$. In particular, taking $A = 1$ shows that $|P^\perp W_k|^2 \to \phi(1) = 1$. Thus $|P^\perp W_k| \to 1$ and so $|P^\perp W_k|$ is invertible for all large k. Assume it is invertible for all k; so $|P^\perp W_k|^{-1} \to 1$. Thus if $P^\perp W_k = V_k|P^\perp W_k|$ is the polar decomposition, V_k is an isometry with range contained in $\mathcal{N}^\perp$. Moreover

$$V_k^*AV_k = |P^\perp W_k|^{-1}(P^\perp W_k)^*A(P^\perp W_k)|P^\perp W_k|^{-1} \to \phi(A)$$

for all A in $\mathcal{A}$. ∎

40.6 Lemma. *If $\{F_n\}$ is a sequence of positive operators such that $\sum_n F_n^2 = 1$ (SOT), then for every operator T, $\sum_n F_nTF_n$ converges in the strong operator topology.*

Proof. Let $\mathcal{H}^{(\infty)} = \mathcal{H} \oplus \mathcal{H} \oplus \cdots$, the infinite inflation of $\mathcal{H}$, and define $V: \mathcal{H} \to \mathcal{H}^{(\infty)}$ by $Vh = F_1h \oplus F_2h \oplus \cdots$. Computing gives that $\|Vh\|^2 = \sum_n \|F_nh\|^2 = \sum_n\langle F_n^2h, h\rangle = \|h\|^2$; so V is an isometry. Let P_n be the projection of $\mathcal{H}^{(\infty)}$ onto the sum of the first n coordinates and put $V_n = P_nV$. It is easily computed that $V_nh = F_1h \oplus \cdots \oplus F_nh$. Another computation shows that $V_n^*(h_1 \oplus \cdots) = \sum_{j=1}^n F_jh_j$.

Let $\tau: \mathcal{B}(\mathcal{H}) \to \mathcal{B}(\mathcal{H}^{(\infty)})$ be the inflationary representation $\tau(T) = T \oplus T \oplus \cdots$. Since $P_n \to 1$ (SOT), $V_n^*\tau(T)V_n = V^*P_n\tau(T)P_nV \to V^*\tau(T)V$. But $V_n^*\tau(T)V_n = F_1TF_1 + \cdots + F_nTF_n$. ∎

40.7 Lemma. *If $\mathcal{C}$ is a separable C^*-algebra containing $\mathcal{B}_0$, $\mathcal{F}$ is a finite subset of $\mathcal{C}$, and $\epsilon > 0$, then there is a sequence of positive finite rank operators $\{F_n\}$ such that $\sum_n F_n^2 = 1$ (SOT), $C - \sum_n F_nCF_n \in \mathcal{B}_0$ for every C in $\mathcal{C}$, and $\|C - \sum_n F_nCF_n\| < \epsilon$ for every C in $\mathcal{F}$.*

Proof. Let $\mathcal{L}$ be a countable dense subset of $\mathcal{C}$ that contains the given set $\mathcal{F}$ and is also a complex-rational linear manifold. Put $\mathcal{L} = \bigcup_n \mathcal{F}_n$, where each $\mathcal{F}_n$ is a finite set, $\mathcal{F}_n \subseteq \mathcal{F}_{n+1}$ for all $n \geq 1$, and $\mathcal{F}_1 = \mathcal{F}$. Using Proposition 36.7, choose a decreasing sequence $\{\delta_n\}$ of positive numbers such that $\delta_n \to 0$ and $\|E^{1/2}C - CE^{1/2}\| < \epsilon/2^n$ whenever $C, E \in \operatorname{ball}\mathcal{C}$, $E \geq 0$, and $\|EC - CE\| < \delta_n$.

Let $\{E_n\}$ be a sequence of finite rank operators that forms a quasicentral approximate identity for $\mathcal{C}$ (36.6). By passing to a subsequence if necessary, it can be assumed that $\|E_nC - CE_n\| < \delta_n/2$ for all C in $\mathcal{F}_{n+1}$. Putting $E_0 = 0$ and $\delta_0 = 0$, this implies that for $n \geq 1$ and C in $\mathcal{F}_n$, $\|(E_n - E_{n-1})C - C(E_n - E_{n-1})\| < \delta_n/2 + \delta_{n-1}/2 \leq \delta_n$. Thus, if $F_n = (E_n - E_{n-1})^{1/2}$, F_n is a positive finite rank operator and $\|F_nC - CF_n\| < \epsilon/2^n$ for all C in $\mathcal{F}_n$.

Since $\{E_n\}$ is an approximate identity, $\sum_n F_n^2 = 1$ (SOT). According to Lemma 40.6, $\Delta(C) \equiv \sum_n F_nCF_n$ converges (SOT) for all C. On the other hand, if $C \in \mathcal{L} = \bigcup_n \mathcal{F}_n$, $\sum_n \|CF_n - F_nC\| < \infty$. Thus for any such C, $C - \Delta(C) = \sum_n(CF_n^2 - F_nCF_n) = \sum_n(CF_n - F_nC)F_n$, where this sum converges in norm. Therefore, $C - \Delta(C) \in \mathcal{B}_0$ for all C in $\mathcal{L}$. Since $C \to C - \Delta(C)$ is a bounded operator on $\mathcal{L}$, it extends to a bounded operator on $\mathcal{C}$. This implies $C - \Delta(C) \in \mathcal{B}_0$ for all C in $\mathcal{C}$. Finally, if $C \in \mathcal{F} = \mathcal{F}_1$, $\|CF_n - F_nC\| < \epsilon/2^n$ for all $n \geq 1$, so $\|C - \sum_n F_nCF_n\| < \epsilon$. ∎

Proof of Voiculescu's Theorem. As noted at the beginning of the section, there is no loss in generality in assuming that $\mathcal{A}$ contains the compact operators. First a partial version of the theorem will be proved, after which only a small trick is required to finish the proof.

40.8 Claim. There is an isometry $V : \mathcal{K} \to \mathcal{H}$ such that $\phi(A) - V^*AV \in \mathcal{B}_0(\mathcal{K})$.

Let $\mathcal{C}$ be the separable C*-algebra $C^*(\phi(\mathcal{A})) + \mathcal{B}_0(\mathcal{K})$ contained in $\mathcal{B}(\mathcal{K})$. By Lemma 40.7 there is a sequence of positive finite rank operators on $\mathcal{K}$, $\{F_n\}$, such that $\sum F_n^2 = 1$ (SOT) and

$$C - \sum_n F_nCF_n \in \mathcal{B}_0(\mathcal{K})$$

for all C in $\mathcal{C}$. Put $E_n = F_1^2 + \cdots + F_n^2$ and let P_n be the projection onto $\operatorname{ran} E_n$. So P_n has finite rank and $P_n \uparrow 1$ (SOT).

Let $\mathcal{F} = \bigcup_n \mathcal{F}_n$, where each $\mathcal{F}_n$ is a finite subset of $\mathcal{A}$ with $\mathcal{F}_n = \mathcal{F}_n^*$, $\mathcal{F}_n \subseteq \mathcal{F}_{n+1}$, $1 \in \mathcal{F}_1$, and $\mathcal{F}$ is dense in $\mathcal{A}$. Now for each $n \geq 1$, $A \to P_n\phi(A)P_n$ is a completely positive map from $\mathcal{A}$ into $\mathcal{B}(P_n)$ that annihilates the compact operators. Using Lemma 40.5 and induction, there is an isometry U_n from P_n into $\mathcal{H}$ such that $\|P_n\phi(A)P_n - U_n^*AU_n\| < 2^{-n}$ for A in $\mathcal{F}_n$

and

$$\operatorname{ran} U_n \perp [\{ A(\operatorname{ran} U_i) : 1 \le i \le n-1, A \in \mathcal{F}_n \}].$$

It follows that

$$\sum_{n=1}^{\infty} \|P_n \phi(A) P_n - U_n^* A U_n\| < \infty$$

for all A in $\mathcal{F}$.

Note that $\operatorname{ran} U_n \perp \operatorname{ran} U_m$ for $n \ne m$ since $1 \in \mathcal{F}_1$. This, combined with the fact that $\|U_n F_n\| \le 1$ for all n, implies that $V \equiv \text{SOT} \sum_{n=1}^{\infty} U_n F_n$ is a bounded operator from $\mathcal{K}$ into $\mathcal{H}$ with $\|V\| \le 1$. For k in $\mathcal{K}$, moreover, $\|Vk\|^2 = \sum \|U_n F_n k\|^2 = \sum \|F_n k\|^2 = \sum \langle F_n^2 k, k\rangle = \|k\|^2$ since SOT $\sum F_n^2 = 1$. Thus $V : \mathcal{K} \to \mathcal{H}$ is an isometry. Put $V_n = \sum_{j=1}^{n} U_j F_j$. Because $V_n \to V$ (SOT), $V_n^* \to V^*$ (WOT). Therefore $V_n^* A V_n \to V^* A V$ (WOT) for all operators A in $\mathcal{B}(\mathcal{H})$. A direct computation shows that

$$V_n^* A V_n = \sum_{j=1}^{n} F_j U_j^* A U_j F_j + \sum_{i,j \le n, i \ne j} F_j U_j^* A U_i F_i.$$

Temporarily fix $p \ge 1$ and let $A \in \mathcal{F}_p$. If $p \le i < j$, then the choice of U_j gives that $A(\operatorname{ran} U_i) \subseteq (\operatorname{ran} U_j)^{\perp} = \ker U_j^*$. Thus $U_j^* A U_i = 0$. Similarly, $p \le j < i$ implies $U_i^* A U_j = 0$. But $\mathcal{F}_p^* = \mathcal{F}_p$, so that

$$U_j^* A U_i = 0 \text{ for all } A \text{ in } \mathcal{F}_p : \ i, j \ge p, i \ne j.$$

If $\rho_p : \mathcal{B}(\mathcal{H}) \to \mathcal{B}(\mathcal{K})$ is defined by $\rho_p(A) = \sum_{i,j<p; i \ne j} F_j U_j^* A U_i F_i$, then for $n > p$ and A in $\mathcal{F}_p$,

$$V_n^* A V_n = \sum_{j=1}^{n} F_j U_j^* A U_j F_j + \rho_p(A).$$

If $n \to \infty$, then this converges in the WOT to

$$V^* A V = \sum_{n=1}^{\infty} F_j U_j^* A U_j F_j + \rho_p(A)$$

whenever $A \in \mathcal{F}_p$.

Therefore for A in $\mathcal{F}_p$,

$$\begin{aligned} \phi(A) - V^* A V &= \phi(A) - \sum_{n=1}^{\infty} F_n U_n^* A U_n F_n - \rho_p(A) \\ &= \phi(A) - \sum_{n=1}^{\infty} F_n \phi(A) F_n \end{aligned}$$

$$+\sum_{n=1}^{\infty} F_n \left[P_n\phi(A)P_n - U_n^* A U_n\right] F_n - \rho_p(A).$$

From the choice of the finite rank operators $\{F_n\}$, it follows that $\phi(A) - \sum_n F_n\phi(A)F_n \in \mathcal{B}_0(\mathcal{K})$. Also the sequence $\{P_n\phi(A)P_n - U_n^*AU_n\}$ is norm summable, so the second summation is a norm convergent series of finite rank operators and, thus, is compact. Finally, $\rho_p(A)$ is finite rank for all A. Hence

$$\phi(A) - V^*AV \in \mathcal{B}_0(\mathcal{K})$$

for all A in $\mathcal{F}_p$. Since p was arbitrary, this membership holds for all A in a dense subset of $\mathcal{A}$. Taking limits establishes Claim 40.8.

Now to finish off the proof. Consider the infinite inflation $\phi^{(\infty)} : \mathcal{A} \to \mathcal{B}(\mathcal{K}^{(\infty)})$ defined by

$$\phi^{(\infty)}(A) = \phi(A)^{(\infty)} = \phi(A) \oplus \phi(A) \oplus \cdots .$$

Verify that $\phi^{(\infty)}$ is completely positive. Clearly $\phi^{(\infty)}$ annihilates the compact operators. Now apply Claim 40.8 to obtain an isometry $V : \mathcal{K}^{(\infty)} \to \mathcal{H}$ such that $\phi^{(\infty)}(A) - V^*AV$ is compact for every A in $\mathcal{A}$. Now V decomposes into a direct sum $V = V_1 \oplus V_2 \oplus \cdots$, where each $V_k : \mathcal{K} \to \mathcal{H}$ is an isometry and $\operatorname{ran} V_n \perp \operatorname{ran} V_m$ for $n \neq m$. Thus $\phi^{(\infty)}(A \oplus A \oplus \cdots) - V^*(A \oplus A \oplus \cdots)V = \bigoplus_n \left[\phi(A) - V_n^*AV_n\right] \in \mathcal{B}_0(\mathcal{K}^{(\infty)})$. This implies that $\phi(A) - V_n^*AV_n \in \mathcal{B}_0(\mathcal{K})$ for each $n \geq 1$ and $\|\phi(A) - V_n^*AV_n\| \to 0$ as $n \to \infty$. ■

Exercises.

1. Show that a sequence of vectors $\{h_n\}$ in $\mathcal{H}$ converges weakly to 0 if and only if $\|Kh_n\| \to 0$ for every compact operator K.

2. Let $\mathcal{A}$ be a separable C*-algebra contained in $\mathcal{B}(\mathcal{H})$ such that $\mathcal{A} \supseteq \mathcal{B}_0$. (a) Show that if $\{V_k\}$ is a sequence of operators from a Hilbert space $\mathcal{K}$ into $\mathcal{H}$ such that $V_k \to 0$ (WOT) and $K \in \mathcal{B}_0$, then $KV_k \to 0$ (SOT). Give the vector space $\mathcal{B}(\mathcal{A}, \mathcal{B}(\mathcal{K}))$ of all bounded linear maps with the topology defined by the semi-norms $\Phi \to \|\Phi(A)\|$, $A \in \mathcal{A}$. (b) If $\mathcal{C}$ denotes the collection of all completely positive maps ϕ from $\mathcal{A}$ into $\mathcal{B}(\mathcal{K})$ such that $\phi(1) = 1$, show that $\mathcal{C}$ is a closed subset of $\mathcal{B}(\mathcal{A}, \mathcal{B}(\mathcal{K}))$ with respect to the topology just defined. (c) If $V : \mathcal{K} \to \mathcal{H}$ is an isometry and $\phi : \mathcal{A} \to \mathcal{B}(\mathcal{K})$ is defined by $\phi(A) = V^*AV$, show that $\phi \in \mathcal{C}$. (d) If $\{V_k\}$ is a sequence of isometries from $\mathcal{K}$ into $\mathcal{H}$ such that $V_k \to 0$ (WOT) and $\phi(A) = \lim_k V_k^*AV_k$ exists for each A in $\mathcal{A}$, show that $\phi \in \mathcal{C}$ and $\phi(\mathcal{A} \cap \mathcal{B}_0) = 0$. (This establishes the converse of Theorem 40.1.)

§41. Approximately equivalent representations

The first definition is reminiscent of Definition 39.9.

41.1 Definition. If $\mathcal{A}$ is a C*-algebra and $(\rho, \mathcal{H})$ and $(\kappa, \mathcal{K})$ are two representations, say that ρ and κ are *approximately equivalent* if there is a sequence of isomorphisms $U_n : \mathcal{K} \to \mathcal{H}$ such that $\|\kappa(a) - U_n^* \rho(a) U_n\| \to 0$ for every a in $\mathcal{A}$. If, in addition, $\kappa(a) - U_n^* \rho(a) U_n \in \mathcal{B}_0(\mathcal{K})$ for each a in $\mathcal{A}$, then say that ρ and κ are *strongly approximately equivalent*. The notation for these two relations is $\rho \cong_a \kappa$ and $\rho \cong_{sa} \kappa$.

It is left to the reader to verify that both of these are equivalence relations. If ρ and κ are approximately equivalent representations, then, in the sense of Definition 39.9, $\rho(a) \cong_a \kappa(a)$ for each a in $\mathcal{A}$. The analogous statement holds if ρ and κ are strongly approximately equivalent.

Now assume that A and B are operators on $\mathcal{H}$ and $\mathcal{K}$, respectively, that are strongly approximately equivalent. Let $U_n : \mathcal{K} \to \mathcal{H}$ be isomorphisms such that $B - U_n^* A U_n \in \mathcal{B}_0(\mathcal{K})$ for each $n \geq 1$ and $\|B - U_n^* A U_n\| \to 0$. If p is any polynomial in two non-commuting symbols, simple algebra shows that $p(B, B^*) - U_n^* p(A, A^*) U_n \in \mathcal{B}_0$ and $\|p(B, B^*) - U_n^* p(A, A^*) U_n\| \to 0$. Note that two operators that are approximately equivalent have the same norm. Therefore $\rho(p(A, A^*)) = p(B, B^*)$ defines an isometry from a dense subalgera of $C^*(A)$ into $C^*(B)$. This extends to a *-isomorphism $\rho : C^*(A) \to C^*(B)$ given by $\rho(T) = \lim U_n^* T U_n$. It follows that $\rho \cong_{sa} \text{id}$, where id is the identity representation on $C^*(A)$. The similar correspondence holds when the operators are approximately equivalent. This is gathered together in the next proposition.

41.2 Proposition. *If A and B are operators and $A \cong_{sa} B$ ($A \cong_a B$), then $\rho(p(A, A^*)) = p(B, B^*)$ extends to a $*$-isomorphism $\rho : C^*(A) \to C^*(B)$ and $\rho \cong_{sa} \text{id}$ ($\rho \cong_a \text{id}$). Moreover if $\{U_n\}$ is the sequence of unitaries such that $U_n^* A U_n \to B$, then ρ is defined by $\rho(T) = \lim U_n^* T U_n$ for all T in $C^*(A)$.*

In the next section on applications, connections of Voiculescu's Theorem with the approximate equivalence of operators will be discussed. The next result is a consequence of Voiculescu's Theorem. It first appeared as Theorem 1.3 in Voiculescu [1976]. Also see Corollary 2 in Arveson [1977].

41.3 Theorem. *If $\mathcal{A}$ is a separable C^*-subalgebra of $\mathcal{B}(\mathcal{H})$ and $\rho \colon \mathcal{A} \to \mathcal{B}(\mathcal{K})$ is a representation such that $\rho(\mathcal{A} \cap \mathcal{B}_0) = 0$, then $\text{id} \cong_{sa} \text{id} \oplus \rho$.*

Proof. Consider the infinite inflation of ρ, $\rho^{(\infty)} \colon \mathcal{A} \to \mathcal{B}(\mathcal{K}^{(\infty)})$. According to Corollary 40.2 there is an isometry $V : \mathcal{K}^{(\infty)} \to \mathcal{H}$ such that $AV - V\rho(A)^{(\infty)} \in \mathcal{B}_0(\mathcal{K}^{(\infty)}, \mathcal{H})$ for all A in $\mathcal{A}$. Put $\mathcal{K}^{(\infty)} = \mathcal{K}_1 \oplus \mathcal{K}_2 \oplus \cdots$, where $\mathcal{K}_n = \mathcal{K}$ for each $n \geq 1$; and let $V = V_1 \oplus V_2 \oplus \cdots$, where V_n is the restriction of V to $\mathcal{K}_n$. Let $\mathcal{H}_n$ denote the range of V_n. So $\mathcal{H}_n \perp \mathcal{H}_m$ for $n \neq m$ and $\mathcal{H} = \mathcal{H}_0 \oplus \mathcal{H}_1 \oplus \cdots$ with $\mathcal{H}_0 = (\operatorname{ran} V)^\perp$. Let J_n be the canonical embedding of $\mathcal{K}$ into $\mathcal{K}_n \leq \mathcal{K}^{(\infty)}$. Put $S_n = \sum_{i=1}^{n-1} J_i J_i^* + \text{SOT} \sum_{i=n}^{\infty} J_{i+1} J_i^* \in \mathcal{B}(\mathcal{K}^{(\infty)})$. It can be checked that S_n is an isometry and, relative to the decomposition $\mathcal{K}^{(\infty)} = \mathcal{K}_1 \oplus \mathcal{K}_2 \oplus \cdots$, has the matrix

$$S_n = \left[\begin{array}{cccc|cccc} 1 & 0 & & & & & & \\ 0 & 1 & & & & & & \\ & & \ddots & 0 & & & 0 & \\ & & 0 & 1 & & & & \\ \hline & & & & 0 & & & \\ & & 0 & & 1 & 0 & & \\ & & & & 0 & 1 & 0 & \\ & & & & & & & \ddots \end{array}\right].$$

That is, S_n is the identity on $\mathcal{K}_1 \oplus \cdots \oplus \mathcal{K}_{n-1}$ and a shift on the orthocomplement of this space. Note that $VS_nV^*\mathcal{H}_0 = 0$, VS_nV^* is the identity on $\mathcal{H}_1 \oplus \cdots \oplus \mathcal{H}_{n-1}$, and it maps $\mathcal{H}_i$ isometrically onto $\mathcal{H}_{i+1}$ for $i \geq n$. Thus $1 - VV^* + VS_nV^*$ maps $\mathcal{H}$ isometrically onto $\mathcal{H}_n^\perp$. Also VJ_n maps $\mathcal{K}$ isometrically onto $\mathcal{H}_n$. Thus $U_n = (1 - VV^* + VS_nV^*) \oplus VJ_n$ is a unitary from $\mathcal{H} \oplus \mathcal{K}$ onto $\mathcal{H}$. It will now be shown that $U_n^* A U_n - A \oplus \rho(A) \in \mathcal{B}_0(\mathcal{H})$ and $\|U_n^* A U_n - A \oplus \rho(A)\| \to 0$. Equivalently, it will be shown that $AU_n - U_n(A \oplus \rho(A))$ is compact and converges to 0 in norm. In preparation for these demonstrations, observe that

$$\begin{aligned} AU_n - & U_n(A \oplus \rho(A)) \\ &= A(1 - VV^* + VS_nV^*) \oplus AVJ_n \\ &\quad - (1 - VV^* + VS_nV^*)A \oplus VJ_n\rho(A) \end{aligned}$$

41.4
$$\begin{aligned} &= [-(AVV^* - VV^*A) + (AVS_nV^* - VS_nV^*A)] \\ &\quad \oplus [AVJ_n - VJ_n\rho(A)]. \end{aligned}$$

Each of the direct summands in (41.4) will be shown to be compact. First

$$\begin{aligned} AVV^* - VV^*A &= [AV - V\rho(A)^{(\infty)}]V^* + V[\rho(A)^{(\infty)}V^* - V^*A] \\ &= [AV - V\rho^{(\infty)}(A)]V^* + V[V\rho(A^*)^{(\infty)} - A^*V]^* \end{aligned}$$

which is compact.

Next

$$S_n\rho(A)^{(\infty)} = \left[\begin{array}{ccc|ccc} 1 & & 0 & & & \\ & \ddots & & & 0 & \\ 0 & & 1 & & & \\ \hline & & & 0 & & \\ & & & 1 & 0 & \\ & & & & 1 & \ddots \end{array}\right] \left[\begin{array}{cc|cc} \rho(A) & & & \\ & \ddots & & \\ \hline & & \ddots & \\ & & & \end{array}\right]$$

$$= \left[\begin{array}{ccc|ccc} \rho(A) & & & & & \\ & \ddots & & & 0 & \\ & & \rho(A) & & & \\ \hline & & & 0 & \cdots & \\ & 0 & & \rho(A) & 0 & \\ & & & & \ddots & \ddots \end{array}\right].$$

A similar matrix multiplication shows that

$$\rho(A)^{(\infty)} S_n = S_n \rho^{(\infty)}(A).$$

So

$$\begin{aligned} &AVS_nV^* - VS_nV^*A \\ &\qquad = [(AV - V\rho(A)^{(\infty)}]S_nV^* + VS_n[\rho(A)^{(\infty)}V^* - V^*A] \\ &\qquad = [AV - V\rho(A)^{(\infty)}]S_nV^* + VS_n[V\rho(A^*)^{(\infty)} - A^*V]^* \end{aligned}$$

which is compact.

Combining these equations gives that the first direct summand in (41.4) becomes

41.5
$$\begin{aligned} -(AVV^* - VV^*A) &+ (AVS_nV^* - VS_nV^*A) \\ &= [AV - V\rho(A)^{(\infty)}](S_n - 1)V^* \\ &\quad + V(S_n - 1)(A^*V - V\rho(A^*)^{(\infty)})^*, \end{aligned}$$

which is compact.

Finally

41.6
$$AVJ_n - VJ_n\rho(A) = [AV - V\rho(A)^{(\infty)}]J_n$$

is also a compact operator. This implies that $AU_n - U_n(A \oplus \rho(A))$ is compact for every A in $\mathcal{A}$ and all $n \geq 1$.

Now to show convergence. Since $(S_n - 1)(\mathcal{K}_1 \oplus \cdots \oplus \mathcal{K}_n \oplus 0 \oplus \cdots) = 0$, it is easy to see that $S_n - 1 \to 0$ (SOT) as $n \to \infty$. Since, in (41.5), these operators are multiplied by compact operators, Exercise 16.5 implies that

$$\| - (AVV^* - VV^*A) + (AVS_nV^* - VS_nV^*A)\| \to 0.$$

Since $AV - V\rho(A)^{(\infty)} \in \mathcal{B}_0(\mathcal{K}^{(\infty)}, \mathcal{H})$, $(AV - V\rho(A)^{(\infty)})|\mathcal{K}_n \in \mathcal{B}_0(\mathcal{K}_n, \mathcal{H})$ for all $n \geq 1$ and $\|(AV - V\rho(A)^{(\infty)})|\mathcal{K}_n\| \to 0$. But $\|(AV - V\rho(A)^{(\infty)})|\mathcal{K}_n\| = \|(AV - V\rho(A)^{(\infty)})|J_n\| = \|AVJ_n - VJ_n\rho(A)\|$ by Equation 41.6. Hence $\|AVJ_n - VJ_n\rho(A)\| \to 0$. This completes the demonstration of convergence and, therefore, $\text{id} \cong_{sa} \text{id} \oplus \rho$. ■

Continue to denote the natural map by $\pi : \mathcal{B}(\mathcal{H}) \to \mathcal{B}(\mathcal{H})/\mathcal{B}_0(\mathcal{H})$.

41.7 Proposition. *If ρ, κ are representations of the separable C^*-algebra $\mathcal{A}$ such that*

$$\ker \rho = \ker \pi\rho = \ker \pi\kappa = \ker \kappa,$$

then $\rho \cong_{sa} \kappa$.

Proof. Assume the condition holds. Since $\ker \rho = \ker \kappa$, there is a unique *-isomorphism $\sigma : \rho(\mathcal{A}) \to \kappa(\mathcal{A})$ such that $\sigma\rho = \kappa$. The assumption that $\ker \rho = \ker \pi\rho$ implies that $\rho(\mathcal{A})$ has no non-zero compact operators. Thus the preceding theorem applies to the representation σ of $\rho(\mathcal{A})$. If id_ρ is the identity representation of $\rho(\mathcal{A})$, then $\mathrm{id}_\rho \cong_{sa} \mathrm{id}_\rho \oplus \sigma$. Therefore $\rho = \mathrm{id}_\rho \rho \cong_{sa} (\mathrm{id}_\rho \oplus \sigma)\rho = \rho \oplus \kappa$. A similar argument shows that $\kappa \cong_{sa} \kappa \oplus \rho$. ■

Now to begin the path to a characterization of approximately equivalent representations. In fact it will be shown that approximate and strong approximate equivalence are the same, thus extending Proposition 39.4. If $\mathcal{A}$ is any C*-algebra and $\rho : \mathcal{A} \to \mathcal{B}(\mathcal{H})$ is a representation, put

$$\mathcal{H}_e \equiv [\{ \rho(a)\mathcal{H} : \rho(a) \in \mathcal{B}_0(\mathcal{H}) \}] .$$

The space $\mathcal{H}_e$ is called the *essential subspace* for the representation. Because $\{ a \in \mathcal{A} : \rho(a) \in \mathcal{B}_0(\mathcal{H}) \} = \ker \pi\rho$ is an ideal of $\mathcal{A}$, $\mathcal{H}_e$ reduces $\rho(\mathcal{A})$. Let $\mathcal{H}_1 = \mathcal{H}_e^\perp$. Thus it is possible to define the subrepresentations $\rho_e \colon \mathcal{A} \to \mathcal{B}(\mathcal{H}_e)$ and $\rho_1 \colon \mathcal{A} \to \mathcal{B}(\mathcal{H}_1)$:

$$\rho_e(a) = \rho(a)|\mathcal{H}_e \qquad \rho_1(a) = \rho(a)|\mathcal{H}_1.$$

The subrepresentation ρ_e is called the *essential part* of ρ and ρ_1 is the *remainder.* Since $\rho_e(\mathcal{A})$ is a C*-algebra of compact operators, the results from §16 apply.

Thus the representation ρ decomposes as the direct sum $\rho = \rho_e \oplus \rho_1$. It is an easy exercise to show that $\rho(a)$ is compact if and only if $\rho_e(a)$ is compact and $\rho_1(a) = 0$. It may be, however, that $\rho_1(A)$ is a non-zero compact operator on $\mathcal{H}_e^\perp$. In this case $\rho_e(a)$ is not compact. The next definition is transitory as this notion of equivalence will eventually be shown to be the same as strong approximate equivalence.

41.8 Definition. Two representations ρ and κ are *weakly approximately equivalent* if there are two sequences of unitaries $\{U_n\}$ and $\{V_n\}$ such that for every a in $\mathcal{A}$, $U_n^*\rho(a)U_n \to \kappa(a)$ and $V_n^*\kappa(a)V_n \to \rho(a)$ in the weak operator topology. This relationship is denoted by $\rho \cong_w \kappa$.

A few observations must be made. First, $\cong_w$ is an equivalence relation. Second, the requirement that the convergence is in the weak operator topology is equivalent to requiring that the convergence be in the strong operator topology. See Exercise 2. Also the existence of the sequence $\{U_n\}$ does not imply the existence of the sequence $\{V_n\}$. See Exercise 3.

41.9 Lemma. *If $(\rho, \mathcal{H})$ and $(\kappa, \mathcal{K})$ are representations of the separable C^*-algebra $\mathcal{A}$ and $\rho \cong_w \kappa$, then $\ker \rho = \ker \kappa$, $\ker \pi\rho = \ker \pi\kappa$, and the essential parts of ρ and κ are unitarily equivalent.*

Proof. Adopt the notation in Definition 41.8. By passing to subsequences if necessary, it can be assumed that there are contractions U and V such that $U_n \to U$ (WOT) and $V_n \to V$ (WOT). Clearly $\ker \rho = \ker \kappa$. Suppose $a \in \ker \pi\rho$; so $\rho(a)$ is compact. Therefore $\rho(a)U_n \to \rho(a)U$ (SOT) (Exercise 16.5) and so $U_n^*\rho(a)U_n \to U^*\rho(a)U$ (WOT). Thus

41.10 $$\kappa(a) = U^*\rho(a)U \in \mathcal{B}_0(\mathcal{K}) \qquad \text{for all } a \text{ in } \ker \pi\rho.$$

This implies $\ker \pi\rho \subseteq \ker \pi\kappa$. The reverse inclusion is proved similarly.

Now to prove that the essential parts are equivalent. Let P_ρ be the projection of $\mathcal{H}$ onto the essential space for ρ. Define P_κ similarly. Let $\{e_n\}$ be a sequential approximate identity for the ideal $\ker \pi\rho$. It is left to the reader to show that $\rho(e_n) \to P_\rho$ (SOT). Since $\ker \pi\rho = \ker \pi\kappa$, it also holds that $\kappa(e_n) \to P_\kappa$ (SOT). By (41.10), $U^*\rho(e_n)U = \kappa(e_n)$ for all n. Thus $U^*P_\rho U = P_\kappa$.

Define $W : P_\kappa \to P_\rho$ by

$$W = P_\rho U | P_\kappa.$$

Because $U^*P_\rho U = P_\kappa$, $W^*W = 1$ on P_κ and so W is an isometry. Now to show that

41.11 $$W\kappa_e(a) = \rho_e(a)W \qquad \text{for all } a \text{ in } \mathcal{A}.$$

Once this is shown, the proof can be completed as follows. In fact, (41.11) shows two things. First it says that $\operatorname{ran} W$ is invariant for $\rho_e(\mathcal{A})$, and thus is reducing. Since $\operatorname{ran} W$ reduces $\rho_e(\mathcal{A})$, it is possible to define the subrepresentation of ρ_e, $\lambda = \rho_e | \operatorname{ran} W$. Since W is an isometry, (41.11) says that $\kappa_e \cong \lambda$. By a similar argument, there is a subrepresentation of κ_e that is equivalent to ρ_e. Since $\kappa_e(\mathcal{A})$ and $\rho_e(\mathcal{A})$ are C*-algebras of compact operators, Corollary 16.22 implies that $\rho_e \cong \kappa_e$.

Now to prove (41.11). First note that if $a \in \mathcal{A}$, then, because $\kappa(a) = \lim_n U_n^*\rho(a)U_n$ (SOT) (Exercise 2), $\rho(a)U_n - U_n\kappa(a) \to 0$ (WOT). So for x in $\ker \pi\rho$, and h in P_κ,

$$\begin{aligned} W\kappa_e(x) = P_\rho U\kappa_e(x)h &= \text{wk} \lim_n P_\rho U_n \kappa_e(x)h \\ &= \text{wk} \lim_n P_\rho \rho_e(x)U_n h \\ &= \rho_e(x)Uh. \end{aligned}$$

Hence $W\kappa_e(x)P_\kappa = \rho_e(x)UP_\kappa$ for all x in $\ker \pi\rho$. Therefore for any a in $\mathcal{A}$ and b in $\ker \pi\rho$, $W\kappa_e(a)[\kappa_e(b)P_\kappa] = W\kappa_e(ab)P_\kappa = \rho_e(ab)UP_\kappa = \rho_e(a)[\rho_e(b)UP_\kappa] = \rho_e(a)W[\kappa_e(b)P_\kappa]$. This proves (41.11). ■

Here is the main result of the section.

41.12 Theorem. *If $\mathcal{A}$ is a separable C^*-algebra and $(\rho, \mathcal{H})$ and $(\kappa, \mathcal{K})$ are two representations of $\mathcal{A}$ on a separable Hilbert space, the following statements are equivalent.*

(a) $\rho \cong_{sa} \kappa$.

(b) $\rho \cong_a \kappa$.

(c) $\rho \cong_w \kappa$.

(d) $\ker \rho = \ker \kappa$, $\ker \pi\rho = \ker \pi\kappa$, *and the essential parts of ρ and κ are unitarily equivalent.*

(e) *For each a in $\mathcal{A}$ the operators $\rho(a)$ and $\kappa(a)$ have the same rank.*

Proof. Clearly (a) implies (b) and this implies (c). Condition (c) implies (d) by the preceding lemma.

(d) *implies* (a). Write each of the representations ρ and κ as the sum of its essential part and the remainder: $\rho = \rho_e \oplus \rho_1$ and $\kappa = \kappa_e \oplus \kappa_1$. Define $\tau : \rho(\mathcal{A}) \to \mathcal{B}(P_\kappa^\perp)$ by $\tau(\rho(a)) = \kappa_1(a)$. Note that τ is well defined since if $\rho(a) = 0$, $a \in \ker \rho = \ker \kappa \subseteq \ker \kappa_1$. It is easily checked that τ is a representation of the algebra $\rho(\mathcal{A})$.

If $\rho(a) \in \mathcal{B}_0$, then $a \in \ker \pi\rho = \ker \pi\kappa$ so that $\kappa_1(a) = 0$. Thus, $\tau(\rho(\mathcal{A}) \cap \mathcal{B}_0) = 0$. By Theorem 41.3 $\mathrm{id} \oplus \tau \cong_{sa} \mathrm{id}$. (id here being the identity representation on $\rho(\mathcal{A})$.) Hence $\rho \oplus \kappa_1 = (\mathrm{id} \oplus \tau) \circ \rho \cong_{sa} \mathrm{id} \circ \rho = \rho$. Similarly $\kappa \oplus \rho_1 \cong_{sa} \kappa$. But, according to condition (d) $\rho_e \cong \kappa_e$. Therefore $\rho \cong_{sa} \rho \oplus \kappa_1 = \rho_e \oplus \rho_1 \oplus \kappa_1 \cong \kappa_e \oplus \kappa_1 \oplus \rho_1 = \kappa \oplus \rho_1 \cong_{sa} \kappa$.

(c) *implies* (e). This follows from Exercise 4.

(e) *implies* (d). One application of (e) shows that $\ker \rho = \ker \kappa$. It is also true that $\rho^{-1}(\mathcal{B}_{00}(\mathcal{H})) = \kappa^{-1}(\mathcal{B}_{00}(\mathcal{K}))$, hence (Exercise 5)

$$\ker \pi\rho = \mathrm{cl}\,[\rho^{-1}(\mathcal{B}_{00}(\mathcal{H}))] = \mathrm{cl}\,[\kappa^{-1}(\mathcal{B}_{00}(\mathcal{K}))] = \ker \pi\kappa.$$

Let $\mathcal{I} = \ker \pi\rho = \ker \pi\kappa$. So $\mathcal{H}_e = [\rho(\mathcal{I})\mathcal{H}]$ and $\rho_e(a) = \rho(a)|\mathcal{H}_e$; $\mathcal{K}_e$ and κ_e are defined similarly. Note that $\rho_e|\mathcal{I}$ is a non-degenerate representation of $\mathcal{I}$. Therefore, according to Corollary 6.12, to show that $\rho_e \cong \kappa_e$, it suffices to show that $\rho_e|\mathcal{I} \cong \kappa_e|\mathcal{I}$. Let $\mathcal{R} = \rho_e(\mathcal{I})$, a C*-algebra of compact operators on $\mathcal{H}_e$. If $x \in \mathcal{I}$ and $\rho_e(x) = 0$, then $\rho(xx^*)\mathcal{H} = \rho_e(x)\rho(x)^*\mathcal{H} = (0)$. So $x \in \ker \rho = \ker \kappa$ and $\kappa_e(x) = 0$. This says that there is a $*$-homomorphism $\tau : \mathcal{R} \to \mathcal{B}(\mathcal{K})$ defined by $\tau(R) = \kappa_e(x)$, whenever $R = \rho_e(x)$

for some x in $\mathcal{I}$. Also note that for $R = \rho_e(x)$ for x in $\mathcal{I}$, $\operatorname{ran}\rho(x) \subseteq \mathcal{H}_e$ so that $\dim \operatorname{cl}[\operatorname{ran} R] = \dim \operatorname{cl}[\operatorname{ran}\rho(x)] = \dim \operatorname{cl}[\operatorname{ran}\kappa(x)] = \dim \operatorname{cl}[\operatorname{ran}\tau(R)]$. According to Proposition 16.23, τ and the identity representation of $\mathcal{R}$ are equivalent. This implies that $\rho_e|\mathcal{I} \cong \kappa_e|\mathcal{I}$. ■

41.13 Corollary. *Let $\mathcal{A}$ be a separable C^*-algebra and let ρ and κ be non-degenerate, infinite dimensional, separable representations of $\mathcal{A}$.*

(a) *If $\ker\rho = \ker\kappa$ and neither $\rho(\mathcal{A})$ nor $\kappa(\mathcal{A})$ contains a non-zero compact operator, then $\rho \cong_{sa} \kappa$.*

(b) *If $\ker\pi\rho \subseteq \ker\kappa$, then $\rho \oplus \kappa \cong_{sa} \rho$.*

Proof. (a) Since neither $\rho(\mathcal{A})$ nor $\kappa(\mathcal{A})$ contains a non-zero compact operator, $\ker\pi\rho = \ker\pi\kappa = (0)$ and the essential parts of the two representations are 0. By the preceding theorem, this implies $\rho \cong_{sa} \kappa$.

(b) It's easy to check that ρ and $\rho \oplus \kappa$ have the same kernels. If $a \in \mathcal{A}$ such that $(\rho \oplus \kappa)(a) = \rho(a) \oplus \kappa(a) \in \mathcal{B}_0$, then $a \in \ker\pi\rho \subseteq \ker\kappa$. That is, $(\rho \oplus \kappa)(\mathcal{A}) \cap \mathcal{B}_0 \subseteq (\rho(\mathcal{A}) \cap \mathcal{B}_0) \oplus (0)$. The reverse inclusion follows similarly so that $(\rho \oplus \kappa)(\mathcal{A}) \cap \mathcal{B}_0 = (\rho(\mathcal{A}) \cap \mathcal{B}_0) \oplus (0)$. This shows two things. First, that $\ker\pi(\rho\oplus\kappa) = \ker\pi\rho$, and, second, that the essential part of $\rho\oplus\kappa$ is just the essential part of ρ. By part (d) of the preceding theorem, this implies $\rho \oplus \kappa \cong_{sa} \rho$. ■

Exercises. 1. Let $(\rho, \mathcal{H})$ and $(\kappa, \mathcal{K})$ be two representations of the C^*-algebra $\mathcal{A}$ and assume that $\rho \cong_{sa} \kappa$. If $\mathcal{C}$ is a C^*-algebra and $\nu : \mathcal{C} \to \mathcal{A}$ is a $*$-homomorphism, show that $\rho\nu \cong_{sa} \kappa\nu$. Prove the same thing with $\cong_{sa}$ replaced by $\cong_a$.

2. Let $\rho, \rho_1, \rho_2, \ldots$ be representations of the C^*-algebra $\mathcal{A}$ into the same Hilbert space. Show that $\rho_n(a) \to \rho(a)$ (WOT) for every a in $\mathcal{A}$ if and only if $\rho_n(a) \to \rho(a)$ (SOT) for every a in $\mathcal{A}$.

3. Consider the C^*-algebra $\mathcal{A} = \mathcal{B}_0 + \mathbb{C}$ and the representation $\kappa(K + \lambda) = \lambda$. Find a sequence of unitaries on $\mathcal{H}$ such that $U_n \to 0$ (WOT). Show that $U_n^* \operatorname{id}(K + \lambda) U_n \to \kappa(K + \lambda)$ for every $K + \lambda$ in $\mathcal{A}$. Show that there is no sequence of unitaries $\{V_n\}$ such that $V_n^* \kappa(K + \lambda) V_n \to K + \lambda$ for every $K + \lambda$ in $\mathcal{A}$.

4. If n is any positive integer, show that the set of operators having rank at most n is WOT closed.

5. Show that if $\mathcal{A}$ is a C^*-algebra and $\rho \colon \mathcal{A} \to \mathcal{B}(\mathcal{H})$ is a representation, then $\ker\pi\rho = \operatorname{cl}\left[\rho^{-1}(\mathcal{B}_0(\mathcal{H}))\right]$.

§42. Some applications

This section will look at a few applications of Voiculescu's Theorem and the preceding section. The first application is a non-abelian version of the Weyl-von Neumann-Berg Theorem. Re-examine the Weyl–von Neumann–Berg Theorem. Let N be a normal operator with $X = \sigma(N)$. Let D be a diagonal operator with $\sigma(D) = X$ and $\dim\ker(D - \lambda) = \dim\ker(N - \lambda)$ for every isolated eigenvalue λ of N. So Proposition 39.10 implies $D \cong_{sa} N$. Thus if ρ and Δ are defined on $C(X)$ by $\rho(f) = f(N)$ and $\Delta(f) = f(D)$, Proposition 41.2 implies $\rho \cong_{sa} \Delta$. Now Δ is the direct sum of one dimensional representations. That is, Δ is the direct sum of irreducible representations of $C(X)$. Moreover this is an equivalent formulation of the Weyl–von Neumann–Berg Theorem, at least in the case of a single normal operator. In this form, the result generalizes to arbitrary separable C*-algebras.

42.1 Theorem. *If $\mathcal{A}$ is a separable C^*-algebra, every separable representation of $\mathcal{A}$ is strongly approximately equivalent to the direct sum of irreducible representations. If ρ is the representation and $\rho(\mathcal{A})$ is infinite dimensional, then ρ is approximately equivalent to an infinite direct sum of irreducible representations.*

Proof. Let $\rho : \mathcal{A} \to \mathcal{B}(\mathcal{H})$ be a representation with $\mathcal{H}$ separable. It is assumed that $\rho(\mathcal{A})$ is infinite dimensional since the proof of the finite dimensional case can be gleaned from this one. As in the preceding section, write ρ as the direct sum $\rho_e \oplus \rho_1$ of its essential part and the remainder. Applying Proposition 32.10 to the algebra $\mathcal{A}/\ker\rho_1$ produces a sequence of irreducible representations $\{\omega_n\}$ of $\mathcal{A}$ such that $\bigcap \ker\omega_n = \ker\rho_1$. Modify this sequence of irreducible representations so that the same representations are used but each ω_n appears infinitely often. Put $\kappa = \bigoplus_n \omega_n$. It will be shown that $\rho \cong_{sa} \rho_e \oplus \kappa$ and $\rho_e \oplus \kappa$ is the direct sum of irreducible representations. Start with the second of these.

If $\mathcal{I} = \ker\pi\rho$, then $\rho_e(\mathcal{I})$ is a C*-algebra of compact operators. Thus the identity representation is the direct sum of irreducible representations by Theorem 16.11. Thus $\rho_e|\mathcal{I}$ is the direct sum of irreducible representations. But each irreducible representation of the ideal $\mathcal{I}$ extends to an irreducible representation of $\mathcal{A}$ (6.11). But since $\rho_e|\mathcal{I}$ is a non-degenerate representation of $\mathcal{I}$, ρ_e is the unique extension of $\rho_e|\mathcal{I}$ to $\mathcal{A}$. Thus ρ_e is the direct sum of irreducible representations. Since κ is defined as a direct sum of irreducible representations, this shows that $\rho_e \oplus \kappa$ can be so expressed.

It will be shown that $\rho \cong_{sa} \rho_e \oplus \kappa$ by applying Theorem 41.12. Since $\ker\kappa = \ker\rho_1$, $\ker(\rho_e \oplus \kappa) = \ker\rho_e \cap \ker\kappa = \ker\rho_e \cap \ker\rho_1 = \ker\rho$. Because each representation ω_n appears infinitely often in the sequence defining κ,

$\kappa(\mathcal{A}) \cap \mathcal{B}_0 = (0)$. Thus $\ker \pi(\rho_e \oplus \kappa) = \ker \pi\rho_e \cap \ker \kappa = \ker \pi\rho_e \cap \ker \rho_1 = \ker \pi\rho$. In addition to saying that the two kernels are equal, this also says that $\rho_e(a) \oplus \kappa(a)$ is compact if and only if $\rho_e(a)$ is compact and $\kappa(a) = 0$. Therefore the essential space of $\rho_e \oplus \kappa$ is the same as the essential space for ρ. That is, $(\rho_e \oplus \kappa)_e = \rho_e$. By Theorem 41.12, $\rho \cong_{sa} \rho_e \oplus \kappa$. ∎

The work of Voiculescu in proving Theorem 40.1 and the preceding theorem was motivated by an attempt to answer a question raised in Halmos [1970]: Is the set of reducible operators dense in $\mathcal{B}(\mathcal{H})$? The answer is yes.

42.2 Theorem. *If* $\dim \mathcal{H} = \infty$, $T \in \mathcal{B}(\mathcal{H})$, *and* $\epsilon > 0$, *then there is an operator* S *such that* $S - T$ *is compact,* $\|S - T\| < \epsilon$, *and* S *is the direct sum of an infinite number of irreducible operators.*

Proof. Consider the identity representation $\text{id} : C^*(T) \to \mathcal{B}(\mathcal{H})$. According to the preceding theorem, there is a representation $\rho : C^*(T) \to \mathcal{B}(\mathcal{H})$ that is the direct sum of an infinite number of irreducible representations of $C^*(T)$ such that $\text{id} \cong_{sa} \rho$. If $\{U_n\}$ is a sequence of unitaries such that $\|A - U_n^*\rho(A)U_n\| \to 0$ and $A - U_n^*\rho(A)U_n \in \mathcal{B}_0$ for all $n \geq 1$, then taking $S = U_n^*TU_n$ for an appropriately large value of n gives the result. ∎

Now for a flirtation with a topic that will be examined in some detail in §58. If $\mathcal{A} \subseteq \mathcal{B}/\mathcal{B}_0$, let $\text{Lat}_e\, \mathcal{A}$ be the collection of all projections p in the Calkin algebra such that $p^\perp a p = 0$ for all a in $\mathcal{A}$. Of course this is in analogy with the lattice of all invariant subspaces of a set of operators. The analogy is somewhat misleading in that $\text{Lat}_e\, \mathcal{A}$ is not always a lattice. The next result is from Arveson [1977]. This is based on a result of Voiculescu [1976], which appears as a corollary of this result. Note that in this theorem it is not assumed that the algebra $\mathcal{A}$ is a C*-algebra.

42.3 Theorem. *If* $\mathcal{A}$ *is a unital separable subalgebra of the Calkin algebra and* $t \in \mathcal{B}/\mathcal{B}_0$, *then there is a* p *in* $\text{Lat}_e\, \mathcal{A}$ *such that*

$$\text{dist}\,(t, \mathcal{A}) = \|p^\perp t p\|.$$

Proof. One inequality is trivial. If $p \in \text{Lat}_e\mathcal{A}$, then for every a in $\mathcal{A}$, $\|p^\perp t p\| = \|p^\perp (t - a) p\| \leq \|t - a\|$. Taking the infimum over all a in $\mathcal{A}$ gives that $\|p^\perp t p\| \leq \text{dist}\,(t, \mathcal{A})$ for every p in $\text{Lat}_e\mathcal{A}$.

Fix t and let $\mathcal{C} = C^*(\mathcal{A}, t)$. Clearly it can be assumed that $t \notin \mathcal{A}$. Thus there is a bounded linear functional L on $\mathcal{C}$ such that $\|L\| = 1$, $L(\mathcal{A}) = (0)$, and $L(t) = \text{dist}\,(t, \mathcal{A})$.

Claim. There is a non-degenerate, separable representation $(\kappa, \mathcal{K})$ of $\mathcal{C}$ and a $\kappa(\mathcal{A})$ invariant subspace E such that

$$\|\pi(E^\perp \kappa(t)E)\| \geq \operatorname{dist}(t, \mathcal{A}).$$

To see this, first obtain a cyclic representation $(\mu, \mathcal{M})$ of $\mathcal{B}$ with unit cyclic vector e and a unit vector k in $\mathcal{K}$ such that $L(c) = \langle \mu(c)e, k\rangle$ for all c in $\mathcal{C}$. In fact this was Exercise 7.8, but here is a proof for the sake of completeness. Let $\tau : \mathcal{C} \to \mathcal{B}(\mathcal{L})$ be the universal representation. So according to Corollary 7.14 there are vectors e and g in $\mathcal{L}$, which can be taken with norm 1, such that $L(c) = \langle \tau(c)e, g\rangle$ for all c in $\mathcal{C}$. Let $\mathcal{M} = [\tau(\mathcal{C})e]$ and let $\mu(c) = \tau(c)|\mathcal{M}$. If k is the projection of g onto $\mathcal{M}$, this is the desired representation of the linear functional L.

Put $F = [\mu(\mathcal{A})e]$. So F is an invariant subspace for $\mu(\mathcal{A})$ that contains e. Since L annihilates $\mathcal{A}$, $k \perp F$. Hence

$$\begin{aligned}\operatorname{dist}(t, \mathcal{A}) = L(t) &= \langle \mu(t)e, k\rangle \\ &= \langle \mu(t)Fe, F^\perp k\rangle \\ &\leq \|F^\perp \mu(t)F\|.\end{aligned}$$

This almost proves the claim in that this is the inequality for the norm in $\mathcal{B}(\mathcal{M})$ rather than the norm in the Calkin algebra. Let $\mathcal{K} = \mathcal{M}^{(\infty)}$, $\kappa = \mu^{(\infty)}$, and $E = F^{(\infty)}$. Now note that $\|\pi(E^\perp\kappa(t)E)\| = \|F^\perp\mu(t)F\|$ and the claim follows.

Consider the separable C*-algebra $\mathcal{D} = \pi^{-1}(\mathcal{C})$ and consider the representation of $\mathcal{D}$, $\mathrm{id} \oplus \kappa\pi : \mathcal{D} \to \mathcal{B}(\mathcal{H}) \oplus \mathcal{B}(\mathcal{K})$. Since $\kappa\pi$ is a representation of $\mathcal{D}$ that annihilates the compact operators, Theorem 41.3 implies $\rho \equiv \mathrm{id} \oplus \kappa\pi \cong_{sa} \mathrm{id}$. Thus there is an isomorphism $U : \mathcal{H} \oplus \mathcal{K} \to \mathcal{H}$ such that $U^*DU - \rho(D) \in \mathcal{B}_0(\mathcal{H} \oplus \mathcal{K})$ for every D in $\mathcal{D}$.

Now if $A \in \pi^{-1}(\mathcal{A}) \subseteq \mathcal{D}$, $\rho(A) = A \oplus \kappa(a)$, where $a = \pi(A)$. Thus for the projection E from the Claim, $0 \oplus E \in \operatorname{Lat}\rho(A)$. Thus $P \equiv U(0 \oplus E)U^* \in \operatorname{Lat} U^*\rho(A)U = \operatorname{Lat} U^*[A \oplus \kappa(a)]U$. Hence there is a compact operator K_1 such that

$$\begin{aligned}P^\perp AP &= U(1 + E^\perp)U^*AU(0 \oplus E)U^* \\ &= K_1 + UU(1 \oplus E^\perp)(A \oplus \kappa(a))U(0 \oplus E)U^* \\ &= K_1 + U\left[0 \oplus E^\perp\kappa(a)E\right]U^* \\ &= K_1.\end{aligned}$$

Thus $p = \pi(P) \in \operatorname{Lat}_e\mathcal{A}$.

If $T \in \mathcal{B}(\mathcal{H})$ such that $\pi(T) = t$, then $T \in \mathcal{D}$. So there is a compact operator K_2 such that $T = K_2 + U\rho(T)U^* = K_2 + U[T \oplus \kappa(t)]U^*$. Therefore

$$\begin{aligned} P^\perp TP &= P^\perp K_2 P + P^\perp U[T \oplus \kappa(t)]U^* P^\perp \\ &= P^\perp K_2 P + u\left[0 \oplus E^\perp \kappa(t) E\right] U^*. \end{aligned}$$

Therefore $\|p^\perp tp\| = \|\pi(E^\perp \kappa(t)E)\| \geq \operatorname{dist}(t, \mathcal{A})$. ■

42.4 Corollary. *If $\mathcal{A}$ is a norm closed, separable unital subalgebra of the Calkin algebra, then $a \in \mathcal{A}$ if and only if $p^\perp ap = 0$ for every p in* $\text{Lat}_e\mathcal{A}$.

Proof. If $p^\perp ap = 0$ for every p in $\text{Lat}_e\mathcal{A}$, then the preceding theorem says that $\operatorname{dist}(a, \mathcal{A}) = 0$. The converse is trivial. ■

42.5 Corollary. *If $\mathcal{A}$ is a separable C^*-algebra contained in the Calkin algebra, then $\mathcal{A}$ equals its double commutant within $\mathcal{B}/\mathcal{B}_0$. That is, $a \in \mathcal{A}$ if and only if $ab = ba$ for ever b in $\mathcal{B}/\mathcal{B}_0$ that commutes with $\mathcal{A}$.*

Proof. The fact that $\mathcal{A}$ is a self-adjoint algebra implies that $\text{Lat}_e\mathcal{A}$ consists of all projections that commute with $\mathcal{A}$. Therefore the preceding corollary implies that every a in $\mathcal{B}/\mathcal{B}_0$ that commutes with $\{ b \in \mathcal{B}/\mathcal{B}_0 : xb = bx \text{ for all } x \text{ in } \mathcal{A} \}$, belongs to $\mathcal{A}$. The other inclusion is trivial. ■

Berger and Coburn [1996] give a non-separable C*-algebra contained in the Calkin algebra that is not equal to its double commutant. On the other hand, Johnson and Parrott [1972] and Popa [1987] show that the image of a von Neumann algebra in the Calkin algebra does equal its double commutant.

42.6 Definition. Two operators A and B are *C^*-equivalent* if there is a $*$-isomorphism ρ: $C^*(A) \to C^*(B)$ such that $\rho(A) = B$. In symbols this is written as $C^*(A) \approx C^*(B)$.

It is easy to see that this is an equivalence relation. In Proposition 41.2 it was shown that if $A \cong_{sa} B$, then $C^*(A) \approx C^*(B)$. Also note that for $1 \leq d \leq \infty$, A and $A^{(d)}$ are C*-equivalent. These next several results appeared in Feldman [1997].

42.7 Theorem. *If $A \in \mathcal{B}(\mathcal{H})$ and $B \in \mathcal{B}(\mathcal{K})$ are operators, then there is a $*$-homomorphism $\rho : C^*(A) \to C^*(B)$ with $\rho(A) = B$ if and only if $A^{(\infty)} \cong_{sa} A^{(\infty)} \oplus B$.*

Proof. Assume such a $*$-homomorphism ρ exists and define $\kappa: C^*(A^{(\infty)}) \to C^*(B)$ by $\kappa(T^{(\infty)}) = \rho(T)$. So κ is a representation of $C^*(A^{(\infty)})$ and $\kappa(C^*(A^{(\infty)}) \cap \mathcal{B}_0) = \kappa(0) = 0$. By Theorem 41.3 id $\cong_{sa}$ id $\oplus\, \kappa$. This implies $A^{(\infty)} \cong_{sa} A^{(\infty)} \oplus B$.

Now assume that $A^{(\infty)} \cong_{sa} A^{(\infty)} \oplus B$. By Proposition 41.2 there is a $*$-homomorphism $\kappa: C^*(A^{(\infty)}) \to C^*(A^{(\infty)} \oplus B)$ such that $\kappa(A^{(\infty)}) = A^{(\infty)} \oplus B$. Define $\rho: C^*(A) \to C^*(B)$ by $\rho(T) = \kappa(T)|(0) \oplus \mathcal{K}$. ■

42.8 Corollary. *Two operators A and B are C^*-equivalent if and only if $A^{(\infty)} \cong_{sa} B^{(\infty)}$.*

Proof. If $A^{(\infty)} \cong_{sa} B^{(\infty)}$, then $C^*(A^{(\infty)}) \approx C^*(B^*)$ by Proposition 41.2. But as was observed, $C^*(A) \approx C^*(A^{(\infty)})$ and $C^*(B) \approx C^*(B^{(\infty)})$.

Conversely, assume $\rho: C^*(A) \to C^*(B)$ is a $*$-isomorphism such that $\rho(A) = B$. If $\kappa: C^*(A) \to C^*(B^{(\infty)})$ is defined by $\kappa(T) = \rho(T)^{(\infty)}$, then κ is a $*$-homomorphism. The theorem implies that $A^{(\infty)} \cong_{sa} A^{(\infty)} \oplus B^{(\infty)}$. If the same reasoning is applied to ρ^{-1}, it follows that $B^{(\infty)} \cong_{sa} B^{(\infty)} \oplus A^{(\infty)}$. ■

42.9 Proposition. *For any operator A, the following statements are equivalent.*

(a) $C^*(A) \cap \mathcal{B}_0 = (0)$.

(b) $S \cong_{sa} S^{(\infty)}$.

(c) $S \cong_{sa} S \oplus S$.

Proof. (a) *implies* (b). Define $\rho: C^*(A) \to \mathcal{B}(\mathcal{H}^{(\infty)})$ by $\rho(T) = T^{(\infty)}$. Since $\rho(C^*(A) \cap \mathcal{B}_0) = (0)$, Theorem 41.3 implies id $\cong_{sa}$ id $\oplus\, \rho \cong \rho$. Hence $A \cong_{sa} A^{(\infty)}$.

(b) *implies* (c). This is clear.

(c) *implies* (a). Let $\{U_n\}$ be a sequence of unitaries such that $U_n^* A U_n - A \oplus A \to 0$ and let $\rho: C^*(A) \to C^*(A \oplus A)$ be the $*$-isomorphism defined be $\rho(T) = \lim_n U_n^* T U_n$ (41.2). Note that the $*$-isomorphism $T \to T \oplus T$ agrees with ρ on the generator A. Thus $T \oplus T = \lim_n U_n^* T U_n$ for all T in $C^*(A)$. If $C^*(A) \cap \mathcal{B}_0 \neq (0)$, then there is a positive compact operator K in $C^*(A)$, $K \neq 0$. Thus $K \cong_{sa} K \oplus K$. But for $\lambda \neq 0$, $\dim\ker(K \oplus K - \lambda) = 2\dim\ker(K - \lambda)$, contradicting Proposition 39.10. ■

Chapter 7

Introduction to von Neumann Algebras

This chapter examines the foundations of the theory of von Neumann algebras. This was begun in §13 and §14, which characterize the maximal abelian von Neumann algebras. This treatment of von Neumann algebras is designed to give the readers an acquaintance with the subject and prepare them for further study. For more complete presentations of this vast subject, the interested reader can see Dixmier [1956], Kadison and Ringrose [1983] and [1986], Pedersen [1979], or Stratila and Zsido [1979]. The book Topping [1971] is another abbreviated treatment and was used in the preparation of this chapter.

One particular facet of a von Neumann algebra is that, unlike a C*-algebra, it is not defined abstractly. A von Neumann algebra is defined as a subalgebra of $\mathcal{B}(\mathcal{H})$ having certain properties. It is possible to give an abstract characterization of a von Neumann algebra, but this will not be carried out here. The development of the theory of von Neumann algebras starting from this abstract, spaceless point of view appears in Sakai [1971]. It is shown there that any C*-algebra that is the dual of a Banach space has a representation as a von Neumann algebra. That a von Neumann algebra is the dual of a Banach space is trivial, since it is a weak* closed subspace of the dual space $\mathcal{B}(\mathcal{H}) = \mathcal{B}_1(\mathcal{H})^*$.

§43. Elementary properties and examples

It has already been seen that a von Neumann algebra is a unital C*-algebra contained in $\mathcal{B}(\mathcal{H})$ that is closed in the weak operator topology. By the Double Commutant Theorem (12.3), a von Neumann algebra is a C*-subalgebra of $\mathcal{B}(\mathcal{H})$ such that $\mathcal{A} = \mathcal{A}''$. Equivalently, it is a C*-subalgebra of $\mathcal{B}(\mathcal{H})$ that contains the identity and is closed in either the strong operator topology or the weak* topology.

Begin by examining order completeness in $\mathcal{B}(\mathcal{H})$.

43.1 Proposition. *If $\{A_i\}$ is an increasing net of hermitian operators on $\mathcal{H}$ such that $\sup_i \|A_i\| < \infty$, then there is an operator A such that the following hold.*

(a) $A = \sup_i A_i$. *That is, $A_i \leq A$ for all i and if B is any other hermitian operator satisfying $A_i \leq B$ for all i, then $A \leq B$.*

(b) $A_i \to A$ *in the weak operator topology.*

(c) $A_i \to A$ *in the strong operator topology.*

(d) $A_i \to A$ *in the weak* topology.*

Proof. Without loss of generality it may be assumed that each operator A_i is positive. Indeed, fixing any index i_0 and proving the proposition for the net $\{A_i - A_{i_0}\}$ will prove the proposition for the original net.

If $h \in \mathcal{H}$, then $\{\langle A_i h, h\rangle\}$ is an increasing net of positive numbers that is bounded by $\alpha\|h\|^2$, where $\alpha = \sup_i \|A_i\|$. Thus $F(h,h) \equiv \lim_i \langle A_i h, h\rangle$ exists and $F(h,h) \leq \alpha\|h\|^2$ for all h in $\mathcal{H}$. Using the polar identity, $F(g,h) \equiv \lim_i \langle A_i g, h\rangle$ exists for all g, h in $\mathcal{H}$ and $|F(g,h)| \leq \alpha \|g\|\,\|h\|$. It is easy to check that F is a sesquilinear form on $\mathcal{H} \times \mathcal{H}$. Therefore there is a bounded operator A on $\mathcal{H}$ such that $\langle Ag, h\rangle = \lim_i \langle A_i g, h\rangle$ for all g, h in $\mathcal{H}$.

Clearly A is positive, $A_i \to A$ (WOT), and $\|A\| \leq \alpha$. Since the net is increasing, $0 \leq A - A_i \leq A$ for all i. Thus $(A - A_i)^{1/2} \leq A^{1/2}$. So if $h \in \mathcal{H}$, $\|(A - A_i)h\|^2 \leq \|(A - A_i)^{1/2}\|^2\, \|(A - A_i)^{1/2}h\|^2 \leq \|A\|\, \|(A - A_i)^{1/2}h\|^2 = \|A\|\, \langle (A - A_i)h, h\rangle \to 0$. This proves (c).

In the course of the proof of the existence of A it was noted that A must be an upper bound for the net $\{A_i\}$. If B is any hermitian operator on $\mathcal{H}$ such that $A_i \leq B$ for all i, then $\langle A_i h, h\rangle \leq \langle Bh, h\rangle$ for all vectors h. Taking limits shows that $A \leq B$. So A is the supremum.

To show that $A_i \to A$ (weak*), consider the net $\{A_i^{(\infty)}\}$ in $\mathcal{B}(\mathcal{H}^{(\infty)})$. This net is also increasing and the norms bounded above by α. Thus, by the preceding part of this proof, it converges to its supremum. It is left to the reader to show that $A^{(\infty)}$ is the supremum of this net, so that $A_i^{(\infty)} \to A^{(\infty)}$ (WOT) in $\mathcal{B}(\mathcal{H}^{(\infty)})$. By Proposition 21.6, $A_i \to A$ (weak*). ∎

What happens if there is a weak* closed C*-algebra of operators that is not assumed to have an identity, let alone contain the identity operator? Though it is easy to manufacture examples where the second situation holds, the first possibility cannot occur. The next proposition shows this and clarifies matters.

43.2 Proposition. *If $\mathcal{A}$ is a C^*-algebra contained in $\mathcal{B}(\mathcal{H})$ that is weak* closed, then there is a projection P in $\mathcal{A}$ such that $A = AP = PA$ for all A in $\mathcal{A}$. So P is an identity for $\mathcal{A}$ and $\mathcal{A}|P$ is a von Neumann algebra in $\mathcal{B}(P)$.*

Proof. Because $\mathcal{A}$ is a C*-algebra, it has an approximate identity $\{A_i\}$. By Proposition 43.1, $P \equiv \sup_i A_i \in \mathcal{A}$ and $A_i \to P$ (weak*); hence $P \in \mathcal{A}$. If $A \in \mathcal{A}$, then $\|AA_i - A\| \to 0$ and $AA_i \to AP$ (weak*). Thus $A = AP$. Similarly $A = PA$. ∎

Here is another way to reduce the study of such algebras to the study of von Neumann algebras. If $\mathcal{A}$ is a weak* closed C*-algebra contained in $\mathcal{B}(\mathcal{H})$ that does not contain the identity, then $\mathcal{A} + \mathbb{C}$ is also weak* closed. With a little thought it follows that $(\mathcal{A} + \mathbb{C})' = \mathcal{A}'$. Therefore $\mathcal{A}'' = (\mathcal{A} + \mathbb{C})'' = \mathcal{A} + \mathbb{C}$. That is, $\mathcal{A} + \mathbb{C}$ is a von Neumann algebra. Indeed, if P is the projection in (43.2), then it follows that $\mathcal{A}+\mathbb{C} = \mathcal{A}+\mathbb{C}P^{\perp}$. With these observations it becomes apparent that assuming that a von Neumann algebra contains the identity is more a convenience than a significant restriction.

For the remainder of the section $\mathcal{A}$ will be used to denote a von Neumann algebra contained in $\mathcal{B}(\mathcal{H})$. A few basic properties of von Neumann algebras were presented in §13. The reader should review these properties as they will be used repeatedly without reference. Two of these are immediately crucial. Recall that every element of a C*-algebra is the linear combination of four unitaries in the algebra (13.3.b). Also for a von Neumann algebra $\mathcal{A}$, $\phi(A) \in \mathcal{A}$ whenever A is a normal operator in $\mathcal{A}$ and ϕ is a bounded Borel function on the spectrum of A. It is also the case that if $A \in \mathcal{A}$, then the parts of the canonical polar decomposition of A belong to $\mathcal{A}$ (13.3.d).

Several examples of von Neumann algebras have already been encountered. The abelian algebras $\{ M_\phi : \phi \in L^\infty(\mu) \}$ have appeared several times and were studied in some detail in §14. Other examples are $\mathcal{B}(\mathcal{H})$ itself, including its finite dimensional specialization. It was also shown in Proposition 13.2 that the intersection of von Neumann algebras is a von Neumann algebra as is the center of a von Neumann algebra $\mathcal{A}$, $\mathcal{Z} = \mathcal{A} \cap \mathcal{A}'$. (Note that

$\mathcal{A}$ and $\mathcal{A}'$ have the same center.) More sophisticated examples will appear in this section and the subsequent material.

Recall that the collection of projections in $\mathcal{B}(\mathcal{H})$ (or the collection of subspaces of $\mathcal{B}(\mathcal{H})$) has a natural lattice structure.

43.3 Proposition. *If $\mathcal{A}$ is a von Neumann algebra and $\mathcal{E}$ is a collection of projections in $\mathcal{A}$, then $\bigvee \mathcal{E}$ and $\bigwedge \mathcal{E}$ belong to $\mathcal{A}$.*

Proof. Because self-adjoint algebras are involved, the projections in $\mathcal{A}$ are precisely the invariant subspaces of the operators in $\mathcal{A}'$. This makes this proposition immediate. ∎

Recall the d-fold inflation of an operator T (6.3), $T^{(d)} \equiv T \oplus T \oplus \cdots$, where there are d direct summands.

43.4 Proposition. *If $\mathcal{A}$ is a von Neumann algebra and $1 \leq d \leq \infty$, $\mathcal{A}^{(d)}$ is a von Neumann algebra and $(\mathcal{A}^{(d)})' = \{T = (T_{ij}) \in \mathcal{B}(\mathcal{H}^{(d)}) : T_{ij} \in \mathcal{A}' \text{ for all } i, j\}$.*

Proof. The fact that $(\mathcal{A}^{(d)})' = \{T = (T_{ij}) \in \mathcal{B}(\mathcal{H}^{(d)}) : T_{ij} \in \mathcal{A}' \text{ for all } i, j\}$ is an interpretation of Proposition 12.2.d. According to part (e) of that same proposition, $\{\mathcal{A}^{(d)}\}'' = [\mathcal{A}'']^{(d)} = \mathcal{A}^{(d)}$ since $\mathcal{A}'' = \mathcal{A}$. Thus $\mathcal{A}^{(d)}$ is a von Neumann algebra. ∎

If the preceding proposition is applied to the abelian algebra $L^\infty(\mu)$ acting on $L^2(\mu)$, an interesting example surfaces. Let $\mathcal{A} = \{M_\phi : \phi \in L^\infty(\mu)\} \subseteq \mathcal{B}(L^2(\mu))$ and consider $\mathcal{A}^{(d)}$ for a finite value of d. According to Proposition 43.4 and the fact that $\mathcal{A}' = \mathcal{A}$, $\mathcal{A}^{(d)'}$ is the set of all $d \times d$ matrices with entries from $\mathcal{A}$. Equivalently,

$$\mathcal{A}^{(d)'} = \{M_\Phi : \Phi \in L^\infty(\mu :, M_d)\},$$

where $L^\infty(\mu; M_d)$ is the algebra of μ-essentially bounded Borel functions with values in M_d, the algebra of $d \times d$ matrices, and M_Φ is defined on $L^2(\mu)^{(d)} = L^2(\mu; \mathbb{C}^d)$ by multiplication. This example and the case where $d = \infty$ are crucial for the theory. These will receive considerable attention later in §52 below.

43.5 Proposition. *If for each i, $\mathcal{A}_i$ is a von Neumann algebra in $\mathcal{B}(\mathcal{H}_i)$ and $\mathcal{H} = \bigoplus_i \mathcal{H}_i$, then*

$$\mathcal{A} = \bigoplus_i \mathcal{A}_i \equiv \left\{ \bigoplus_i A_i : A_i \in \mathcal{A}_i \text{ and } \sup_i \|A_i\| < \infty \right\}$$

is a von Neumann algebra in $\mathcal{B}(\mathcal{H})$ and $\mathcal{A}' = \bigoplus_i \mathcal{A}_i'$.

Proof. This is rather straightforward since $\mathcal{A}$ contains the projections of $\mathcal{H}$ onto the coordinate spaces $\mathcal{H}_i$. The details are left to the reader. ■

43.6 Definition. If $\mathcal{Z}$ is the center of $\mathcal{A}$ and $A \in \mathcal{A}$, the *central support* or *central cover* of A is the projection

$$C_A \equiv \inf\{\, C : C \text{ is a projection in } \mathcal{Z} \text{ and } AC = A\,\}.$$

Since $\mathcal{Z}$ is itself a von Neumann algebra, Proposition 43.3 implies that C_A is a projection in $\mathcal{Z}$. It is also clear that $C_A A = AC_A = A$. Introduce the notation

$$\operatorname{cran} A \equiv \operatorname{cl}[\operatorname{ran} A]$$

for the closure of the range of the operator A. Remember that $\operatorname{cran} A \in \mathcal{A}$ whenever $A \in \mathcal{A}$, and, in general, $\operatorname{cran} A = (\ker A^*)^\perp$.

43.7 Proposition. (a) *If* $A \in \mathcal{A}$, $E = (\ker A)^\perp$ *and* $F = \operatorname{cran} A$, *then* $C_A = C_{A^*} = C_{A^*A} = C_{AA^*} = C_E = C_F$.

(b) *If E is a projection in $\mathcal{A}$, then*

$$C_E = [Ah : h \in E, A \in \mathcal{A}]\,.$$

Proof. (a) The fact that $C_A = C_{A^*}$ is left to the reader to verify. If C is a central projection such that $AC = A$, then clearly $A^*AC = A^*A$. Conversely, if $A^*AC = A^*A$, then, by taking square roots, $|A|C = |A|$. Using the polar decomposition of A, it follows that $AC = A$. This establishes that $C_A = C_{A^*A}$. By symmetry, $C_{AA^*} = C_{A^*} = C_A$.

Now $E \in \mathcal{A}$, so if C is a central projection such that $AC = A$, then $C(\ker A) \subseteq \ker A$ so that $EC = CE = E$. That is, $E \le C$ and, hence, $E \le C_A$. Therefore $C_E \le C_A$. On the other hand, if C is a central project such that $E \le C$, then $AC = AEC = AE = A$. Thus $C_A \le C_E$. By symmetry, $C_{A^*} = C_F$, so this completes the proof of (a).

(b) Put $P = [Ah : h \in E, A \in \mathcal{A}]$. Since P is invariant for $\mathcal{A}$, $P \in \mathcal{A}'$. Also, if $A' \in \mathcal{A}'$, then for all h in E and A in $\mathcal{A}$, $A'h \in E$ so that $A'(Ah) = A(A'h) \in P$. Thus $A'P \subseteq P$ for all A' in $\mathcal{A}'$ and, hence, $P \in \mathcal{A}'' = \mathcal{A}$. Therefore $P \in \mathcal{A} \cap \mathcal{A}' = \mathcal{Z}$. Clearly $E \le P$, so that $C_E \le P$. On the other hand, if C is any central projection such that $EC = E$, then for A in $\mathcal{A}$ and h in E, $C(Ah) = ACEh = AEh = Ah$. Thus $CP = P = PC$, so that $P \le C$. Therefore $P \le C_E$. ■

If $\mathcal{A}$ is a von Neumann algebra and E is a any projection onto a subspace of $\mathcal{H}$, consider the collection of all the compressions of operators in $\mathcal{A}$ to the subspace E, $\mathcal{A}_E \equiv \{ A_E : A \in \mathcal{A} \}$, where A_E is defined on E by $A_E h = EAh$ for all h in E. In general $\mathcal{A}_E$ may not be an algebra. But if $E \in \mathcal{A}$, it is easy to see that there is a natural identification of $\mathcal{A}_E$ with the subalgebra $E\mathcal{A}E$ of $\mathcal{A}$. Indeed, it is easy to verify that the map $T \to T_E$ is a well defined $*$-isomorphism from $E\mathcal{A}E$ onto $\mathcal{A}_E$. Hence $\mathcal{A}_E$ is a *-algebra. If $E \in \mathcal{A}'$ it is even easier to see that $\mathcal{A}_E$ is a C*-algebra since the map $A \to A|E$ is a $*$-homomorphism.

43.8 Proposition. (a) *If E is a projection in $\mathcal{A}'$, then $\mathcal{A}_E$ is a von Neumann algebra in $\mathcal{B}(E)$ and $(\mathcal{A}_E)' = (\mathcal{A})'_E$.*

(b) *If E is a projection in $\mathcal{A}$, then $\mathcal{A}_E$ is a von Neumann algebra in $\mathcal{B}(E)$ and $(\mathcal{A}_E)' = (\mathcal{A})'_E$.*

(c) *If $\mathcal{Z}$ is the center of $\mathcal{A}$ and E is a projection in either $\mathcal{A}$ or $\mathcal{A}'$, then the center of $\mathcal{A}_E$ is $\mathcal{Z}_E$.*

Proof. The proof of (a), being easier than that of (b), is left to the reader. To prove (b), assume that $E \in \mathcal{A}$. From the remarks preceding the proposition it follows that $\mathcal{A}_E$ and $(\mathcal{A}')_E$ are *-algebras and it is clear that they contain the identity on the Hilbert space E. If $A' \in \mathcal{A}'$ and $A \in \mathcal{A}$, then for each h in E, $A'_E A_E h = EA'EAh = EAEA'h = A_E A'_E h$. Thus $(\mathcal{A}')_E \subseteq (\mathcal{A}_E)'$. Now to show that $(\mathcal{A}_E)' \subseteq (\mathcal{A})'_E$.

Suppose U is a unitary in the von Neumann algebra $(\mathcal{A}_E)' \subseteq \mathcal{B}(E)$. If $A_1, \ldots, A_n \in \mathcal{A}$ and $h_1, \ldots, h_n \in E$, then

$$\begin{aligned}
\left\| \sum_{j=1}^{n} A_j U h_j \right\|^2 &= \sum_{i,j=1}^{n} \langle A_i U h_i, A_j U h_j \rangle \\
&= \sum_{i,j=1}^{n} \langle (E A_j^* A_i E) U h_i, U h_j \rangle \\
&= \sum_{i,j=1}^{n} \langle U (E A_j^* A_i E) h_i, U h_j \rangle \\
&= \sum_{i,j=1}^{n} \langle E A_j^* A_i E h_i, h_j \rangle \\
&= \left\| \sum_{j=1}^{n} A_j h_j \right\|^2 .
\end{aligned}$$

Hence if W is defined on $\mathcal{H}$ by setting $W(\sum_j A_j h_j) = \sum_j A_j U h_j$ on $C = [Ah : A \in \mathcal{A}, h \in E]$ and $W = 0$ on $C^\perp$, then W defines a partial isometry on $\mathcal{H}$. But Proposition 43.7 implies $C = C_E$. Therefore for any A in $\mathcal{A}$, $WAC^\perp = WC^\perp A = 0 = AWC^\perp$; while on C

$$WA\sum_j A_j h_j = W\sum_j AA_j h_j = A\sum_j A_j U h_j = AW\sum_j A_j h_j.$$

Thus $W \in \mathcal{A}'$. If $h \in E$, then $Wh = Uh$ by definition, so that $W_E = U$. That is, $U \in (\mathcal{A}')_E$ whenever U is a unitary in $(\mathcal{A}_E)'$. Since $(\mathcal{A}_E)'$ is a C*-algebra and every element can be written as a linear combination of four unitaries, $(\mathcal{A}_E)' \subseteq (\mathcal{A}')_E$.

Now let $T \in (\mathcal{A}_E)''$ and extend T to be defined on all of $\mathcal{H}$ by setting $T = 0$ on $E^\perp$. If $A' \in \mathcal{A}'$, then the first part of the proof says that A'_E is the generic element of $(\mathcal{A}_E)'$. For $h \in E$, $TA'h = (TE)A'h = TA'_E h = A'_E Th = A'Th$. On the other hand, $TA'E^\perp = TE^\perp A' = 0 = A'TE^\perp$. Thus $T \in \mathcal{A}'' = \mathcal{A}$. So $\mathcal{A}_E = \mathcal{A}''_E$ and $\mathcal{A}_E$ is a von Neumann algebra.

(c) Using the first two parts of this proposition, it follows that the center of $\mathcal{A}_E$ is $\mathcal{A}_E \cap (\mathcal{A}_E)' = \mathcal{A}_E \cap \mathcal{A}'_E$, which the reader will quickly verify to be $\mathcal{Z}_E$. ■

In light of the preceding proposition, there is no ambiguity when $\mathcal{A}'_E = (\mathcal{A}')_E = (\mathcal{A}_E)'$ is written.

The remainder of this section is devoted to defining an important collection of examples of von Neumann algebras. Let G be a group and set $\mathcal{H} = \ell^2(G)$. That is, $\mathcal{H}$ consists of all the functions $f\colon G \to \mathbb{C}$ such that $f(a) = 0$ for all but a countable number of group elements a and $\sum_{a\in G} |f(a)|^2 < \infty$. Of course, $\ell^2(G)$ is separable if and only if G is countable.

If $a \in G$, define $R_a\colon \ell^2(G) \to \ell^2(G)$ and $L_a\colon \ell^2(G) \to \ell^2(G)$ by

$$R_a f(x) = f(xa),$$
$$L_a f(x) = f(a^{-1}x).$$

It is easy to check that both R_a and L_a are isometries on $\ell^2(G)$ and that $R_a R_b = R_{ab}$ and $L_a L_b = L_{ab}$. (The reason for using a^{-1} in the definition of L_a was to get this last equation to hold.) Thus R_a and L_a are unitaries and $R_a^* = R_{a^{-1}}$, $L_a^* = L_{a^{-1}}$. In summary, the maps $a \to R_a$ and $a \to L_a$ are unitary representations of the group G.

Define the von Neumann algebras

$$\mathcal{R}(G) = W^*(\{R_a : a \in G\}),$$
$$\mathcal{L}(G) = W^*(\{L_a : a \in G\}).$$

The remainder of this section is devoted to an exploration of the properties of these algebras.

If $f, g \in \ell^2(G)$, the convolution of f and g is defined as:

43.9 $$f * g(a) \equiv \sum_{x\in G} f(x)g(x^{-1}a) = \sum_{x \in G} f(ax^{-1})g(x).$$

It is easy to see that $f * g \in \ell^\infty(G)$ and by the CBS Inequality, $|(f*g)(a)| \le \|f\|_2 \, \|g\|_2$. So even though it is generally not the case that $f * g$ belongs to $\ell^2(G)$, it is the case that $f * g \in \ell^\infty(G)$ and $\|f * g\|_\infty \le \|f\|_2 \, \|g\|_2$. (Consult the exercises for additional properties of the convolution.) A relation between convolution and the algebras $\mathcal{R}(G)$ and $\mathcal{L}(G)$ is on the agenda.

Recall that a vector h is separating for an algebra $\mathcal{A}$ if $A = 0$ whenever $A \in \mathcal{A}$ and $Ah = 0$. If h is a cyclic vector for $\mathcal{A}$, then h is a separating vector for $\mathcal{A}'$. If $\mathcal{A}$ is a $*$-algebra and h is separating for $\mathcal{A}'$, then h is cyclic for $\mathcal{A}$ (14.3). Using the notation of Exercise 5, if e is the identity of G, then it is easy to see that ϵ_e is a cyclic vector for both $\mathcal{R}(G)$ and $\mathcal{L}(G)$.

43.10 Proposition. *If $T \in \mathcal{R}(G)'$, there is a unique function f in $\ell^2(G)$ such that $Th = f * h$ for all h in $\ell^2(G)$. Similarly, if $S \in \mathcal{L}(G)'$, there is a unique function g in $\ell^2(G)$ such that $Sh = h * g$ for all h in $\ell^2(G)$.*

Proof. Only the first statement will be proved, the proof of the second being analogous. Adopt the notation of Exercise 6. If e is the identity of G, put $f = T(\epsilon_e)$. Note that $T \to T(\epsilon_e)$ is a linear map from $\mathcal{R}(G)'$ into $\ell^2(G)$. If $a \in G$, $T(\epsilon_a) = TR_{a^{-1}}\epsilon_e = R_{a^{-1}}f = f * \epsilon_a$. If $\mathcal{D}$ is the linear span of $\{\epsilon_a : a \in G\}$, this shows that $Tk = f * k$ for every k in $\mathcal{D}$.

But now if $h \in \ell^2(G)$, there is a sequence $\{k_n\}$ in $\mathcal{D}$ such that $\|k_n - h\|_2 \to 0$. Thus $\|T(h) - f * k_n\|_2 = \|T(h) - T(k_n)\|_2 \to 0$. In particular, $T(h)(x) = \lim_n (f * k_n)(x)$ for each x in G . But, on the other hand, $f*h, f*k_n \in \ell^\infty(G)$ and $\|f * h - f * k_n\|_\infty = \|f * (h - k_n)\|_\infty \le \|f\|_2 \, \|h - k_n\|_2 \to 0$. Thus $T(h)(x) = (f * h)(x)$ for all x in G and h in $\ell^2(G)$.

To prove uniqueness, observe that because ϵ_e is a cyclic vector for $\mathcal{R}(G)$, it is separating for $\mathcal{R}(G)'$. Thus if $f = 0$, then $T = 0$ and uniqueness follows. ■

In relation to the preceding proposition, see Exercise 9.

43.11 Theorem. *For any group G, $\mathcal{R}(G)' = \mathcal{L}(G)$ and $\mathcal{L}(G)' = \mathcal{R}(G)$.*

Proof. By the Double Commutant Theorem, it suffices to prove the first equality. By Exercise 8, $\mathcal{L}(G) \subseteq \mathcal{R}(G)'$. To show the reverse inclusion it suffices to show that if $T \in \mathcal{R}(G)'$ and $S \in \mathcal{L}(G)'$, then $TS = ST$. But the preceding proposition implies there are f, g in $\ell^2(G)$ such that $T(h) = f * h$ and $S(h) = h * g$ for all vectors h. Using the fact that these convolutions always belong to $\ell^2(G)$, this implies $TSh = T(h*g) = f*(h*g) = (f*h)*g = STh$. (Justify these equalities.) ■

The only operators that are guaranteed to belong to the center of an arbitrary von Neumann algebra are the scalar multiples of the identity. The algebras for which this is the entirety of the center are singled out.

43.12 Definition. A *factor* is a von Neumann algebra whose center consists of the scalar multiples of the identity.

One easy example comes quickly to mind: $\mathcal{B}(\mathcal{H})$ is a factor. Also the commutant of a factor is easily seen to also be a factor. Here are a few more examples.

43.13 Theorem. *If G is a group, the following statements are equivalent.*

(a) *$\mathcal{R}(G)$ is a factor.*

(b) *$\mathcal{L}(G)$ is a factor.*

(c) *If $a \in G$ and a is not the identity, the conjugacy class of a is infinite.*

Proof. Since $\mathcal{L}(G) = \mathcal{R}(G)'$, (a) and (b) are equivalent.

Let T belong to the center of $\mathcal{R}(G)$; thus $T \in \mathcal{R}(G) \cap \mathcal{L}(G)$. By Proposition 43.10, there is an f in $\ell^2(G)$ such that $Th = h * f$ for all h in $\ell^2(G)$. But for any a and b in G, $f(a) = \epsilon_e * f(a) = T(\epsilon_e)(a)$. Since $T \in \mathcal{R}(G)'$, $f(a) = R_b T R_b^{-1}(\epsilon_e)(a) = R_b T(\epsilon_b)(a) = T(\epsilon_b)(ab) = (\epsilon_b * f)(ab) = f(b^{-1}ab)$. Thus f must be constant on the conjugacy class of a. If (c) holds, then the fact that $f \in \ell^2(G)$ implies that $f = \alpha\epsilon_e$ for some scalar α. Thus $T = \alpha$ and so $\mathcal{R}(G)$ is a factor.

Now assume that (c) is false and fix an element a in G such that its conjugacy class, C, is finite. Let $T = \sum_{c \in C} R_c^{-1}$, so $T \in \mathcal{R}(G)$ and T is not a multiple of the identity. It is left as an exercise for the reader to show that $Th = h * f$, where f is the characteristic function of C. For any x in G, $R_x T R_x^{-1} = \sum_{c \in C} R_x R_c^{-1} R_x^{-1} = \sum_{c \in C} R_{xcx^{-1}}^{-1} = \sum_{c \in C} R_c^{-1} = T$, since C is a conjugacy class. Thus T belongs to the center of $\mathcal{R}(G)$ and so $\mathcal{R}(G)$ is not a factor. ■

Exercises.

1. If $\mathcal{S} \subseteq \mathcal{B}(\mathcal{H})$, show that $(\mathcal{S} \cap \mathcal{S}')' = (\mathcal{S}'' \cup \mathcal{S}')''$.

2. If H is a subset of $\mathcal{H}$, let $F = [Ah : h \in H, A \in \mathcal{A}]$. (a) Show that F is a projection in $\mathcal{A}'$. (b) Show that

$$[A'g : A' \in \mathcal{A}', g \in F] = [Z'h : Z' \in \mathcal{Z}', h \in H]$$

and this is a central projection. (c) Show that the projection $E = [A'h : A' \in \mathcal{A}, h \in H]$ in $\mathcal{A}$ and the projection F in $\mathcal{A}'$ satisfy $C_E = C_F = [Z'h : Z' \in \mathcal{Z}', h \in H]$.

3. If E is a projection in $\mathcal{A}'$, show that $\mathcal{A}_E$ is a von Neumann algebra in $\mathcal{B}(E)$ and compute its commutant. Show that the $*$-homomorphism $A \to A_E$ has kernel $\mathcal{A}(1-C_E)$. Consequently this $*$-homomorphism is an isomorphism if and only if $C_E = 1$.

4. Prove that $AC_A = A = C_A A$. If G is a central projection, show that $C_{AG} = GC_A$.

In the remaining exercises, G will always denote a group.

5. Show that if $f, g \in \ell^2(G)$, then

$$f * g(a) = \sum_{x \in G} f(ax)g(x^{-1}) = \sum_{x \in G} f(ax^{-1})g(x).$$

6. For any element a of G, let ϵ_a denote the characteristic function of the singleton $\{a\}$. Show that if $a, b \in G$, the $\epsilon_a * \epsilon_b = \epsilon_{ab}$.

7. Show that for each f in $\ell^2(G)$ and each a, $R_a f = f * \epsilon_{a^{-1}}$ and $L_a f = \epsilon_a * f$.

8. For all a, b in G show that $R_a L_b = L_b R_a$.

9. Show that if $f \in \ell^2(G)$ such that $h * f \in \ell^2(G)$ for all h in $\ell^2(G)$, then $R_f(h) \equiv h * f$ is a bounded operator on $\ell^2(G)$ and $\mathcal{R}(G) = \{R_f : f \in \ell^2(G)$ and $h * f \in \ell^2(G)$ whenever $h \in \ell^2(G)\}$. State and prove the analogous result for $\mathcal{L}(G)$.

10. Using the notation of the preceding exercise, show that if $R_f \in \mathcal{R}(G)$, then $R_f^* = R_{f^*}$, where $f^*(x) \equiv \overline{f(x^{-1})}$ for all x in G.

11. Show that T belongs to the center of $\mathcal{R}(G)$ if and only if there is an f in $\ell^2(G)$ that is constant on all conjugacy classes of G and $Th = h * f$ for all h in $\ell^2(G)$.

12. Prove that the following statements are equivalent. (a) G is abelian. (b) $\mathcal{R}(G)$ is abelian. (c) $\mathcal{L}(G)$ is abelian. (d) $\mathcal{R}(G) = \mathcal{L}(G)$.

§44. The Kaplansky Density Theorem

If $\mathcal{X}$ is a Banach space and $\mathcal{Y}$ is a subspace of its dual space, $\mathcal{X}^*$, then it may not be the case that ball $\mathcal{Y}$ is weak* dense in the weak* closure of $\mathcal{Y}$. For a discussion of this phenomenon, see §V.12 in [ACFA]. Now if $\mathcal{A}$ is a von Neumann algebra, $\mathcal{A}$ is the dual of a Banach space. Indeed, $\mathcal{A}$ is naturally isomorphic to $\mathcal{B}_1(\mathcal{H})/\mathcal{A}_\perp$. This leads to the following query. Suppose $\mathcal{C}$ is a unital C*-algebra contained in $\mathcal{B}(\mathcal{H})$ and $\mathcal{A}$ is the SOT closure of $\mathcal{C}$. So $\mathcal{A}$ is a von Neumann algebra, and $\mathcal{C}$ is weak* dense in $\mathcal{A}$. Is ball $\mathcal{C}$ weak* dense in ball $\mathcal{A}$? There is no reason to conclude from general Banach space theory that this is the case. (Recall that the SOT closure of a convex subset of $\mathcal{B}(\mathcal{H})$ is the same as its WOT closure and that on bounded sets of operators the WOT and weak* topologies agree.) However this is true and is the content of the main result of this section.

44.1 Kaplansky Density Theorem. *If $\mathcal{C}$ is a unital C^*-subalgebra of $\mathcal{B}(\mathcal{H})$ and $\mathcal{A}$ is its* SOT *closure, then:*

(a) ball $\mathcal{A}$ *is the* SOT *closure of* ball $\mathcal{C}$*;*

(b) ball Re $\mathcal{A}$ *is the* SOT *closure of* ball Re $\mathcal{C}$*;*

(c) $(\text{ball}\,\mathcal{A})_+$ *is the* SOT *closure of* $(\text{ball}\,\mathcal{C})_+$*;*

(d) *The* SOT *closure of the unitaries in $\mathcal{C}$ contains the unitaries of $\mathcal{A}$.*

This requires a preliminary result. Before stating this lemma it might be wise to underscore the difference between condition (d) and the first three parts of the theorem. Of course it follows from (d) that the SOT closure of the set of unitaries in $\mathcal{A}$ is the SOT closure of the unitaries in $\mathcal{C}$. But the set of unitary operators in a von Neumann algebra is not necessarily SOT closed. Indeed, if $\{U_n\}$ is a sequence of unitaries in $\mathcal{A}$ and $U_n \to U$ (SOT), then $U_n^* \to U^*$ (WOT). It follows that $U_n^*U_n \to U^*U$ (WOT), so that $U^*U = 1$. That is, U is an isometry. However it does not follow that $\{U_nU_n^*\}$ converges (WOT) to UU^*. Thus the SOT closure of the unitary operators in $\mathcal{A}$ is contained in the set of isometries. (In fact, the SOT closure of the set of unitaries equals the set of isometries. See Exercise 1.) As a curiosity, it might be pointed out that "most of the time" the WOT closure of the unitaries in a von Neumann algebra is the entire unit ball. See Dye [1953] and Conway and Szüch [1973].

The next lemma has some interest independent of its use in the proof of the theorem.

44.2 Lemma. *If $f\colon \mathbb{R} \to \mathbb{R}$ is a continuous function with $f(0) = 0$ and there are positive constants a, b such that $|f(t)| \le a|t| + b$ for all t in $\mathbb{R}$, then $f(T_i) \to f(T)$* (SOT) *whenever $\{T_i\}$ is a net of operators such that $T_i \to T$* (SOT).

Proof. The proof is somewhat "soft" in that it introduces a space of functions to which it will be shown that such a function f belongs. First note that via the functional calculus for hermitian operators, every continuous function $g\colon \mathbb{R} \to \mathbb{R}$ defines a function $g\colon \operatorname{Re}\mathcal{B}(\mathcal{H}) \to \operatorname{Re}\mathcal{B}(\mathcal{H})$. Let S denote the vector space over $\mathbb{R}$ of real-valued continuous functions g on $\mathbb{R}$ such that

$g\colon (\operatorname{Re}\mathcal{B}(\mathcal{H}), \mathrm{SOT}) \to (\operatorname{Re}\mathcal{B}(\mathcal{H}), \mathrm{SOT})$ is continuous. It will be shown that any function f satisfying the hypothesis belongs to S.

Let S^b denote the bounded functions in S. Observe that S^b is uniformly closed and if $g \in S^b$ and $h \in S$, then $gh \in S$. (Verify!) In particular, S^b is a real Banach subalgebra of the real-valued bounded continuous functions on $\mathbb{R}$. Now to give some examples of functions in S^b other than the constants.

Let $e(t) = t/(1+t^2)$. If A and B are hermitian operators,

$$\begin{aligned} e(A) - e(B) &= (1+A^2)^{-1}\left[A(1+B^2) - (1+A^2)B\right](1+B^2)^{-1} \\ &= (1+A^2)^{-1}A(1+B^2)^{-1} + (1+A^2)^{-1}AB^2(1+B^2)^{-1} \\ &\quad - (1+A^2)^{-1}B(1+B^2)^{-1} - (1+A^2)^{-1}A^2B(1+B^2)^{-1} \\ &= (1+A^2)^{-1}(A-B)(1+B^2)^{-1} \\ &\quad + (1+A^2)^{-1}A(B-A)B(1+B^2)^{-1}. \end{aligned}$$

But $(1+A^2)^{-1}$ and $(1+A^2)^{-1}A$ are both contractions. Hence for every vector h in $\mathcal{H}$,

$$\|[e(A)-e(B)]h\| \le \|(A-B)(1+B^2)^{-1}h\| + \|(B-A)B(1+B^2)^{-1}h\|.$$

So $e \in S^b$.

Put $e_\alpha(t) = \alpha t/(1+\alpha^2 t^2)$. Since S is closed under the composition of functions, $e_\alpha \in S^b$ for each $\alpha > 0$. Since $\{e_\alpha : \alpha > 0\}$ separates the points of $\mathbb{R}\setminus\{0\}$, the uniformly closed algebra generated by $\{e_\alpha\}_{\alpha>0}$ is $C_0(\mathbb{R}\setminus\{0\})$. Since S^b is a uniformly closed algebra, $C_0(\mathbb{R}\setminus\{0\}) \subseteq S^b$.

Now let f be as in the statement of the lemma. Thus $t \to f(t)/(1+t^2)$ belongs to $C_0(\mathbb{R}\setminus\{0\})$, and hence to S^b. Since the identity function belongs to S, $t \to tf(t)/(1+t^2)$ belongs to S and therefore to S^b. Therefore $t \to t^2 f(t)/(1+t^2)$ belongs to S. But $f(t) = [t^2 f(t)/(1+t^2)] + [f(t)/(1+t^2)]$, so $f \in S$. ■

Proof of The Kaplansky Density Theorem. Fix a hermitian operator A in the unit ball of $\mathcal{A}$ and fix a net $\{C_i\}$ in $\operatorname{Re}\mathcal{C}$ such that $C_i \to A$ (SOT).

If $f(t) = (t \wedge 1) \vee (-1)$, then f satisfies the hypothesis of the preceding lemma and so $f(C_i) \to f(A) = A$ (SOT). But $f(C_i) \in \operatorname{ball}\operatorname{Re}\mathcal{C}$ for each i, so that (b) is proved.

If $A \in (\operatorname{ball}\mathcal{A})_+$ and $\{C_i\}$ is as before, let $f(t) = (t \wedge 1) \vee 0$. Again the lemma implies that $f(C_i) \to f(A) = A$ (SOT), proving (c).

To prove (d), let U be a unitary operator in $\mathcal{A}$. Using the functional calculus, there is an operator $A \in \operatorname{Re}\mathcal{A}$ such that $U = \exp(iA) = \cos A + i\sin A$. Let $\{C_i\}$ be a net in $\operatorname{Re}\mathcal{C}$ such that $C_i \to A$ (SOT). Now both $\sin t$

and $\cos t - 1$ satisfy the hypothesis of the preceding lemma, so $U_i \equiv e^{iC_i} = \cos C_i + i \sin C_i \to e^{iA} = U$ (SOT). Since each U_i is a unitary in $\mathcal{C}$, this proves (d).

Now to prove (a). Fix an operator A in ball $\mathcal{A}$, and consider the algebras $M_2(\mathcal{C})$ and $M_2(\mathcal{A})$ of all 2×2 matrices with entries from the appropriate algebra acting on $\mathcal{H} \oplus \mathcal{H}$. It is left to the reader to check that these are both C*-algebras and that $M_2(\mathcal{A})$ is the SOT closure of $M_2(\mathcal{C})$. But

$$\begin{bmatrix} 0 & A \\ A^* & 0 \end{bmatrix} \in \operatorname{ball} \operatorname{Re} M_2(\mathcal{A}).$$

Therefore part (b) implies there is a net $\{T_i\}$ in ball Re $M_2(\mathcal{C})$ that converges (SOT) to this matrix. Put

$$T_i = \begin{bmatrix} X_i & C_i \\ C_i^* & Y_i \end{bmatrix},$$

with X_i, Y_i, C_i in $\mathcal{C}$. Since $\|T_i\| \le 1$, $\|C_i\| \le 1$. Clearly $C_i \to A$ (SOT). ∎

Exercises.

1. Let $\mathcal{A}$ be a von Neumann algebra. (a) Show that the set $\mathcal{I}$ of isometries in $\mathcal{A}$ is SOT closed. (b) If W is an isometry in $\mathcal{A}$ and $W = S \oplus U$ is its von Neumann–Wold decomposition with S a shift on $\mathcal{H}_0$ and U a unitary on $\mathcal{H}_1$, show that the projections onto $\mathcal{H}_0$ and $\mathcal{H}_1$ belong to $\mathcal{A}$. Call these projections E_0 and E_1, respectively. (c) Show that SE_0 and UE_1 belong to $\mathcal{A}$. (d) Show that there is a sequence of unitaries $\{V_n\}$ on $\mathcal{H}_0$ such that $V_n E_0 \in \mathcal{A}$ for all $n \ge 1$ and $V_n \to S$ (SOT). (e) Show that $U_n = V_n \oplus U$ is a unitary in $\mathcal{A}$ and $U_n \to W$ (SOT). Thus the SOT closure of the unitary operators in $\mathcal{A}$ is its set of isometries.

2. If $\mathcal{A}$ is a von Neumann algebra and $n \in \mathbb{N}$, show that the algebra $M_n(\mathcal{A})$ of all $n \times n$ matrices with entries from $\mathcal{A}$ acting on $\mathcal{H}^{(n)}$ is a von Neumann algebra.

§45. The Pedersen Up–Down Theorem

This section will prove a result from Pedersen [1972] concerning the natural ordering on the real linear space of hermitian operators in a von Neumann algebra. This result will have a familiar ring to the reader when it is interpreted in the commutative case. (See Exercises 1 and 2.)

45.1 Definition. For any subset $\mathcal{S}$ of $\operatorname{Re}\mathcal{B}(\mathcal{H})$ define the following sets of hermitian operators.

$$\begin{aligned}\mathcal{S}^\sigma \equiv \{T :\ & \text{there is an increasing sequence } \{S_n\} \text{ in } \mathcal{S} \\ & \text{such that } S_n \to T(\text{SOT})\,\},\\ \mathcal{S}_\sigma \equiv \{T :\ & \text{there is a decreasing sequence } \{S_n\} \text{ in } \mathcal{S} \\ & \text{such that } S_n \to T(\text{SOT})\,\}.\end{aligned}$$

Note that $\mathcal{S} \subseteq \mathcal{S}^\sigma$ and $\mathcal{S}_\sigma = -(-\mathcal{S})^\sigma$.

45.2 The Pedersen Up–Down Theorem. *If $\mathcal{H}$ is separable, $\mathcal{C}$ is a unital C^*-algebra contained in $\mathcal{B}(\mathcal{H})$, and $\mathcal{A}$ is the strong closure of $\mathcal{C}$, then* $\operatorname{ball}\mathcal{A}_+ = ((\operatorname{ball}\mathcal{C}_+)^\sigma)_\sigma$ *and* $\operatorname{Re}\mathcal{A} = ((\operatorname{Re}\mathcal{C})^\sigma)_\sigma$.

The reader should check that the "Down-Up" Theorem is also true. Indeed this follows from (45.2) by using the observation that if $\{C_n\}$ is a increasing sequence in $\operatorname{ball}\mathcal{C}_+$, then $\{1 - C_n\}$ is a decreasing sequence in $\operatorname{ball}\mathcal{C}_+$.

45.3 Lemma. *If $\mathcal{C}$ is a C^*-algebra, $\mathcal{A}$ is the strong closure of $\mathcal{C}$, and P is a projection in $\mathcal{A}$, then for each sequence of unit vectors $\{h_n\}$ in $\mathcal{H}$ there is an operator A in $((\operatorname{ball}\mathcal{C}_+)^\sigma)_\sigma$ such that*

$$A(1-P)h_n = 0 = (1-A)Ph_n$$

for all $n \geq 1$.

Proof. This proof will make heavy use of Exercise 3. According to the Kaplansky Density Theorem there is a sequence $\{C_n\}$ in $\operatorname{ball}\mathcal{C}_+$ that converges strongly to P. (A sequence can be found because the underlying Hilbert space is separable and the SOT is metrizable on bounded sets.) Passing to a subsequence if necessary, it can be assumed that $\|C_n(1-P)h_j\| < (n2^n)^{-1}$ and $\|(1-C_n)Ph_j\| < n^{-1}$ for $1 \leq j \leq n$. For $n < m$ define

$$A_{nm} \equiv \left(1 + \sum_{k=n}^{m} kC_k\right)^{-1} \sum_{k=n}^{m} kC_k.$$

Note that each $A_{nm} \in \operatorname{ball}\mathcal{C}_+$ and $A_{nm} \leq \sum_{k=n}^{m} kC_k$. Hence for $j \leq n$

45.4 $$\langle A_{nm}(1-P)h_j, (1-P)h_j\rangle \leq \sum_{k=n}^{m} \frac{1}{2^k} < \frac{1}{2^{n-1}}.$$

Also $\sum_{k=n}^{m} kC_k \geq mC_m$, so $A_{nm} \geq (1 + mC_m)^{-1}(mC_m)$ by Exercise 3. Therefore $1 - A_{nm} \leq 1 - (1+mC_m)^{-1}(mC_m) = (1+mC_m)^{-1}$. But whenever

$0 \le t \le 1$, $(1+mt)^{-1} \le (1+m)^{-1}(1+m(1-t))$. Thus the fact that each C_m belongs to ball $\mathcal{C}_+$ implies that $1 - A_{nm} \le (1+m)^{-1}[1+m(1-C_m)]$. This gives

$$\begin{aligned} \langle (1-A_{nm})Ph_j, Ph_j\rangle &\le \frac{1}{m+1}\langle [1+m(1-C_m)]Ph_j, Ph_j\rangle \\ &\le \frac{1}{m+1}[\|Ph_j\| + m\|(1-C_m)Ph_j\|] \\ &\le \frac{2}{m+1}. \end{aligned} \tag{45.5}$$

Now if n is fixed, Exercise 3 implies that $\{A_{nm} : m > n\}$ is an increasing sequence. Since this sequence is also bounded, Proposition 43.1 implies there is an operator A_n such that $A_{nm} \to A_n$ (SOT) as $m \to \infty$. Therefore $A_n \in (\text{ball}\, \mathcal{C}_+)^\sigma$.

On the other hand, when $n+1 < m$, $A_{n+1,m} \le A_{nm}$, so $A_{n+1} \le A_n$. Thus there is an operator A such that $A_n \to A$ (SOT). By definition, $A \in ((\text{ball}\, \mathcal{C}_+)^\sigma)_\sigma$. Now by (45.4) and (45.5),

$$\begin{aligned} \langle A_n(1-P)h_j, (1-P)h_j\rangle &\le \frac{1}{2^{n-1}} \quad \text{for } j \le n, \\ \langle (1-A_n)Ph_j, Ph_j\rangle &\le 0 \quad \text{for all } j \ge 1. \end{aligned}$$

Letting $n \to \infty$ shows that

$$\begin{aligned} \langle A(1-P)h_j, (1-P)h_j\rangle &\le 0, \\ \langle (1-A)Ph_j, Ph_j\rangle &\le 0 \end{aligned}$$

for all $j \ge 1$. But $0 \le A \le 1$, so this implies $A(1-P)h_j = (1-A)Ph_i = 0$ for all j. ■

Proof of Theorem 45.2. For any projection P in $\mathcal{A}$ apply the preceding lemma with a dense sequence in $\{h \in \mathcal{H} : \|h\| = 1\}$ to get an operator A in $((\text{ball}\, \mathcal{C}_+)^\sigma)_\sigma$ such that $A(1-P) = 0 = (1-A)P$. Thus $P = A \in ((\text{ball}\, \mathcal{C}_+)^\sigma)_\sigma$. That is, each projection in $\mathcal{A}$ belongs to $((\text{ball}\, \mathcal{C}_+)^\sigma)_\sigma$.

Let $A \in \text{ball}\, \mathcal{A}_+$. If $A = \int_0^1 t\, dE(t)$ is the spectral decomposition and for each $k \ge 1$

$$P_k = E\left(\bigcup_{j=1}^{2^k-1}\left(\frac{j}{2^k}, \frac{j+1}{2^k}\right]\right),$$

then Exercise 10.5 implies that $A = \sum_{k=1}^\infty 2^{-k}P_k$, where the convergence is in the norm.

For each k, the preceding lemma implies there is a decreasing sequence $\{Z_{kn}\}$ in $(\text{ball}\,\mathcal{C}_+)^\sigma$ such that $Z_{kn} \searrow P_k$ (SOT) as $n \to \infty$. Define

$$T_n = \sum_{k=1}^{n} \frac{1}{2^k} Z_{kn} + \frac{1}{2^n} \geq A.$$

Since $(\text{ball}\,\mathcal{C}_+)^\sigma$ is convex (Why?), each T_n belongs to $(\text{ball}\,\mathcal{C}_+)^\sigma$. Because $Z_{n+1,n+1} \leq 1$, $2^{-n} - (2^{-(n+1)} Z_{n+1,n+1} + 2^{-(n+1)}) \geq 0$ and so

$$\begin{aligned} T_n - T_{n+1} &= \sum_{k=1}^{n} \frac{1}{2^k}[Z_{kn} - Z_{k,n+1}] + \frac{1}{2^n} - \left(\frac{1}{2^{n+1}} Z_{n+1,n+1} + \frac{1}{2^{n+1}}\right) \\ &\geq 0. \end{aligned}$$

Thus $\{T_n\}$ is a decreasing sequence. It must be that $\{T_n\}$ converges strongly to some operator T in $((\text{ball}\,\mathcal{C}_+)^\sigma)_\sigma$ and $T \geq A$.

Since $0 \leq Z_{kn} - P_k \leq 1$, the series $\sum_k 2^{-k}(Z_{kn} - P_k)$ is convergent. If h is a unit vector and $\epsilon > 0$, choose K such that $2^{-K} < \epsilon/2$. But then there is an n_0 such that $0 \leq \langle (Z_{kn} - P_k)h, h\rangle < \epsilon/2$ when $1 \leq k \leq K$ and $n \geq n_0$. It follows that

$$\lim_{n\to\infty} \sum_{k=1}^{\infty} \frac{1}{2^k} \langle (Z_{kn} - P_k)h, h\rangle = 0$$

for every vector h in $\mathcal{H}$. But

$$0 \leq T_n - A = T_n - \sum_{k=1}^{\infty} \frac{1}{2^k} P_k \leq \frac{1}{2^n} + \sum_{k=1}^{\infty} \frac{1}{2^k}(Z_{kn} - P_k).$$

Therefore $T_n \to A$ (WOT). But it is known that $T_n \searrow T$ (SOT). Thus $A = T \in ((\text{ball}\,\mathcal{C}_+)^\sigma)_\sigma$.

To prove the second equality, fix A in $\text{Re}\,\mathcal{A}$ and put $\alpha = -\|A\|$. So $A + \alpha \geq 0$. Let $\beta = \|A + \alpha\|$, so that $T = \beta^{-1}(A + \alpha) \in \text{ball}\,\mathcal{A}_+$ and $A = \beta T - \alpha$. From the first part of the proof, $\beta T \in ((\text{Re}\,\mathcal{C})^\sigma)_\sigma$. But $-\alpha \in \text{Re}\,\mathcal{C} \subseteq ((\text{Re}\,\mathcal{C})^\sigma)_\sigma$. Thus $A \in ((\text{Re}\,\mathcal{C})^\sigma)_\sigma$. ∎

The definition of a C*-algebra is an abstraction and, as has been seen, leads to the fact that every C*-algebra is isomorphic to a C*-subalgebra of $\mathcal{B}(\mathcal{H})$. Similar questions arise for von Neumann algebras, as discussed in the introduction of this chapter. Kadison [1956] gives a characterization in terms of the order completeness property of a von Neumann algebra. The next theorem is a special case of this result as it is assumed that the C*-algebra in question is already a subalgebra of $\mathcal{B}(\mathcal{H})$, and not an abstract C*-algebra as in Kadison [1956]. (For an historical account of the abstract characterization of von Neumann algebras, see Kadison [1985].) Also, for technical reasons, it is not assumed that the C*-algebra is unital. In the proof, however, it will surface that this case follows from the case where it is assumed that $\mathcal{A}$ is unital.

45.6 Theorem. *If $\mathcal{A}$ is a C^*-algebra contained in $\mathcal{B}(\mathcal{H})$, then $\mathcal{A}$ is weak* closed if and only if it contains the supremum of every bounded increasing net of hermitian operators in the algebra.*

Proof. It is known that every weak* closed algebra has this property (Proposition 43.1). So assume that $\mathcal{A}$ is a C*-subalgebra of $\mathcal{B}(\mathcal{H})$ and $A \in \mathcal{A}$ whenever $\{A_i\}$ is an increasing net in $\operatorname{Re}\mathcal{A}$ and $A_i \nearrow A$ (SOT). The proof of Proposition 43.2 shows that there is a projection P in $\mathcal{A}$ that is an identity for $\mathcal{A}$. By considering $\mathcal{A}|P \subseteq \mathcal{B}(P)$, it follows that without loss of generality it can be assumed that $1 \in \mathcal{A}$.

By considering the negatives of decreasing nets, it is also true that if $\{A_i\}$ is a decreasing net in $\operatorname{Re}\mathcal{A}$ and $A_i \searrow A$ (SOT), then $A \in \mathcal{A}$. In particular, $\operatorname{Re}\mathcal{A} = ((\operatorname{Re}\mathcal{A})^\sigma)_\sigma$. If it were assumed that the underlying Hilbert space is separable, the Up–Down Theorem would complete the proof. Here is a proof that is valid in the non-separable case.

Let $\mathcal{B}$ be the strong closure of $\mathcal{A}$; it must be shown that $\mathcal{A} = \mathcal{B}$. To do this, it suffices to show that $\mathcal{A}$ contains every projection in $\mathcal{B}$ (13.3.e). So let P be a projection in $\mathcal{B}$ and let $g \in P$ and $h \in P^\perp$. By Lemma 45.3 there is an operator T in $\mathcal{A}_+$ such that $Tg = g$ and $Th = 0$. If R is the projection onto the closure of the range of T, then the Spectral Theorem and the hypothesis imply that $R \in \mathcal{A}$. Also $Rg = g$ and $Rh = 0$.

In other words, for every pair of vectors (g, h) with g in P and h in $P^\perp$, there is a projection $R_{g,h}$ in $\mathcal{A}$ such that $R_{g,h}\, g = g$, $R_{g,h}\, h = 0$. Fix g in P and order the finite subsets of $P^\perp$ by inclusion. For any finite subset $F = \{h_1, \ldots, h_m\}$ of $P^\perp$, $R_F = R_{g,h_1} \wedge \cdots \wedge R_{g,h_m} \in \mathcal{A}$ (Exercise 4). But $\{R_F : F \text{ a finite subset of } P^\perp\}$ is a decreasing net of projections in $\mathcal{A}$. If R_g is the strong limit of this net, then, by hypothesis, R_g is a projection in $\mathcal{A}$.

Note that $R_g h = 0$ for all h in $P^\perp$. That is, for all g in P, $R_g \leq P$ and $R_g\, g = g$. Consider the net consisting of all projections $R_{g_1} \vee \cdots \vee R_{g_m}$ corresponding to finite subsets $\{g_1, \ldots, g_m\}$ of P. This is an increasing net in $\mathcal{A}$ that is bounded above by P. By hypothesis, it converges strongly to a projection R in $\mathcal{A}$ and $R \leq P$. But $Rg = g$ for each g in P, so $P = R \in \mathcal{A}$.
■

Exercises.

1. If X is a locally compact space, and $u\colon X \to \mathbb{R}$ is a lower semicontinuous function, show that $u(x) = \sup\{f(x) : f\colon X \to \mathbb{R}$ is continuous and $f(x) \leq u(x)$ for all $x\}$.

2. If μ is a σ-finite measure on the locally compact space X, $\mathcal{L}$ denotes the collection of real-valued, bounded lower semicontinuous functions on X,

and $\phi \in \operatorname{Re} L^\infty(\mu)$, show that in the ordered space $\operatorname{Re} L^\infty(\mu)$, $\phi = \inf\{u : u \in \mathcal{L}\}$.

3. Prove that if A and B are positive operators and $A \leq B$, then $(1+A)^{-1}A \leq (1+B)^{-1}B$. (See Exercise VIII.3.11 in [ACFA].)

4. Let E, F be two projections. Show that $A_n = (EFE)^n$ is a decreasing sequence of hermitian operators that strongly converges to the projection $E \wedge F$, the projection onto the intersection of the ranges of E and F. What about the sequence $(FEF)^n$?

§46. Normal homomorphisms and ideals

This section will focus attention on the ideals in a von Neumann algebra and morphisms between them. The twist is that the weak topology of the von Neumann algebra is incorporated in these notions. So the objects for consideration here are not the ideals that are just norm closed; these are part of the theory of C*-algebras. Only ideals that are weakly closed are considered.

The starting point is an examination of positive linear maps between von Neumann algebras. Again, in order to remain in the theory of von Neumann algebras, something about the positive maps other than their boundedness must be assumed. Tradition has dealt with the following concept, which will later be seen is actually a familiar one.

46.1 Definition. If $\mathcal{A}$ and $\mathcal{B}$ are von Neumann algebras and $\rho\colon \mathcal{A} \to \mathcal{B}$ is a positive linear map, then ρ is *normal* if $\rho(A_i) \to \rho(A)$ (SOT) for any increasing net $\{A_i\}$ in $\mathcal{A}$ that converges strongly to A.

If $\mathcal{A}$ is a von Neumann algebra and E is a projection in $\mathcal{A}$, then the map $A \to EAE$ is a normal positive map. Of course any positive linear map between von Neumann algebras that is strongly continuous is normal. The converse is not necessarily true.

46.2 Example. Let $\mathcal{A} = \mathcal{B}(\mathcal{H})$, $\mathcal{K} = \mathcal{H}^{(\infty)}$, and $\mathcal{B} = \mathcal{B}(\mathcal{K})$. Define $\rho\colon \mathcal{A} \to \mathcal{B}$ by $\rho(A) = A^{(\infty)}$, the infinite inflation of A. It is left as an exercise for the reader to show that ρ is normal. On the other hand, if $\{e_n\}$ is a basis for $\mathcal{H}$ and $k = (e_1, 2^{-1}e_2, 3^{-1}e_3, \dots) \in \mathcal{K}$, then $A \to \langle \rho(A)k, k\rangle$ is not SOT continuous on $\mathcal{B}(\mathcal{H})$. Hence ρ is not SOT continuous.

The astute reader will have noticed that the map in the preceding example is weak* continuous. It will be proved soon that a positive map between von Neumann algebras is normal if and only if it is weak* continuous. First it is necessary to handle the case for linear functionals, from which the statement for positive maps will follow.

46.3 Lemma. *If ϕ and ψ are positive linear functionals on a von Neumann algebra $\mathcal{A}$ such that there is an operator A in $\mathcal{A}_+$ with $\phi(A) < \psi(A)$, then there is an operator B in $\mathcal{A}_+$ such that $B \leq A$ and $\phi(T) < \psi(T)$ for all T in $\mathcal{A}$ with $0 < T \leq B$. If A is a projection, then B can be chosen to be a projection.*

Proof. Consider the set $\mathcal{C}$ of all operators C in $\mathcal{A}_+$ with $C \leq A$ and $\phi(C) \geq \psi(C)$. Give $\mathcal{C}$ the usual ordering. Let $\{C_i\}$ be a linearly ordered family in $\mathcal{C}$ and put $C = \sup_i C_i$. So $C \in \mathcal{A}_+$ and $C \leq A$. Because ϕ and ψ are normal, $C \in \mathcal{C}$. Therefore Zorn's Lemma implies there is a maximal element C in $\mathcal{C}$. Put $B = A - C$. If $T \in \mathcal{A}_+$ such that $0 < T \leq B$, then the maximality of C implies that $T \notin \mathcal{C}$ so that $\phi(T) < \psi(T)$.

Now assume that A is a projection and let B be the positive operator obtained in the preceding paragraph. The condition $B \leq A$ implies $\operatorname{cran} B \leq A$. If $B = \int_0^1 t\, dP(t)$ is the spectral decomposition of B, then for any $\epsilon > 0$, $P \equiv P[\epsilon, 1]$ is a non-zero projection contained in $\operatorname{cran} B$. Thus $P \leq A$ and the lemma holds if B is replaced by P. ∎

If $\mathcal{A}$ is a von Neumann algebra in $\mathcal{B}(\mathcal{H})$, then, as a weak* closed subspace of $\mathcal{B}(\mathcal{H})$, $\mathcal{A}$ is the dual of a Banach space. Denote by $\mathcal{A}_*$ the space of all weak* continuous linear functionals on $\mathcal{A}$. So $\mathcal{A} = [\mathcal{A}_*]^*$ by general Banach space theory. In fact $\mathcal{A}_*$ can be represented as $\mathcal{B}_1(\mathcal{H})/\mathcal{A}_\perp$, where $\mathcal{A}_\perp$ is the set of trace class operators that annihilate $\mathcal{A}$. Thus for every weak* continuous linear functional L on $\mathcal{A}$, there is a trace class operator T such that $L(A) = \operatorname{tr}(AT)$ for all A in $\mathcal{A}$. It will be shown that if the linear functional L is positive, the trace class operator can be chosen to be positive. In addition it will follow that a positive linear functional is normal if and only if it is weak* continuous.

46.4 Theorem. *If ψ is a positive linear functional on the von Neumann algebra $\mathcal{A}$, then the following statements are equivalent.*

(a) *ψ is normal.*

(b) *If $\{E_i\}$ is a pairwise orthogonal family of projections in $\mathcal{A}$, then $\psi(\sum_i E_i) = \sum_i \psi(E_i)$.*

(c) *ψ is weak* continuous.*

(d) *There is a positive trace class operator C such that $\psi(A) = \operatorname{tr}(AC)$ for all A in $\mathcal{A}$.*

Proof. There is no loss in generality in assuming that $\psi(1) = 1$.

(a) *implies* (b). If (a) holds and $\{E_i\}$ is a pairwise orthogonal family of projections in $\mathcal{A}$, then $\{\sum_{i\in F} E_i : F$ is a finite subset of the index set$\}$ is an increasing net that converges SOT to $\sum_i E_i$. Since ψ is normal, (b) holds.

(b) *implies* (c). Assume ψ satisfies (b). Since $\psi \geq 0$, $0 \leq \psi(E) \leq 1$ for any non-zero projection E . Thus there is a non-zero vector h such that $\psi(E) < \langle Eh, h\rangle$. By the preceding lemma there is a projection F satisfying $0 < F \leq E$ such that $\psi(T) < \langle Th, h\rangle$ whenever $0 \leq T \leq F$. Note that if $A \in \mathcal{A}_+$, then $FAF \leq \|A\|F$, so that $\psi(FAF) < \|A\| \langle FAFh, h\rangle$. If $[\cdot,\cdot]$ is the semi-inner product defined on $\mathcal{A}$ by ψ, then $|\psi(AF)|^2 = |[AF, 1]|^2 \leq [AF, AF]\,[1, 1] = \psi(FA^*AF) \leq \|A\|^2\langle FA^*AFh, h\rangle = \|A\|^2\,\|AFh\|^2$ for any A in $\operatorname{ball}\mathcal{A}$. Therefore $A \to \psi(AF)$, denoted by $\psi(\cdot F)$, is SOT continuous on $\mathcal{A}$.

Now let $\{E_i\}$ be a maximal family of pairwise orthogonal projections in $\mathcal{A}$ such that $\psi(\cdot E_i)$ is SOT continuous on $\mathcal{A}$. Let $E = \sum_i E_i$. If $E \neq 1$, then the preceding paragraph shows there is a projection $F \leq E^\perp$ such that $\psi(\cdot F)$ is SOT continuous, contradicting the maximality of $\{E_i\}$. Thus $\sum_i E_i = 1$ and so $\sum_i \psi(E_i) = \psi(1) = 1$ by (b). Therefore if $\epsilon > 0$, then there is a finite set of indices I_0 such that if J is any finite set containing I_0 and $P_J \equiv \sum_{i\in J} E_i$, then $\psi(P_J^\perp) = 1 - \psi(P_J) < \epsilon$. Hence for J containing I_0 and A in $\operatorname{ball}\mathcal{A}$, $|\psi(AP_J^\perp)|^2 = |[P_J^\perp, A^*]|^2 \leq \psi(AA^*)\psi(P_J^\perp) < \epsilon$. That is, $\|\psi - \psi(\cdot P_J)\| = \sup\{\,|\psi(AP_J^\perp)| : A \in \operatorname{ball}\mathcal{A}\,\} < \sqrt{\epsilon}$ whenever $J \supseteq I_0$. Since each $\psi(\cdot P_J)$ belongs to $\mathcal{A}_*$, so does ψ.

(c) *implies* (d). Begin with a special case. Assume there are vectors g and h in $\mathcal{H}$ such that $\psi(A) = \langle Ag, h\rangle$. It will be shown that there is a vector f in $\mathcal{H}$ such that $\psi(A) = \langle Af, f\rangle$ for all A in $\mathcal{A}$. In fact if $A \in \mathcal{A}_+$, then the fact that $\langle Ag, h\rangle \geq 0$ implies that $4\langle Ag, h\rangle = 2\langle Ag, h\rangle + 2\overline{\langle Ag, h\rangle} = 2\langle Ag, h\rangle + 2\langle Ah, g\rangle = \langle A(g+h), (g+h)\rangle - \langle A(g-h), (g-h)\rangle \leq \langle A(g+h), (g+h)\rangle$. That is $\langle Ag, h\rangle \equiv \psi(A) \leq \frac{1}{4}\langle A(g+h), (g+h)\rangle \equiv \phi(A)$. By Proposition 32.1 there is an operator T in the commutant of $\mathcal{A}$ such that $0 \leq T \leq 1$ and $\psi(A) = \langle AT(g+h), (g+h)\rangle$. Letting $f = T^{1/2}(g+h)$ completes the proof of the special case.

Now for the arbitrary case. Since ψ is weak* continuous, the Hahn–Banach Theorem implies there is a trace class operator D such that $\psi(A) = \operatorname{tr}(AD)$ for all A in $\mathcal{A}$. By Proposition 21.3.a there are vectors g, h in $\mathcal{H}^{(\infty)}$ such that $\operatorname{tr}(TD) = \langle T^{(\infty)}g, h\rangle$ for all T in $\mathcal{B}(\mathcal{H})$. Applying the preceding special case to the von Neumann algebra $\mathcal{A}^{(\infty)}$, there is a vector f in $\mathcal{H}^{(\infty)}$ such that for all A in $\mathcal{A}$, $\psi(A) = \operatorname{tr}(AD) = \langle A^{(\infty)}g, h\rangle = \langle A^{(\infty)}f, f\rangle$. Again Proposition 21.3.b implies there is a trace class operator C on $\mathcal{H}$ such that $\langle T^{(\infty)}f, f\rangle = \operatorname{tr}(TC)$ for all T in $\mathcal{B}(\mathcal{H})$. But this implies that $\operatorname{tr}(TC) \geq 0$ for all positive operators T on $\mathcal{H}$, so that $C \geq 0$, and clearly $\psi(A) = \operatorname{tr}(AC)$ for all A in $\mathcal{A}$.

(d) *implies* (a). If C is a positive trace class operator such that $\psi(A) = \operatorname{tr}(AC)$ for every A in $\mathcal{A}$, then it is clear that ψ is weak* continuous. Thus if

$\{A_i\}$ is an increasing net that converges to A (SOT), then $A_i \to A$ weak*. Thus $\psi(A_i) \to \psi(A)$. ■

In light of this theorem one is tempted to abandon the term "normal" in favor of weak* continuous. But this would conflict with a tradition. In this case, tradition wins.

46.5 Corollary. *If $\mathcal{A}$ and $\mathcal{B}$ are von Neumann algebras and $\rho\colon \mathcal{A} \to \mathcal{B}$ is a positive linear map, then ρ is normal if and only if it is weak* continuous.*

Proof. That weak* continuity implies normality is manifest from Proposition 43.1. On the other hand if ρ is normal and $\phi \in \mathcal{B}_*$ such that $\phi \geq 0$, then $\phi \circ \rho$ is a normal linear functional on $\mathcal{A}$. By the preceding theorem, $\phi \circ \rho \in \mathcal{A}_*$. But every weak* continuous linear functional on $\mathcal{B}$ is the linear combination of four positive elements of $\mathcal{B}_*$ (Exercise 3). By Banach space theory this implies that ρ is weak* continuous. ■

To expand on a point and enhance the reader's understanding of general functional analysis, this is a good place to emphasize the limitations of the Hahn–Banach Theorem. Powerful as this result is, it cannot do two things at once. If $\mathcal{X}$ is a Banach space, $\mathcal{Z}$ is a weak* closed subspace of $\mathcal{X}^*$, and $L\colon \mathcal{Z} \to \mathbb{C}$ is a weak* continuous linear functional, then the Hahn–Banach Theorem says that L can be extended to $\mathcal{X}^*$ preserving the norm or there is an extension that is weak* continuous. It is not always possible to do both (Exercise 2). Similarly, if $\mathcal{X}$ has an order structure and L is positive as well as weak* continuous, the Hahn–Banach Theorem does not imply the existence of a weak* continuous extension that is also positive. Theorem 46.4 gives a particular circumstance where an extension that simultaneously preserves the two properties is possible.

46.6 Proposition. *Every $*$-isomorphism between von Neumann algebras is normal.*

Proof. Let $\rho\colon \mathcal{A} \to \mathcal{B}$ be a $*$-isomorphism. If $\{A_i\}$ is an increasing net of hermitian operators in $\mathcal{A}$ with $A = \sup_i A_i$, then $\{\rho(A_i)\}$ is an increasing net in $\operatorname{Re}\mathcal{B}$ with $\rho(A_i) \leq \rho(A)$ for all A_i. Thus $B \equiv \sup_i \rho(A_i) \leq \rho(A)$. But then $A_i = \rho^{-1}\rho(A_i) \leq \rho^{-1}(B)$ for all i, so that $A \leq \rho^{-1}(B)$. Therefore $B = \rho(A)$ and ρ is proved to be normal. ■

A study of normal $*$-homomorphisms between von Neumann algebras implies an interest in examining the kernel and the induced map on the quotient algebra. To do this several pieces of information about $*$-homomorphisms and weak* closed ideals in a von Neumann algebra are needed.

A somewhat more general result will be proved. Recall the correspondence between hereditary subalgebras of a C*-algebra and closed left ideals (Proposition 5.3).

46.7 Proposition. *Let $\mathcal{A}$ be a von Neumann algebra.*

(a)) $\mathcal{B}$ is a weak* closed hereditary subalgebra of $\mathcal{A}$ if and only if there is a unique projection P in $\mathcal{A}$ such that $\mathcal{B} = P\mathcal{A}P$.

(b) $\mathcal{L}$ is a weak* closed left ideal of $\mathcal{A}$ if and only if there is a unique projection P in $\mathcal{A}$ such that $\mathcal{L} = \mathcal{A}P$.

(c) $\mathcal{I}$ is a weak* closed ideal in $\mathcal{A}$ if and only if there is a unique central projection P such that $\mathcal{I} = P\mathcal{A} = \mathcal{A}P$.

Proof. (a) Clearly if P is a projection in $\mathcal{A}$, then $P\mathcal{A}P$ is a weak* closed hereditary algebra. Conversely, if $\mathcal{B}$ is weak* closed, Proposition 43.2 implies there is a projection P in $\mathcal{B}$ such that $BP = PB = B$ for all B in $\mathcal{B}$. Clearly $\mathcal{B} \subseteq P\mathcal{A}P$. If $A \in \mathcal{A}_+$, then $PAP \leq \|A\|P$, an element of $\mathcal{B}$. Since $\mathcal{B}$ is hereditary, $PAP \in \mathcal{B}$. Thus $P\mathcal{A}P = \mathcal{B}$. The uniqueness of P follows from the fact that any such projection is an identity for the algebra $\mathcal{B}$.

(b) If $\mathcal{L}$ is a weak* closed left ideal, $\mathcal{B} = \mathcal{L} \cap \mathcal{L}^*$ is a weak* closed hereditary subalgebra of $\mathcal{A}$ (5.3). By the first part there is a projection P with $\mathcal{B} = P\mathcal{A}P$. Thus $P \in \mathcal{B} \subseteq \mathcal{L}$, and so $\mathcal{A}P \subseteq \mathcal{L}$. If $T \in \mathcal{L}$, then $T^*T \in \mathcal{L}_+ = \mathcal{B}_+$. Thus $T^*TP = T^*T$ and so $|T|P = |T|$. Using the polar decomposition, $T = TP \in \mathcal{A}P$ so that $\mathcal{L} = \mathcal{A}P$. But $\mathcal{B} = \mathcal{L} \cap \mathcal{L}^* = \mathcal{A}P \cap P\mathcal{A} = P\mathcal{A}P$, and so the uniqueness of P follows from the uniqueness part of (a). The converse is clear.

(c) If $\mathcal{I}$ is a weak* closed bilateral ideal, then $\mathcal{I}^* = \mathcal{I}$ and so $\mathcal{I}$ is also a hereditary subalgebra of $\mathcal{A}$. By parts (a) and (b) there is a unique projection P in $\mathcal{A}$ such that $\mathcal{I} = P\mathcal{A} = \mathcal{A}P$. So for any A in $\mathcal{A}$, $PA = P(PA) = P(AP) = (PA)P = AP$. Therefore $P \in \mathcal{A} \cap \mathcal{A}'$. Again the converse is clear. ■

Note that the hereditary subalgebras of $\mathcal{A}$ are precisely those subalgebras that are $*$-isomorphic to the compression algebras $\mathcal{A}_P$, with P in $\mathcal{A}$. Also note that part (c) of this last proposition implies that every weak* closed ideal splits the algebra $\mathcal{A}$. Indeed, if P is a central projection in $\mathcal{A}$, then $\mathcal{A} = P\mathcal{A} + P^\perp\mathcal{A}$, $P\mathcal{A} \cap P^\perp\mathcal{A} = (0)$, and for every A in $\mathcal{A}$, $\|A\| = \max\{\|PA\|, \|P^\perp A\|\}$. Moreover. $\mathcal{A}/\mathcal{I} \cong \mathcal{A}P^\perp$.

In the next theorem there is no assumption that the $*$-homomorphism takes the identity of $\mathcal{A}$ to the identity of $\mathcal{B}$. The proof, however, reduces to this case.

46.8 Theorem. *If $\mathcal{A}$ and $\mathcal{B}$ are von Neumann algebras and $\rho\colon \mathcal{A} \to \mathcal{B}$ is a normal $*$-homomorphism, then* $\ker \rho$ *is weak* closed and $\rho(\mathcal{B})$ is weak* closed in $\mathcal{B}$. Moreover there is a central projection P such that* $\ker \rho = P^{\perp}\mathcal{A}$ *and ρ is a $*$-isomorphism of $P\mathcal{A}$ onto* $\operatorname{ran}\rho$.

Proof. Note that the last part of the theorem follows from the first part and Proposition 46.7.c. In fact that result gives a central projection such that $\ker\rho = P^{\perp}\mathcal{A}$ and it becomes clear that ρ is a $*$-isomorphism of $P\mathcal{A}$ onto its range. So it suffices to prove the first part.

From C*-algebra theory it is known that $\ker\rho$ and $\operatorname{ran}\rho$ are norm closed. If $F = \rho(1)$, then F acts as an identity for the weak* closure of the range of ρ. Thus, replacing $\mathcal{B}$ by $\mathcal{B}_F$, it can be assumed that $\rho(1) = 1$.

Theorem 45.6 is used to show that $\operatorname{ran}\rho$ and $\ker\rho$ are weak* closed. If $\{A_i\}$ is a bounded increasing net in $\operatorname{Re}(\ker\rho)$, then put $A = \sup_i A_i$. Since ρ is normal, $\rho(A) = \sup_i \rho(A_i) = 0$. So $A \in \ker\rho$. By Theorem 45.6, $\ker\rho$ is weak* closed.

Now let $\{B_i\}$ be a bounded, increasing net in $\operatorname{Re}(\operatorname{ran}\rho)$. Let $B = \sup_i B_i \in \mathcal{B}$. Show that $B \in \operatorname{ran}\rho$. For each i, let $A \in \mathcal{A}$ such that $\rho(A) = B_i$. Replacing A by $|A|$, it can be assumed that $A \in \mathcal{A}_+$. Let $\Lambda \equiv \{\, A \in \mathcal{A}_+ : \rho(A) \in \{B_i\}\,\}$. It is claimed that Λ is directed upward.

Suppose $A_i, A_j \in \Lambda$ with $\rho(A_i) = B_i, \rho(A_j) = B_j$. Pick a B_k that dominates both B_i and B_j. Since $B_k - B_i \geq 0$, there is an operator T in $\mathcal{A}_+$ with $\rho(T) = B_k - B_i$. If $C_i = A_i + T$, then $C_i \geq A_i$ and $\rho(C_i) = B_k$. Similarly there is a C_j in $\mathcal{A}_+$ with $C_j \geq A_j$ and $\rho(C_j) = B_k$. Now $C_i - C_j \in \ker\rho$; so if $A = C_i + |C_i - C_j|$, then $\rho(A) = B_k$. Hence $A \in \Lambda$. But $A \geq C_i$ and $A \geq C_i + (C_j - C_i) = C_j$. This proves that Λ is directed.

If it was known that Λ was bounded above, the proof would come to a quick conclusion. Alas, from the way Λ is defined, it is clearly not bounded. One could have chosen Λ to be a bounded set of positive operators that map onto the net $\{B_i\}$, but then this would have shifted the trouble to showing it is directed. Improvisation is needed.

Fix $\epsilon > 0$. As in the proof of Exercise 45.3, $A \to (1+\epsilon A)^{-1}A$ is an increasing function on $\mathcal{B}(\mathcal{H})_+$. Thus $\{\,(1+\epsilon A)^{-1}A : A \in \Lambda\,\}$ is an increasing net in $\mathcal{A}$ whose norms are bounded by ϵ^{-1}. Put $A_\epsilon = \sup\{\,(1+\epsilon A)^{-1}A : A \in \Lambda\}$; so $\|A_\epsilon\| \leq \epsilon^{-1}$. If $A \in \Lambda$ and $\rho(A) = B_i$, then $\rho((1+\epsilon A)^{-1}A) = (1+\epsilon B_i)^{-1}B_i$, and this converges to $(1+\epsilon B)^{-1}B$ (weak*). Because ρ is normal, $\rho(A_\epsilon) = (1+\epsilon B)^{-1}B$ and so $(1+\epsilon B)^{-1}B \in \operatorname{ran}\rho$. But $\operatorname{ran}\rho$ is norm closed and $\|(1+\epsilon B)^{-1}B - B\| \to 0$ as $\epsilon \to 0$. Thus $B \in \operatorname{ran}\rho$. ■

46.9 Example. If (X, Ω, μ) is a σ-finite measure space and $\Delta \in \Omega$, then

$$\chi_\Delta L^\infty(\mu) = \{\, \phi \in L^\infty(\mu) : \phi(x) = 0 \text{ a.e. on } X \setminus \Delta\,\}.$$

is a weak* closed ideal of $L^\infty(\mu)$. Moreover all weak* closed ideals of $L^\infty(\mu)$ are of this form. Indeed, if $L^\infty(\mu)$ is identified with the algebra of multiplication operators on $L^2(\mu)$, the characteristic functions are the central projections.

46.10 Example. The maximal weak* closed left ideals of $\mathcal{B}(\mathcal{H})$ have the form $\mathcal{L}_h \equiv \{ T \in \mathcal{B}(\mathcal{H}) : Th = 0 \}$, where h is some fixed non-zero vector in $\mathcal{H}$. In fact, if $\mathcal{L}$ is any weak* closed left ideal there is a projection P with $\mathcal{L} = \mathcal{B}(\mathcal{H})P = \{T \in \mathcal{B}(\mathcal{H}) : TP^\perp = 0\}$. Clearly $\mathcal{L}$ is maximal when the projection P is maximal, or $P^\perp$ is minimal. This is the case precisely when $P^\perp$ has rank 1.

The left ideal $\mathcal{B}(\mathcal{H})P$ is minimal precisely when P has rank one. But rank one projections have the form $e \otimes e$ for some unit vector e. If $T \in \mathcal{B}(\mathcal{H})$, then $Te \otimes e = (Te) \otimes e$. So the minimal weak* closed left ideals of $\mathcal{B}(\mathcal{H})$ are those of the form $\{ h \otimes e : h \in \mathcal{H} \}$, where e is some fixed vector.

Exercises.

In these exercises $\mathcal{A}$ always denotes a von Neumann algebra acting on the Hilbert space $\mathcal{H}$.

1. Show that a positive linear map between von Neumann algebras is normal if and only if its restriction to bounded sets is strongly continuous.

2. Give an example of a measure space (X, Ω, μ), a weak* closed subspace $\mathcal{Y}$ of $L^\infty(\mu)$, and a weak* continuous linear functional on $\mathcal{Y}$ that does not have a weak* continuous extension to $L^\infty(\mu)$ with the same norm as L.

3. Show that every weak* continuous linear functional on $\mathcal{A}$ is the linear combination of four positive, weak* continuous linear functionals. (Hint: Represent the linear functional as a trace class operator.)

4. If $\mathcal{I}_0$ is a closed subset of $\mathcal{A}_+$, there is a closed ideal $\mathcal{I}$ of $\mathcal{A}$ such that $\mathcal{I}_0 = \mathcal{I}_+$ if and only if the following conditions hold: (a) $\mathcal{I}_0$ is a cone; (b) $\mathcal{I}_0$ is hereditary; that is, if $A \in \mathcal{I}_0$ and $T \in \mathcal{A}_+$ such that $T \leq A$, then $T \in \mathcal{I}_0$; (c) $UAU^* \in \mathcal{I}_0$ whenever $A \in \mathcal{I}_0$ and U is a unitary in $\mathcal{A}$. What happens if $\mathcal{I}_0$ is not assumed to be closed?

5. If $\mathcal{I}$ is a weak* closed ideal of $\mathcal{A}$ and N is a normal operator in $\mathcal{I}$ with spectral decomposition $N = \int z\, dE(z)$, show that $E(\Delta) \in \mathcal{I}$ for every Borel set Δ with $\inf\{ |z| : z \in \Delta \} > 0$.

6. Let (X, Ω, μ) be a σ-finite measure space. If $\Delta \in \Omega$, show that $\chi_\Delta L^\infty(\mu)$ is a minimal weak* closed ideal if and only if Δ is an atom of the measure space. Show that there is a one-to-one correspondence between weak* continuous homomorphisms from $L^\infty(\mu)$ into $\mathbb{C}$ and atoms of the measure μ.

7. If $\rho\colon \mathcal{A} \to \mathcal{B}$ is a normal $*$-homomorphism, $A \in \mathcal{A}_+$, and $B \in \mathcal{B}_+$ such that $B \le \rho(A)$, show that there is an operator A_1 in $\mathcal{A}_+$ with $A_1 \le A$ such that $\rho(A_1) = B$.

8. Let $X = \mathbb{N}$, $\Omega = 2^{\mathbb{N}}$, and let μ be counting measure. So $L^\infty(\mu) = \ell^\infty \cong C(\beta\mathbb{N})$, where $\beta\mathbb{N}$ is the Stone–Čech compactification of the natural numbers. (See Exercise 2.10.) With this identification of ℓ^∞ and $C(\beta\mathbb{N})$, show that a homomorphism from $C(\beta\mathbb{N})$ into $\mathbb{C}$ is weak* continuous if and only if it is given by evaluation at a point of $\mathbb{N}$. Show that if $p \in \beta\mathbb{N} \setminus \mathbb{N}$, then $\{ f \in \ell^\infty : \widehat{f}(p) = 0 \}$ is a maximal ideal of the von Neumann algebra ℓ^∞ that is weak* dense.

9. Let (X, Ω, μ) be a σ-finite measure space and consider the von Neumann algebra $\mathcal{A} = L^\infty(\mu; M_2)$ acting on $L^2(\mu) \oplus L^2(\mu)$. What are the weak* closed left ideals of $\mathcal{A}$? What are the minimal and maximal weak* closed left ideals?

10. Continuing the preceding exercise for the measure space of Exercise 8, show that $L^\infty(\mu; M_2)$ can be identified with $\ell^\infty(M^2)$, the space of all bounded sequences of 2×2 matrices. Show that there is a natural isomorphism $f \to \widehat{f}$ from $\ell^\infty(M_2)$ onto $C(\beta\mathbb{N}, M_2)$. What are the weak* closed left ideals of $\ell^\infty(M_2)$? For p in $\beta\mathbb{N}$, discuss the nature of the set $\{ f \in \ell^\infty(M_2) : \widehat{f}(p) = 0 \}$.

§47. Equivalence of projections

This section defines an equivalence relation on the set of projections in a von Neumann algebra $\mathcal{A}$, which, when $\mathcal{A} = \mathcal{B}(\mathcal{H})$, is the relation that the ranges of the projections have the same dimension. This definition of relative dimension will later lead to the classification of von Neumann algebras into types. Throughout the section $\mathcal{A}$ is a von Neumann algebra acting on $\mathcal{H}$.

47.1 Definition. If E and F are projections in $\mathcal{A}$, say that E and F are *equivalent* if there is a partial isometry W in $\mathcal{A}$ such that $W^*W = E$ and $WW^* = F$. In symbols, equivalence of projections is denoted by $E \sim F$. If there is a need to emphasize the role of the partial isometry W, write $E \sim_W F$.

Write $E \precsim F$ if there is a projection G in $\mathcal{A}$ with $G \le F$ and $E \sim G$. In this case say that E is *subequivalent* to F or that F *relatively dominates* E.

So two projections are equivalent exactly when there is a partial isometry with one projection as the initial space and the other projection as the final space. The first thing to check is that this is truly an equivalence relation on the set of projections in $\mathcal{A}$. (The reader who has not done so should do the exercises on partial isometries at the end of §3. These will be used without reference.) Reflexivity for this relation follows because a projection is also a partial isometry. Symmetry follows since W^* is a partial isometry whenever W is. Proving transitivity is a good way for readers to check out their facility with partial isometries.

47.2 Examples. (a) In an abelian algebra, two projections are equivalent if and only if they are equal.

(b) Two projections in $\mathcal{B}(\mathcal{H})$ are equivalent if and only if they have the same dimension.

If $\{e_n\}$ is a basis for $\mathcal{H}$ and $P = [e_1, e_3, \dots]$, then $P \sim 1$. So equivalence of a pair of projections does not imply anything about their orthogonal compliments. (See Exercise 1.) In light of Example (b) above, it is sometimes said that equivalent projections have the same relative dimension. (The word "relative" meaning relative to the algebra $\mathcal{A}$.)

Whenever W is a partial isometry in $\mathcal{A}$, $W^*W \sim WW^*$. The most obvious way to get a partial isometry in a von Neumann algebra is to look at the polar decomposition of an operator in $\mathcal{A}$.

47.3 Proposition. *If $A \in \mathcal{A}$, then* $\operatorname{cran} A \sim \operatorname{cran} A^*$.

Proof. Let $A = W|A|$ be the polar decomposition of A. Now $W^*W = \operatorname{cran}|A| = \operatorname{cran} A^*$ and $WW^* = \operatorname{cran} A$. ■

In order to explore this concept of equivalence, a few elementary facts about projections are needed. These may be considered a review for some readers. Recall that the lattice operations on subspaces (projections) is defined by $E \vee F = \operatorname{cl}[E+F] = \operatorname{cran}(E+F)$ and $E \wedge F = E \cap F$. (See Exercise 45.4.)

47.4 Proposition. *For projections E and F:*

(a) $\operatorname{cran}(EF) = E \wedge (E \wedge F^\perp)^\perp = E - E \wedge F^\perp$;

(b) $\operatorname{cran}(EF^\perp) = E - E \wedge F$;

(c) $\operatorname{cran}(F^\perp E) = E \vee F - F$.

Proof. (a) This is shown by proving that $\ker FE = E^\perp + (E \wedge F^\perp)$. Once this is shown, then $\operatorname{cran} EF = (\ker FE)^\perp = [E^\perp + (E \wedge F^\perp)]^\perp = E \wedge (E \wedge F^\perp)^\perp = E - (E \wedge F^\perp)$. If $h \in \ker FE$, then write $h = g + f$, where $g \in E^\perp$ and $f \in E$. So $0 = FEh = Ff$, and, hence, $f \in F^\perp$. That is, $f \in E \wedge F^\perp$ so that $\ker FE \leq E^\perp + (E \wedge F^\perp)$. Conversely, if $h = g + f$ with g in $E^\perp$ and f in $E \wedge F^\perp$, then $FEh = Ff = 0$.

Part (b) follows from (a) by making appropriate substitutions. For (c) note that (a) implies that $\operatorname{cran}(F^\perp E) = F^\perp - F^\perp \wedge E^\perp$. It is elementary to check that this is $E \vee F - F$. ■

Often in discussing equivalence of projections or trying to gain intuition, it helps to think of E and F as sets and draw Venn diagrams. This is the abelian case where equivalence means equality. If two expressions involving sets are equal, the corresponding expressions in projections may be equivalent. But that's a big maybe. These matters are fraught with the wrecks of statements that hold when the algebra is abelian but fail in general.

47.5 Proposition. *If E and F are projections in $\mathcal{A}$, then $E \vee F - F \sim E - E \wedge F$ and $E - E \wedge F^\perp \sim F - E^\perp \wedge F$.*

Proof. By Proposition 47.3, $\operatorname{cran} F^\perp E \sim \operatorname{cran}(F^\perp E)^* = \operatorname{cran} EF^\perp$. But the preceding proposition gives that $\operatorname{cran} F^\perp E = E \vee F - F$ and $\operatorname{cran} EF^\perp = E - E \wedge F$. To show the second equivalence, once again invoke Proposition 47.3 to get that $\operatorname{cran} EF \sim \operatorname{cran} FE$. Now the preceding proposition implies $\operatorname{cran}(EF) = E - E \wedge F^\perp$ and, by symmetry, $\operatorname{cran}(FE) = F - F \wedge E^\perp$. ■

The next result will sometimes be called the "additivity of equivalence."

47.6 Proposition. *If $\{E_i\}$ and $\{F_i\}$ are two sets of pairwise orthogonal projections such that $E_i \sim_{W_i} F_i$ for each i, then $\sum_i E_i \sim_W \sum_i F_i$, where $W = \sum_i W_i$.*

Proof. Put $E = \sum_i E_i$ and $F = \sum_i F_i$. By hypothesis for each i, $W_i^* W_i = E_i$ and $W_i W_i^* = F_i$. Now for any vector h in $\mathcal{H}$

$$\begin{aligned}\|Eh\|^2 = \langle Eh, h\rangle &= \sum_i \langle E_i h, h\rangle \\ &= \sum_i \langle W_i^* W_i h, h\rangle = \sum_i \|W_i h\|^2.\end{aligned}$$

Since $\{W_i h\}$ is an orthogonal family in $\mathcal{H}$, this proves that $\sum_i W_i$ converges strongly; put $W = \sum_i W_i$. Because the projections $\{E_i\}$ are pairwise orthogonal, it follows that for any finite set of indices I, $(\sum_{i\in I} W_i)^*(\sum_{i\in I} W_i) = \sum_{i\in I} E_i$. It is now left to the reader to supply the details that $W^*W = \sum_i W_i^* W_i = \sum_i E_i = E$. Similarly, $\sum_i W_i^*$ converges strongly, this limit is W^*, and $WW^* = F$. ■

In some ways the next result is a lemma. But it has more interest than one usually associates with such an appellation. There is also the author's belief that no result that gives a necessary and sufficient condition should be called a lemma. This proposition will also underscore one role of the central cover of a projection.

47.7 Proposition. *If E and F are projections in $\mathcal{A}$, then the following statements are equivalent.*

(a) *There are non-zero projections E_1 and F_1 in $\mathcal{A}$ such that $E_1 \leq E$, $F_1 \leq F$, and $E_1 \sim F_1$.*

(b) $C_E C_F \neq 0$.

(c) $E\mathcal{A}F \neq (0)$.

Proof. (a) *implies* (b). Since $E_1 \sim F_1$, $C_{E_1} = C_{F_1}$ (Exercise 5). But $C_{E_1} \leq C_E$ and $C_{F_1} \leq C_F$, whence (b).

(b) *implies* (c). Recall (43.7) that $C_E = [Ah : h \in E, A \in \mathcal{A}] = [\mathcal{A}E]$ and $C_F = [\mathcal{A}F]$. So if $E\mathcal{A}F = (0)$, then $EC_F = 0$. Thus $C_E \leq C_F^{\perp}$ by the definition of C_E, contradicting (b).

(c) *implies* (a). If $A \in \mathcal{A}$ and $EAF \neq 0$, let $E_1 = \operatorname{cran}(EAF)$ and $F_1 = \operatorname{cran}(EAF)^* = \operatorname{cran}(FA^*E)$. By Proposition 47.3, $E_1 \sim F_1$ and clearly $E_1 \leq E$ and $F_1 \leq F$. ∎

The justification for the title of the next result might be more easily gleaned from an interpretation of the result in the case of a factor. In that situation it says that if E and F are two projections in a factor, either $E \precsim F$ or $F \precsim E$. When these ideas were first developed in Murray and von Neumann [1936] and [1937], their approach was to prove results for factors, then use direct integral theory (see §52) to express every von Neumann algebra as a direct integral of factors, and then deduce the properties of the arbitrary algebras. As the subject progressed, this approach lost currency in favor of a more direct one such as that given in this book.

47.8 The Comparison Theorem. *If E and F are two projections in $\mathcal{A}$, then there is a central projection Z such that $EZ \precsim FZ$ and $FZ^{\perp} \precsim EZ^{\perp}$.*

Proof. By Zorn's Lemma there is a maximal family $\{(E_i, F_i)\}$ of pairs of projections in $\mathcal{A}$ such that for all i, $E_i \sim F_i$, $E_i \leq E$, $F_i \leq F$, and $E_i \perp E_j$, $F_i \perp F_j$ when $j \neq i$. By the additivity of equivalence, $E_1 \equiv \sum_i E_i \sim \sum_i F_i \equiv F_1$. Put $E_2 = E - E_1$ and $F_2 = F - F_1$. By the maximality of $\{(E_i, F_i)\}$, E_2 and F_2 have no non-zero subprojections in $\mathcal{A}$ that are equivalent. By Proposition 47.7 $C_{E_2} C_{F_2} = 0$; let $Z = C_{F_2}$.

Also let W be a partial isometry in $\mathcal{A}$ such that $W^*W = E_1$ and $WW^* = F_1$. Since Z is a central projection, $(WZ)^*(WZ) = E_1 Z$ and $(WZ)(WZ)^* = F_1 Z$. Therefore $EZ = E_1 Z + E_2 Z = E_1 Z + E_2 C_{E_2} C_{F_2} = E_1 Z \sim F_1 Z \leq FZ$. That is, $EZ \precsim FZ$. Similarly $FZ^{\perp} = F_1 Z^{\perp} + F_2 Z^{\perp} = F_1 Z^{\perp} + F_2 C_{F_2}^{\perp} = F_1 Z^{\perp} \sim E_1 Z^{\perp} \leq EZ^{\perp}$, so that $FZ^{\perp} \precsim EZ^{\perp}$. ∎

47.9 Corollary. *If $\mathcal{A}$ is a factor and E and F are two projections in $\mathcal{A}$, then either $E \precsim F$ or $F \precsim E$.*

The next result is a Schröder–Bernstein type theorem for the notion of the relative dominance of projections. The result dates back to Murray and von Neumann, but the proof here is due to Lebow [1968] and is based on the following easy fixed point theorem for order preserving maps defined on a complete lattice. In fact, if $\mathcal{L}$ is a complete lattice and $\tau: \mathcal{L} \to \mathcal{L}$ is an order preserving map, then put $T = \{x \in \mathcal{L} : x \leq \tau(x)\}$ and $x_0 = \sup T$. If $x \in T$, then $x \leq x_0$ so $x \leq \tau(x) \leq \tau(x_0)$; thus $x_0 \leq \tau(x_0)$ by the definition of x_0. But order preservation implies that $\tau(x_0) \leq \tau(\tau(x_0))$ so that $\tau(x_0) \in T$ and so $\tau(x_0) \leq x_0$. (Cute, huh?)

47.10 Theorem. *If E and F are projections in $\mathcal{A}$ such that $E \precsim F$ and $F \precsim E$, then $E \sim F$.*

Proof. Let W and V be partial isometries in $\mathcal{A}$ such that $WW^* = E$, $W^*W \leq F$, $VV^* = F$, and $V^*V \leq E$. Let $\mathcal{L}$ be the set of all projections G in $\mathcal{A}$ such that $G \leq F$. Note that $\mathcal{L}$ is a complete lattice. Define $\tau: \mathcal{L} \to \mathcal{L}$ by

$$\tau(G) = F - W^*(E - V^*GV)W.$$

As soon as the initial and final subspaces are discerned, it is straightforward to check that τ is order preserving on $\mathcal{L}$. Therefore τ has a fixed point P. That is, there is a projection P in $\mathcal{A}$ such that $P \leq F$ and $P = F - W^*(E - V^*PV)W$. If V_1 and W_1 are the partial isometries PV and $(E - V^*PV)W$, respectively, then:

$$V_1V_1^* = PV^*VP = P \qquad V_1^*V_1 = V^*PV,$$
$$W_1^*W_1 = W^*(E - V^*PV)W = F - P \qquad W_1W_1^* = E - V^*PV.$$

Thus $P \sim V^*PV$ and $F - P \sim E - V^*PV$ so that, by the additivity of equivalence, $E \sim F$. ∎

There are some who might not like the preceding proof because it seems to hide what is happening. It is slick and falls into the category of proofs that could never have been the original. In Exercise 8 another proof is sketched that is messier but, perhaps for some, more revealing.

Exercises.

1. If E and F are two projections in $\mathcal{A}$, show that $E \sim F$ and $E^\perp \sim F^\perp$ if and only if there is a unitary W in $\mathcal{A}$ such that $W^*EW = F$.

2. If E and F are equivalent projections in $\mathcal{A}$ and Z is a central projection, show that $EZ \sim FZ$.

3. If E and F are projections in $\mathcal{A}$, show that $E \precsim F$ if and only if there is a projection G in $\mathcal{A}$ with $E \leq G$ and $G \sim F$.

4. If $h \in \mathcal{H}$, $F = [\mathcal{A}'h]$, and $E \precsim F$, show that there is a vector g such that $E = [\mathcal{A}'g]$.

5. Show that if E and F are projections in a von Neumann algebra $\mathcal{A}$ and $E \precsim F$, then $C_E \leq C_F$. Is the converse true?

6. Let $\mathcal{I}$ be an ideal of $\mathcal{A}$. (a) Show that if E is a projection in $\mathcal{I}$ and $F \precsim E$, then $F \in \mathcal{I}$. (b) If E and F are two projections in $\mathcal{I}$, then $E \vee F \in \mathcal{I}$.

7. If E and F are projections in $\mathcal{A}$ such that $E \sim E^\perp$ and $F \sim F^\perp$, show that $E \sim F$.

8. Here is another proof of Theorem 47.10. Let W and V be partial isometries in $\mathcal{A}$ such that $V^*V = E$, $VV^* = F_1 \leq F$, $W^*W = F$, and $WW^* = E_1 \leq E$. (a) Show that VE_1 and WF_1 are partial isometries. (b) Show that there are sequences of projections $\{E_n\}$ and $\{F_n\}$ in $\mathcal{A}$ such that for $n \geq 1$: (i) $E \geq E_1 \geq E_2 \geq \cdots$; (ii) $F \geq F_1 \geq F_2 \geq \cdots$; (iii) V maps E_n isometrically onto F_{n+1}; (iv) W maps F_n isometrically onto E_{n+1}. Let $E_0 = E$ and $F_0 = F$. (c) Show that: V maps $E_n \ominus E_{n+1}$ isometrically onto $F_{n+1} \ominus F_{n+2}$ and W maps $F_n \ominus F_{n+1}$ isometrically onto $E_{n+1} \ominus E_{n+2}$. (d) Show that $E_n \ominus E_{n+1} \sim E_{n+2} \ominus E_{n+3}$ for $n \geq 0$. Put $E_\infty = \bigwedge_n E_n$. Using (d) only for the even values of $n \geq 0$, show that $E \sim E_1 \sim F$.

§48. Classification of projections

Here there will be a discussion of the various types of von Neumann algebras. It will be shown that every von Neumann algebra is the direct sum of algebras, each of which is one of the types. The first step is to introduce the notion of a finite and an infinite algebra and then the concepts of discrete and continuous algebras. These are done by first defining what it means for a projection to be of this type.

To set notation in this section, $\mathcal{A}$ is always a von Neumann algebra in $\mathcal{B}(\mathcal{H})$, and E and F will be projections in $\mathcal{A}$.

48.1 Definition. A projection E in $\mathcal{A}$ is *finite* if the only projection F in $\mathcal{A}$ such that $F \leq E$ and $F \sim E$ is $F = E$. A projection is *infinite* if it is not finite. A von Neumann algebra is *finite* or *infinite* depending on whether the identity is a finite or infinite projection.

Note that a projection E is finite or infinite exactly when the von Neumann algebra $\mathcal{A}_E$ is finite or infinite. For the von Neumann algebra $\mathcal{B}(\mathcal{H})$, the finite projections are precisely those with finite rank. In an abelian algebra, all projections are finite.

48.2 Proposition.

(a) *If E is finite and $F \precsim E$, then F is finite.*

(b) *If $\mathcal{A}$ is a finite algebra and E is a projection in $\mathcal{A}$, then $\mathcal{A}_E$ is a finite algebra.*

(c) *If $\mathcal{A}$ is abelian, then $\mathcal{A}$ is a finite von Neumann algebra.*

(d) *If $\{E_i\}$ is a family of finite projections in $\mathcal{A}$ such that $C_{E_i} \perp C_{E_j}$ for $i \neq j$, then $\sum_i E_i$ is a finite projection.*

(e) *If $\{Z_i\}$ is a family of finite central projections, then $Z = \bigvee_i Z_i$ is a finite central projection.*

(f) *If $\rho\colon \mathcal{A} \to \mathcal{B}$ is a $*$-isomorphism and E is a finite projection in $\mathcal{A}$, then $\rho(E)$ is a finite projection in $\mathcal{B}$.*

Proof. (a) First assume that $F \leq E$. If $F_1 \leq F$ and $F_1 \sim F$, then $F_1 + (E - F) \sim E$, so that, by finiteness, $F_1 + (E - F) = E$. Hence $F_1 = F$ and F is a finite projection. In light of this, to prove (a) it suffices to show that if E is finite and $F \sim E$, then F is finite. Let W be the partial isometry such that $W^*W = F$ and $WW^* = E$. If $F_1 \leq F$ and $F_1 \sim F$, then $E_1 = WF_1$ satisfies $E_1 \sim F_1 \sim F \sim E$. Since E is finite, $E = E_1$ so that $F = W^*E = W^*WE_1 = F_1$, and F must be finite.

(b) This statement is a consequence of (a).

(c) This is clear from the definition.

(d) Adopt the notation in the statement of (d) and put $E = \sum_i E_i$. Let F be a projection in $\mathcal{A}$ such that $F \leq E$ and $F \sim_W E$. For any i, $E - E_i \leq \sum_{j \neq i} C_{E_j}$, and so $(E - E_i)C_{E_i} = 0$ by hypothesis. That is, $E_i = EC_{E_i}$. But then if $W_i = WC_{E_i}$, $FC_{E_i} \sim_{W_i} EC_{E_i} = E_i$. Also $FC_{E_i} \leq EC_{E_i} = E_i$. Hence the finiteness of E_i implies $FC_{E_i} = E_i$. Therefore $E = \sum_i E_i = \sum_i FC_{E_i} \leq F \leq E$. Thus $F = E$ and E must be finite.

(e) Let $Z = \bigvee_i Z_i$; thus Z is a central projection. If F is any (not necessarily central) projection in $\mathcal{A}$ with $F \leq Z$ and $F \sim Z$, then $Z_iF \sim Z_iZ = Z_i$. Since Z_i is finite, $Z_iF = Z_i$. Since i was arbitrary, $F = Z$.

(f) This proof is a routine exercise in the definitions. ∎

A consideration of $\mathcal{B}(\mathcal{H})$ shows that part (d) in the preceding proposition is not valid without the restriction on the central covers. It should also be noted that in light of Theorem 46.8 and part (b) of the preceding proposition, part (f) can be extended to the case that ρ is a normal, surjective $*$-homomorphism. See Exercises 1 and 2.

The preceding proposition displays the counterintuitive fact that an infinite collection of finite projections can be put together (with some restriction) and still produce a finite projection. In particular, part (d) implies the following corollary.

48.3 Corollary. *If $\{ \mathcal{A}_i \}$ is a collection of von Neumann algebras, then $\bigoplus_i \mathcal{A}_i$ is a finite von Neumann algebra if and only if each $\mathcal{A}_i$ is finite.*

The next result gives a further justification of the definition of finiteness that connects with the concept of a finite dimensional space in linear algebra.

48.4 Proposition. *$\mathcal{A}$ is finite if and only if every left (right) invertible element of $\mathcal{A}$ is invertible.*

Proof. Suppose every left invertible element of $\mathcal{A}$ is invertible. If $E \sim_W 1$, then $W^*W = E$ and $WW^* = 1$. So W^* is left invertible and, hence, invertible. Thus $E = 1$ and $\mathcal{A}$ is finite. Now assume that $\mathcal{A}$ is finite and $A, B \in \mathcal{A}$ such that $AB = 1$. Thus $\operatorname{ran} A = 1$. But $\operatorname{cran} A^* \sim \operatorname{cran} A = 1$. Since A is finite, $\operatorname{cran} A^* = 1$ so that $\ker A = (\operatorname{ran} A^*)^\perp = 0$. That is A is bijective. Therefore A is invertible and this implies that B is invertible. ∎

Here is another property of finite algebras that has its analogue in finite dimensional space.

48.5 Proposition. *Let $\mathcal{A}$ be a finite von Neumann algebra.*

(a) *If E, E_1, F, and F_1 are projections in $\mathcal{A}$, $E_1 \leq E$, $F_1 \leq F$, $E \sim F$, and $E_1 \sim F_1$, then $E - E_1 \sim F - F_1$.*

(b) *If E and F are equivalent projections in $\mathcal{A}$, then there is a unitary U in $\mathcal{A}$ such that $U^*EU = F$.*

Proof. (a) By the Comparison Theorem there is a central projection Z such that $(E - E_1)Z \precsim (F - F_1)Z$ and $(F - F_1)Z^\perp \precsim (E - E_1)Z^\perp$. Let $F_2 \leq (F - F_1)Z$ such that $(E - E_1)Z \sim F_2$. Thus $F_2 + F_1Z \leq FZ$ and $F_2 + F_1Z \sim (E - E_1)Z + E_1Z = EZ \sim FZ$. Since FZ is finite, $F_2 + F_1Z = FZ$, hence $F_2 = (F - F_1)Z$. That is, $(E - E_1)Z \sim (F - F_1)Z$. Similarly, $(E - E_1)Z^\perp \sim (F - F_1)Z^\perp$, so that (a) is proved.

(b) Apply part (a) with the identity as the dominating projection to obtain that $E^\perp \sim F^\perp$. So there are partial isometries V and W such that $W^*W = E$, $WW^* = F$, $V^*V = E^\perp$, and $VV^* = F^\perp$. It follows that $U = W + V$ is unitary and $U^*EU = F$. (See Exercise 47.1.) ∎

48.6 Definition. E is an *abelian* projection if $E\mathcal{A}E \approx \mathcal{A}_E$ is an abelian algebra. An algebra $\mathcal{A}$ is *discrete* if for every non-zero central projection

Z there is a non-zero abelian projection F with $F \leq Z$. An algebra $\mathcal{A}$ is *continuous* if it contains no non-zero abelian projections. A projection E in $\mathcal{A}$ is *discrete* or *continuous* when the compression algebra $\mathcal{A}_E$ is discrete or continuous.

By (48.2.b) every abelian projection is finite and also every abelian projection is discrete. In $\mathcal{B}(\mathcal{H})$ the only abelian projections are those of rank one. Furthermore $\mathcal{B}(\mathcal{H})$ is a discrete algebra. Finding an example of a continuous algebra is more difficult and this is postponed until §53.

At this point it is possible to proceed directly to the classification of von Neumann algebras. But this objective is put on pause to establish some elementary facts about abelian, discrete, and continuous projections.

48.7 Proposition. *Let $\mathcal{A}$ be a von Neumann algebra.*

(a) *A projection E in $\mathcal{A}$ is discrete if and only if for every central projection Z in $\mathcal{A}$ with $ZE \neq 0$, there is an abelian projection F in $\mathcal{A}$ with $F \leq ZE$.*

(b) *A projection E in $\mathcal{A}$ is continuous if and only if there is no non-zero abelian projection F in $\mathcal{A}$ satisfying $F \leq E$.*

(c) *If $\mathcal{B}$ is a von Neumann algebra, $\rho\colon \mathcal{A} \to \mathcal{B}$ is a normal, surjective $*$-homomorphism, and E is a projection in $\mathcal{A}$ that is abelian (respectively, discrete or continuous), then $\rho(E)$ is abelian (respectively, discrete or continuous).*

(d) *If E is abelian (respectively, discrete or continuous) and $F \precsim E$, then F is abelian (respectively, discrete or continuous).*

(e) *If $\{E_i\}$ is a collection of abelian (respectively, discrete or continuous) projections such that $C_{E_i} \perp C_{E_j}$ for $i \neq j$, then $E = \sum_i E_i$ is abelian (respectively, discrete or continuous).*

Proof. To prove (a) note that if the center of $\mathcal{A}$ is $\mathcal{Z}$, then the center of $\mathcal{A}_E$ is $\mathcal{Z}_E$ (43.8.c). Now just interpret the definition that $\mathcal{A}_E$ is discrete and (a) is proved. The proof of (b) is similar. The proof of (c) proceeds rather routinely once it is realized that by Theorem 46.8 it suffices to assume that ρ is a $*$-isomorphism .

(d). First observe that (c) can be invoked to reduce proving the statement to proving the analogous statement where it is assumed that $F \leq E$ instead of $F \precsim E$. With this, assume that E is abelian. Since $F \leq E$, $F\mathcal{A}F \subseteq E\mathcal{A}E$, so F is abelian. The proof for the continuous case is also easy.

Now assume that E is discrete and $F \leq E$. Replacing $\mathcal{A}$ by $\mathcal{A}_E$, it follows that it suffices to assume that $E = 1$. That is, assume that $\mathcal{A}$ is a discrete algebra. Part (a) will be used to show that F is discrete. So let Z be a central projection with $ZF \neq 0$. Recall (Exercise 43.4) that $C_{ZF} = ZC_F$. Since $C_{ZF} \neq 0$, the fact that $\mathcal{A}$ is discrete implies there is an abelian projection G with $G \leq ZC_F$. Now $C_G C_{ZF} = C_{ZF} \neq 0$, so by (47.7) there are projections $G_1 \leq G$ and $F_1 \leq ZF$ with $G_1 \sim F_1$. That is, $F_1 \precsim G$, so that F_1 is abelian and, hence, F is discrete.

(e) Assume the projections $\{E_i\}$ are abelian. Because $\sum_i C_{E_i} = C_E$, if $A \in \mathcal{A}$, then $EAE = \sum_i (EAE)C_{E_i} = \sum_i E_i A E_i$. Thus for A, B in $\mathcal{A}$

$$\begin{aligned}(EAE)(EBE) &= \left(\sum_i E_i A E_i\right)\left(\sum_i E_i B E_i\right)\\ &= \sum_i (E_i A E_i)(E_i B E_i)\\ &= \left(\sum_i E_i B E_i\right)\left(\sum_i E_i A E_i\right)\\ &= (EBE)(EAE).\end{aligned}$$

Hence E is abelian.

Now assume each E_i is discrete. If Z is a central projection with $ZE \neq 0$, then $ZE_i \neq 0$ for some indices i. When this occurs, let G_i be an abelian projection with $G_i \leq ZE_i$. If $ZE_i = 0$, let $G_i = 0$. Put $G = \sum_i G_i$. Now $C_{G_i} \leq C_{E_i}$, so the first part of this proof implies that G is abelian and so E is discrete.

Finally assume each E_i is continuous. Suppose there is a non-zero abelian projection $G \leq E$. If $GC_{E_i} = 0$ for all E_i, then $G = GC_E = 0$, a contradiction. So there is an i with $GC_{E_i} \neq 0$. But $GC_{E_i} = GEC_{E_i} = GC_{E_i}\sum_j E_j = GE_i$. But then GE_i is a non-zero abelian projection that is dominated by E_i, contradicting the fact that E_i is discrete. ■

48.8 Corollary. *If $\{\mathcal{A}_i\}$ is a collection of von Neumann algebras and $\mathcal{A} = \bigoplus_i \mathcal{A}_i$, then $\mathcal{A}$ is discrete (respectively, continuous) if and only if each $\mathcal{A}_i$ is discrete (respectively, continuous).*

48.9 Lemma. *If E is an abelian projection and $F \leq E$, then $F = C_F E$.*

Proof. Since E is abelian, $E\mathcal{A}E$ is an abelian algebra. It follows that $F\mathcal{A}(E-F) = (0)$. In fact, if $A \in \mathcal{A}$, then $FA(E-F) = F(EAE)(E-F) = F(E-F)(EAE) = 0$. By Proposition 43.7, $C_F = [\mathcal{A}F] \perp C_{E-F} = [\mathcal{A}(E-F)]$. In particular, $0 = C_F(E-F)$, or $F = C_F E$. ■

48.10 Proposition. *If E and F are abelian projections, then $F \precsim E$ if and only if $C_F \leq C_E$.*

Proof. If $F \precsim E$, then $C_F \leq C_E$ independent of whether the projections are abelian (Exercise 47.5.) Now assume that $C_F \leq C_E$. By the Comparison Theorem there is a central projection Z such that $ZE \precsim ZF$ and $Z^\perp F \precsim Z^\perp E$. Without loss of generality it can be assumed that $ZE \leq ZF$. Note that ZE is also abelian. Thus using the preceding lemma and Exercise 43.4 it follows that $ZE = ZFC_{ZE} = ZFC_E = ZF$. Therefore, $F = ZF + Z^\perp F = ZE + Z^\perp F \precsim E$. ∎

Now for the classification scheme for von Neumann algebras, though, as mentioned, finding examples of the various types must wait.

48.11 Definition. A von Neumann algebra $\mathcal{A}$ is:

(a) *Type* I if it is discrete.

(b) *Type* II if it is continuous and every non-zero central projection dominates a non-zero finite projection.

(c) *Type* III if it contains no non-zero finite projections.

(d) *Type* I_n if it is Type I and finite.

(e) *Type* I_∞ if it is Type I and infinite.

(f) *Type* II_1 if it is Type II and finite.

(g) *Type* II_∞ if it is Type II and infinite.

Of course the last four parts of the definition are refinements of the first three. Here is a small table to see what has happened.

	I	II	III
Finite	I_n	II_1	*
Infinite	I_∞	II_∞	III

* –Impossible

There are additional terms that are used in discussing types of von Neumann algebras. Some further definitions connected with the classification will appear shortly, but for the moment be aware that the term *purely infinite* is a synonym for Type III.

48.12 Examples. (a) $\mathcal{B}(\mathcal{H})$ is Type I. If $\dim \mathcal{H} = \infty$, then $\mathcal{B}(\mathcal{H})$ is Type I_∞; if $\dim \mathcal{H} = n < \infty$, then $\mathcal{B}(\mathcal{H})$ is Type I_n.

(b) If (X, Ω, μ) is a σ-finite measure space, $n < \infty$, and $L^\infty(\mu; M_n)$ acting on $L^2(\mu; \mathbb{C}^n)$ is defined as is done following Proposition 43.4, then $L^\infty(\mu; M_n)$ is Type I_n.

48.13 Proposition. *Let $\mathcal{A}$ be a von Neumann algebra and let E be a projection in $\mathcal{A}$.*

(a) *If $\mathcal{A}$ is Type I (respectively, II, III), then $\mathcal{A}_E$ is Type I (respectively, II, III).*

(b) *If $\mathcal{A}$ is Type I_n, then $\mathcal{A}_E$ is Type I_n.*

(c) *If $\mathcal{A}$ is Type II_1, then $\mathcal{A}_E$ is Type II_1.*

Proof. (a) The Type I statement is a rephrasing of (48.7.d). If $\mathcal{A}$ is Type II, then $\mathcal{A}_E$ is continuous by (48.7.d). Now let Z be a central projection in $\mathcal{A}$ such that $ZE \neq 0$. Since $\mathcal{A}$ is Type II, there is a non-zero finite projection F with $F \leq Z$. Thus $EZF = EZ \neq 0$. But $EZF \in (EZ)\mathcal{A}F$, By Proposition 47.7 there are projections E_1 and F_1 with $E_1 \leq EZ$, $F_1 \leq F$, and $E_1 \sim F_1$. Thus E_1 is finite and $\mathcal{A}_E$ is Type II. If $\mathcal{A}$ is Type III, then again $\mathcal{A}_E$ is continuous. But $\mathcal{A}_E$ cannot contain a finite projection since this projection would be a finite projection in $\mathcal{A}$. Thus $\mathcal{A}$ is Type III.

(b) From (a) it is known that $\mathcal{A}_E$ is Type I. By (48.2.b), $\mathcal{A}_E$ is also finite. The proof of (c) is similar. ■

If $\mathcal{A}$ is Type I_∞ or II_∞, then there is a finite projection E in $\mathcal{A}$ and, in this case, $\mathcal{A}_E$ will be Type I_n or II_1.

Just as the various combinations of the pairs (finite, infinite) and (discrete, continuous) are patched together with suitable modifiers to obtain the definitions of the various types, so too can the various preceding results (in particular, (48.3) and (48.8)) be patched together to arrive at the following. The details are left to the reader.

48.14 Proposition. *Let $\{\mathcal{A}_i\}$ be a collection of von Neumann algebras and put $\mathcal{A} = \bigoplus_i \mathcal{A}_i$.*

(a) *$\mathcal{A}$ is Type I_n if and only if each $\mathcal{A}_i$ is Type I_n.*

(b) *$\mathcal{A}$ is Type I_∞ if and only if each $\mathcal{A}_i$ is Type I and at least one of the algebras is Type I_∞.*

(c) *$\mathcal{A}$ is Type II_1 if and only if each $\mathcal{A}_i$ is Type II_1.*

(d) *$\mathcal{A}$ is Type II_∞ if and only if each $\mathcal{A}_i$ is Type II and at least one of the algebras is Type II_∞.*

(e) *$\mathcal{A}$ is Type III if and only if each $\mathcal{A}_i$ is Type III.*

48.15 Proposition. *The various types are preserved under a $*$-isomorphism.*

48.16 Theorem. *If $\mathcal{A}$ is a von Neumann algebra, then there are unique pairwise orthogonal central projections $Z_1, \ldots, Z_5$ such that $Z_1 + \cdots + Z_5 = 1$ and:*

(a) *$\mathcal{A}Z_1$ is Type I_n.*

(b) *$\mathcal{A}Z_2$ is Type I_∞;*

(c) *$\mathcal{A}Z_3$ is Type II_1;*

(d) *$\mathcal{A}Z_4$ is Type II_∞;*

(e) *$\mathcal{A}Z_5$ is Type III.*

Proof. Define the central projections

$$H = \sup\{\, Z : Z \text{ is a central projection and } \mathcal{A}Z \text{ is Type I} \,\},$$
$$K = \sup\{\, Z : Z \text{ is a central projection and } \mathcal{A}Z \text{ is Type II} \,\}.$$

It is easy to see that $\mathcal{A}H$ is Type I and $\mathcal{A}K$ is Type II (Exercise 6) and, moreover, $HK = 0$. Put $Z_5 = 1 - (H + K)$. It will be shown that $\mathcal{A}Z_5$ is Type III.

Begin by showing that $\mathcal{A}Z_5$ is continuous. In fact, suppose there is an abelian projection P in $\mathcal{A}Z_5$ and consider its central cover, C_P. Clearly $C_P \leq Z_5$. If G is any non-zero central projection in $\mathcal{A}C_P$, then G is in the center of $\mathcal{A}$. $GP \neq 0$, and $(GP)\mathcal{A}(GP) = G(P\mathcal{A}P)$ is abelian. That is, every non-zero central projection of $\mathcal{A}C_P$ dominates a non-zero abelian projection. In other words, $\mathcal{A}C_P$ is Type I. Thus, by definition, $C_P \leq H$. Since $C_P \leq Z_5 \perp H$, this implies that $C_P = 0$, and thus $P = 0$. Since P was an arbitrary abelian projection in $\mathcal{A}Z_5$, this shows that $\mathcal{A}Z_5$ is continuous.

Now to finish the proof that $\mathcal{A}Z_5$ is Type III. Indeed, suppose there is a non-zero finite projection P in $\mathcal{A}Z_5$. By the preceding paragraph, $\mathcal{A}C_P$ is continuous. If G is any central projection of $\mathcal{A}C_P$, then $G \in \mathcal{Z}$ and $G \leq C_P$. Therefore $GP \neq 0$ and GP is finite. That is, every non-zero central projection of $\mathcal{A}C_P$ dominates a non-zero finite projection. This implies that $\mathcal{A}C_P$ is Type II and so $C_P \leq K \leq Z_5^\perp$. But $P \leq Z_5$ implies that $C_P \leq Z_5$, so $P = 0$.

Now define the central projection

$$F = \sup\{\, Z : Z \text{ is a finite central projection} \,\}.$$

Put $Z_1 = HF$, $Z_2 = HF^\perp$, $Z_3 = KF$, and $Z_4 = KF^\perp$. It is left to the reader to verify that the desired properties are achieved. ∎

48.17 Corollary. *If $\mathcal{A}$ is a factor, then $\mathcal{A}$ is exactly one of the types I_n, I_∞, II_1, II_∞, or III.*

Exercises.

1. If $\mathcal{A}$ is a finite von Neumann algebra and $\rho\colon \mathcal{A} \to \mathcal{B}$ is a surjective, normal $*$-homomorphism, show that $\mathcal{B}$ is finite.

2. If $\mathcal{A}$ is an infinite von Neumann algebra and $\rho\colon \mathcal{A} \to \mathcal{B}$ is a $*$-isomorphism, show that $\mathcal{B}$ is infinite. If ρ is only assumed to be a normal, surjective $*$-homomorphism, does the conclusion remain valid?

3. Show that E is an abelian projection in $\mathcal{A}$ if and only if $\mathcal{A}_E = \mathcal{Z}_E$, where $\mathcal{Z}$ is the center of $\mathcal{A}$.

4. Prove that a von Neumann algebra $\mathcal{A}$ is finite if and only if the set of unitaries in $\mathcal{A}$ is SOT closed. (See Exercise 44.1.)

5. If $\mathcal{A}$ is a von Neumann algebra of Type I_n (resp. I_∞, II_1, II_∞, III) and Z is a central projection, then $\mathcal{A}_Z$ is Type I_n (resp. I_∞, II_1, II_∞, III).

6. Using the notation from the proof of Theorem 48.16, show that $\mathcal{A}H$ is Type I and $\mathcal{A}K$ is Type II.

7. Is the following statement true or false? $\mathcal{A}$ is Type II if it is continuous and has a non-zero finite projection. Is the converse true or false?

8. Show that if $\mathcal{A}$ is finite and E, E_1, F, F_1 are projections in $\mathcal{A}$ such that $E_1 \leq E$, $F_1 \leq F$, $E \precsim F$, and $F_1 \precsim E_1$, then $E - E_1 \precsim F - F_1$.

9. Suppose $\mathcal{A}$ and $\mathcal{B}$ are von Neumann algebras and $\rho\colon \mathcal{A} \to \mathcal{B}$ is a normal, surjective $*$-homomorphism. If $\mathcal{A}$ is Type I_n (respectively, I_∞, II_1, II_∞, III), what can be said about $\mathcal{B}$?

§49. Properties of projections

This section establishes some additional properties of projections and thus more about the structure of the various types of von Neumann algebras.

The next proposition, sometimes called the"Halving Lemma," says that in a certain sense a continuous projection satisfies what might intuitively be consider an inherent property of a continuous object — that it satisfies something like the intermediate value property. Recall that if (X, Ω, μ) is a measure space and $\Delta \in \Omega$ such that Δ contains no atoms of the measure μ, then for any number α with $0 \leq \alpha \leq \mu(\Delta)$ there is a measurable set $\Delta_1 \subseteq \Delta$ such that $\mu(\Delta_1) = \alpha$. In the notion of equivalence of projections there is no scalar-valued function attached to the concept, but the next proposition says that a continuous projection can always be divided into two halves. The corollary to this result shows that such a property characterizes continuous projections.

49.1 Lemma. *If E is not an abelian projection, then there are projections E_1, E_2 in $\mathcal{A}$ that are dominated by E and such that $E_1 \perp E_2$ and $E_1 \sim E_2$.*

Proof. Because E is not abelian, there is a projection $F \le E$ that does not belong to the center of $E\mathcal{A}E = \mathcal{Z}E$. It is claimed that $F\mathcal{A}(E-F) \neq (0)$. To show this it suffices to show that $F\text{Re}\,\mathcal{A}(E-F) \neq (0)$. Suppose, to the contrary, that $FA(E-F) = 0$ for every hermitian operator A in $\mathcal{A}$. Thus $FAE = FAF$. Taking adjoints gives that $EAF = FAF$. Therefore $F(EAE) = FAE = FAF = EAF = (EAE)F$. Since A was arbitrary, this implies that $F \in (E\mathcal{A}E) \cap (E\mathcal{A}E)' = \mathcal{Z}E$, contradicting the fact that F is not central in $E\mathcal{A}E$. But since $F\mathcal{A}(E-F) \neq (0)$, Proposition 47.7 implies there are projections $E_1 \le F$ and $E_2 \le E - F$ such that $E_1 \sim E_2$. ■

49.2 Proposition. *If E is continuous, then there are projections E_1, E_2 in $\mathcal{A}$ such that $E_1 \perp E_2$, $E_1 \sim E_2$, and $E = E_1 + E_2$.*

Proof. Consider the collection of all sets of pairs $\{(E_{i1}, E_{2i})\}$ such that for all i,

$$E_{1i} \sim E_{2i}\,; \quad E_{1i}, E_{2i} \le E\,; \quad \text{and } E_{ni} \perp E_{mj}$$

for $i \neq j$ and $m, n = 1, 2$. According to the lemma, this collection of pairs is non-void. By Zorn's Lemma there is a maximal family of such pairs $\{(E_{i1}, E_{2i})\}$. For $k = 1, 2$ put $E_k = \sum_i E_{ik}$. By the additivity of equivalence, $E_1 \sim E_2$. Clearly $E_1 \perp E_2$. If $E' \equiv E - (E_1 + E_2) \neq 0$, then E' is not abelian and so the lemma produces a pair of equivalent orthogonal projections E_1', E_2' under E'. This violates the maximality of $\{(E_{i1}, E_{2i})\}$. ■

49.3 Corollary. *The projection E is continuous if and only if for every $F \le E$ there are projections F_1, F_2 in $\mathcal{A}$ such that $F_1 \perp F_2$, $F_1 \sim F_2$, and $F = F_1 + F_2$.*

Proof. If E is continuous, then so is any subprojection. Hence the proof of necessity follows from the proposition. On the other hand, if E is not continuous and F is a non-zero abelian projection under E, F cannot be divided into the sum of two equivalent subprojections. ■

It is possible to have a projection that is not continuous but which dominates two equivalent orthogonal projections. For example, let $\mathcal{A} = \mathcal{B}(\mathcal{H}) \oplus \mathcal{B}(\mathcal{H})$ and take $E = 1$. Thus in the preceding corollary it is required that every subprojection of E have the halving property.

The next proposition says that an infinite projection has a property that conforms to one of the preconceived notions of the infinite. The perspicacious reader will notice that this proof considers the von Neumann–Wold decomposition of the isometry $W|E$.

49.4 Proposition. *A projection E is infinite if and only if there is a sequence $\{E_n\}$ of pairwise orthogonal subprojections of E that are pairwise equivalent.*

Proof. Assume E is infinite; so there is a projection $F \leq E$ such that $F \sim_W E$ and $F \neq E$. So $W^*W = E$ and $WW^* = F$. For each $n \geq 1$, let $F_n = W^n(F)$. Thus $F_1 \geq F_2 \geq \cdots$; put $E_n = F_n - F_{n+1}$. Clearly $E_n \leq E$ for all $n \geq 1$ and the projections $\{E_n\}$ are pairwise orthogonal. Also $E_n \sim E_{n+1}$, where the appropriate partial isometry is $W|E_n$.

Conversely, if E has such a sequence of subprojections, put $E_\infty = \sum_{n=1}^\infty E_n$. If $F = E_\infty^\perp + \sum_{n=2}^\infty E_n$, then $F \sim E$ and $F \neq E$. Hence E is infinite. ∎

Is it possible to choose the projections E_n in the preceding proposition to be equivalent to E? No. For example, let $\mathcal{A} = \mathcal{B}(\mathcal{H}) \oplus L^\infty(\mu)$ and take $E = 1$. If $\{ E_n \}$ is a sequence of pairwise orthogonal, pairwise equivalent projections in $\mathcal{A}$, then $E_n = F_n \oplus \chi_{\Delta_n}$ for a sequence of pairwise orthogonal, pairwise equivalent projections $\{ F_n \}$ in $\mathcal{B}(\mathcal{H})$ and a sequence of pairwise disjoint sets $\{ \Delta_n \}$ with $\chi_{\Delta_n} \sim \chi_{\Delta_m}$ for all n, m. But this implies that simultaneously, $\Delta_n \cap \Delta_m = \emptyset$ and $\Delta_n = \Delta_m$ for all $n \neq m$. This can only happen if $\Delta_n = \emptyset$ for all $n \geq 1$. Thus no $E_n = F_n \oplus 0$ can be equivalent to 1. The difficulty here is the presence of the finite summand $L^\infty(\mu)$.

49.5 Definition. A von Neumann algebra is *properly infinite* if it contains no non-zero central projections that are finite. A projection E in $\mathcal{A}$ is properly infinite if $\mathcal{A}_E$ is a properly infinite algebra.

A factor is properly infinite if it isn't finite. So a properly infinite algebra can have some finite projections as does $\mathcal{B}(\mathcal{H})$. $\mathcal{A}$ is properly infinite if and only if the central projections Z_1 and Z_3 in the decomposition in Theorem 48.16 are both 0. So a Type III algebra, sometimes called purely infinite, is properly infinite. Every direct summand of a properly infinite algebra is properly infinite (Exercise 1).

49.6 Lemma. *If C is a central projection that is properly infinite, then there is a non-zero central projection Z with $Z \leq C$ and a sequence $\{E_n\}$ of pairwise orthogonal projections such that $Z = \sum_n E_n$ and $E_n \sim Z$ for all $n \geq 1$.*

Proof. Because C is central, the center of $\mathcal{A}_C = \mathcal{Z}_C \subseteq \mathcal{Z}$. Thus by restricting attention to the algebra $\mathcal{A}_C$, it can be assumed that $C = 1$ and $\mathcal{A}$ is properly infinite. Since 1 is an infinite projection, Proposition 49.4 and Zorn's Lemma imply that there is a maximal family $\{ E_i : i \in I \}$ of pairwise orthogonal, pairwise equivalent projections in $\mathcal{A}$ such that the index set I is infinite. Put $E = \sum_i E_i$.

Claim. There is a central projection Z and a family of pairwise orthogonal, pairwise equivalent projections $\{ H_i : i \in I \}$ such that $Z = \sum_i H_i$ and $H_i \sim ZE_i$ for all i in I.

In fact, fix a j in I. By the maximality of $\{E_i\}$, E_j is not equivalent to a subprojection of $E^\perp$. On the other hand, the Comparison Theorem implies there is a central projection Z such that $ZE^\perp \precsim ZE_j$ and $Z^\perp E_j \precsim Z^\perp E^\perp$. Note that $Z = ZE^\perp + \sum_i ZE_i$.

Since the projections $\{ZE_i\}$ are pairwise equivalent, there are orthogonal projections F_i and G_i such that $ZE_i = F_i + G_i$ and $F_i \sim ZE^\perp$. Because I is an infinite index set, there is a bijection τ from I onto $I \cup \{0\}$. Putting $F_0 = ZE^\perp$, let $H_i = G_i + F_{\tau(i)}$. Thus $\{H_i\}$ is a family of pairwise orthogonal projections and, for each i in I, $H_i \sim G_i + ZE^\perp \sim G_i + F_i = ZE_i$. So the projections $\{H_i\}$ are pairwise equivalent. Finally, $\sum_i H_i = \sum_i G_i + \sum_i F_i + ZE^\perp = Z$. This proves the claim.

Now write I as the union of a sequence of pairwise disjoint sets $\{I_n\}$, each of which has the same cardinality as I. Put $E_n = \sum\{ H_i : i \in I_n \}$. Since I and I_n have the same cardinality, $E_n \sim Z$ for all $n \geq 1$. Clearly $Z = \sum_n E_n$. ■

49.7 Theorem. *The projection E is properly infinite if and only if there is a sequence of pairwise orthogonal subprojections $\{E_n\}$ such that $E = \sum_n E_n$ and $E_n \sim E$ for all $n \geq 1$.*

Proof. First assume that $\{ E_n \}$ is as in the statement of the theorem. Suppose Z is a central projection such that ZE is finite. Now $ZE = \sum_n ZE_n$, and $ZE_n \sim ZE$ for all $n \geq 1$. Using an elementary countability argument, for any $m \geq 1$, $\sum_{n \neq m} ZE_n \sim \sum_n ZE_n = ZE$. Because ZE is finite, it must be that $ZE_m = 0$. Since m was arbitrary, $ZE = 0$. Thus E is properly infinite.

Now assume that E is properly infinite. Replacing $\mathcal{A}$ with the algebra $\mathcal{A}_E$, it can be assumed, with no loss in generality, that 1 is a properly infinite projection. Apply Lemma 49.6 to obtain a central projection Z with the stated property. This is the beginning of an argument using Zorn's Lemma to obtain a maximal family of orthogonal central projections $\{ Z_i : i \in I \}$ such that each $Z_i = \sum_{n=1}^\infty E_{in}$, where $\{E_{in}\}_n$ is a sequence of pairwise orthogonal projections in $\mathcal{A}$ such that $E_{in} \sim Z_i$ for all $n \geq 1$ and all i. If $Z = 1 - \sum_i Z_i \neq 0$, then Z is a properly infinite central projection. Applying Lemma 49.6 again produces a contradiction to the maximality of

$\{Z_i\}$. Thus $\sum_i Z_i = 1$. Put $E_n = \sum_i E_{in}$. So $\sum_n E_n = 1$ and, by the additivity of equivalence, $E_n = \sum_i E_{in} \sim \sum_i Z_i = 1$. ■

Exercises.

1. If $\mathcal{A}$ is a properly infinite algebra, show that every non-zero central projection is properly infinite.

2. (a) Show that if $\{Z_i\}$ is a collection of central projections such that $\mathcal{A}_{Z_i}$ is properly infinite for each i and $Z = \sup_i Z_i$, then $\mathcal{A}_Z$ is properly infinite. (b) Show that there is a central projection Z such that $\mathcal{A}_Z$ is properly infinite and $Z^\perp$ dominates no properly infinite central projections. (c) If Z is as in part (b), show that $Z = Z_2 + Z_4 + Z_5$, where Z_k is as defined in Theorem 48.16.

§50. The structure of Type I algebras

In this section the structure of Type I von Neumann algebras is determined. The starting point is a concept valid for all von Neumann algebras.

50.1 Definition. A projection is *faithful* if its central cover is the identity.

In an abelian algebra, no projection other than the identity is faithful. In a factor, every non-zero projection is faithful. In particular, in $\mathcal{B}(\mathcal{H})$ every rank 1 projection is faithful; they are also abelian. This is no accident.

50.2 Proposition. *A von Neumann algebra is Type I if and only if it has a faithful abelian projection.*

Proof. One way is easy. If E is a faithful abelian projection in $\mathcal{A}$ and G is any non-zero central projection, then $EG \neq 0$. But $(EG)\mathcal{A}(EG) = (E\mathcal{A}E)G$, which is abelian. Since $EG \leq G$, this shows that $\mathcal{A}$ is Type I.

Now suppose $\mathcal{A}$ is Type I. By Zorn's Lemma there is a maximal family $\{E_i\}$ of abelian projections such that $C_{E_i} \perp C_{E_j}$ if $i \neq j$; set $Z = \sum_i C_{E_i}$. By Proposition 48.13, $\mathcal{A}Z^\perp$ is Type I. So if $Z^\perp \neq 0$, this furnishes a contradiction to the maximality of $\{E_i\}$. Thus $\sum_i C_{E_i} = 1$. If $E = \sum_i E_i$, then E is abelian by Proposition 48.7.e and $C_E = \sum_i C_{E_i} = 1$. That is, E is a faithful abelian projection. ■

Note that Proposition 48.10 implies that any two faithful abelian projections are equivalent. The next result is a rephrasing of Exercise 43.3. The proof remains the responsibility of the reader. Also see Exercise 1 below.

50.3 Lemma. *A projection E is faithful if and only if the $*$-homomorphism $A' \to A'E$ of $\mathcal{A}'$ onto $\mathcal{A}'_E$ is an isomorphism.*

The next result is one of the principal structure results of this section. Later a more detailed structure theorem for Type I algebras will be presented in Theorem 50.19.

50.4 Theorem. *The following are equivalent statements.*

(a) $\mathcal{A}$ *is a Type I von Neumann algebra.*

(b) $\mathcal{A}'$ *is a Type I von Neumann algebra.*

(c) $\mathcal{A}$ *is* $*$*-isomorphic to a von Neumann algebra with abelian commutant.*

Proof. (a) *implies* (b). First consider the case that $\mathcal{A}$ is abelian. In this case $\mathcal{A}$ is its own center. Let G be a non-zero projection in $\mathcal{A}$, and pick any non-zero vector h in G. Put $E = [\mathcal{A}h]$; so $E \neq 0$, $E \in \mathcal{A}'$, and $E \leq G$. Since $A \to AE$ is a homomorphism, $\mathcal{A}_E$ is abelian in $\mathcal{B}(E)$. But h is a cyclic vector for $\mathcal{A}_E$, so $\mathcal{A}_E$ is a maximal abelian von Neumann algebra (14.5)*. Hence $\mathcal{A}_E = \mathcal{A}'_E \approx E\mathcal{A}'E$ is abelian. Thus, E is an abelian projection in $\mathcal{A}'$. Since G was an arbitrary central projection of $\mathcal{A}'$, this establishes that $\mathcal{A}'$ is Type I.

For the general case, Proposition 50.2 implies there is a faithful abelian projection F in $\mathcal{A}$. But then the preceding lemma implies that $\mathcal{A}' \approx \mathcal{A}'_F = (\mathcal{A}_F)'$. But since $\mathcal{A}_F$ is abelian, the first case says that $\mathcal{A}'_F$ is Type I. Therefore $\mathcal{A}'$ is Type I.

(b) *implies* (c). Since $\mathcal{A}'$ is Type I, it has a faithful, abelian projection Q (50.2). Because Q is faithful, $\mathcal{A} \approx \mathcal{A}_Q$ (50.3). Also $(\mathcal{A}_Q)' = (\mathcal{A}')_Q$ is abelian.

(c) *implies* (a). If $\mathcal{A} \approx \mathcal{B}$ and $\mathcal{B}'$ is abelian, then $\mathcal{B}'$ is Type I. Since it is already known that (a) implies (b), this implies that $\mathcal{B}$ is Type I. Thus $\mathcal{A}$ is Type I. ■

The following corollary concerns the structure of algebras that are not Type I, but it is an easy consequence of the theorem.

50.5 Corollary. *If* $\mathcal{A}$ *is a continuous algebra, then* $\mathcal{A}'$ *is also continuous.*

Proof. If $\mathcal{A}'$ is not continuous, then there is a non-zero central projection Z such that $\mathcal{A}'_Z$ is Type I. Thus $\mathcal{A}_Z = (\mathcal{A}'_Z)'$ is Type I by the theorem, a contradiction. ■

*In Theorem 14.5 the underlying Hilbert space is assumed separable for measure theoretic reasons. The fact that an abelian von Neumann algebra is maximal if it has a cyclic vector remains valid when the Hilbert space is not assumed separable.

50.6 Proposition. *If $\mathcal{A}$ is abelian and Q is a faithful, abelian projection in $\mathcal{A}'$, then $\mathcal{A}$ is isomorphic to $\mathcal{A}_Q$ and $\mathcal{A}_Q$ is a maximal abelian von Neumann algebra in $\mathcal{B}(Q)$.*

Proof. Since $\mathcal{A}$ is abelian, the center of $\mathcal{A}'$ is $\mathcal{A}$. Because Q is a faithful projection in $\mathcal{A}'$, there is no projection Z in $\mathcal{A}$ such that $ZQ = 0$. Thus the map $Z \to ZQ$ from $\mathcal{A}$ onto $\mathcal{A}_Q$ is a $*$-isomorphism. To show maximality, it suffices to show that $\mathcal{A}_Q = \mathcal{A}'_Q$. But since $\mathcal{A}_Q$ is abelian, $\mathcal{A}_Q \subseteq (\mathcal{A}_Q)' = \mathcal{A}'_Q$. On the other hand, the fact that Q is an abelian projection in $\mathcal{A}'$ implies $\mathcal{A}'_Q$ is abelian. Hence $\mathcal{A}'_Q \subseteq (\mathcal{A}'_Q)' = \mathcal{A}_Q$. ∎

The existence of a projection Q such that $\mathcal{A}_Q$ is maximal abelian in $\mathcal{B}(Q)$ could be obtained by using the results of §14 if the underlying Hilbert space is separable. Indeed, if h is a separating vector for $\mathcal{A}$ and $Q = [\mathcal{A}h]$ (14.6), then $Q \in \mathcal{A}'$ and $\mathcal{A}_Q$ is maximal by Theorem 14.5. The fact that h is a separating vector for $\mathcal{A}$ implies it is cyclic for $\mathcal{A}'$. This, in turn, implies that $C_Q = 1$.

50.7 Proposition. *A projection E in $\mathcal{A}$ is minimal if and only if $E\mathcal{A}E$ is one dimensional. If E is a minimal projection, then E is abelian, C_E is a minimal projection in the center of $\mathcal{A}$, $\mathcal{A}_{C_E}$ is a factor, and $\mathcal{A}'_E = \mathcal{B}(E)$.*

Proof. If $E\mathcal{A}E$ is one dimensional, then $E\mathcal{A}E = \mathbb{C}E$, so there are no non-zero projections in $\mathcal{A}$ that are dominated by E. Conversely, if E is minimal, then the fact that the von Neumann algebra $E\mathcal{A}E$ is generated by its projections implies that it must be one dimensional.

Continue to assume that E is minimal. The fact that E is abelian is clear. If Z is a central projection with $Z \leq C_E$, then $ZE \leq E$. Thus, by minimality, either $ZE = 0$ or $ZE = E$. If $ZE = 0$, then, by definition of the central cover, $C_E \leq Z^\perp$, so that $Z = 0$. If $ZE = E$, then $Z = C_E$. Thus C_E is minimal in $\mathcal{Z}$. Also $\mathcal{A}'_E = (\mathcal{A}_E)' = (\mathbb{C}E)' = \mathcal{B}(E)$. Applying Exercise 43.3 to the algebra $\mathcal{A}_{C_E}$, it follows that $\mathcal{A}'_{C_E} \approx \mathcal{A}'_E = \mathcal{B}(E)$. Thus $\mathcal{A}'_{C_E}$, and hence $\mathcal{A}_{C_E}$, is a factor. ∎

50.8 Theorem. *$\mathcal{A}$ is a Type I factor if and only if there is a Hilbert space $\mathcal{K}$ such that $\mathcal{A}$ is $*$-isomorphic to $\mathcal{B}(\mathcal{K})$.*

Proof. If $\mathcal{A} \approx \mathcal{B}(\mathcal{K})$ for some Hilbert space $\mathcal{K}$, then $\mathcal{A}$ is clearly a Type I factor. Conversely assume that $\mathcal{A}$ is a Type I factor. By Theorem 50.4, $\mathcal{A}'$ is also a Type I factor. Let E be an abelian projection in $\mathcal{A}'$. Since $\mathcal{A}'_E$ is abelian, the fact that $\mathcal{A}'$ is a factor implies that $\mathcal{A}'_E = \mathcal{Z}_E = \mathbb{C}E$. Thus E is a minimal projection in $\mathcal{A}'$. According to the preceding proposition, $\mathcal{A}_E = \mathcal{B}(E)$. But again the fact that $\mathcal{A}$ is a factor implies that $C_E = 1$. So Exercise 43.3 implies $\mathcal{A} \approx \mathcal{A}_E = \mathcal{B}(E)$. ∎

Note that a Type I factor need not be spatially isomorphic to $\mathcal{B}(\mathcal{K})$ for some Hilbert space $\mathcal{K}$. For example if d is a cardinal number greater than $\dim \mathcal{K}$ and $\mathcal{H} = \mathcal{K}^{(d)}$, then $\mathcal{A} = \mathcal{B}(\mathcal{K})^{(d)}$ is a Type I factor that is isomorphic but not spatially isomorphic to $\mathcal{B}(\mathcal{K})$.

Now the goal is a finer classification of Type I algebras. That is, a collection of objects is sought that can be attached to each Type I algebra so that two such algebras are isomorphic (or spatially isomorphic) if and only if the objects satisfy some condition of equivalency.

50.9 Definition. A von Neumann algebra is *homogeneous* if there is a family of pairwise orthogonal abelian projections that are mutually equivalent and whose sum is the identity.

Note that a homogeneous algebra is Type I (Exercise 3) and that if $\{ E_i \}$ is a collection of pairwise orthogonal, mutually equivalent projections in $\mathcal{A}$ whose sum is 1, then each E_i is faithful independent of whether it is abelian (Exercise 4). It will be shown that every Type I algebra is the direct sum of homogeneous algebras. This will lead to complete information on the structure of Type I algebras.

To generate examples of homogeneous von Neumann algebras revisit some results from §12. Recall that Proposition 12.2.d showed that for any subset $\mathcal{S}$ of $\mathcal{B}(\mathcal{H})$ and $1 \leq n \leq \infty$, $T = [T_{ij}] \in \{ \mathcal{S}^{(n)} \}' \subseteq \mathcal{B}(\mathcal{H}^{(n)})$ if and only if $T_{ij} \in \mathcal{S}'$ for all i, j. In part (e) of that same proposition it was proved that $[\mathcal{S}^{(n)}]'' = [\mathcal{S}'']^{(n)}$. It is left to the reader to extend the notion of the n-inflation of a set of operators to the context where n is assumed to be an arbitrary cardinal number and to show that these results extend to that context.

50.10 Definition. If $\mathcal{A}$ is a von Neumann algebra and n is some cardinal number, let

$$M_n(\mathcal{A}) = \left\{ (\mathcal{A}')^{(n)} \right\}'.$$

50.11 Proposition. *If $\mathcal{A}$ is a von Neumann algebra, then $M_n(\mathcal{A})$ is a von Neumann algebra, $M_n(\mathcal{A})' = \mathcal{A}^{(n)}$, and its center is $\mathcal{Z}^{(n)}$, where $\mathcal{Z}$ is the center of $\mathcal{A}$. If $\mathcal{A}$ is a factor, so is $M_n(\mathcal{A})$. If $\mathcal{A}$ is abelian and n is finite, then $M_n(\mathcal{A})$ is a finite von Neumann algebra.*

Proof. The fact that $M_n(\mathcal{A})$ is a von Neumann algebra with the stated commutant is a consequence of the Double Commutant Theorem and the fact that $(\mathcal{A}')^{(n)}$ is a von Neumann algebra. The form of the center is apparent. This, in turn, implies that $M_n(\mathcal{A})$ is a factor whenever $\mathcal{A}$ is.

Now assume that $n < \aleph_0$ and $\mathcal{A}$ is abelian. Thus it suffices to assume that there is a compact space X with $\mathcal{A} = C(X)$. Consider the normalized trace on the matrices, $\omega\colon M_n \to \mathbb{C}$, defined by $\omega(T) = n^{-1}\mathrm{tr}\,(T)$ (18.10) and define the map $\tau : M_n(C(X)) \to C(X)$ by $\tau([f_{ij}])(x) = \omega([f_{ij}(x)]) = n^{-1}\sum_{i=1}^{n} f_{ii}(x)$. It will be shown that τ has the following properties.

50.12
$$\begin{cases} \text{(a)} & \tau \text{ is linear;} \\ \text{(b)} & \tau(1) = 1; \\ \text{(c)} & \tau \text{ is positive and if } F \in M_n(C(X)), F \geq 0, \text{ and} \\ & \tau(F) = 0, \text{ then } F = 0; \\ \text{(d)} & \tau(FG) = \tau(GF) \text{ for all } F, G \text{ in } C(X). \end{cases}$$

Parts (a), (b), and (d) follow easily from the properties of the trace. The same can be said for (c), though here are the details. If $F = [f_{ij}] \in M_n(C(X))$ and $F \geq 0$, then $[f_{ij}(x)] \geq 0$ in M_n for each x in X. Thus $\tau(F)(x) = \omega([f_{ij}(x)]) \geq 0$ for every x. Thus $\tau(F) \geq 0$. If $\tau(F) = 0$, then for each x in X, $\omega(F(x)) = 0$, so that $F(x) = 0$. Hence $F = 0$.

Now if E is a projection in $M_n(C(X))$ and $E \sim 1$, then there is a partial isometry W in $M_n(C(X))$ with $W^*W = E$ and $WW^* = 1$. Thus $\tau(E) = \tau(W^*W) = \tau(WW^*) = \tau(1) = 1$. Hence $\tau(1 - E) = 0$, But $1 - E \geq 0$, so $E = 1$ by (50.12.d). ∎

If $\mathcal{A}$ is a von Neumann algebra with center $\mathcal{Z}$, a map $\tau\colon \mathcal{A} \to \mathcal{Z}$ satisfying (50.12) is called a *faithful, center-valued trace* on $\mathcal{A}$. So the previous argument proves the following corollary, which is worth recording for future reference.

50.13 Corollary. *If $\mathcal{A}$ is a von Neumann algebra with a faithful, center-valued trace, then $\mathcal{A}$ is finite.*

The converse of this is true: a finite von Neumann algebra has a faithful, center-valued trace. The proof of this is postponed to §55.

50.14 Example. If $\mathcal{A}$ is an abelian von Neumann algebra, then $M_n(\mathcal{A})$ is homogeneous. In fact, if P_i is the projection onto the i-th coordinate of $\mathcal{H}^{(n)}$, then P_i is an abelian projection in $M_n(\mathcal{A})$ and the P_i are pairwise equivalent.

Another way to define $M_n(\mathcal{A})$ is to define the tensor product of von Neumann algebras. In this setting $M_n(\mathcal{A})$ is precisely $\mathcal{A} \otimes \mathcal{B}(\mathcal{K})$, where $\mathcal{K}$ is a Hilbert space with dimension n. In fact the Hilbert space $\mathcal{H} \otimes \mathcal{K}$ can be identified with $\mathcal{H}^{(n)}$. With this identification, $\mathcal{A}' \otimes \mathbb{C} \subseteq \mathcal{B}(\mathcal{H} \otimes \mathcal{K})$ is identified with $\mathcal{A}'^{(n)}$ and $M_n(\mathcal{A})$ with $\mathcal{A} \otimes \mathcal{B}(\mathcal{K})$. The reader can see the references for an exposition of the tensor product of von Neumann algebras.

Now to show that every homogeneous algebra has the form in Example 50.14. The starting point is a somewhat more general result that will be used later in discussing the structure of algebras that are not Type I.

50.15 Proposition. *Let $\mathcal{A}$ be a von Neumann algebra and let $\{E_i\}$ be a collection of pairwise orthogonal, mutually equivalent projections in $\mathcal{A}$ with $\sum_i E_i = 1$. If $\{E_i\}$ has cardinality n and $\mathcal{B} = \mathcal{A}_{E_1}$, then $\mathcal{A}$ is spatially isomorphic to $M_n(\mathcal{B})$.*

Proof. By assumption there is a family of partial isometries $\{W_i\}$ in $\mathcal{A}$ with $W_i^*W_i = E_1$ and $W_iW_i^* = E_i$. For all i and j and all A in $\mathcal{A}$, $W_i^*AW_j = E_1(W_i^*AW_j) = (W_i^*AW_j)E_1$, so $W_i^*AW_j \in E_1\mathcal{A}E_1$, which is identified with $\mathcal{B}$. Let $\mathcal{K} = E_1$ and define $U\colon \mathcal{K}^{(n)} \to \mathcal{H}$ by $U(h_i) = \sum_i W_ih_i$; so U is an isomorphism. Furthermore, if $A \in \mathcal{A}$, then $U[W_i^*AW_j]U^{-1} = A$. Thus $A \to U^{-1}AU$ maps $\mathcal{A}$ into $M_n(\mathcal{B})$.

To show that this map is surjective, fix an operator T in $M_n(\mathcal{B})$. If $A' \in \mathcal{A}'$, then $A'E_i = E_iA'$ for all i. Also if $B' \equiv A'|\mathcal{K}$, then $B' \in \mathcal{B}'$ and $U^{-1}A'U = B'^{(n)} \in (\mathcal{B}')^{(n)} = M_n(\mathcal{B})'$ (Verify!). Thus

$$\begin{aligned}(UTU^{-1})A' &= U[T(U^{-1}A'U)]U^{-1} = U[(U^{-1}A'U)T]U^{-1}\\ &= A'(UTU^{-1}).\end{aligned}$$

This says that $UTU^{-1} \in \mathcal{A}'' = \mathcal{A}$, so that $A \to U^{-1}AU$ is surjective. ■

50.16 Corollary. *If $\mathcal{A}$ is a homogeneous algebra, then there is a cardinal number n and an abelian algebra $\mathcal{B}$ that is isomorphic to the center of $\mathcal{A}$ such that $\mathcal{A}$ is spatially isomorphic to $M_n(\mathcal{B})$. If $\mathcal{Z}$ is the center of $\mathcal{A}$, then $\mathcal{A}$ is $*$-isomorphic to $M_n(\mathcal{Z})$.*

Proof. By definition there is a family $\{E_i\}$ of pairwise orthogonal abelian projections that are mutually equivalent and satisfy $\sum_i E_i = 1$. If $\mathcal{B} = \mathcal{A}_{E_1}$, the preceding proposition says that $\mathcal{A}$ is spatially isomorphic to $M_n(\mathcal{B})$. $\mathcal{B}$ is abelian since E_1 is an abelian projection. So $\mathcal{Z}$ is spatially isomorphic to $\mathcal{B}^{(n)}$, which is isomorphic to $\mathcal{B}$. An elementary argument will now show that $\mathcal{A}$ and $M_n(\mathcal{Z})$ are $*$-isomorphic. ■

50.17 Example. Let $\mathcal{H}$ be any Hilbert space and n any cardinal number. Consider the algebra $\mathcal{A} = \mathcal{B}(\mathcal{H})^{(n)}$. This algebra is homogeneous. Indeed, let $\{E_i\}$ be a collection of pairwise orthogonal, rank 1 projections in $\mathcal{B}(\mathcal{H})$ with $\sum_i E_i = 1$. Then $\{E_i^{(n)}\}$ is a collection of pairwise orthogonal, mutually equivalent abelian projections in $\mathcal{A}$ with sum equal to the identity. If this collection is used and Proposition 50.15 applied, it follows that $\mathcal{A}$ is spatially isomorphic to $M_n(\mathbb{C})$, which is clear without the proposition.

50.18 Lemma. *If $\mathcal{A}$ is Type I and E is a non-zero projection in $\mathcal{A}$, then there is an abelian projection P with $P \leq E$ and $C_P = C_E$.*

Proof. Since $\mathcal{A}$ is Type I, for any non-zero projection F, $\mathcal{A}_F$ is also Type I (48.13). Thus if Z is any central projection in $\mathcal{A}$ and $ZF \neq 0$, there is a non-zero abelian projection that is dominated by ZF.

Now fix the non-zero projection E and let $\{ P_i \}$ be a maximal family of abelian projections in $\mathcal{A}$ such that for all $i \neq j$, $P_i \leq E$ and $C_{P_i} \perp C_{P_j}$. Put $P = \sum_i P_i$. By (48.7.e) P is abelian. Clearly $P \leq E$ and so $C_P \leq C_E$. On the other hand, if $C_P \neq C_E$, then $E(C_E - C_P) \neq 0$, and, by the first paragraph of this proof, there is a non-zero abelian projection contained in $E(C_E - C_P)$. This contradicts the maximality of $\{ P_i \}$. Thus $C_E = C_P$. ■

50.19 Theorem. *If $\mathcal{A}$ is Type I, then there is a family of pairwise orthogonal central projections $\{ Z_i \}$ such that $\sum_i Z_i = 1$ and each $\mathcal{A}_{Z_i}$ is homogeneous. Consequently, if $\mathcal{A}$ is Type I, then there is a collection of abelian algebras $\{ \mathcal{B}_i \}$ and a collection of distinct cardinal numbers $\{ n_i \}$ such that $\mathcal{A}$ is spatially isomorphic to*

50.20 $$\bigoplus_i M_{n_i}(\mathcal{B}_i) .$$

Proof. First establish that any Type I algebra contains a non-zero central projection Z such that $\mathcal{A}_Z$ is homogeneous. Indeed, if $\mathcal{A}$ is Type I, there is a maximal family $\{ E_i \}$ of pairwise orthogonal, faithful, abelian projections. By Proposition 48.10, $E_i \sim E_j$ for all i, j. Put $E = \sum_i E_i$. If $E = 1$, then $\mathcal{A}$ is homogeneous. Assume $E^\perp \neq 0$. By maximality, $E^\perp$ contains no non-zero faithful, abelian projection. Now the preceding lemma implies there is an abelian projection P contained in $E^\perp$ with $C_P = C_{E^\perp}$. So the maximality of $\{ E_i \}$ implies $C_{E^\perp} \neq 1$. Put $Z = 1 - C_{E^\perp}$. Hence $ZE^\perp = 0$ and so $Z = ZE = \sum_i ZE_i$. Also $ZE_i \leq E_i$ so that each ZE_i is abelian. Thus $\{ ZE_i \}$ is a collection of pairwise orthogonal, mutually equivalent abelian projections with $\sum_i ZE_i = Z$. Therefore $\mathcal{A}_Z$ is homogeneous.

Apply Zorn's Lemma to find a maximal family $\{ Z_i \}$ of pairwise orthogonal central projections such that each $\mathcal{A}_{Z_i}$ is homogeneous. Put $Z = \sum_i Z_i$. If $Z \neq 1$, then $\mathcal{A}Z^\perp$ is Type I and by the first part of this proof there is a non-zero central projection G in $\mathcal{A}_{Z^\perp}$ such that $(\mathcal{A}_{Z^\perp})_G$ is homogeneous. But G is also a central projection in $\mathcal{A}$ and so this contradicts the maximality of $\{ Z_i \}$. Thus $\sum_i Z_i = 1$.

To obtain (50.20) apply Theorem 50.16 to the summands $\mathcal{A}_{Z_i}$. If some of the cardinal numbers $\{ n_i \}$ are equal, then apply Exercise 5 to collapse these summands into one. ■

At this point there is one more crucial step remaining in the discussion of Type I algebras: to give a complete set of spatially isomorphic invariants for the algebras. This is done in two stages and will be taken up in the next section when the underlying Hilbert space is separable. The non-separable case, in its full glory, is technically more complex and would violate the nature of an introduction.

Exercises.

1. If E is a projection in $\mathcal{A}$, show that the following statements are equivalent. (a) E is faithful. (b) There is no central projection G such that $EG = 0$. (c) The $*$-homomorphism $Z \to ZE$ is injective on $\mathcal{Z}$.

2. Say that a von Neumann algebra $\mathcal{A}$ is *semi-finite* if it has no Type III summand. That is, there is no central projection G such that $\mathcal{A}G$ is Type III. Show that $\mathcal{A}$ is semi-finite if and only if it has a faithful finite projection.

3. Show that a homogeneous algebra is Type I.

4. If $\{ E_i \}$ is a collection of pairwise orthogonal, mutually equivalent projections in $\mathcal{A}$ whose sum is 1, show that each E_i is faithful.

5. If n is some cardinal number and $\{ \mathcal{B}_i \}$ is a collection of abelian algebras, show that $\bigoplus_i M_n(\mathcal{B}_i) = M_n(\bigoplus \mathcal{B}_i)$.

6. If $\mathcal{A}$ is Type I algebra with $\mathcal{A} = \bigoplus_i M_{n_i}(\mathcal{B}_i)$ as in (50.19), show that for any cardinal number n, $\mathcal{A}^{(n)}$ is Type I and compute the (50.19) type decomposition for $\mathcal{A}^{(n)}$.

§51. The classification of Type I algebras

The material in this section can be approached in the non-separable case, but this would require technicalities that seem best avoided in an introduction. Therefore the following overriding assumption is now made and will remain in force for the remainder of the book.

All Hilbert spaces are now assumed to be separable.

Begin with the proof of the existence of a well defined cardinal number associated with each homogeneous algebra.

51.1 Proposition. *If $\mathcal{A}$ is a homogeneous von Neumann algebra and $\{ E_i : i \in I \}$ and $\{ F_k : k \in K \}$ are two collections of pairwise orthogonal, mutually equivalent abelian projections in $\mathcal{A}$ with $\sum_i E_i = \sum_k F_k = 1$, then I and K have the same cardinality.*

Proof. If $\operatorname{card} I = n < \infty$ and $\mathcal{B} = \mathcal{A}_{E_1}$, then $\mathcal{A} \cong M_n(\mathcal{B})$ and so, by Proposition 50.12, $\mathcal{A}$ is finite. If K is infinite, then shifting the projections shows that $1 = \sum_{k=1}^{\infty} F_k \sim \sum_{k=2}^{\infty} F_k$ in $\mathcal{A}$. Since $\mathcal{A}$ is finite and $F_1 \neq 0$, this is impossible. Therefore K is also finite.

Thus either both I and K are finite or both are infinite. If they are both infinite, then they have the same cardinality, $\aleph_0$. Assume they are both finite: say $n = \operatorname{card} I$ and $m = \operatorname{card} K$. Note that $C_{E_i} = 1$ for each $i = 1, \ldots, n$ (Exercise 50.4). Similarly, $C_{F_k} = 1$ for each k. Therefore Proposition 48.10 implies $E_i \sim F_k$ for all i and k. So if $n < m$, $1 = \sum_{i=1}^{n} E_i \sim \sum_{k=1}^{n} F_k < 1$, an impossibility in the finite algebra $\mathcal{A}$. Hence $n = m$. ■

When $\mathcal{A}$ is homogeneous, the *multiplicity* of $\mathcal{A}$ is the unique cardinal number n such that $\mathcal{A}$ is spatially isomorphic to $M_n(\mathcal{B})$ for an abelian algebra. Interpreting Theorem 50.19 with this language, every Type I von Neumann algebra $\mathcal{A}$ is spatially isomorphic to $\mathcal{A}_\infty \oplus \mathcal{A}_1 \oplus \mathcal{A}_2 \oplus \cdots$, where $\mathcal{A}_n$ is homogeneous with multiplicity n, $1 \leq n \leq \infty$.

51.2 Theorem. *Let $\mathcal{A}$ and $\mathcal{B}$ be two Type I von Neumann algebras acting on separable Hilbert spaces, and let $\{ A_n : 1 \leq n \leq \infty \}$ and $\{ B_n : 1 \leq n \leq \infty \}$ be pairwise orthogonal central projections in these algebras whose sum is 1 and such that $\mathcal{A}_{A_n}$ and $\mathcal{B}_{B_n}$ are homogeneous with multiplicity n. The algebras $\mathcal{A}$ and $\mathcal{B}$ are spatially isomorphic* (respectively, $*$-*isomorphic*) *if and only if $\mathcal{A}_{A_n} \cong \mathcal{B}_{B_n}$* (respectively, $\mathcal{A}_{A_n} \approx \mathcal{B}_{B_n}$) *for $1 \leq n \leq \infty$.*

Proof. If the condition holds, then clearly $\mathcal{A}$ and $\mathcal{B}$ are isomorphic. Now let $\rho \colon \mathcal{A} \to \mathcal{B}$ be a $*$-isomorphism that is implemented by a unitary. (The case where ρ is only assumed to be a $*$-isomorphism is identical.) So $\rho(A_n)$ is a central projection in $\mathcal{B}$ and $\mathcal{A}_{A_n} \cong \mathcal{B}_{\rho(A_n)}$. In particular, $\mathcal{B}_{\rho(A_n)}$ is homogeneous with multiplicity n. By Exercise 1, for $1 \leq m \leq \infty$, $\mathcal{B}_{B_m \rho(A_n)} = (\mathcal{B}_{\rho(A_n)})_{B_m}$ has multiplicity n and $\mathcal{B}_{B_m \rho(A_n)} = (\mathcal{B}_{B_m})_{\rho(A_n)}$ has multiplicity m. Since the multiplicity is unique, this implies that $\rho(A_n) \leq B_m^{\perp}$ for $m \neq n$ and so $\rho(A_n) \leq B_n$. Applying the same type of reasoning to ρ^{-1} shows that $\rho^{-1}(B_n) \leq A_n$ and so $\rho(A_n) = B_n$ for $1 \leq n \leq \infty$. Therefore $\mathcal{A}_{A_n} \cong \mathcal{B}_{B_n}$. ■

51.3 Corollary. *If $\mathcal{A}$ and $\mathcal{B}$ are abelian von Neumann algebras and $1 \leq n, m \leq \infty$, then $M_n(\mathcal{A})$ and $M_m(\mathcal{B})$ are spatially isomorphic* (respectively, $*$-*isomorphic*) *if and only if $n = m$ and $\mathcal{A}$ and $\mathcal{B}$ are spatially isomorphic* (respectively, $*$-*isomorphic*).

The preceding theorem furnishes a complete set of isomorphism invariants for Type I algebras acting on a separable Hilbert space. Namely, a sequence of cardinal numbers $\{ n_k \}$ between 1 and $\aleph_0$ and a sequence of abelian von Neumann algebras $\{ Z_k \}$ is associated with each Type I algebra. The isomorphism class of the algebra is completely determined by these two sequences. The effect of this is to reduce the study of Type I algebras to the study of the abelian ones. A close examination of the abelian algebras begins with the next theorem.

51.4 Theorem. *If $\mathcal{A}$ is an abelian von Neumann algebra on a separable Hilbert space, then there are compact metric spaces $\{X_\infty, X_1, \dots\}$ (some of which may be void) and measures $\{\mu_\infty, \mu_1, \dots\}$ supported on these metric spaces such that the following hold.*

(a) $\mathcal{A} \cong L^\infty(\mu_\infty)^{(\infty)} \oplus L^\infty(\mu_1) \oplus L^\infty(\mu)^{(2)} \oplus \dots$ *acting on* $L^2(\mu_\infty)^{(\infty)} \oplus L^2(\mu_1) \oplus L^2(\mu)^{(2)} \oplus \dots$;

(b) $\mathcal{A}' \cong M_\infty(L^\infty(\mu_\infty)) \oplus L^\infty(\mu_1) \oplus M_2(L^\infty(\mu_2)) \oplus \cdots$

Proof. Since $\mathcal{A}$ is abelian, its commutant is Type I. Therefore there are central projections $Z_\infty, Z_1, Z_2 \dots$ such that $\mathcal{A}'Z_n$ is homogeneous of multiplicity n. Combining Proposition 50.6 with Corollary 50.16, it follows that $\mathcal{A}'Z_n \cong M_n(\mathcal{Z}_n)$, where $\mathcal{Z}_n$ is a maximal abelian von Neumann algebra. By Theorem 14.5, There is a compact metric space X_n and a measure μ_n on X_n such that $\mathcal{Z}_n \cong L^\infty(\mu_n)$ acting on $L^2(\mu_n)$. Therefore

$$\begin{aligned}\mathcal{A}' &= \mathcal{A}'Z_\infty \oplus \mathcal{A}'Z_1 \oplus \mathcal{A}'Z_2 \oplus \cdots \\ &\cong M_\infty(L^\infty(\mu_\infty)) \oplus L^\infty(\mu_1) \oplus M_2(L^\infty(\mu_2)) \oplus \cdots\end{aligned}$$

Taking the commutant gives part (a). ■

51.5 Corollary. *If $\mathcal{A}$ is a Type I von Neumann algebra acting on a separable space, then there are compact metric spaces $\{X_\infty, X_1, \dots\}$ (some of which may be void) and measures $\{\mu_\infty, \mu_1, \dots\}$ supported on these metric spaces such that $\mathcal{A}$ is $*$-isomorphic to*

$$M_\infty(L^\infty(\mu_\infty)) \oplus L^\infty(\mu_1) \oplus M_2(L^\infty(\mu_2)) \oplus \cdots$$

Proof. By Theorem 50.4, $\mathcal{A}$ is $*$-isomorphic to an algebra with abelian commutant. Now apply the preceding theorem to that abelian commutant. ■

The uniqueness result for the representation in Theorem 51.4 can now be stated. The proof is little more than a piecing together of previous results: a "collage" rather than a proof.

51.6 Theorem. *Let $\mathcal{A}$ be an abelian von Neumann algebra acting on a separable Hilbert space and let $\{X_n, \mu_n\}$ be as in* Theorem 51.4. *If $\{Y_n : 1 \le n \le \infty\}$ is a collection of compact metric spaces and $\{\nu_n\}$ a collection* *of measures supported on the metric spaces* Y_n *such that* $\mathcal{A} \cong L^\infty(\nu_\infty)^{(\infty)} \oplus L^\infty(\nu_1) \oplus L^\infty(\nu_2)^{(2)} \oplus \ldots$ *and* $\mathcal{A}' \cong M_\infty(L^\infty(\nu_\infty)) \oplus L^\infty(\nu_1) \oplus M_2(L^\infty(\nu_2)) \oplus \cdots$, *then for* $1 \le n \le \infty$, $L^\infty(\mu_n)$ *acting on* $L^2(\mu_n)$ *is spatially isomorphic to* $L^\infty(\nu_n)$ *acting on* $L^2(\nu_n)$.

Proof. Let Z_n be the central projection in $\mathcal{A}$ with $\mathcal{A}'_{Z_n} \cong M_n(L^\infty(\mu_n)^{(n)})$. If Z is any central projection in $\mathcal{A}$ such that $\mathcal{A}'_Z$ is homogeneous with multiplicity n, then $Z \le Z_n$ (Exercise 1). Thus it is also true that $\mathcal{A}'_{Z_n} \cong M_n(L^\infty(\nu_n))^{(n)}$ and it follows that $L^\infty(\mu_n) \cong L^\infty(\nu_n)$ by Corollary 51.5. ∎

For measures μ and ν, what are necessary and sufficient conditions for $L^\infty(\mu)$ acting on $L^2(\mu)$ to be spatially equivalent to $L^\infty(\nu)$ acting on $L^2(\nu)$? It turns out that this solely depends on whether μ and ν simultaneously have a purely non-atomic restriction and whether they have the same number of atoms. In fact, if $L^2(\mu)$ is separable and μ has no atoms, then there are Borel sets Δ_1 in X and Δ_2 in $[0,1]$ and a bijection $\tau : \Delta_1 \to \Delta_2$ such that: (a) $\mu(X \setminus \Delta_1) = 0 = m([0,1] \setminus \Delta_2)$; (b) For a subset Δ of Δ_1, Δ is μ-measurable if and only if $\tau(\Delta)$ is Lebesgue measurable; (c) If Δ is a Borel set contained in Δ_1, $\mu(\Delta) = 0$ if and only if $m(\tau(\Delta)) = 0$. (m is Lebesgue measure on $[0,1]$.)

This can be found in Halmos and von Neumann [1942]. (Also see Billingsley [1965].)

Define a measure η on $[0,1]$ by $\eta(\Lambda) = \mu(\tau^{-1}(\Lambda))$ for all Borel sets Λ contained in the interval. Two things emerge from the conditions on τ. First, if $W : L^2(\eta) \to L^2(\mu)$ is defined by $Wf = f \circ \tau$, then U is an isomorphism of these Hilbert spaces and for each ϕ in $L^\infty(\eta)$, $WM_\phi W^{-1} = M_{\phi\circ\tau}$. Second, the measures η and m are mutually absolutely continuous. Thus if $\psi = \sqrt{dm/d\eta}$ and $V : L^2([0,1]) \to L^2(\eta)$ is defined by $Vh = \psi h$, then V is a Hilbert space isomorphism. Thus $U = VW : L^2([0,1]) \to L^2(\mu)$ is an isomorphism and for each ϕ in $L^\infty([0,1])$, $UM_\phi U^{-1} = M_{\phi\circ\tau}$. Thus $L^\infty(\mu)$ acting on $L^2(\mu)$ is spatially isomorphic to $L^\infty([0,1])$ acting on $L^2([0,1])$. Putting this argument together with some additional material produces the following.

51.7 Theorem. *For any measure* μ *with* $L^2(\mu)$ *separable, let* $\kappa(\mu) = (\epsilon, n)$, *where* $\epsilon = 0$ *if* μ *is purely atomic and* $\epsilon = 1$ *otherwise and where* n *is the number of atoms of* μ, $1 \le n \le \infty$. *For any two such measure* μ *and* ν, $L^\infty(\mu)$ *acting on* $L^2(\mu)$ *is spatially isomorphic to* $L^\infty(\nu)$ *acting on* $L^2(\nu)$ *if and only if* $\kappa(\mu) = \kappa(\nu)$.

The next section develops the nature of the algebra $M_n(L^\infty(\mu))$ more thoroughly. This especially needs to be done when $n = \infty$.

This section concludes with a small detour to apply the results of this section to the particular case when $\mathcal{A}$ is generated by a single normal operator. This result appears in [ACFA] as Theorem IX.10.16. Recall the definition of the prototypical star-cyclic normal operator N_μ for a compact supported measure μ in the plane (2.5).

51.8 Theorem. *If N is a normal operator acting on a separable Hilbert space, then there are mutually singular, compactly supported measures μ_∞, $\mu_1, \mu_2, \ldots$ on the plane such that*

51.9 $$N \cong N_{\mu_\infty}^{(\infty)} \oplus N_{\mu_1} \oplus N_{\mu_2}^{(2)} \oplus \cdots .$$

If M is another normal operator with corresponding measures $\nu_\infty, \nu_1, \nu_2, \ldots$, then $M \cong N$ if and only if $[\nu_n] = [\mu_n]$ for $1 \leq n \leq \infty$.

Proof. Apply Theorem 51.4 to the abelian von Neumann algebra $\mathcal{A} = W^*(N)$. Use the notation from that theorem and let Z_n be the central projection such that $\mathcal{A}'_{Z_n} \cong M_n(L^\infty(\mu_n))$ and $\mathcal{A}_{Z_n} \cong L^\infty(\mu_n)^{(n)}$. The fact that $N|Z_n$ is the generator of $\mathcal{A}_{Z_n}$ implies that the metric space X_n can be taken to be a subset of the plane, and, under this spatial isomorphism, the operator $N|Z_n$ corresponds to the element $z^{(n)}$ in $L^\infty(\mu_n)^{(n)}$. Once these identifications are made, (51.9) quickly follows.

The fact that the measures $\{\mu_n\}$ are mutually singular follows from the fact that $W^*(N)' = M_\infty(L^\infty(\mu_\infty)) \oplus L^\infty(\mu_1) \oplus M_2(L^\infty(\mu_2)) \oplus \cdots$ and Exercise 15.6.

Now let M and $\{\nu_n\}$ be as described in the statement of the theorem. Applying Theorem 51.6 to the present situation shows that $L^\infty(\mu_n) \cong L^\infty(\nu_n)$. Using the particular situation at hand, $N_{\mu_n} \cong N_{\nu_n}$. According to Theorem 11.4, $[\mu_n] = [\nu_n]$. The converse is also a consequence of Theorem 11.4. ∎

Exercises.

1. If $\mathcal{A}$ is a homogeneous von Neumann algebra with multiplicity n and Z is a central projection, show that $\mathcal{A}_Z$ is a homogeneous algebra with multiplicity n.

2. If $\mathcal{A}$ is Type I and $2 \leq n \leq \infty$, show that $\mathcal{A}^{(n)}$ is Type I. If $\mathcal{A} = \bigoplus_{1 \leq k \leq \infty} \mathcal{A}_k$ is the decomposition of $\mathcal{A}$ into the direct sum of homogeneous algebras of multiplicity k, find the corresponding direct sum decomposition of $\mathcal{A}^{(n)}$.

3. Prove that (51.1) and (51.2) remain true in the non-separable case. Let N and M be two sets of distinct cardinal numbers and for each n in N and m in M, let $\mathcal{A}_n$ and $\mathcal{B}_m$ be abelian algebras. Show that $\bigoplus_n M_n(\mathcal{A}_n)$

and $\bigoplus_m M_m(\mathcal{B}_m)$ are spatially isomorphic if and only if $N = M$ and $\mathcal{A}_n$ and $\mathcal{B}_m$ are spatially isomorphic for all n in N.

4. If $\mathcal{A}$ is an abelian von Neumann algebra and $2 \leq n \leq \infty$, find the decomposition obtained in Theorem 51.4 for the algebra $\mathcal{A}^{(n)}$.

5. If $\mathcal{A}$ is a Type I von Neumann algebra, give a necessary and sufficient condition that $\mathcal{A} \cong \mathcal{A}^{(\infty)}$.

6. Use Theorem 51.8 to show that if N is a normal operator acting on a separable Hilbert space, then there is a sequence of compactly supported measures $\{\mu_n\}$ on $\mathbb{C}$ such that $\mu_{n+1} << \mu_n$ for all n and $N \cong N_{\mu_1} \oplus N_{\mu_2} \oplus \cdots$. If $\{\nu_n\}$ is another such sequence of measures with $\nu_{n+1} << \nu_n$ for all n, then $N \cong N_{\nu_1} \oplus N_{\nu_2} \oplus \cdots$ if and only if $[\mu_n] = [\nu_n]$ for all n. (This is Theorem IX.10.1 in [ACFA]. The interested reader could work the exercises at the end of §IX.10 there.)

7. Use Theorem 51.8 to show that if N is a normal operator acting on a separable Hilbert space, then there is a compactly supported measure μ on the plane and a measurable function $m: \mathbb{C} \to \mathbb{N} \cup \{0, \infty\}$ such that $m = 0$ off the support of μ and the following hold: (a) if $\Delta_n = m^{-1}(n)$ for $1 \leq n \leq \infty$ and $\mu_n \equiv \mu|\Delta_n$, then

$$N \cong N_{\mu_\infty}^{(\infty)} \oplus N_{\mu_1} \oplus N_{\mu_2}^{(2)} \oplus \cdots .$$

The measure μ is, of course, the scalar-valued spectral measure for N and m is called the *multiplicity function* for N. Show that two normal operators are unitarily equivalent if and only if their scalar-valued spectral measures are equivalent and their multiplicity functions agree a.e.

8. If N is a normal operator and $2 \leq n \leq \infty$, find the scalar-valued spectral measure and multiplicity function for $N^{(n)}$ (in terms of the same objects for N). Show that for any normal operator N, the multiplicity function for $N^{(\infty)}$ is identically ∞ on the spectrum of N and 0 elsewhere.

9. Suppose that N is a normal operator with scalar-valued spectral measure μ and multiplicity function m. Prove that the following statements are equivalent. (a) $m \equiv \infty$ on support μ. (b) $N \cong N^{(\infty)}$. (c) $N \cong N_\mu^\infty$.

§52. Operator-valued measurable functions

Remember that all Hilbert spaces considered here are assumed to be separable.

52.1 Definition. If (X, Ω) is a measurable space, a function $f : X \to \mathcal{H}$ is *measurable* if for every vector h in $\mathcal{H}$ the function $x \to \langle F(x), h\rangle$ is measurable.

The functions defined here could be called *weakly measurable* and a function $f\colon X \to \mathcal{H}$ could be termed *norm measurable* if $f^{-1}(U) \in \Omega$ for every norm open set U in $\mathcal{H}$. In fact the two concepts are equivalent as part (c) of the next result will show.

52.2 Proposition. *Fix a measurable space* (X, Ω).

(a) *The set of all measurable functions from* (X, Ω) *into* $\mathcal{H}$ *is a vector space if the operations are defined pointwise.*

(b) *If* $\{ e_n \}$ *is any orthonormal basis for* $\mathcal{H}$, *a function* $f \to \mathcal{H}$ *is measurable if and only if* $x \to \langle f(x), e_n \rangle$ *is measurable for all* n.

(c) *If* f *and* g *are measurable functions from* X *into* $\mathcal{H}$, *then the function* $x \to \langle f(x), g(x) \rangle$ *is a measurable scalar-valued function. In particular,* $x \to \|f(x)\|$ *is measurable.*

Proof. The proof of (a) follows from the same fact about scalar-valued functions. If f is measurable and $\{ e_n \}$ is a basis, then each $x \to \langle f(x), e_n \rangle$ is measurable by definition. Conversely, if each $x \to \langle f(x), e_n \rangle$ is measurable and $h \in \mathcal{H}$, then $\langle f(x), h \rangle$ is the convergent sum of the measurable functions, $\sum_n \langle f(x), e_n \rangle \langle e_n, h \rangle$. Hence (b) follows.

(c) If $\{ e_n \}$ is a basis for $\mathcal{H}$, then

$$\langle f(x), g(x) \rangle = \sum_n \langle f(x), e_n \rangle \langle e_n, g(x) \rangle.$$

Each of the functions $\langle f(\cdot), e_n \rangle$ and $\langle e_n, g(\cdot) \rangle$ are measurable by definition, so their product, and hence sum, is also measurable. Since $\|f(x)\| = \sqrt{\langle f(x), f(x) \rangle}$, $\|f(\cdot)\|$ is measurable. ■

So part (c) of this proposition says that a weakly measurable function is norm-measurable. If it is assumed that f is norm-measurable, then it is clear that it must be weakly measurable since $\langle f(\cdot), h \rangle$ is then the composition of a measurable function and a continuous one. In light of this proposition, the next definition is justified.

52.3 Definition. If (X, Ω, μ) is a measure space and $p < \infty$, $L^p(\mu; \mathcal{H}) = \{ f : X \to \mathcal{H} : f$ is measurable and $\int \|f(x)\|^p \, d\mu(x) < \infty \}$.

One defines the norm in the usual way and proves that $L^p(\mu; \mathcal{H})$ is a Banach space. The main interest here is when $p = 2$ and in this case $L^2(\mu; \mathcal{H})$ is a Hilbert space. In fact, the inner product is given by

$$\langle f, g \rangle = \int \langle f(x), g(x) \rangle \, dm(x)$$

for functions f and g in $L^2(\mu; \mathcal{H})$. The reader will be asked to prove the needed measure and integration theory connected with these spaces. Now for another collection of measurable functions.

52.4 Definition. If (X, Ω) is a measurable space, a function $\Phi : X \to \mathcal{B}(\mathcal{H})$ is *measurable* if for all vectors h and g in $\mathcal{H}$, the function $x \to \langle \Phi(x)h, g\rangle$ is measurable.

52.5 Proposition. *Fix a measurable space* (X, Ω).

(a) *A function* $\Phi : X \to \mathcal{B}(\mathcal{H})$ *is measurable if and only if for every vector* h *in* $\mathcal{H}$ *the function* $x \to \Phi(x)h$ *is a measurable function from* X *in* $\mathcal{H}$.

(b) *If* $\Phi : X \to \mathcal{B}(\mathcal{H})$ *is measurable, then the function* $x \to \Phi(x)^*$ *is measurable.*

(c) *If* $\Phi : X \to \mathcal{B}(\mathcal{H})$ *and* $f: X \to \mathcal{H}$ *are measurable functions, then* $x \to \Phi(x)f(x)$ *is a measurable function from* X *into* $\mathcal{H}$.

(d) *The set of all measurable functions from* (X, Ω) *into* $\mathcal{B}(\mathcal{H})$ *is an algebra if the operations are defined pointwise.*

(e) *If* $\Phi : X \to \mathcal{B}(\mathcal{H})$ *is a measurable function, then* $x \to \|\Phi(x)\|$ *is measurable.*

Proof. Part (a) is clear from the definition, and part (b) is an easy exercise. To prove (c), let $h \in \mathcal{H}$ and note that

$$\langle \Phi(x)f(x), h\rangle = \langle f(x), \Phi(x)^*h\rangle.$$

By part (b) and (52.2.c), this proves part (c).

For part (d), the fact that the set of measurable operator-valued functions is a vector space is straightforward. To prove that products are measurable, let Φ and Ψ be two such functions. If $\{ e_n \}$ is a basis for $\mathcal{H}$ and $h, g \in \mathcal{H}$, then

$$\langle \Phi(x)\Psi(x)h, g\rangle = \sum_n \langle \Psi(x)h, e_n\rangle \langle e_n, \Phi(x)^*g\rangle.$$

Since each of the functions $\langle \Psi(\cdot)h, e_n\rangle$ and $\langle e_n, \Phi(\cdot)^*g\rangle$ are measurable, so is their product. Hence the sum of these products is measurable and (d) follows.

Finally, if $\{ h_n \}$ is a countable dense subset of ball $\mathcal{H}$, then $\|\Phi(x)\| = \sup\{ \|\Phi(x)h_n\| : n \geq 1 \}$. So $\|\Phi(\cdot)\|$ is the sup of a countable number of measurable functions. ■

52.6 Definition. If (X, Ω, μ) is a measure space, then

$$L^\infty(\mu; \mathcal{B}(\mathcal{H})) \equiv \{\Phi : X \to \mathcal{B}(\mathcal{H}) : \Phi \text{ is a measurable function and } \Phi \text{ is bounded off a set of } \mu \text{ measure } 0\}.$$

Functions that agree a.e. are identified and for each Φ in $L^\infty(\mu; \mathcal{H})$ the norm is defined as the μ-essential supremum of the function $\|\Phi(\cdot)\|$. That is, $\|\Phi\|_\infty \equiv \| \, \|\Phi(\cdot)\| \, \|_\infty$.

52.7 Theorem. *Let (X, Ω, μ) be a σ-finite measure space.*

(a) *If for each Φ in $\Phi \in L^\infty(\mu; \mathcal{B}(\mathcal{H}))$*

$$\Phi^*(x) \equiv \Phi(x)^*,$$

then $L^\infty(\mu; \mathcal{B}(\mathcal{H}))$ is a C^*-algebra.

(b) *If for each f in $L^2(\mu; \mathcal{H})$*

$$(M_\Phi f)(x) \equiv \Phi(x) f(x),$$

then M_Φ is a bounded operator on $L^2(\mu; \mathcal{H})$. The map $\Phi \to M_\Phi$ is a $*$-isomorphism from $L^\infty(\mu; \mathcal{B}(\mathcal{H}))$ into $\mathcal{B}(L^2(\mu; \mathcal{H}))$.

Proof. The proof of (a) is left to the reader. To prove (b) apply standard measure theory to find a set Δ in Ω with $\mu(\Delta) = 0$ such that $\|\Phi\|_\infty = \sup\{ \|\Phi(x)\| : x \notin \Delta \}$. Therefore if $x \notin \Delta$ and $f \in L^2(\mu; \mathcal{H})$, then $\|\Phi(x) f(x)\| \leq \|\Phi\|_\infty \|f(x)\|$. Thus

$$\|M_\Phi f\|^2 = \int_{X \setminus \Delta} \|\Phi(x) f(x)\|^2 \, d\mu(x) \leq \|\Phi\|_\infty^2 \, \|f\|^2.$$

So $M_\Phi \in \mathcal{B}(L^2(\mu; \mathcal{H}))$ and $\|M_\Phi\| \leq \|\Phi\|_\infty$. The fact that the map $\Phi \to M_\Phi$ is linear, multiplicative, and star preserving can be checked by the reader.

It remains to show that the map is an isometry. Since C*-algebras are involved, it suffices to show that the map is injective. Suppose $\Phi \in L^\infty(\mu; \mathcal{B}(\mathcal{H}))$ such that $M_\Phi f = \Phi f = 0$ for all f in $L^2(\mu; \mathcal{H})$. If $f \in L^2(\mu)$ and $h \in \mathcal{H}$, define $f \otimes h$ to be the function in $L^2(\mu; \mathcal{H})$ with $(f \otimes h)(x) = f(x) h$ for all x in X. Thus $M_\Phi(f \otimes h) = 0$. Thus for almost all x, $0 = \Phi(x)[f(x)h] = f(x)\Phi(x)h$. This holds for every f in $L^2(\mu)$ so clearly this implies that for each h in $\mathcal{H}$, $\Phi(x)h = 0$ a.e. Let $\{ h_n \}$ be a countable dense subset of $\mathcal{H}$. So for each n there is a measurable set Δ_n with $\mu(\Delta_n) = 0$ such that $\Phi(x)$ is a bounded operator and $\Phi(x) h_n = 0$ if $x \notin \Delta_n$. Letting $\Delta = \bigcup_n \Delta_n$, it follows that $\mu(\Delta) = 0$. If $x \notin \Delta$, then $\Phi(x)$ is a bounded operator and $\Phi(x)h_n = 0$ for all n. By density, $\Phi(x) = 0$ for all x not belonging to Δ. That is, $\Phi = 0$. ∎

Note that it is possible to canonically embed $L^\infty(\mu)$ into $L^\infty(\mu; \mathcal{B}(\mathcal{H}))$ by letting each ϕ in $L^\infty(\mu)$ be identified with the function whose value at x is $\phi(x)$ times the identity operator. That is, for ϕ in $L^\infty(\mu)$ and f in $L^2(\mu)$, define $(M_\phi f)(x) = \phi(x)f(x)$. This map, $\phi \to M_\phi$ is also a $*$-isomorphism.

52.8 Theorem. *If (X, Ω, μ) is a σ-finite measure space, $\mathcal{A} = \{ M_\phi : \phi \in L^\infty(\mu) \}$ acting on $L^2(\mu)$, $\mathcal{A}_\mathcal{H} = \{ M_\phi : \phi \in L^\infty(\mu) \}$ acting on $L^2(\mu; \mathcal{H})$, and $\mathcal{C}_\mathcal{H} = \{ M_\Phi : \Phi \in L^\infty(\mu; \mathcal{B}(\mathcal{H})) \}$, then:*

(a) *$\mathcal{A}_\mathcal{H}$ and $\mathcal{C}_\mathcal{H}$ are von Neumann algebras;*

(b) *$\mathcal{A}'_\mathcal{H} = \mathcal{C}_\mathcal{H}$ and $\mathcal{C}'_\mathcal{H} = \mathcal{A}_\mathcal{H}$;*

(c) *If $n = \dim \mathcal{H}$, then there is a Hilbert space isomorphism $U : L^2(\mu)^{(n)} \to L^2(\mu; \mathcal{H})$ such that*

$$U\mathcal{A}^{(n)}U^{-1} = \mathcal{A}_\mathcal{H}$$

and

$$UM_n(\mathcal{A})U^{-1} = \mathcal{C}_\mathcal{H}.$$

Proof. The preceding results establish that $\mathcal{A}_\mathcal{H}$ and $\mathcal{C}_\mathcal{H}$ are C*-algebras. Clearly $\mathcal{C}_\mathcal{H} \subseteq \mathcal{A}'_\mathcal{H}$. The proof starts by showing the existence of an isomorphism $U : L^2(\mu)^{(n)} \to L^2(\mu; \mathcal{H})$ such that $U\mathcal{A}^{(n)}U^{-1} = \mathcal{A}_\mathcal{H}$. (The proof will be accomplished by establishing various parts of statements (a), (b), and (c).) Let $\{ e_i \}$ be a basis for $\mathcal{H}$ and define $U : L^2(\mu)^{(n)} \to L^2(\mu; \mathcal{H})$ by

$$U(\xi_1, \xi_2, \dots) = \sum_i \xi_i \otimes e_i,$$

where for a function ξ in $L^2(\mu)$ and a vector h in $\mathcal{H}$, $\xi \otimes h$ is the function in $L^2(\mu; \mathcal{H})$ whose value at x is $\xi(x)h$. Note that

$$\begin{aligned}\|U(\xi_i)\|^2 &= \int \|\sum_i \xi_i(x)e_i\|^2 \, d\mu(x) = \int \sum_i |\xi_i(x)|^2 \, d\mu(x)\\ &= \sum_i \|\xi_i\|^2 = \|(\xi_i)\|^2.\end{aligned}$$

So U is an isometry. It is left to the reader to check that U is invertible with $U^{-1}f = (\langle f(\cdot), e_1\rangle, \langle f(\cdot), e_2\rangle, \dots)$. If $\phi \in L^\infty(\mu)$, $f \in L^2(\mu; \mathcal{H})$, and $f_i(x) = \langle f(x), e_i\rangle$, then $f = \sum_i f_i \otimes e_i$ and

$$UM_\phi^{(n)}U^{-1} = UM_\phi^{(n)}(f_i)$$

$$
\begin{aligned}
&= U(\phi f_i) \\
&= \sum_i (\phi f_i) \otimes e_i \\
&= \phi f.
\end{aligned}
$$

This proves the first half of (c).

Now to prove the first half of (b). Let $T \in \mathcal{A}'_{\mathcal{H}}$.

Case 1. Assume $\mu(X) < \infty$.

In this case for every vector h in $\mathcal{H}$, the function on X that is constantly equal to h belongs to $L^2(\mu; \mathcal{H})$. Using the notation from the first part of this proof, this is the function $1 \otimes h$. So for any h in $\mathcal{H}$, $T(1 \otimes h) \in L^2(\mu; \mathcal{H})$. Thus for h and g in $\mathcal{H}$, $x \to \langle T(1 \otimes h)(x), g\rangle_{\mathcal{H}}$ is a measurable function. Because $|\langle T(1 \otimes h)(x), g\rangle| \le \|T(1 \otimes h)(x)\| \, \|g\|$, this function belongs to $L^2(\mu)$; since the measure space is finite, it also belongs to $L^1(\mu)$.

Let $\phi \in L^\infty(\mu)$ and write $\phi = \phi_1 \overline{\phi_2}$, where $\phi_1 = |\phi|^{1/2}$ and $\phi_2 = |\phi|^{1/2} u$ with $|u| = 1$. Therefore

$$
\begin{aligned}
&\left| \int \phi(x) \langle T(1 \otimes h)(x), g\rangle \, d\mu(x) \right| \\
&\qquad = \left| \int \langle \phi_1(x) T(1 \otimes h)(x), \phi_2(x) g\rangle \, d\mu(x) \right| \\
&\qquad = \left| \int \langle M_{\phi_1} T(1 \otimes h)(x), \phi_2 \otimes g(x)\rangle \, d\mu(x) \right| \\
&\qquad = |\langle (T M_{\phi_1}(1 \otimes h))(x), \phi_2 \otimes g(x)\rangle \, d\mu(x)| \\
&\qquad = |\langle T(\phi_1 \otimes h), \phi_2 \otimes g\rangle| \\
&\qquad \le \|T\| \, \|\phi_1\|_2 \, \|\phi_2\|_2 \, \|h\| \, \|g\|.
\end{aligned}
$$

But $\|\phi_1\|_2 = \|\phi_2\|_2 = \sqrt{\|\phi\|_1}$. This shows that for every ϕ in $L^\infty(\mu)$

$$
\left| \int \phi(x) \langle T(1 \otimes h)(x), g\rangle \, d\mu(x) \right| \le \|T\| \, \|h\| \, \|g\| \, \|\phi\|_1.
$$

Since $L^\infty(\mu)$ is dense in $L^1(\mu)$, this implies $\tau_{h,g}(x) = \langle T(1 \otimes h)(x), g\rangle \in L^\infty(\mu)$ with $\|\tau_{h,g}\|_\infty \le \|T\| \, \|h\| \, \|g\|$.

Let $\mathcal{L}$ be a countable dense subset of $\mathcal{H}$, and, to maintain a comfort zone, assume that $\mathcal{L}$ is a complex-rational vector space. Because $\mathcal{L}$ is countable there is a measurable set Δ with $\mu(\Delta) = 0$ such that for all x in $X \setminus \Delta$ and all g, h in $\mathcal{L}$, $|\langle T(1 \otimes h)(x), g\rangle| \le \|T\| \, \|h\| \, \|g\|$. Momentarily fix x in $X \setminus \Delta$. So $[h, g] \equiv \langle T(1 \otimes h)(x), g\rangle$ is a bounded sesquilinear form on $\mathcal{L}$. Since $\mathcal{L}$ is dense, there is a bounded operator $\Phi(x)$ on $\mathcal{H}$ such that $\langle T(1\otimes h)(x), g\rangle = \langle \Phi(x)h, g\rangle$. This says that Φ is defined almost everywhere and is measurable. Also $|\langle \Phi(x)h, g\rangle| \leq \|T\|\,\|h\|\,\|g\|$. Hence $\|\Phi(x)\| \leq \|T\|$ for all x in $X \setminus \Delta$ so that $\Phi \in L^\infty(\mu; \mathcal{B}(\mathcal{H}))$ and $\|\Phi\| \leq \|T\|$. Using the methods of the last paragraph it follows that if $\xi_1, \xi_2 \in L^2(\mu)$ and $h, g \in \mathcal{H}$, then

$$\langle T(\xi_1 \otimes h), \xi_2 \otimes g\rangle = \langle M_\Phi(\xi_1 \otimes h), \xi_2 \otimes g\rangle.$$

Hence $\langle Tf_1, f_2\rangle = \langle M_\Phi f_1, f_2\rangle$ for all f_1 and f_2 in the linear span of $\{\,\xi \otimes h : \xi \in L^2(\mu) \text{ and } h \in \mathcal{H}\,\}$. But this linear span is dense in $L^2(\mu; \mathcal{H})$ (Exercise 1). Therefore $T = M_\Phi$.

Case 2. μ is arbitrary.

For every measurable set Δ with finite measure, let $\mathcal{K}_\Delta = L^2(\mu|\Delta; \mathcal{H})$. Clearly $\mathcal{K}_\Delta = M_{\chi_\Delta} L^2(\mu; \mathcal{H})$ so that $\mathcal{K}_\Delta$ reduces $\mathcal{A}_\mathcal{H}$, $\mathcal{C}_\mathcal{H}$, and T. Let $\mathcal{A}_\Delta$, $\mathcal{C}_\Delta$, and T_Δ be the restrictions of these entities to $\mathcal{K}_\Delta$. It is routine to show that $T_\Delta \in \mathcal{A}'_\Delta$. Therefore by Case 1 there is a function Φ_Δ in $L^\infty(\mu|\Delta; \mathcal{B}(\mathcal{H}))$ such that $T_\Delta = M_{\Phi_\Delta}$. If Δ and Λ are two sets having finite measure, then $T_\Delta|\mathcal{K}_{\Delta\cap\Lambda} = T_\Lambda|\mathcal{K}_{\Delta\cap\Lambda} = T_{\Delta\cap\Lambda}$. This implies that $\Phi_\Delta|\Lambda = \Phi_\Lambda|\Delta$. That is, the functions Φ_Δ are consistent. Because the measure space is σ-finite, this implies there is a function Φ in $L^\infty(\mu; \mathcal{B}(\mathcal{H}))$ such that $\Phi|\Delta = \Phi_\Delta$ for every measurable set Δ having finite measure. Thus $T|\mathcal{K}_\Delta = M_\Phi|\mathcal{K}_\Delta$ for all sets having finite measure. A measure-theoretic argument shows that $T = M_\Phi$.

Now to tie up the loose ends and finish the proof. Since $M_n(\mathcal{A}) = (\mathcal{A}^{(n)})'$, it follows that $UM_n(\mathcal{A})U^{-1} = (U\mathcal{A}^{(n)}U^{-1})' = \mathcal{A}'_\mathcal{H} = \mathcal{C}_\mathcal{H}$. This finishes (c). From here it is possible to also polish off (b). In fact, $\mathcal{C}'_\mathcal{H} = (UM_n(\mathcal{A})U^{-1})' = UM_n(\mathcal{A})'U^{-1} = U(\mathcal{A}^{(n)})'U^{-1} = \mathcal{A}_\mathcal{H}$. Since both $\mathcal{A}_\mathcal{H}$ and $\mathcal{C}_\mathcal{H}$ are C*-algebras, it is immediate that they are von Neumann algebras, giving (a). ■

This completes the description of Type I algebras. So the homogeneous algebras are precisely the algebras $\{\, M_\Phi : \Phi \in L^\infty(\mu; \mathcal{B}(\mathcal{H}))\,\}$. Hence if $\{(X_n, \Omega_n, \mu_n)\}$ is a family of finite measure spaces, and $\mathcal{H}_n$ is a Hilbert space with dimension n, $1 \leq n \leq \infty$, then

$$L^\infty(\mu_\infty; \mathcal{H}_\infty) \oplus L^\infty(\mu_1) \oplus L^\infty(\mu_2; \mathcal{H}_2) \oplus \cdots$$

is a Type I algebra and every Type I algebra is spatially equivalent to such an algebra.

Exercises.

1. If (X, Ω, μ) is a σ-finite measure space, show that the linear span of $\{\,\xi \otimes h : \xi \in L^2(\mu) \text{ and } h \in \mathcal{H}\,\}$ is dense in $L^2(\mu; \mathcal{H})$.

2. Using Theorem 51.7 as a guide, parametrize the spatial isomorphism classes of Type I algebras.

§53. Some structure theory for continuous algebras

In this section a few results about the structure of continuous algebras are obtained. Information about these algebras does not approach the state of completeness as can be obtained in the Type I case, though much work has been done in recent times. The starting point is the existence of Type II factors.

Recall the definition of the algebras $\mathcal{R}(G)$ and $\mathcal{L}(G)$ for a group G (just prior to 43.9). Also recall (43.13) that $\mathcal{R}(G)$ and $\mathcal{L}(G)$ are factors if and only if the group satisfies the condition: save for the identity, every element of the group has an infinite conjugacy class. Since only separable Hilbert spaces are being considered, an example of a countable group having this property must be found. The reader is asked in Exercise 1 below to verify that such an example exists.

53.1 Theorem. *If G is a countable group with the property that every non-trivial conjugacy class is infinite, then $\mathcal{R}(G)$ and $\mathcal{L}(G)$ are Type II_1 factors.*

Proof. Only the result for $\mathcal{R}(G)$ will be proved, the corresponding result for $\mathcal{L}(G)$ being proved in a similar manner. Begin by showing that $\mathcal{R}(G)$ is finite. The plan is to use Corollary 50.13 and show that there is a faithful, complex-valued trace τ defined on the factor $\mathcal{R}(G)$.

For a in G, let ϵ_a be the characteristic function of the singleton $\{a\}$; so $\epsilon_a \in \ell^2(G)$. Define a map $\tau : \mathcal{R}(G) \to \mathbb{C}$ by

$$\tau(T) = T(\epsilon_e)(e) = \langle T(\epsilon_e), \epsilon_e\rangle,$$

where e is the identity of G. It is easy to conclude that τ is linear and $\tau(1) = 1$. Also $\tau(T^*T) = \langle T^*T\epsilon_e, \epsilon_e\rangle = \|T\epsilon_e\|^2 \geq 0$, so that τ is a positive map. On the other hand, Proposition 43.10 implies there is an f in $\ell^2(G)$ such that $Th = h * f$ for all h in $\ell^2(G)$. Thus $\tau(T^*T) = \|T(\epsilon_e)\|^2 = \|f\|^2$. So if $\tau(T^*T) = 0$, $T = 0$ and τ is a faithful map.

Again use Proposition 43.10 and Exercise 43.9 to establish that $\mathcal{R}(G) = \{\, R_f : f \in \ell^2(G) \text{ and } h * f \in \ell^2(G) \text{ whenever } h \in \ell^2(G)\}$. If $R_f, R_g \in \mathcal{R}(G)$, then $R_f R_g = R_{g*f}$. So $\tau(R_f R_g) = g * f(e) = \sum_y g(y)f(y^{-1})$. Similarly, $\tau(R_g R_f) = f * g(e) = \sum_y f(y)g(y^{-1})$. Substituting y^{-1} for y in one of these sums shows they are equal, and so $\tau(ST) = \tau(TS)$ for all T and S in $\mathcal{R}(G)$. That is, τ is a faithful trace. By Corollary 50.13, $\mathcal{R}(G)$ is finite.

Since $\mathcal{R}(G)$ is a factor, this means that either $\mathcal{R}(G)$ is Type I_n or Type II_1. But if it were Type I_n, it would be $*$-isomorphic to M_n, the algebra of $n \times n$ matrices, and hence finite dimensional. Since this is not the case, $\mathcal{R}(G)$ is a Type II_1 factor. ■

Recall that a von Neumann algebra $\mathcal{A}$ is properly infinite if it contains no non-zero finite, central projection (49.5). Equivalently, $\mathcal{A}$ is properly infinite if $\mathcal{A} = \mathcal{A}_1 \oplus \mathcal{A}_2 \oplus \mathcal{A}_3$, where $\mathcal{A}_1$ is Type I_∞, $\mathcal{A}_2$ is Type II_∞, and $\mathcal{A}_3$ is Type III (Exercise 49.2).

53.2 Proposition. *If $\mathcal{A}$ is a properly infinite von Neumann algebra, then there is a projection E in $\mathcal{A}$ such that $E \sim E^\perp \sim 1$.*

Proof. According to Theorem 49.7 there is a sequence $\{ E_n \}$ of pairwise orthogonal projections in $\mathcal{A}$ such that $1 = \sum_n E_n$ and $E_n \sim 1$ for all $n \geq 1$. Put $E = \sum_k E_{2k}$; so $E^\perp = \sum_k E_{2k-1}$. Since $1 = \sum_n E_n$ it is easy to see that $E \sim 1 \sim E^\perp$. ■

53.3 Corollary. *If $\mathcal{A}$ is properly infinite, $1 \leq n \leq \infty$, and E is any projection in $\mathcal{A}$ such that $1 \sim E \sim E^\perp$, then $\mathcal{A}$ is spatially isomorphic to $M_n(\mathcal{A}_E)$.*

Proof. Put $E_1 = E$. If V is an isometry in $\mathcal{A}$ with $VV^* = E^\perp$, then $A \to VAV^*$ is a spatial isomorphism of $\mathcal{A}$ onto $\mathcal{A}_{E^\perp}$. Thus $\mathcal{A}_{E^\perp}$ is properly infinite. The preceding proposition implies there is a projection $E_2 \leq E^\perp$ such that $E_2 \sim E^\perp \sim E^\perp - E^2$. Continue this line of argument to obtain projections $\{ E_k \}$ that are pairwise orthogonal, $E_1 = E$, $E_k \sim E$ for all k, and $\sum_k E_k = 1$. Now let U_k be a partial isometry in $\mathcal{A}$ with $U_k U_k^* = E_k$ and $U_k^* U_k = E = E_1$. It is left to the reader to verify that $A \to [(U_i V)^* A (U_j V)]_{i,j=1}^n$ is the required spatial isomorphism. ■

53.4 Corollary. *If $\mathcal{A}$ is properly infinite and $1 \leq n \leq \infty$, then $\mathcal{A}$ is spatially isomorphic to $M_n(\mathcal{A})$.*

53.5 Proposition. *If $\mathcal{A}$ is a von Neumann algebra and $1 \leq n \leq \infty$, then the following statements are equivalent.*

(a) *$\mathcal{A}$ is continuous.*

(b) *$M_n(\mathcal{A})$ is continuous.*

(c) *$\mathcal{A}^{(n)}$ is continuous.*

Proof. Since $\mathcal{A}$ and $\mathcal{A}^{(n)}$ are $*$-isomorphic, (a) and (c) are equivalent. If (b) holds, then $(\mathcal{A}')^{(n)} = M_n(\mathcal{A})'$ is continuous by Corollary 50.5. Since $\mathcal{A}'$ and $(\mathcal{A}')^{(n)}$ are isomorphic, $\mathcal{A}'$ is continuous. Once again, Corollary 50.5 implies that $\mathcal{A}$ is continuous. This proves that (b) implies (a); it remains to show that (a) implies (b). But if $\mathcal{A}$ is assumed to be continuous, so is $\mathcal{A}'$. Since (a) and (c) are known to be equivalent, this implies that $(\mathcal{A}')^{(n)}$ and hence $M_n(\mathcal{A}) = [(\mathcal{A}')^{(n)}]'$ are continuous. ■

53.6 Theorem. *Let $\mathcal{A}$ be a von Neumann algebra and let $1 \leq n \leq \infty$.*

(a) *$\mathcal{A}$ is Type I if and only if $M_n(\mathcal{A})$ is Type I.*

(b) *$\mathcal{A}$ is Type II if and only if $M_n(\mathcal{A})$ is Type II.*

(c) *$\mathcal{A}$ is Type III if and only if $M_n(\mathcal{A})$ is Type III.*

(d) *If $\mathcal{A}$ is Type II_1, then $M_n(\mathcal{A})$ is Type II_1 when $n < \infty$ and $M_\infty(\mathcal{A})$ is Type II_∞.*

(e) *If $\mathcal{A}$ is Type II_∞, then $M_n(\mathcal{A})$ is Type II_∞.*

Proof. (a) This follows immediately from the work on classifying the Type I algebras and is only stated here for balance.

(c) If $\mathcal{A}$ is Type III, then Corollary 53.4 implies $\mathcal{A} \cong M_n(\mathcal{A})$, so that $M_n(\mathcal{A})$ is Type III. Now assume that $M_n(\mathcal{A})$ is Type III and that E is a finite projection in $\mathcal{A}$; show that $E = 0$. Now $E_1 = E \oplus 0 \oplus \cdots$ is a projection in $M_n(\mathcal{A})$. It will be shown that E_1 is a finite projection in $M_n(\mathcal{A})$, so that $E_1 = 0$ and, hence, $E = 0$. So assume there is a projection F_1 in $M_n(\mathcal{A})$ with $F_1 \leq E_1$ and $F_1 \sim E_1$. Because $F_1 \leq E_1$, it follows that there is a projection F in $\mathcal{A}$ such that $F_1 = F \oplus 0 \oplus \cdots$ and $F \leq E$. Let V_1 be a partial isometry in $M_n(\mathcal{A})$ with $V_1^* V_1 = E_1$ and $V_1 V_1^* = F_1$. Looking at the matrix representation $V_1 = [V_{ij}]$ in $M_n(\mathcal{A})$ and performing the matrix multiplication in the equation $V_1 E_1 = V_1$, it follows that $V_{ij} = 0$ when $j \geq 2$. Now performing the matrix multiplication in the equation $V_1 = F_1 V_1$ shows that $V_1 = V \oplus 0 \oplus \cdots$, where V is a partial isometry in $\mathcal{A}$, $V^*V = E$, and $VV^* = F$. That is, $E \sim F$ in $\mathcal{A}$, so that $F = E$ since E is assumed to be finite. Thus $F_1 = E_1$ and E_1 is finite in $M_n(\mathcal{A})$. Therefore $E = 0$.

(b) Assume that $\mathcal{A}$ is Type II. By Proposition 53.5, $M_n(\mathcal{A})$ is continuous. Let Z_1 be a non-zero central projection in $M_n(\mathcal{A})$. So there is a central project Z in $\mathcal{A}$ such that $Z_1 = Z^{(n)}$. Since $\mathcal{A}$ is Type II, there is a non-zero finite projection E in $\mathcal{A}$ with $E \leq Z$. Put $E_1 = E \oplus 0 \oplus \cdots$. Thus E_1 is a projection in $M_n(\mathcal{A})$ and $E_1 \leq Z_1$. It will be shown that E_1 is a finite projection in $M_n(\mathcal{A})$. In fact, suppose F_1 is any projection in $M_n(\mathcal{A})$ with $F_1 \leq E_1$ and $F_1 \sim E_1$. It follows that $F_1 = F \oplus 0 \oplus \cdots$ for some projection F in $\mathcal{A}$ with $F \leq E$. If V_1 is a partial isometry in $M_n(\mathcal{A})$ with $V_1^* V_1 = E_1$ and $V_1 V_1^* = F_1$, then matrix calculations similar to those carried out in the proof of (c) show that $V_1 = V \oplus 0 \oplus \cdots$ for some partial isometry V in $\mathcal{A}$, and $V^*V = E$ and $VV^* = F$. That is, $F \sim E$ in $\mathcal{A}$. So $F = E$ since E is finite in $\mathcal{A}$. Therefore $F_1 = E_1$ and E_1 is finite in $M_n(\mathcal{A})$. By definition, $M_n(\mathcal{A})$ is Type II.

Now assume that $M_n(\mathcal{A})$ is Type II. By the general theory, $M_n(\mathcal{A}) = M_n(\mathcal{A}_1) \oplus M_n(\mathcal{A}_\infty)$, where $M_n(\mathcal{A}_1)$ is Type II_1 and $M_n(\mathcal{A}_\infty)$ is Type II_∞. So it suffices to consider the two separate cases where $M_n(\mathcal{A})$ is Type II_1

and II_∞. First assume that $M_n(\mathcal{A})$ is Type II_1. By Corollary 53.5, $\mathcal{A}$ is continuous. To show that $\mathcal{A}$ is finite, assume that E is a projection in $\mathcal{A}$ with $E \sim 1$ and let V be a partial isometry in $\mathcal{A}$ with $V^*V = 1$ and $VV^* = E$. If $E_1 = E \oplus 0 \oplus \cdots$ and $V_1 = V \oplus 0 \oplus \cdots$, then E_1 is a projection in $M_n(\mathcal{A})$, V_1 is a partial isometry in $M_n(\mathcal{A})$, $V_1^*V_1 = 1$, and $V_1V_1^* = E_1$. Since $M_n(\mathcal{A})$ is finite, this implies that $E_1 = 1$ and so $E = 1$ and $\mathcal{A}$ is finite.

Now assume that $M_n(\mathcal{A})$ is Type II_∞. Again $\mathcal{A}$ is continuous by (53.5). Thus $\mathcal{A} = \mathcal{A}_1 \oplus \mathcal{A}_2 \oplus \mathcal{A}_3$, where $\mathcal{A}_1$ is Type II_1, $\mathcal{A}_2$ is Type II_∞, and $\mathcal{A}_3$ is Type III. It follows that $M_n(\mathcal{A}) = M_n(\mathcal{A}_1) \oplus M_n(\mathcal{A}_2) \oplus M_n(\mathcal{A}_3)$. But $M_n(\mathcal{A}_3)$ is Type II by (c), so it must be that $\mathcal{A}_3 = 0$. Thus $\mathcal{A} = \mathcal{A}_1 \oplus \mathcal{A}_2$ is Type II.

(d) Assume that $\mathcal{A}$ is Type II_1. To show that $M_\infty(\mathcal{A})$ is Type II_∞, show that no non-zero central projection in $M_\infty(\mathcal{A})$ is finite. In fact, if Z_1 is a central projection in $M_\infty(\mathcal{A})$, then there is a central projection Z in $\mathcal{A}$ with $Z_1 = Z^{(\infty)}$. For $1 \le n < \infty$, let E_n be the projection in $M_\infty(\mathcal{A})$ with Z in the n-th coordinate and 0 elsewhere. It is easy to verify that $E_n \sim E_m$ for all n and m. Thus $Z_1 = \bigoplus_{n=1}^\infty E_n \sim \bigoplus_{n=2}^\infty E_n$. Thus Z_1 is not finite and $M_\infty(\mathcal{A})$ must be Type II_∞.

Now assume that $1 \le n < \infty$. By part (b), $M_n(\mathcal{A})$ is Type II. To show that $M_n(\mathcal{A})$ is finite, use is made of Corollary 55.9 below that a von Neumann algebra is finite if it has a faithful, center-valued trace. (The proof of Theorem 55.9 does not use this result and the present result appears here so that it is grouped with its cousins.) If $\tau : \mathcal{A} \to \mathcal{Z}$ is a faithful, center-valued trace on $\mathcal{A}$, define $\widetilde{\tau} : M_n(\mathcal{A}) \to \mathcal{Z}^{(n)}$ by

$$\widetilde{\tau}([A_{ij}]) = \tau\left[\frac{1}{n}\sum_{i=1}^n A_{ii}\right]^{(n)}.$$

To facilitate the proof also define $\rho : M_n(\mathcal{A}) \to \mathcal{A}$ by

$$\rho([A_{ij}]) = \frac{1}{n}\sum_{i=1}^n A_{ii}.$$

So $\widetilde{\tau}([A_{ij}]) = \tau(\rho([A_{ij}]))^{(n)}$. It is easy to see that $\widetilde{\tau}$ is linear, positive, and $\widetilde{\tau}(1) = 1$.

If $A = [A_{ij}], B = [B_{ij}] \in M_n(\mathcal{A})$, then two equations emerge:

$$\rho(AB) = \frac{1}{n}\sum_{i=1}^n\sum_{k=1}^n A_{ik}B_{ki},$$

$$\rho(BA) = \frac{1}{n}\sum_{i=1}^n\sum_{k=1}^n B_{ik}A_{ki}.$$

Thus, by changing the variables of summation

$$\tau(\rho(AB)) = n^{-1}\sum_i\sum_k \tau(A_{ik}B_{ki}) = n^{-1}\sum_i\sum_k \tau(B_{ki}A_{ik}) = \tau(\rho(BA)).$$

Therefore $\widetilde{\tau}$ has the trace property. It remains to show that $\widetilde{\tau}$ is faithful.

If $A \in M_n(\mathcal{A})$ and $\widetilde{\tau}(A^*A) = 0$, then $\rho(A^*A) = 0$. In the last displayed equations, taking $B = A^*$ so that $B_{ij} = A_{ji}^*$ implies that $0 = \rho(A^*A) = n^{-1}\sum_i\sum_k A_{ki}^*A_{ki}$, which can only happen if $A_{ki} = 0$ for all i, k.

(e) This is left as an exercise for the reader. ■

53.7 Corollary. *If $\mathcal{A}$ is a Type II_1 factor, then $M_\infty(\mathcal{A})$ is a Type II_∞ factor.*

With the example of a Type II_1 factor from Theorem 53.1, an example of each type of factor has been obtained save for one that is Type III. Getting such an example would involve additional material and this is not presented here. The interested reader can see the references. Much work has been done in the last 25 years on Type III algebras. The reader can see the expositions by van Daele [1978] and Wright [1989].

Exercise.

1. Let G be the group of all maps $x \to ax + b$ of $\mathbb{R}$ into itself under composition where a and b are rational and a is positive. Show that the conjugacy class of every non-zero element of G is infinite. Show that the free group on two generators is another example of a group having this property.

§54. Weak-star continuous linear functionals revisited

In this section some results are presented on the predual of a von Neumann algebra. These could have appeared earlier in this text, before the classification of projections. But while they are certainly basic, they are the type of information that might be skipped in an introductory treatment. They appear here because they are required for the proof of the existence of a center-valued trace on a finite von Neumann algebra, given in the following section. The starting point is a collection of results on extreme points of some sets in a C*-algebra.

54.1 Lemma. *If $\mathcal{A}$ is a C^*-algebra, then the identity is an extreme point of* ball $\mathcal{A}$.

Proof. Assume there are operators A, B in ball $\mathcal{A}$ such that $1 = (A+B)/2$. Since the identity is hermitian, it is easy to see that the operators A and B can be replaced in this equation by their real parts; hence it can be assumed that $A, B \in \operatorname{Re}\mathcal{A}$. Since $B = 2 - A$, A and B commute. Let $\mathcal{C}$ be the C*-algebra generated by A, B, and the identity. Since $\mathcal{C}$ is abelian, there is a compact space X and functions a and b in ball $\operatorname{Re} C(X)$ such that $1 = [a(x) + b(x)]/2$ for all x. But 1 is an extreme point of $[-1, 1]$ and $-1 \leq a(x), b(x) \leq 1$, so $a = b = 1$. ■

54.2 Proposition. *Let $\mathcal{A}$ be a C^*-algebra.*

(a) *The extreme points of* $(\text{ball}\,\mathcal{A})_+$ *are the projections in* $\mathcal{A}$.

(b) *The set of extreme points in* ball Re $\mathcal{A}$ *is the set of hermitian unitary elements in* $\mathcal{A}$.

(c) *Every extreme point of* ball $\mathcal{A}$ *is a partial isometry.*

Proof. (a) First assume that $\mathcal{A}$ is abelian; that is, assume that $\mathcal{A} = C(X)$ for some compact set X. If p is a projection in $C(X)$ and $f, g \in (\text{ball}\, C(X))_+$ such that $p = (f+g)/2$, then $p(x) = [f(x) + g(x)]/2$ for every x in X. But since p is a projection, it is a characteristic function. So $p(x) = 0$ or 1 for all x. On the other hand, $0 \leq f(x), g(x) \leq 1$ for all x. But 0 and 1 are the extreme points of $[0, 1]$, so for each x, $f(x) = g(x) = p(x)$. That is, $p = f = g$.

Now assume that p is an extreme point in $(\text{ball}\, C(X))_+$. If there is an x such that $0 < p(x) < 1$, there is an open set U such that $0 < p(x) < 1$ for x in U. Urysohn's lemma implies there is a continuous function f that vanishes outside U and such that $0 \leq p(x) \pm f(x) \leq 1$ on U. Therefore $p \pm f \in (\text{ball}\, C(X))_+$ and $2^{-1}[(p+f) + (p-f)] = p$, contradicting the fact that p is an extreme point. Thus $p(x)$ is either 0 or 1 for all x, implying that p is a characteristic function.

Now for the general case. If P is a projection and $A, B \in (\text{ball}\,\mathcal{A})_+$ such that $P = (A+B)/2$, then $2^{-1}B = P - 2^{-1}A \leq P$. Thus B and P commute. Therefore A, B, and P all commute and so the C*-algebra generated by these three operators is abelian. From the proof of the abelian case it is known that $P = A = B$, and P is an extreme point. Conversely, assume that P is an extreme point in $(\text{ball}\,\mathcal{A})_+$ and let $\mathcal{B}$ be the C*-algebra generated by P and the identity. It follows that P is an extreme point of $\mathcal{B}$, and, since $\mathcal{B}$ is abelian, the special case shows that P is a projection.

(b) Assume that U is a hermitian unitary in $\mathcal{A}$. Note that the map $A \to UA$ is a linear isometry of $\operatorname{Re}\mathcal{A}$ onto itself. Since the identity is an extreme point of ball $\mathcal{A}$ (54.1), it is an extreme point of ball $\operatorname{Re}\mathcal{A}$. Hence its image under this isometry, U, is also an extreme point of ball $\operatorname{Re}\mathcal{A}$.

Conversely assume that A is an extreme point of ball $\operatorname{Re}\mathcal{A}$ and write $A = A_+ - A_-$. It will be shown that A_+ and A_- are extreme points of $(\operatorname{ball}\mathcal{A})_+$ and, thus, are projections by (a); this will show that A is a hermitian unitary.

Suppose there are operators B, C in $(\operatorname{ball}\mathcal{A})_+$ such that $A_+ = (B+C)/2$. Now $0 = A_-A_+A_- = (A_-BA_- + A_-CA_-)/2$, and both A_-BA_- and A_-CA_- are positive. Hence $0 = A_-BA_- = (A_-B^{\frac{1}{2}})(A_-B^{\frac{1}{2}})^*$. So $A_-B^{\frac{1}{2}} = 0$ from which it follows that $0 = A_-B = BA_-$. Similarly $A_-C = CA_- = 0$. Thus $B - A_-, C - A_- \in \operatorname{ball}\operatorname{Re}\mathcal{A}$. Since $A = [(B-A_-)+(C-A_-)]/2$ and A is extreme, $A = B - A_- = C - A_-$. By the uniqueness of the Jordan decomposition, this implies that $A_+ = B = C$.

(c) Let A be an extreme point of ball $\mathcal{A}$. If A^*A is not a projection, then using the functional calculus for A^*A shows that there exists an element B in $\mathcal{A}_+$ such that $A^*AB = BA^*A \neq 0$ and $\|A(1 \pm B)\|^2 = \|A^*A(1 \pm B)^2\| \leq 1$. Since, $A = [(A + AB) + (A - AB)]/2$ and A is an extreme point, $AB = 0$, a contradiction. ∎

It can be shown without too much additional effort that an element A in a C*-algebra is an extreme point of ball $\mathcal{A}$ if and only if $(1 - A^*A)\mathcal{A}(1 - AA^*) = (0)$. (For example, see Sakai [1971], p. 12.) The examination of extreme points stops here to derive the Jordan decomposition of weak* continuous linear functionals, where the information on extreme points just obtained will be used.

The first step is to obtain the decomposition of an arbitrary linear functional into its real and imaginary parts. If ϕ is a linear functional on $\mathcal{A}$, define $\phi^* : \mathcal{A} \to \mathbb{C}$ by $\phi^*(A) = \overline{\phi(A^*)}$. The next proposition gathers together the basic properties of this associated functional and the joy of executing its proof is saved for the reader.

54.3 Proposition. *Let ϕ be a bounded linear functional on the C^*-algebra $\mathcal{A}$.*

(a) $\|\phi^*\| = \|\phi\|$.

(b) *If $\phi = \phi^*$, then $\|\phi\| = \|\phi|\operatorname{Re}\mathcal{A}\|$.*

(c) *If ϕ is weak* continuous, then so is ϕ^*.*

(d) *There are bounded linear functionals ϕ_1 and ϕ_2 on $\mathcal{A}$ such that $\phi_1^* = \phi_1$, $\phi_2^* = \phi_2$, and $\phi = \phi_1 + i\phi_2$.*

(e) *If $\mathcal{A}$ is a von Neumann algebra and ϕ is weak* continuous, then so are the functionals ϕ_1 and ϕ_2 in (d).*

A linear functional ϕ satisfying $\phi^* = \phi$ is called *hermitian*.

54.4 Definition. If ϕ and ψ are bounded linear functionals on a C*-algebra, then ϕ and ψ are said to be *orthogonal* if $\|\phi - \psi\| = \|\phi\| + \|\psi\|$. Symbolically this is denoted by $\phi \perp \psi$.

If $\mathcal{A} = C(X)$ for some compact space X, then it is not difficult to see that two functionals on $C(X)$ are orthogonal if and only if the corresponding measures are mutually singular.

Now let $\mathcal{A}$ be a von Neumann algebra and let ϕ be a positive weak* continuous linear functional on $\mathcal{A}$. The proof of the GNS construction (7.7) established that $\mathcal{L} = \{ X \in \mathcal{A} : \phi(X^*X) = 0 \}$ is a left ideal of $\mathcal{A}$. Because $\phi \in \mathcal{A}_*$, $\mathcal{L}$ is weak* closed. Therefore Proposition 46.7 implies there is a projection P in $\mathcal{A}$ such that $\mathcal{L} = \mathcal{A}P$.

54.5 Definition. If ϕ is a positive weak* continuous linear functional on the von Neumann algebra $\mathcal{A}$, then the *support* of ϕ is the projection $P^\perp$, where $\{ X \in \mathcal{A} : \phi(X^*X) = 0 \} = \mathcal{A}P$. The support of ϕ is denoted by $S(\phi)$.

If $\mathcal{A} = L^\infty(\mu)$ and the linear functional is given by the function f in $L^1(\mu)$, then its support is χ_Δ, where $\Delta = \{ x : f(x) \neq 0 \}$. The proof of the next proposition is left to the reader.

54.6 Proposition. *If $\phi \in \mathcal{A}_*$, $\phi \geq 0$, and P is its support, then:*

(a) $P^\perp = \sup\{ E : E$ *is a projection in* $\mathcal{A}$ *and* $\phi(E) = 0 \}$*;*

(b) *For all A in $\mathcal{A}$, $\phi(AP) = \phi(PA) = \phi(A)$.*

Now for a result that fully exhibits the connection between this support and the support of a measure.

54.7 Proposition. *If ϕ is a positive linear functional on the von Neumann algebra $\mathcal{A}$, $Q = S(\phi)$, and $A \in \mathcal{A}_+$ such that $A = QA = AQ$ and $\phi(A) = 0$, then $A = 0$.*

Proof. Consider the spectral representation $A = \int_0^a t\,dE(t)$. The assumption that $AQ = A = QA$ implies that $Q^\perp \leq \ker A$. If $A \neq 0$, then there is an $\epsilon > 0$ such that $E_\epsilon = E([\epsilon, a]) \neq 0$. Thus $A \geq \int_\epsilon^a t\,dE(t) \geq \epsilon E_\epsilon$. Thus $\phi(E_\epsilon) = 0$. But $0 < E_\epsilon \leq (\ker A)^\perp \leq S(\phi)$, which is a contradiction. ∎

Now for one of the important results of this section.

54.8 Theorem. *If $\mathcal{A}$ is a von Neumann algebra and ϕ is a hermitian, weak* continuous linear functional on $\mathcal{A}$, then there are unique orthogonal, positive, weak* continuous linear functionals ϕ_1 and ϕ_2 on $\mathcal{A}$ such that $\phi = \phi_1 - \phi_2$.*

Proof. Without loss of generality it can be assumed that $\|\phi\| = 1$. Since $\operatorname{ball}\operatorname{Re}\mathcal{A}$ is weak* compact, (54.3.b) implies $\mathcal{X} = \{A \in \operatorname{ball}\operatorname{Re}\mathcal{A} : \phi(A) = 1\}$ is a non-empty set. It is easy to see that it is weak* compact and convex. Let U be an extreme point of this set $\mathcal{X}$; it follows that U is an extreme point of $\operatorname{ball}\operatorname{Re}\mathcal{A}$. By (54.2) U is a hermitian unitary. Thus there is a projection P such that $U = P - P^{\perp}$. Define ϕ_1 and ϕ_2 by $\phi_1(A) = \phi(PA)$ and $\phi_2(A) = -\phi(P^{\perp}A)$ for all A in $\mathcal{A}$. It must be shown that ϕ_1 and ϕ_2 are positive.

Now $(\phi_1 + \phi_2)(1) = \phi(P - P^{\perp}) = \phi(U) = 1$. Also for every A, $|(\phi_1 + \phi_2)(A)| = |\phi(UA)| \leq \|A\|$. So $\|\phi_1 + \phi_2\| \leq 1$. This implies that $\|\phi_1 + \phi_2\| = (\phi_1 + \phi_2)(1) = 1$, and so $\phi_1 + \phi_2 \geq 0$.

Claim. $\|\phi_1 | P\mathcal{A}P\| = \phi_1(P)$ and, hence, $\phi_1 | P\mathcal{A}P \geq 0$.

Suppose not; so there is an operator X in $P\mathcal{A}P$ with $\|X\| = 1$ and $\phi_1(X) > \phi_1(P)$. Since $XP^{\perp} = P^{\perp}X = 0$, $\|X - P^{\perp}\| = 1$. Hence $1 \geq \phi(X - P^{\perp}) = \phi_1(X) - \phi(P^{\perp}) > \phi_1(P) - \phi(P^{\perp}) = \phi(U) = 1$, a contradiction. This establishes the claim, which implies that $\phi_1 | P\mathcal{A}P \geq 0$.

Now $\phi_1 = 2^{-1}[\phi + (\phi_1 + \phi_2)]$, so ϕ_1 is hermitian. Therefore for an operator A in $\mathcal{A}$, $\phi(AP) = \phi((PA^*)^*) = \overline{\phi(PA^*)} = \overline{\phi_1(A^*)} = \phi_1(A)$. When combined with the claim, this implies $\phi_1 \geq 0$. Similarly $\phi_2 \geq 0$. Clearly ϕ_1 and ϕ_2 are orthogonal and weak* continuous. This completes the existence part of the proof.

To prove uniqueness, assume that ψ_1 and ψ_2 are orthogonal, positive, weak* continuous linear functionals on $\mathcal{A}$ such that $\phi_1 - \phi_2 = \psi_1 - \psi_2$. From the equations $\phi_1(1) - \phi_2(1) = \psi_1(1) - \psi_2(1)$ and $\phi_1(1) + \phi_2(1) = 1 = \psi_1(1) + \psi_2(1)$, it follows that $\phi_1(1) = \psi_1(1)$ and $\phi_2(1) = \psi_2(1)$. Let E be the support of ϕ_1. So $\psi_1(E) - \psi_2(E) = \phi(E) = \phi_1(E) = \phi_1(1) = \psi_1(1) = \psi_1(E) + \psi_1(E^{\perp})$. Hence $-\psi_2(E) = \psi_1(E^{\perp}) \geq 0$, so $\psi_2(E) = \psi_1(E^{\perp}) = 0$. Therefore, $S(\psi_1) \leq E \leq P$ and $S(\psi_2) \leq E^{\perp}$. But this implies that for any A in $\mathcal{A}$, $\phi_1(A) = \phi(EA) = \psi_1(EA) - \psi_2(EA) = \psi_1(EA) = \psi_1(A)$. Similarly it also follows that $\psi_2 = \phi_2$. ∎

54.9 Corollary. *If ϕ is a weak* continuous linear functional on the von Neumann algebra $\mathcal{A}$, then there are positive, weak* continuous linear functionals ϕ_j, $1 \leq j \leq 4$, such that $\phi = \phi_1 - \phi_2 + i[\phi_3 - \phi_4]$, $\phi_1 \perp \phi_2$, and $\phi_3 \perp \phi_4$.*

A weaker form of the preceding corollary, with the orthogonality restriction removed, was obtained in Exercise 46.3. Combine this corollary with Theorem 46.4 to obtain the following.

54.10 Corollary. *If ϕ is a bounded linear functional on $\mathcal{A}$, the following statements are equivalent.*

(a) *ϕ is weak* continuous.*

(b) *For every sequence $\{E_n\}$ of pairwise orthogonal projections in $\mathcal{A}$, $\phi(\sum_n E_n) = \sum_n \phi(E_n)$.*

(c) *There is a trace class operator C on $\mathcal{H}$ such that $\phi(A) = \text{tr}\,(AC)$ for all A in $\mathcal{A}$.*

Now the promised polar decomposition is delivered.

54.11 Theorem. *If ω is a weak* continuous linear functional on the von Neumann algebra $\mathcal{A}$, then there is a positive, weak* continuous linear functional ϕ on $\mathcal{A}$ satisfying $\|\phi\| = \|\omega\|$ and a partial isometry W with $W^*W = S(\phi)$ such that*

$$\omega(A) = \phi(AW)$$

for all A in $\mathcal{A}$. This decomposition of ω is unique.

Proof. It can be assumed that $\|\omega\| = 1$. Note that due to the weak* continuity of ω, the set $\{A \in \text{ball}\,\mathcal{A} : \omega(A) = 1\}$ is a non-empty, weak* compact, convex subset of ball $\mathcal{A}$. Therefore it has an extreme point U. An easy argument shows that U is also an extreme point of ball $\mathcal{A}$ so that U is a partial isometry (54.2.c). Let $\omega_1(A) = \omega(AU)$ for all A in $\mathcal{A}$. So ω_1 is a weak* continuous linear functional and it is easy to see that $\|\omega_1\| \leq 1$. On the other hand, $\omega_1(1) = \omega(U) = 1$, so $\omega_1 \geq 0$. Also $\omega_1(UU^*) = \omega(UU^*U) = \omega(U) = 1 = \|\omega_1\|$, so $S(\omega_1) \leq UU^*$. Let $W = U^*S(\omega_1)$. It follows that W is a partial isometry with initial space $W^*W = S(\omega_1)$. Also for any A in $\mathcal{A}$, $\omega_1(A) = \omega_1(AS(\omega_1)) = \omega(AS(\omega_1)U) = \omega(AW^*)$. That is,

$$\omega(AU) = \omega(AW^*)$$

for all A in $\mathcal{A}$.

Let $P = WW^*$, the final space of W.

Claim. $\omega(A) = \omega(AP)$ for all A in $\mathcal{A}$.

To see this fix A in ball $\mathcal{A}$ and let $\beta = \omega(AP^\perp)$. For any positive integer n, $\|nW^* + AP^\perp\|^2 = \|[nW^* + AP^\perp][nW + P^\perp A^*]\| = \|n^2 S(\omega_1) + AP^\perp A^*\| \leq n^2 + 1$. Also

$$\begin{aligned}\omega(nW^* + AP^\perp) &= n\omega(W^*) + \omega(AP^\perp)\\ &= n + \beta.\end{aligned}$$

Now if $\beta \neq 0$, then replacing A by a unimodular multiple of itself shows that it can be assumed that $\beta > 0$. Hence for sufficiently large n, $n + \beta >$

$(n^2+1)^{1/2}$. That is, for sufficiently large n, $\omega(nW^*+AP^\perp) > \|nW^*+AP^\perp\|$, contradicting the fact that $\|\omega\| = 1$. Thus $\omega(AP^\perp) = 0$, or $\omega(A) = \omega(AP)$.

Put $\phi(A) = \omega(AW^*)$. So $\phi(A) = \omega(AS(\omega_1)U) = \omega_1(AS(\omega_1)) = \omega_1(A)$. That is, $\phi = \omega_1$, so ϕ is positive and $S(\phi) = W^*W$. Also $\phi(AW) = \omega(AWW^*) = \omega(AP) = \omega(A)$. This completes the proof of existence.

To prove uniqueness assume that ψ is a positive, weak* continuous linear functional and V is a partial isometry with $\omega(A) = \psi(AV)$ and $V^*V = S(\psi)$. Note that for any A in $\mathcal{A}$, $\phi(A) = \phi(AS(\phi)) = \phi(AW^*W) = \omega(AW^*) = \psi(AW^*V) = \psi(S(\psi)AW^*V)$. Putting $A = S(\psi)^\perp$ shows that $\phi(S(\psi)^\perp) = 0$, so that $S(\phi) \leq S(\psi)$. The reverse inequality is established in a similar manner, so ϕ and ψ have the same support. Put $Q = S(\phi) = W^*W = S(\psi) = V^*V$.

Note that $V^*W \in Q\mathcal{A}Q$, so that $V^*W = A + iB$, where $A, B \in \operatorname{Re}(Q\mathcal{A}Q)$. So $\phi(A) + i\phi(B) = \phi(V^*W) = \omega(V^*) = \psi(V^*V) = \psi(1) = 1$. Hence $\phi(A) = 1$. But A, being the real part of a contraction, is a contraction. So $Q - A \geq 0$ and $Q(Q-A) = (Q-A)Q = Q - A$. Since $\phi(Q-A) = 0$, Proposition 54.7 implies $A = Q$. This implies $B = 0$, or $V^*W = Q$. So for each h in $Q = \operatorname{initial} W$, $\|V^*Wh\| = \|h\| = \|Wh\|$. Therefore $\operatorname{final} W \subseteq \operatorname{initial} V^* = \operatorname{final} V$. Using the equation $W^*V = Q$ shows that $\operatorname{final} V \subseteq \operatorname{final} W$. It follows that $V = W$. Thus $\phi(A) = \omega(AW^*) = \psi(A)$. ■

If $\omega \in \mathcal{A}_*$ and $\omega(\cdot) = \phi(\cdot W)$ as in the preceding theorem, then denote the functional ϕ by $\phi = |\omega|$.

§55. The center-valued trace

Throughout this section $\mathcal{A}$ is a von Neumann algebra acting on the separable Hilbert space $\mathcal{H}$. The main result of this section, that $\mathcal{A}$ is finite if and only if it possesses a faithful, center-valued trace, goes back to the origin of the subject (Murray and von Neumann [1936], [1937]). The proof given here, however, is due to Yeadon [1971] and is based on the exposition of this proof found in Ringrose [1971b]. The fundamental ingredient is an application of the Ryll-Nardzewski Fixed Point Theorem ([ACFA], p. 153).

55.1 Definition. If $\mathcal{A}$ and $\mathcal{B}$ are von Neumann algebras, a bounded linear map $\phi\colon \mathcal{A} \to \mathcal{B}$ is called a *trace* if $\phi(AB) = \phi(BA)$ for all A and B in $\mathcal{A}$.

In this setting the main concern is with a trace whose range is either the center of the algebra $\mathcal{A}$ or the complex numbers. This notion was encountered in (50.13), where it was shown that if $\mathcal{A}$ is a von Neumann algebra with a faithful, positive, center-valued trace, then $\mathcal{A}$ is finite. The starting point here is with linear functionals having the trace property. Before obtaining the main result, it will be established that trace functionals exist on a finite algebra.

55.2 Lemma. *If $\phi : \mathcal{A} \to \mathbb{C}$ is a bounded linear functional, then ϕ is a trace if and only if $\phi(U^*AU) = \phi(A)$ for all A in $\mathcal{A}$ and all unitaries U in $\mathcal{A}$.*

Proof. If ϕ is a trace, it is clear that the condition holds. Now assume the condition is satisfied. First observe that if U is a unitary, then $\phi(AU) = \phi(U^*(UAU)) = \phi((UAU)U^*) = \phi(UA)$ for all A in $\mathcal{A}$. Because every element in $\mathcal{A}$ can be written as a linear combination of four unitaries, it follows that ϕ is a trace. ■

In the proof of the next lemma, substantial use is made of Proposition 48.5.

55.3 Lemma. *If $\mathcal{A}$ is finite and F, $E_1, E_2, \ldots$ are projections in $\mathcal{A}$ such that $E_n \leq E_{n+1}$ and $E_n \precsim F$, then $E = \sup_n E_n \precsim F$.*

Proof. Put $E_0 = 0$.

Claim. There are pairwise orthogonal projections $\{F_n\}$ in $\mathcal{A}$ such that $F_n \leq F$ and $F_n \sim E_n - E_{n-1}$ for all $n \geq 1$.

The proof is by induction. Let F_1 be a projection in $\mathcal{A}$ such that $F_1 \leq F$ and $E_1 \sim F_1$. Now suppose $F_1, \ldots, F_n$ have been chosen with the stated properties. Because $\mathcal{A}$ is finite and $E_{n+1} \precsim F$, $F^\perp \precsim E_{n+1}^\perp$ (48.5). Also $E_n = \sum_k^n F_k$, so

$$\begin{aligned}(E_{n+1} - E_n)^\perp &= E_{n+1}^\perp + E_n \\ &= E_{n+1}^\perp + \sum_{k=1}^n (E_k - E_{k-1}) \\ &\sim E_{n+1}^\perp + \sum_{k=1}^n F_k \\ &\succsim F^\perp + \sum_{k=1}^n F_k.\end{aligned}$$

Again the fact that $\mathcal{A}$ is finite implies that $E_{n+1} - E_n \precsim (F^\perp + \sum_{k=1}^n F_k)^\perp = F - \sum_{k=1}^n F_k$. Thus it is possible to choose $F_{n+1} \leq F - \sum_{k=1}^n F_k$ with $E_{n+1} - E_n \sim F_{n+1}$. This establishes the claim.

Now the lemma follows since $E = \sum_{n=1}^\infty (E_n - E_{n-1}) \sim \sum_{n=1}^\infty F_n \leq F$. ■

Denote the cone of positive weak* continuous (normal) linear functionals on $\mathcal{A}$ by $(\mathcal{A}_*)_+$.

55.4 Proposition. *If K is a uniformly bounded subset of $(\mathcal{A}_*)_+$ and for every sequence $\{E_n\}$ of pairwise orthogonal projections and for every $\epsilon > 0$ there is an integer N such that $\omega(E_n) \leq \epsilon$ for all ω in K and all $n \geq N$, then the closure of K in $\mathcal{A}_*$ is weakly compact.*

Proof. It is given that $K \subseteq (\mathcal{A}_*)_+ \subseteq (\mathcal{A}^*)_+$; let K_1 be the weak* $(\sigma(\mathcal{A}^*, \mathcal{A}))$ closure of K in $\mathcal{A}^*$. Because K_1 is bounded, it is weak* compact. Since the relative weak* topology of $\mathcal{A}^*$ on $\mathcal{A}_*$ is $\sigma(\mathcal{A}_*, \mathcal{A})$, the weak topology on $\mathcal{A}_*$, it suffices to show that K_1 is, in fact, a subset of $\mathcal{A}_*$. So it must be shown that every linear functional in K_1 is weak* continuous on $\mathcal{A}$. To do this apply Theorem 46.4.

Let $\{E_k\}$ be a sequence of pairwise orthogonal projections in $\mathcal{A}$, and put $E = \sum_k E_k$.

Claim. If $\epsilon \geq 0$, then there is an integer N such that

$$0 \leq \omega(E) - \sum_{k=1}^{n} \omega(E_k) < \epsilon$$

for all $n \geq N$ and all ω in K.

In fact if this is not the case, then there is an $\epsilon > 0$ such that for every integer n, $\omega(E) - \sum_{k=1}^{n} \omega(E_k) > \epsilon$ for at least one ω in K. In particular, there is an ω_1 in K and an integer N_1 such that $\omega_1(E) - \sum_{k=1}^{N_1} \omega_1(E_k) \geq \epsilon$. Since ω_1 is weak* continuous, there is an $N_2 \geq N_1 + 1$ such that $\omega_1(E) - \sum_{k=1}^{N_2} \omega_1(E_k) < \epsilon/2$. Putting $F_1 = \sum_{k=N_1+1}^{N_2} E_k$ implies that F_1 is a projection in $\mathcal{A}$ and $\omega_1(F_1) \geq \epsilon/2$.

The assumption that the claim is false now produces an ω_2 in K such that $\omega_2(E) - \sum_{k=1}^{N_2} \omega_2(E_k) \geq \epsilon$. Continue arguing as before. This produces an increasing sequence of integers $\{N_n\}$, a sequence $\{\omega_n\}$ in K, and a sequence of pairwise orthogonal projections $\{F_n\}$ in $\mathcal{A}$ such that $\omega_n(F_n) \geq \epsilon/2$ for all $n \geq 1$. This contradicts the hypothesis of the proposition, and so the claim is true.

Now let $\{\omega_\alpha\}$ be a net in K and assume that $\omega_\alpha \to \omega$ weak* in $\mathcal{A}^*$. That is, assume that $\omega_\alpha(A) \to \omega(A)$ for all A in $\mathcal{A}$. Clearly ω is a positive linear functional on $\mathcal{A}$. To show that ω is weak* continuous, let $\{E_n\}$ be a sequence of pairwise orthogonal projections in $\mathcal{A}$ and let $\epsilon > 0$. By the claim there is an integer N such that

$$\omega_\alpha(E) - \sum_{k=1}^{n} \omega_\alpha(E_k) < \epsilon$$

for all $n \geq N$ and all α. Fix $n \geq N$; so there is an index α_0 such that

$$\left|\omega_\alpha\left(\sum_{k=1}^{n} E_k\right) - \omega\left(\sum_{k=1}^{n} E_k\right)\right| < \epsilon$$

and

$$|\omega_\alpha(E) - \omega(E)| < \epsilon$$

for all $\alpha \geq \alpha_0$. A judicious use of the triangle inequality shows that for $n \geq N$, $\omega(E) - \sum_{k=1}^{n} \omega(E_k) < 3\epsilon$, so ω is weak* continuous by Theorem 46.4. ∎

Note that the preceding proposition says that whenever the convergence $\omega(\sum_n E_n) = \sum_n \omega(E_n)$, established in (46.4) for each element of K, is uniform over K, then K has weakly compact closure. This condition is necessary and sufficient for the weak compactness of any subset of $\mathcal{A}_*$, not just a set of positive normal functionals. This is due to Akemann [1967]; also see Ringrose [1971b].

If U is a unitary in $\mathcal{A}$, define the operator $R_U : \mathcal{A}_* \to \mathcal{A}_*$ by $R_U\omega(A) = \omega(U^*AU)$. Note that each R_U is a surjective isometry, $R_U\omega \geq 0$ whenever $\omega \geq 0$, and $R_U R_V = R_{UV}$. According to Lemma 55.2, to show the existence of a scalar-valued trace on $\mathcal{A}$ it suffices to show that the set of operators $\{R_U\}$ has a fixed point when it operates on the states in $\mathcal{A}_*$. This will be done using the Ryll-Nardzewski Fixed Point Theorem ([ACFA], p. 151). Now to lay the groundwork for that application.

55.5 Lemma. *If $\mathcal{A}$ is finite, $\omega \in \mathcal{A}_*$, and Q is the closed convex hull of $K = \{R_U\omega : U \text{ is unitary}\}$, then Q is weakly compact in $\mathcal{A}_*$.*

Proof. It suffices to assume that $\omega \geq 0$. Indeed, Corollary 54.9 implies an arbitrary ω in $\mathcal{A}_*$ can be written as $\omega = \omega_1 - \omega_2 + i(\omega_3 - \omega_4)$, where each $\omega_i \geq 0$. If Q_i is the weakly closed convex hull of $\{R_U\omega_i : U \text{ is unitary}\}$ and it is given that each Q_i is weakly compact, then $\widetilde{Q} \equiv Q_1 - Q_2 + i(Q_3 - Q_4)$ is weakly compact since it is the continuous image of a weakly compact set. But it is easy to see that the convex hull of K is contained in $\widetilde{Q}$, so $Q \subseteq \widetilde{Q}$, and it follows that Q is weakly compact.

So assume that $\omega \geq 0$. By the Krein–Smulian Theorem, Q is weakly compact if and only if the weak closure of K is weakly compact. Assume that this is not the case. According to the preceding proposition there is an $\epsilon > 0$, a sequence $\{E_n\}$ of pairwise orthogonal projections in $\mathcal{A}$, and a sequence $\{\omega_n\}$ in K such that $\omega_n(E_n) \geq \epsilon$ for all $n \geq 1$. From the definition of K, for each $n \geq 1$ there is a unitary U_n in $\mathcal{A}$ such that $\omega_n = R_{U_n}\omega$. Put $F_n = U_n^* E_n U_n$; so $F_n \sim E_n$ and $\omega(F_n) \geq \epsilon$ for all $n \geq 1$.

Let $P_n = E_n + E_{n+1} + \cdots$ and $G_n = \sup\{ F_n, F_{n+1}, \dots \}$. So $P_n \geq P_{n+1}$ and $G_n \geq G_{n+1}$ for all $n \geq 1$. Let $G = \inf\{ G_n : n \geq 1 \}$. Now fix n, and, for each $k \geq 0$, put $R_k = \sup\{ F_i : n \leq i \leq n+k \}$.

Claim. For all $k \geq 0$, $R_k \precsim \sum_{i=n}^{n+k} E_i \leq P_n$.

The last inequality is trivial. The proof of the rest of the claim is by induction on k. For $k = 0$ this is clear. Assume $R_{k-1} \precsim \sum_{i=n}^{n+k-1} E_i$. An application of Proposition 47.5 shows that

$$\begin{aligned} R_k &= R_{k-1} \oplus [(R_{k-1} \vee F_{n+k}) - R_{k-1}] \\ &\sim R_{k-1} \oplus [F_{n+k} - (F_{n+k} \wedge R_{k-1})] \\ &\leq R_{k-1} \vee F_{n+k} \\ &\precsim \sum_{i=k}^{n+k} E_i. \end{aligned}$$

Now combine the claim and Lemma 55.3 to get that $G_n = \sup\{ R_k : k \geq 0 \} \precsim P_n$ for all $n \geq 1$. Because $\mathcal{A}$ is finite this implies that $P_n^\perp \precsim G_n^\perp \leq G^\perp$. Since $P_n^\perp \leq P_{n+1}^\perp$ for all n, Lemma 55.3 can again be applied to conclude that $\sup_n P_n^\perp \precsim G^\perp$. But $\sup_n P_n^\perp = 1$, so $G = 0$. That is, $G_n \downarrow 0$ (weak*). Since $F_n \leq G_n$, this implies that $F_n \to 0$ weak*. Since ω is weak* continuous, $\omega_n(E_n) = \omega(F_n) \to 0$, contradicting the choice of ω_n and E_n. ■

55.6 Theorem. *If $\mathcal{A}$ is a finite von Neumann algebra with center $\mathcal{Z}$ and $f \in \mathcal{Z}_*$, then there is a unique linear functional ϕ in $\mathcal{A}_*$ such that ϕ is a norm preserving extension of f and*

$$\phi(A) = \phi(U^* A U)$$

for all A in $\mathcal{A}$ and every unitary operator U in $\mathcal{A}$. If f is positive on $\mathcal{Z}$, then ϕ is positive on $\mathcal{A}$.

Proof. Let $\mathcal{U}$ be the group of unitary operators in $\mathcal{A}$. Since $\mathcal{A}$ with its weak* topology is a locally convex space, there is an ω in $\mathcal{A}_*$ such that $\omega|\mathcal{Z} = f$. If $f \geq 0$, then ω can be chosen to be positive (46.4). Let Q be the closed convex hull of $\{ R_U \omega : U \in \mathcal{U} \}$. By the preceding lemma, Q is weakly compact in the Banach space $\mathcal{A}_*$. Also note that $R_U Q \subseteq Q$ for every unitary U. Furthermore, $\{ R_U : U \in \mathcal{U} \}$ is non-contracting on Q ([ACFA], V.10.6). Indeed this is immediate from the following.

Claim. If $\phi \in \mathcal{A}_*$ and $\phi \neq 0$, then $0 \notin \overline{\text{co}}\{ R_U \phi \}$.

In fact, if $0 \in \overline{\text{co}}\{ R_U \phi \}$, then for every $\epsilon > 0$ there is a unitary U in $\mathcal{U}$ such that $\|R_U \phi\| < \epsilon$. But it is easy to see that $\|R_U \phi\| = \|\phi\|$, so this would violate the assumption that $\phi \neq 0$.

(A point deserves emphasis here. In the Ryll-Nardzewski Theorem it is permitted that the convex set Q is weakly compact but the set of maps $\{ R_U : U \in \mathcal{U} \}$ is non-contracting in the norm topology!)

Now apply the Ryll-Nardzewski Fixed Point Theorem to conclude that there is a ϕ in Q such that $R_U\phi = \phi$ for all U in $\mathcal{U}$. It is easy to check that every element of Q is an extension of f, so, in particular, ϕ is such an extension. Because ϕ is a fixed point of the mappings R_U, it also has the required invariance property. Furthermore, if f is positive and ω was chosen positive, so is every element of Q, so that $\phi \geq 0$. It remains to establish uniqueness and the equality of norms.

Suppose ψ is any functional in $\mathcal{A}_*$ such that $\psi|\mathcal{Z} = f$ and $R_U\psi = \psi$ for every U in $\mathcal{U}$. Let $\psi(\cdot) = |\psi|(\cdot W)$ be the polar decomposition of ψ (54.11). So for U in $\mathcal{U}$ and A in $\mathcal{A}$, $\psi(A) = \psi(U^*AU) = |\psi|(U^*AUW) = (R_U)|\psi|(A(UWU^*)$. From Theorem 54.11 it follows that $W^*W = S(|\psi|)$. It follows that $S(R_U|\psi|) = US(|\psi|)U^* = UW^*WU = (UWU^*)^*(UWU^*)$. So the uniqueness of the polar decomposition implies that $|\psi| = R_U|\psi|$ and $W = UWU^*$ for every unitary U. But this last equation implies that W commutes with every unitary, so that $W \in \mathcal{Z}$. Therefore $\|\psi\| = \psi(W^*) = f(W^*) \leq \|f\|$. Since ψ is an extension of f, $\|\psi\| = \|f\|$.

Now that it is known that the extension is norm preserving, suppose that ϕ and ψ are two extensions of f that have the invariance property. Then $\phi - \psi$ is an extension of the zero functional with the invariance property, so that $\|\phi - \psi\| = 0$. That is the extension is unique. ■

What has just been done? To each f in $\mathcal{Z}_*$ has been associated an extension ϕ in $\mathcal{A}_*$ in a unique way. This means that a function $\rho : \mathcal{Z}_* \to \mathcal{A}_*$ has been defined having the properties:

55.7
$$\begin{cases} \text{(a)} & \rho(f)|\mathcal{Z} = f \text{ for every } f \text{ in } \mathcal{Z}_*; \\ \text{(b)} & R_U\rho(f) = \rho(f) \text{ for every } f \text{ in } \mathcal{Z}_*; \\ \text{(c)} & \rho \text{ is a positive, linear isometry.} \end{cases}$$

The properties (a) and (b) are clear from the theorem. The fact that ρ is positive and an isometry also follows from (55.6); the fact that it is linear is a consequence of the uniqueness of the extension. Note that the map ρ is itself unique. Taking the adjoint of ρ produces the desired centered valued trace.

55.8 Theorem. *If $\mathcal{A}$ is a finite von Neumann algebra with center $\mathcal{Z}$, then there is a unique contractive, linear map $\tau : \mathcal{A} \to \mathcal{Z}$ having the following properties.*

(a) *τ is weak* continuous.*

(b) $\tau(Z) = Z$ *for all* Z *in* $\mathcal{Z}$.

(c) $\tau(U^*AU) = \tau(A)$ *for every* A *in* $\mathcal{A}$ *and every unitary* U *in* $\mathcal{A}$.

This map also satisfies the following.

(d) τ *is positive.*

(e) τ *is faithful.*

(f) $\tau(AZ) = Z\tau(A)$ *for all* A *in* $\mathcal{A}$ *and all* Z *in* $\mathcal{Z}$.

Proof. Let $\rho : \mathcal{Z}_* \to \mathcal{A}_*$ be the map described after Theorem 55.6 and satisfying (55.7). Let $\tau = \rho^* : \mathcal{A} \to \mathcal{Z}$. The fact that τ is contractive and satisfies condition (a) is automatic. If $f \in \mathcal{Z}_*$ and $Z \in \mathcal{Z}$, then $\langle f, \tau(Z)\rangle = \langle \rho(f), Z\rangle = \langle f, Z\rangle$ since $\rho(f)$ is an extension of f. Again the Hahn–Banach Theorem implies that (b) is verified. Property (c) follows from (55.7.b).

Now to show that τ is the unique map satisfying (a), (b), and (c). Indeed, if $\eta : \mathcal{A} \to \mathcal{Z}$ is another such contractive, linear map and $f \in \mathcal{Z}_*$, then $\psi = f \circ \eta$ is weak* continuous and extends f by (b). Property (c) implies that $\psi(U^*AU) = \psi(A)$ for all A in $\mathcal{A}$ and all unitaries U in $\mathcal{A}$. Since η is contractive, $\|\psi\| \le \|f\|$. But the fact that ψ is an extension of f implies that the norms are equal. According to the uniqueness part of Theorem 55.6, $f \circ \eta = \rho(f) = f \circ \tau$. Since f was arbitrary, $\eta = \tau$.

Now to establish properties (d), (e), and (f). Because ρ is positive, (d) holds. To prove (f), let Z be a unitary in $\mathcal{Z}$ and define $\eta : \mathcal{A} \to \mathcal{Z}$ by $\eta(A) = Z^*\tau(ZA)$. Clearly η is a contraction that satisfies (a) and (b). If U is any unitary in $\mathcal{A}$, then $\eta(U^*AU) = Z^*\tau(U^*(ZA)U) = \eta(A)$. By the uniqueness of τ, $\eta = \tau$. Thus (f) holds whenever Z is a unitary in the center. Since each central element is the linear combination of four central unitaries, (f) is verified.

It remains to prove that τ is faithful. Note that it suffices to show that $\tau(E) \neq 0$ for every non-zero projection E in $\mathcal{A}$. Indeed, if $A \in \mathcal{A}_+$ and $A \neq 0$, then there is a positive scalar α and a non-zero projection E such that $A \ge \alpha E$. Thus $\tau(A) \ge \alpha\tau(E) > 0$.

Assume there is a non-zero projection E such that $\tau(E) = 0$. By Zorn's lemma there is a maximal family of non-zero, pairwise orthogonal projections $\{ E_i \}$ such that $\tau(E_i) = 0$ for all i. (By the separability of the underlying Hilbert space, this is necessarily a sequence; but this doesn't play a role here.) Let $E = \sum_i E_i$. Since τ is weak* continuous, $\tau(E) = 0$. By the maximality of $\{ E_i \}$, $\tau(F) > 0$ for every non-zero projection F with $F \le E^\perp$.

By the Comparison Theorem there is a central projection Z such that

$$ZE \precsim ZE^\perp \qquad \text{and} \qquad Z^\perp E^\perp \precsim Z^\perp E.$$

Note that because τ is a trace, it takes the same value at two equivalent projections. Now $\tau(ZE) = Z\tau(E) = 0$, so if $ZE \neq 0$, there is a non-zero projection that is dominated by $ZE^\perp \leq E^\perp$ that belongs to the kernel of τ. This contradicts the maximality of $\{E_i\}$. Thus $ZE = 0$, or $E \leq Z^\perp$. Similarly, $\tau(Z^\perp E^\perp) \leq \tau(Z^\perp E) = 0$ and the maximality of $\{E_i\}$ implies $Z^\perp E^\perp = 0$. That is, $Z^\perp \leq E$. Therefore $Z^\perp = E$. But then $0 = \tau(E) = \tau(Z^\perp) = Z^\perp$ by (b). Thus $Z = 1$ so $E = Z^\perp = 0$, a contradiction. Thus τ must be faithful. ∎

55.9 Corollary. *A von Neumann algebra is finite if and only if it has a faithful, finite, center-valued trace.*

Proof. The preceding theorem shows that every finite algebra has such a trace. The converse was shown in Corollary 50.13. ∎

This section closes by mentioning that every semi-finite algebra has a "trace," though it is not defined everywhere. In fact, it isn't defined everywhere but only of a weak* dense ideal. This is analogous to the fact that $\mathcal{B}(\mathcal{H})$ has a trace, though it is defined only on the trace class. See any of the references for this material.

Exercises.

1. For a bounded linear functional $\phi\colon \mathcal{A} \to \mathbb{C}$, show that ϕ is a trace if and only if either of the following two conditions is satisfied: (a) $\phi(A^*A) = \phi(AA^*)$ for all A in $\mathcal{A}$; (b) $\phi(E) = \phi(F)$ for any two equivalent projections E and F in $\mathcal{A}$.

2. Show that Lemma 55.3 does not hold for $\mathcal{A} = \mathcal{B}(\mathcal{H})$.

Chapter 8

Reflexivity

This chapter explores relations between sets of operators and their common invariant subspaces. Specifically the topic is reflexive and hyperreflexive subspaces of operators. Some partial surveys of this topic have appeared, but no monograph on the subject has been written. Hadwin [1994] gives an overview of reflexivity. Ptak [1997] discusses the reflexivity of n-tuples of operators, which will not be touched in this section.

§56. Fundamentals and examples

Throughout this section, $\mathcal{S}$ will denote a linear subspace of $\mathcal{B}(\mathcal{H})$. Recall the definition of the attached subspace and its basic properties (8.4 and 8.5). The reader might also review §22 concerning algebras and lattices.

56.1 Definition. A linear subspace $\mathcal{S}$ of $\mathcal{B}(\mathcal{H})$ is *reflexive* if $\mathcal{S} = \operatorname{Ref} \mathcal{S}$. A single operator is *reflexive* if the WOT closed algebra generated by the operator and the identity is reflexive.

This explains to some extent the choice of the notation $\operatorname{Ref} \mathcal{S}$. Recall (22.3.e) that for an algebra $\mathcal{A}$ containing the identity, $\operatorname{Ref} \mathcal{A} = \operatorname{Alg} \operatorname{Lat} \mathcal{A}$. So an algebra of operators containing the identity is reflexive if and only if $\operatorname{Alg} \operatorname{Lat} \mathcal{A} = \mathcal{A}$. Reflexivity for algebras was first introduced in Radjavi and Rosenthal [1969]. One also defines reflexivity for lattices of subspaces: a lattice $\mathcal{L}$ is *reflexive* if $\mathcal{L} = \operatorname{Lat} \operatorname{Alg} \mathcal{L}$. The study of reflexive lattices has a life of its own. A related question that has occupied several mathematicians is the following: given an abstract lattice $\mathcal{L}$, is there a collection of operators $\mathcal{X}$ such that $\operatorname{Lat} \mathcal{X} = \mathcal{L}$. Such problems can be quite difficult. Note that a lattice arising in this way is reflexive. Before proceeding with the results, the reader might examine Corollary 42.4, which says that, in an analogous sense, separable C*-algebras contained in the Calkin algebra are reflexive.

56.2 Proposition. (a) *If* $\{\mathcal{S}_n\}$ *is a sequence of reflexive subspaces, then* $\bigoplus_n \mathcal{S}_n$ *is reflexive.*

(b) *If* $\{\mathcal{S}_i\}$ *is any collection of reflexive subspaces,* $\bigcap_i \mathcal{S}_j$ *is reflexive.*

Proof. Note that if $\mathcal{S} = \bigoplus_n \mathcal{S}_n$ on $\mathcal{H} = \bigoplus_n \mathcal{H}_n$, then $[\mathcal{S}h_n] \subseteq \mathcal{H}_n$ for any h_n in $\mathcal{H}_n$. So if $T \in \operatorname{Ref}\mathcal{S}$, $T\mathcal{H}_n \subseteq \mathcal{H}_n$ for each $n \geq 1$. Therefore T decomposes as a direct sum, $T = \bigoplus_n T_n$. It is apparent that $T_n \in \operatorname{Ref}\mathcal{S}_n$ for each $n \geq 1$; since $\mathcal{S}_n$ is reflexive, $T_n \in \mathcal{S}_n$. Thus $T \in \mathcal{S}$. To prove part (b) note that if $\mathcal{S} = \cap_i \mathcal{S}_i$ and $T \in \operatorname{Ref}\mathcal{S}$, then $Th \in [\mathcal{S}h] \subseteq [\mathcal{S}_i h]$ for every vector h. Thus $T \in \operatorname{Ref}\mathcal{S}_i = \mathcal{S}_i$ for every i. ∎

It is easy to see that a reflexive subspace must be WOT closed (8.5); and a reflexive algebra must contain the identity. It is also easy to see that an operator is reflexive if and only if its adjoint is reflexive. Most of the easy ways of combining operators do not preserve reflexivity. This is the distinction between dealing with reflexivity for operators and reflexivity for subspaces or algebras. For example, for the direct sum of two algebras $\mathcal{A}$ and $\mathcal{B}$ containing the identity, $1 \oplus 0 \in \mathcal{A} \oplus \mathcal{B}$. On the other hand, for the direct sum of two operators A and B, it may not be that $1 \oplus 0 \in W(A \oplus B)$. Indeed, this occurs if and only if the algebra $W(A \oplus B)$ splits (22.8). It is possible for the direct sum of two reflexive operators not to be reflexive. (See Example 7 in Wogen [1987]. This paper is partially discussed in §61 below, though an example of a pair of reflexive operators whose direct sum is not reflexive is not presented in that section.)

There is a connection between reflexive subspaces and reflexive algebras that is helpful. If $\mathcal{S}$ is a linear subspace of $\mathcal{B}(\mathcal{H})$, then

56.3 $$\mathcal{A} = \left\{ \begin{bmatrix} \alpha & S \\ 0 & \beta \end{bmatrix} : S \in \mathcal{S} \text{ and } \alpha, \beta \in \mathbb{C} \right\}$$

is easily seen to be a subalgebra of $\mathcal{B}(\mathcal{H} \oplus \mathcal{H})$. It can also be checked that $\mathcal{S}$ is WOT (resp., weak*) closed if and only if $\mathcal{A}$ is. There is an intimate connection between the subspace $\mathcal{S}$ and the corresponding algebra $\mathcal{A}$. The following is an indication of this.

56.4 Proposition. *If* $\mathcal{S}$ *is a linear subspace of* $\mathcal{B}(\mathcal{H})$ *and* $\mathcal{A}$ *is the algebra in* (56.3), *then* $\mathcal{S}$ *is reflexive if and only if* $\mathcal{A}$ *is reflexive. Moreover, a closed subspace of* $\mathcal{H} \oplus \mathcal{H}$ *belongs to* $\operatorname{Lat}\mathcal{A}$ *if and only if* $\mathcal{M} = \mathcal{M}_1 \oplus \mathcal{M}_2$, *where* $\mathcal{M}_1$ *and* $\mathcal{M}_2$ *are subspaces of* $\mathcal{H}$ *and* $\mathcal{S}\mathcal{M}_2 \subseteq \mathcal{M}_1$.

Proof. Set the notation that if $T \in \mathcal{B}(\mathcal{H})$,

$$\widetilde{T} \equiv \begin{bmatrix} 0 & T \\ 0 & 0 \end{bmatrix}.$$

Begin with the characterization of $\operatorname{Lat} \mathcal{A}$. First observe that $1 \oplus 0$ and $0 \oplus 1$ both belong to $\mathcal{A}$. So if $\mathcal{M} \in \operatorname{Lat} \mathcal{A}$, $\mathcal{M} = \mathcal{M}_1 \oplus \mathcal{M}_2$ for two closed subspaces $\mathcal{M}_1, \mathcal{M}_2$ of $\mathcal{H}$. Since $\widetilde{S}\mathcal{M} \subseteq \mathcal{M}$ for S in $\mathcal{S}$, it follows that $\mathcal{S}\mathcal{M}_2 \subseteq \mathcal{M}_1$. The converse is clear.

Assume that $\mathcal{A}$ is reflexive and let $T \in \operatorname{Ref} \mathcal{S}$. If $\mathcal{M} = \mathcal{M}_1 \oplus \mathcal{M}_2 \in \operatorname{Lat} \mathcal{A}$, then $\widetilde{T}\mathcal{M} = T\mathcal{M}_2 \oplus (0)$. But for any h_2 in $\mathcal{M}_2$, $Th_2 \in [\mathcal{S}h_2] \subseteq [\mathcal{S}\mathcal{M}_2] \subseteq \mathcal{M}_1$. That is, $T\mathcal{M}_2 \subseteq \mathcal{M}_1$ and so $\widetilde{T}\mathcal{M} \subseteq \mathcal{M}$. By definition, $\widetilde{T} \in \operatorname{Alg} \operatorname{Lat} \mathcal{A} = \mathcal{A}$. This implies that $T \in \mathcal{S}$ and so $\mathcal{S}$ is reflexive.

Now assume that $\mathcal{S}$ is reflexive and let $A \in \operatorname{Alg} \operatorname{Lat} \mathcal{A}$. Represent A as a 2×2 matrix with operator entries: $A = (A_{ij})$. If $\mathcal{N}$ is any subspace of $\mathcal{H}$, then $\mathcal{N} \oplus (0) \in \operatorname{Lat} \mathcal{A}$. Since this is left invariant by A, it follows that $A_{21} = 0$ and $A_{11} = \alpha$ for some scalar α. Consideration of the spaces $[\mathcal{S}h] \oplus \mathbb{C}h$ in $\operatorname{Lat} \mathcal{A}$ for an arbitrary vector h in $\mathcal{H}$, shows that $A_{22} = \beta$ for some scalar β and $A_{12}h \in [\mathcal{S}h]$. That is $A_{12} \in \operatorname{Ref} \mathcal{S} = \mathcal{S}$. Thus $A \in \mathcal{A}$. ∎

56.5 Proposition. *Every one dimensional subspace of $\mathcal{B}(\mathcal{H})$ is reflexive.*

Proof. Let $\mathcal{S} = \mathbb{C}T$. If $A \in \operatorname{Ref} \mathcal{S}$, then for every vector h, $Ah \in [\mathcal{S}h] = \mathbb{C}Th$. Hence there is a scalar $\alpha(h)$ such that $Ah = \alpha(h)Th$. By comparing Ag, Ah, and $A(g+h)$ it is apparent that there is a single scalar α such that $\alpha(h) = \alpha$ for all h. That is, $A = \alpha T$. ∎

56.6 Proposition. *Every von Neumann algebra is reflexive.*

Proof. If $\mathcal{A}$ is a von Neumann algebra, then it was observed in Example 22.4.b that $\operatorname{Lat} \mathcal{A} = \{P : P \text{ a projection in } \mathcal{A}'\}$. Thus $P^\perp \in \operatorname{Lat} \mathcal{A}$ whenever $P \in \operatorname{Lat} \mathcal{A}$. So if $T \in \operatorname{Alg} \operatorname{Lat} \mathcal{A}$, T is reduced by every subspace in $\operatorname{Lat} \mathcal{A}$. Since $\mathcal{A}'$ is generated by its projections, this implies that $T \in \operatorname{Alg} \operatorname{Lat} T$ if and only if $T \in \mathcal{A}'' = \mathcal{A}$. ∎

56.7 Proposition. *The unilateral shift is reflexive.*

Proof. Let S be the unilateral shift acting on H^2. That is, $Sf = zf$ for every function f in H^2. Recall (26.1) that $W(S) = P^\infty(S) = \{T_\phi : \phi \in H^\infty\}$, the algebra of analytic Toeplitz operators.

Now suppose that $A \in \operatorname{Alg} \operatorname{Lat} S$. For $|\lambda| < 1$, let $k_\lambda = (1 - \bar{\lambda}z)^{-1}$ be the reproducing kernel for H^2. So the one dimensional space spanned by k_λ is invariant for S^*, and thus for A^*. Hence there is a scalar $\phi(\lambda)$ such that $A^*k_\lambda = \overline{\phi(\lambda)}k_\lambda$. Therefore, $\langle A(1), k_\lambda \rangle = \phi(\lambda)$. That is, $\phi = A(1)$, a function in H^2. So ϕ is analytic. On the other hand, for each λ in $\mathbb{D}$, $\phi(\lambda) \in \sigma(A)$. Thus $|\phi(\lambda)| \leq \|A\|$ and so $\phi \in H^\infty$. Finally, for any f in H^2, $(Af)(\lambda) = \langle Af, k_\lambda \rangle = \langle f, A^*k_\lambda \rangle = \phi(\lambda)f(\lambda)$. That is, A is the analytic Toeplitz operator with symbol ϕ and hence belongs to $W(S)$. ∎

The proof that the unilateral shift is reflexive can be used to show many other cyclic subnormal operators are reflexive. In particular, the same proof will show that the Bergman operator is reflexive. It is true that every subnormal operator is reflexive, though the proof of this (due to Olin and Thomson [1980]) is very complicated. Also see Conway [1991], Theorem VII.8.5. Another proof using Thomson's result on the existence of bounded point evaluations (Thomson [1991]) is in McCarthy [1993].

The pensive reader might have observed a phenomena in the last proposition. Namely, for the unilateral shift S, $W(S) = P^\infty(S)$ and the WOT and the weak* topologies agree there. This is true for many other operators, as will be seen later in (59.2). The fact that $W(S) = P^\infty(S)$ for the unilateral shift is, however, not something that is true for all operators. Wogen [1986] and [1987] has given an example of an operator for which these two algebras differ. Later (61.8) this as well as several other examples relative to these topics will be presented.

56.8 Proposition. *If $\mathcal{T}$ is the subspace of $\mathcal{B}(H^2)$ consisting of all the Toeplitz operators, then* $\operatorname{Ref}\mathcal{T} = \mathcal{B}(H^2)$ *and so $\mathcal{T}$ is not reflexive.*

Proof. If $h \in H^2$, let $h = ug$, where u is inner and g is outer. So $[\mathcal{T}h] \supseteq \operatorname{cl}\{T_{\phi\bar{u}}h : \phi \in H^\infty\} = \operatorname{cl}\{\phi g : \phi \in H^\infty\} = H^2$ if $g \neq 0$. ∎

This result has an interesting generalization in Azoff and Ptak [1998]. They show that if $\mathcal{S}$ is a weak* closed subspace of Toeplitz operators, either $\mathcal{S}$ is reflexive or $\operatorname{Ref}\mathcal{S} = \mathcal{B}(H^2)$ as in the preceding proposition.

This section concludes with a characterization of reflexivity in terms of the preannihilator of $\mathcal{S}$ in the trace class. Recall that $\mathcal{S}_\perp = \{C \in \mathcal{B}_1 : \operatorname{tr}(CS) = 0 \text{ for all } S \text{ in } \mathcal{S}\}$. The first step is to notice that $\mathcal{S}_\perp$ and $(\operatorname{Ref}\mathcal{S})_\perp$ have the same rank one operators. In fact, $g \otimes h \in \mathcal{S}_\perp$ if and only if $h \perp [\mathcal{S}g]$. But $[\mathcal{S}g] = [\operatorname{Ref}\mathcal{S}g]$, and, thus, $g \otimes h \in \mathcal{S}_\perp$ if and only if $g \otimes h \in (\operatorname{Ref}\mathcal{S})_\perp$.

56.9 Theorem. *If $\mathcal{S}$ is any linear subspace of $\mathcal{B}(\mathcal{H})$,* $(\operatorname{Ref}\mathcal{S})_\perp$ *is the closed linear span of the rank one operators it contains. Consequently, a weak* closed subspace of $\mathcal{B}(\mathcal{H})$ is reflexive if and only if its preannihilator is the closed linear span of the rank one operators it contains.*

Proof. First note that the second statement follows from the first. Indeed, using the first part of the proposition it follows that the second part says that $\mathcal{S}$ is reflexive if and only if $\mathcal{S}_\perp = (\text{Ref}\,\mathcal{S})_\perp$. If $\mathcal{S}$ is weak* closed, this condition is the same as $\mathcal{S} = \mathcal{S}_\perp^\perp = (\text{Ref}\,\mathcal{S})_\perp^\perp = \text{Ref}\,\mathcal{S}$, the definition of reflexivity.

Fix the linear space $\mathcal{S}$ and let $\mathcal{X}$ be the closed linear span of the rank one operators contained in $\mathcal{S}_\perp$. So $\mathcal{X} \subseteq (\text{Ref}\,\mathcal{S})_\perp$. The Hahn–Banach Theorem says to show that $\mathcal{X} = (\text{Ref}\,\mathcal{A})_\perp$ it suffices to show that $\mathcal{X}^\perp \subseteq ((\text{Ref}\,\mathcal{S})_\perp)^\perp = \text{Ref}\,\mathcal{S}$. Let $T \in \mathcal{X}^\perp$ and let g be any vector in $\mathcal{H}$. If $h \perp [\mathcal{S}g]$, then $g \otimes h \in \mathcal{S}_\perp = (\text{Ref}\,\mathcal{S})_\perp$. Thus $0 = \text{tr}\,(T(g \otimes h)) = \langle Tg, h\rangle$. Since h was an arbitrary vector in $[\mathcal{S}g]^\perp$, this implies that $Tg \in [\mathcal{S}g]$. Thus $T \in \text{Ref}\,\mathcal{S}$ and the proof is complete. ■

The preceding proposition was shown by Arveson [1975].

Exercises.

1. Prove that $\mathcal{B}(\mathcal{H})$ and $\mathbb{C}$ are reflexive.

2. Let Γ be a finite Jordan system. That is, $\Gamma = \{\gamma_0, \ldots, \gamma_m\}$, where each γ_j is a Jordan curve, γ_0 is positively oriented, and for $1 \le j \le m$, γ_j is negatively oriented, contained inside γ_0, and contained in the outside of γ_i for $i \neq j$ and $1 \le i \le m$. Let G be the region bounded by this system of curves and let $Sf = zf$ for all f in the Bergman space for G, $L^2_a(G)$. Show that S is reflexive. (Hint: If μ is the restriction of area measure to G, first observe that the weak* closure of the polynomials in $L^\infty(\mu)$ is isomorphic to $H^\infty(G_0)$, where G_0 is the inside of γ_0 and $W(S) = P^\infty(S) = \{M_\phi : \phi \in H^\infty(G_0)\}$.)

§57. Reflexive operators on finite dimensional spaces

This section will present the characterization of the operators on $\mathbb{C}^d$ that are reflexive, due to Deddens and Fillmore [1975]. Observe that reflexivity is invariant under similarity, so it is natural that the characterization of the reflexive transformations on a finite dimensional space be characterized in terms of Jordan canonical forms.

Le J_0 be the 0 operator acting on the zero dimensional space and, if $n \ge 1$, let J_n be the Jordan cell

$$J_n = \begin{pmatrix} 0 & 1 & 0 & \ldots & 0 \\ 0 & 0 & 1 & \ldots & 0 \\ \vdots & \ddots & \ddots & & \vdots \\ 0 & 0 & \ldots & 0 & 1 \\ 0 & 0 & \ldots & 0 & 0 \end{pmatrix}$$

The standard basis on $\mathcal{H} = \mathbb{C}^n$ is denoted by $\{e_1, \ldots, e_n\}$, and so $J_n e_1 = 0, J_n e_2 = e_1, \ldots, J_n e_n = e_{n-1}$.

It is useful to note several properties of Jordan cells. First, an operator on an n-dimensional space is similar to J_n if and only if it is nilpotent with index of nilpotency n. (Recall that the *index of nilpotency* of a nilpotent linear transformation T is the smallest integer k such that $T^k = 0$.) Equivalently, a linear transformation is similar to J_n if and only if it is nilpotent, acts on an n-dimensional space, and is cyclic. In Example 22.4.a it was pointed out that

$$\textbf{57.1} \qquad \begin{cases} \operatorname{Lat} J_n = \{(0), \vee\{e_1\}, \vee\{e_1.e_2\}, \ldots, \vee\{e_1, \ldots, e_{n-1}\}, \mathcal{H}\}, \\ \operatorname{Alg}\operatorname{Lat} J_n = \{\text{upper triangular matrices}\}. \end{cases}$$

It will be useful to set some notation. If T is a linear transformation on the d-dimensional space $\mathcal{H}$, let $\lambda_1, \ldots, \lambda_p$ be the distinct eigenvalues of T and let $\mathcal{H}_1, \ldots, \mathcal{H}_p$ be the corresponding generalized eigenspaces. So if $d_i = \dim \mathcal{H}_i$, $d_1 + \cdots + d_p = d$. Note that the spaces $\mathcal{H}_i$ are also the Riesz spaces corresponding to the clopen subset $\{\lambda_i\}$ of the spectrum of T. Thus $\mathcal{H} = \mathcal{H}_1 \dotplus \cdots \dotplus \mathcal{H}_p$, where this is not necessarily an orthogonal direct sum. If $T_i = T|\mathcal{H}_i$, then $\sigma(T_i) = \{\lambda_i\}$. Therefore T is similar to

$$T_1 \oplus \cdots \oplus T_p,$$

where for $1 \leq i \leq p$, there are integers $n_1^i \geq \cdots \geq n_{m_i}^i$ such that

$$T_i - \lambda_i \approx J_{n_1^i} \oplus \cdots \oplus J_{n_{m_i}^i}.$$

This is the Jordan (canonical) form of T and the numbers $\{n_k^i\}$, $1 \leq i \leq p$, $1 \leq k \leq m_i$, together with the spectrum of T form a complete set of similarity invariants for T. With this notation, the main theorem of this section is the following.

57.2 Theorem. (Deddens and Fillmore [1975]) *With the preceding notation, T is reflexive if and only if whenever* $\dim \mathcal{H}_i > 1$, *$m_i \geq 2$ and either $n_1^i = n_2^i$ or $n_1^i = n_2^i + 1$.*

57.3 Corollary. *If T is any linear transformation on a finite dimensional space and $d \geq 2$, $T^{(d)}$ is reflexive.*

The first step in the proof of Theorem 57.2 is easy. According to Proposition 22.9, $\operatorname{Lat} T = \operatorname{Lat} T_1 \dotplus \cdots \dotplus \operatorname{Lat} T_p$, $P(T) = P(T_1) \dotplus \cdots \dotplus P(T_p)$, and $\operatorname{Alg}\operatorname{Lat} T = \operatorname{Alg}\operatorname{Lat} T_1 \dotplus \cdots \dotplus \operatorname{Alg}\operatorname{Lat} T_p$. Thus it follows that T is reflexive if and only if each T_i is reflexive. Of course, T_i is reflexive if and only if $T_i - \lambda_i$ is reflexive, so it suffices to characterize the reflexive nilpotent operators.

If $p(z)$ is the polynomial of degree $\leq n-1$, $p(z) = \lambda_0 + \lambda_1 z + \cdots + \lambda_{n-1} z^{n-1}$, then

$$p(J_n) = \begin{pmatrix} \lambda_0 & \lambda_1 & \lambda_2 & \dots & \lambda_{n-1} \\ 0 & \lambda_0 & \lambda_1 & \dots & \lambda_{n-2} \\ \vdots & \ddots & \ddots & & \vdots \\ 0 & 0 & \dots & \lambda_0 & \lambda_1 \\ 0 & 0 & \dots & 0 & \lambda_0 \end{pmatrix}.$$

Hence $P(J_n)$ consists of all the upper triangular matrices that are constant on diagonals. Thus (57.1) implies that a single Jordan cell is not reflexive unless $n = 1$.

To characterize all the invariant subspaces of a nilpotent operator Q on $\mathbb{C}^d$ is too difficult. However, it is possible to characterize the operators in $\operatorname{Alg}\operatorname{Lat} Q$ in a way that suffices to decide which nilpotents Q are reflexive.

If Q is nilpotent with index n, the Jordan form for Q is

57.4 $$Q \approx J_{n_1} \oplus \cdots \oplus J_{n_m}$$

where $n = n_1 \geq n_2 \geq \cdots \geq n_m$. Denote this direct sum of spaces by $\mathcal{H} = \mathcal{H}_1 \oplus \cdots \oplus \mathcal{H}_m$. So $J_{n_1} = J_n$ is a cyclic nilpotent on $\mathcal{H}_1$. If e is the cyclic vector for J_{n_1}, then $\mathcal{N} = \mathcal{H}_2 \oplus \cdots \oplus \mathcal{H}_m$ is a complementary invariant subspace for Q.

57.5 Theorem. *Let Q be a nilpotent linear transformation on the finite dimensional space $\mathcal{H}$ and let n be its index. If e is a vector such that $Q^{n-1}e \neq 0$ and $\mathcal{N}$ is an invariant subspace for Q that is complementary to $[P(Q)e]$ with $r =$ the index of $Q|\mathcal{N}$, then $R \in \operatorname{Alg}\operatorname{Lat} Q$ if and only if $R = A + B$, where $A \in P(Q)$ and B is a linear transformation such that $B\mathcal{N} = (0)$ and for $0 \leq k < n$*

$$BQ^k e \in [P(Q)Q^{r+k}e].$$

Proof. Let $R \in \operatorname{Alg}\operatorname{Lat} Q$. Since $Re \in [P(Q)e]$, there is a polynomial p such that $Re = p(Q)e$. Put $A = p(Q)$ and $B = R - A$. Note that $A \in P(Q)$, $B \in \operatorname{Alg}\operatorname{Lat} Q$, and $Be = 0$. Show that B has the desired properties.

First fix a vector x in $\mathcal{N}$ and show that $Bx = 0$. So $[P(Q)x] \subseteq \mathcal{N}$ and hence $[P(Q)x] \cap [P(Q)e] = (0)$. Clearly $[P(Q)(x+e)] + [P(Q)x] \subseteq [P(Q)e] + [P(Q)x]$. On the other hand, $Q^k e + Q^j x = Q^k(e+x) - Q^k x + Q^j x \in [P(Q)x] + [P(Q)(e+x)]$. Thus

$$[P(Q)x] + [P(Q)(x+e)] = [P(Q)e] + [P(Q)x].$$

Also $Q^{n-1}(x+e) = Q^{n-1}x + Q^{n-1}e \neq 0$ since $\mathcal{N} \cap [P(Q)e] = (0)$. Thus $\dim [P(Q)(e+x)] = n$ and so $[P(Q)x] \cap [P(Q)(x+e)] = (0)$. But $Bx = B(x+e) \in [P(Q)x] \cap [P(Q)(e+x)]$, so $B\mathcal{N} = (0)$.

Now to show that B has the other property. If $r = 0$, there is nothing to prove. So assume $r > 0$ and let $h \in \mathcal{N}$ such that $Q^{r-1}h \neq 0$. Because $h \in \mathcal{N}$, $Bh = 0$ and so $y = BQ^k e = B(Q^k e + h) \in [P(Q)Q^k e] \cap [P(Q)(Q^k e + h)]$. Thus there are polynomials $f(z), g(z)$ with $\deg f(z) \leq n - k - 1$ such that $y = f(Q)Q^k e = g(Q)(Q^k e + h)$. Hence $(f(Q) - g(Q))Q^k e = g(Q)h \in [P(Q)e] \cap \mathcal{N} = (0)$. Because $g(Q)h = 0$ and $Q^{r-1}h \neq 0$, it must be that $g(z)$ has the form $g(z) = \alpha_r z^r + \cdots$. On the other hand, the fact that $(f(Q) - g(Q))Q^k e = 0$ implies that $f(z)$ and $g(z)$ have the same first $n - k$ coefficients. The degree restriction on $f(z)$ implies that $f(z) = \alpha_r z^r + \cdots + \alpha_{n-k-1}z^{n-k-1}$. Thus $BQ^k e = y = f(Q)Q^k e \in [P(Q)Q^{r+k}e]$.

For the converse, assume that $R = A + B$ as in the statement of the theorem. It suffices to show that $B \in \operatorname{Alg}\operatorname{Lat} Q$. To do this it suffices to show that for each x in $\mathcal{H}$, $Bx \in [P(Q)x]$. If x is fixed, then $x = \sum_{j=0}^{n-1} \alpha_j Q^j e + y$, where $y \in \mathcal{N}$. Thus $Q^r x = \sum_{j=0}^{n-r-1} \alpha_j Q^{r+j}e$. Let k be the first integer such that $\alpha_k \neq 0$. So $Q^r x = \alpha_k Q^{r+k}e + \cdots + \alpha_{n-r-1}Q^{n-1}e \in [P(Q)x]$. Thus $Q^{n-1-r-k}Q^r x = \alpha_k Q^{n-1}e \in [P(Q)x]$. Since $\alpha_k \neq 0$, $Q^{n-1}e \in [P(Q)x]$. Continue and it follows that $[P(Q)Q^{r+k}e] \subseteq [P(Q)x]$. Thus the assumed properties of B imply that $Bx = \alpha_k BQ^k e + \cdots + \alpha_{n-1}Q^{n-1}e \in [P(Q)Q^{r+k}e] \subseteq [P(Q)x]$. ■

57.6 Corollary. (Brickman and Fillmore [1967]) *For any operator on a finite dimensional space,* $\{T\}' \cap \operatorname{Alg}\operatorname{Lat} T = P(T)$.

Proof. Clearly $\{T\}' \cap \operatorname{Alg}\operatorname{Lat} T \supseteq P(T)$, so it suffices to establish the reverse inclusion. It is left to the reader to show that it suffices to prove this for the case of a nilpotent operator Q. If $R \in \{Q\}' \cap \operatorname{Alg}\operatorname{Lat} Q$, then $R = p(Q) + B$ as in the theorem. Thus $B \in \{Q\}'$. But as in the proof of the theorem, $Be = 0$. Thus $BQ^j e = Q^j Be = 0$. Since $B\mathcal{N} = (0)$, $R = p(Q) \in P(Q)$. ■

This result is false in infinite dimensions, though the counter example was long in coming (Wogen [1987]). This appears later in (61.12).

57.7 Corollary. *If $\mathcal{A}$ is a reflexive abelian algebra of linear transformations on a finite dimensional space and $A \in \mathcal{A}$, then A is a reflexive linear transformation.*

Proof. If $T \in \operatorname{Alg}\operatorname{Lat} A$, then $T \in \operatorname{Alg}\operatorname{Lat} \mathcal{A} = \mathcal{A}$. Since $\mathcal{A}$ is abelian, $T \in \{A\}'$ so that $T \in P(A)$ by the preceding corollary. ■

Deddens and Fillmore [1975] give an example of an abelian algebra of linear transformations on a 4-dimensional space that is not reflexive but such that every element of the algebra is reflexive.

Whether the preceding corollary is true in infinite dimensions is, I believe, unknown.

Proof of Theorem 57.2. As discussed above, it suffices to characterize the reflexive nilpotent linear transformations. Adopt the notation of (57.4) with Q equal to the direct sum rather than being merely similar to it. It must be shown that Q is reflexive if and only if $m \geq 2$ and either $n_1 = n_2$ or $n_1 = n_2 + 1$.

Let $R \in \operatorname{Alg}\operatorname{Lat} Q$ and assume that $m \geq 2$. According to Theorem 57.5, $R = p(Q) + B$, where B has the properties as in the statement of that result. Maintaining the notation of Theorem 57.5, $n = n_1$ and $r = n_2$. If $n = r$, then the condition on B in (57.5) implies that $B = 0$ so that $R = p(Q) \in P(Q)$. If $n = r + 1$, then $Be = \alpha Q^{n-1}e$ for some σ and $BQ^k e = 0$ for $0 < k < n$. Since $B\mathcal{N} = (0)$, this implies that $B = \alpha Q^{n-1}$. Thus $R = q(Q)$ for the polynomial $q(z) = p(z) + \alpha z^{n-1}$.

If $m = 1$, Lemma 57.1 shows that Q is not reflexive. So assume that $m \geq 2$ and that $n > r + 1$. Define the linear transformation B on $\mathcal{H}$ by setting $B\mathcal{N} = (0)$, $Be = Q^r e$, and $BQ^k e = 0$ for $0 < k < n$. According to Theorem 57.5, $B \in \operatorname{Alg}\operatorname{Lat} Q$. However, $QBe = Q^{r+1}e \neq 0$ while $BQe = 0$. Thus $QB \neq BQ$ and so it cannot be that $B \in P(Q)$. Thus Q is not reflexive.
■

§58. Hyperreflexive subspaces

If $\mathcal{S}$ is a subspace of $\mathcal{B}(\mathcal{H})$, define the quantity

$$\alpha(T, \mathcal{S}) = \sup\{\, \|Q^\perp T P\| : P, Q \text{ are projections and} Q^\perp \mathcal{S} P = (0)\,\}.$$

It is left as an exercise to show that if P and Q are projections, $Q^\perp \mathcal{S} P = (0)$ if and only if $g \otimes h \in \mathcal{S}_\perp$ whenever $g \in P$ and $h \in Q^\perp$. (Exercise 1.)

58.1 Proposition. *If $\mathcal{S}$ is a subspace of $\mathcal{B}(\mathcal{H})$, the following properties hold for the quantity $\alpha(T, \mathcal{S})$ just defined.*

(a) $\alpha(T, \mathcal{S}) \leq \operatorname{dist}(T, \mathcal{S})$ *for every operator T.*

(b) $\alpha(T, \mathcal{S}) = \alpha(T, \operatorname{cl}\mathcal{S})$ *for every operator T.*

(c) $T \to \alpha(T, \mathcal{S})$ *defines a seminorm on $\mathcal{B}(\mathcal{H})$.*

(d) *If $T \in \mathcal{B}(\mathcal{H})$, then $\alpha(T, \mathcal{S}) = 0$ if and only if $T \in \operatorname{Ref}\mathcal{S}$.*

(e) $\alpha(T, \mathcal{S}) = \sup\{\, |\langle Tg, h\rangle| : g \otimes h \in \operatorname{ball}\mathcal{S}_\perp \,\}$.

(f) *If $\mathcal{A}$ is a subalgebra of $\mathcal{B}(\mathcal{H})$ containing the identity, then for every operator T,*

$$\alpha(T, \mathcal{A}) = \sup\{\, \|P^\perp T P\| : P \in \operatorname{Lat}\mathcal{A}\,\}.$$

Proof. (a) If $S \in \mathcal{S}$ and P and Q are projections with $Q^\perp \mathcal{S} P = (0)$, then for any operator T, $\|Q^\perp T P\| = \|Q^\perp (T - S)P\| \le \|T - S\|$. Thus $\alpha(T, \mathcal{S}) \le \|T - S\|$ for every S in $\mathcal{S}$. Part (a) now follows.

(e) Let $\beta(T, \mathcal{S})$ be the number on the right hand side of the equation in (e). Note that if $\|g \otimes h\| = \|g\| \, \|h\| \le 1$, it can be assumed that $\|g\| = \|h\| \le 1$. In light of Exercise 1, whenever $Q^\perp \mathcal{S} P = (0)$, it follows that $\|Q^\perp T P\| = \sup\{|\langle Tg, h\rangle| : g \in \operatorname{ball} P, h \in \operatorname{ball} Q^\perp\} \le \beta(T, \mathcal{S})$. Thus $\alpha(T, \mathcal{S}) \le \beta(T, \mathcal{S})$. On the other hand, if $g \otimes h \in \mathcal{S}_\perp$ with $\|g\| = \|h\| \le 1$ and P, Q are the projections $P = \mathbb{C} g$ and $Q = [\mathcal{S} g]$, then $Q^\perp \mathcal{S} P = (0)$ and $|\langle Tg, h\rangle| \le \|Q^\perp T P\| \le \alpha(T, \mathcal{S})$. Thus $\beta(T, \mathcal{S}) \le \alpha(T, \mathcal{S})$.

The proofs of parts (b) and (c) are routine and left for the reader as an exercise. In light of Theorem 56.9 and part (e), $\alpha(T, \mathcal{S}) = 0$ if and only if $T \in ((\operatorname{Ref} \mathcal{S})_\perp)^\perp = \operatorname{Ref} \mathcal{S}$, proving (d).

(f) Denote by $\delta(T, \mathcal{A})$ the quantity on the right hand side of the equation in the statement of part (d). If $P \in \operatorname{Lat} \mathcal{A}$, then $P^\perp \mathcal{A} P = (0)$ so that $\delta(T, \mathcal{A}) \le \alpha(T, \mathcal{A})$. Suppose Q, P are projections satisfying $Q^\perp \mathcal{A} P = (0)$ and define $\widehat{P} = [\mathcal{A} P \mathcal{H}]$. Thus $Q^\perp \widehat{P} = 0$, so that $\widehat{P} \le Q$. The fact that $\mathcal{A}$ is an algebra implies that $\widehat{P} \in \operatorname{Lat} \mathcal{A}$ and the fact that $1 \in \mathcal{A}$ implies that $P \le \widehat{P}$. For any vector h in $\mathcal{H}$,

$$\begin{aligned}
\|Q^\perp T P h\|^2 &= \langle Q^\perp T P h, T P h\rangle \le \langle \widehat{P}^\perp T P h, T P h\rangle \\
&= \langle \widehat{P}^\perp T \widehat{P} P h, T \widehat{P} P h\rangle = \|\widehat{P}^\perp T \widehat{P} P h\|^2 \\
&\le \|\widehat{P}^\perp T \widehat{P}\|^2 \, \|P h\|^2.
\end{aligned}$$

Hence $\|\widehat{P}^\perp T \widehat{P}\| \ge \|Q^\perp T P\|$ so that $\delta(T, \mathcal{A}) \ge \alpha(T, \mathcal{A})$. ∎

58.2 Definition. A linear subspace $\mathcal{S}$ of $\mathcal{B}(\mathcal{H})$ is *hyperreflexive* if there is a constant a such that for each T in $\mathcal{B}(\mathcal{H})$

$$\operatorname{dist}(T, \mathcal{S}) \le a\, \alpha(T, \mathcal{S}).$$

To make the role of the constant explicit, say that $\mathcal{S}$ is hyperreflexive with constant a if the above inequality holds for all operators T. Let $\kappa(\mathcal{S})$ be the infimum of the collection of all constants a such that this inequality holds. Observe that the inequality remains valid with $M = \kappa(\mathcal{S})$. An operator T is *hyperreflexive* if $W(T)$ is hyperreflexive.

Note that Theorem 42.3 says that a separable C*-subalgebra of the Calkin algebra is, in a certain sense, hyperreflexive with constant 1. The following proposition justifies the use of the prefix "hyper" in the terminology.

58.3 Proposition. *A hyperreflexive subspace is reflexive. In particular, every hyperreflexive subspace is* WOT *closed.*

Proof. If $T \in \operatorname{Ref}\mathcal{S}$, then $\alpha(T,\mathcal{S}) = 0$ by (58.1). If $\mathcal{S}$ is hyperreflexive, then this implies that $0 = \operatorname{dist}(T,\mathcal{S})$ so that $T \in \mathcal{S}$. ∎

58.4 Proposition. *Let $\mathcal{S}$ be a closed subspace of $\mathcal{B}(\mathcal{H})$ and let $\mathcal{A}$ be the associated algebra in $\mathcal{B}(\mathcal{H}\oplus\mathcal{H})$ defined in* (56.3). *If $\mathcal{A}$ is hyperreflexive with constant κ, then $\mathcal{S}$ is hyperreflexive with constant κ.*

Proof. For any operator T in $\mathcal{B}(\mathcal{H})$, define the operator $\widetilde{T}$ on $\mathcal{H}\oplus\mathcal{H}$ as in the proof of (56.4). It is left to the reader to show that $\operatorname{dist}(T,\mathcal{S}) = \operatorname{dist}(\widetilde{T},\mathcal{A})$.

According to Proposition 56.4, $R \in \operatorname{Lat}\mathcal{A}$ if and only if $R = Q\oplus P$, where P and Q are projections on $\mathcal{H}$ and $Q^{\perp}\mathcal{S}P = (0)$. Matrix multiplication implies that $R^{\perp}\widetilde{T}R = \widetilde{Q^{\perp}TP}$ for any operator T on $\mathcal{H}$. This implies that $\alpha(\widetilde{T},\mathcal{A}) = \alpha(T,\mathcal{S})$. So for any operator T on $\mathcal{H}$, $\operatorname{dist}(T,\mathcal{S}) = \operatorname{dist}(\widetilde{T},\mathcal{A}) \leq M\,\alpha(\widetilde{T},\mathcal{A}) = M\,\alpha(T,\mathcal{S})$. ∎

A brief encounter with the definitions of $\operatorname{dist}(X,\mathcal{A})$ and $\alpha(X,\mathcal{A})$ for an arbitrary operator $X = (X_{ij})$ in $\mathcal{B}(\mathcal{H}\oplus\mathcal{H})$ will quickly convince the reader that the converse of the preceding proposition is at best hard and possibly false. No example showing it to be false is known.

The main result of this section is the following.

58.5 Theorem. *If $\mathcal{S}$ is a norm closed subspace of $\mathcal{B}(\mathcal{H})$, the following statements are equivalent.*

(a) *$\mathcal{S}$ is hyperreflexive with constant a.*

(b) *There is a constant $b > 0$ such that the closed convex hull of $\{g\otimes h : g\otimes h \in \operatorname{ball}\mathcal{S}_{\perp}\}$ contains $b(\operatorname{ball}\mathcal{S}_{\perp})$.*

(c) *For each C in $\mathcal{S}_{\perp}$ there is a sequence $\{g_n\otimes h_n\}$ in $\mathcal{S}_{\perp}$ such that $\sum_n \|g_n\otimes h_n\| < \infty$ and $C = \sum_n g_n\otimes h_n$.*

(d) *There is a constant $d > 0$ such that for each C in $\mathcal{S}_{\perp}$ there is a sequence $\{g_n\otimes h_n\}$ in $\mathcal{S}_{\perp}$ such that $\sum_n \|g_n\otimes h_n\| \leq d\|C\|_1$ and $C = \sum_n g_n\otimes h_n$.*

(e) *If $\mathcal{B}(\mathcal{H})/\mathcal{S}$ has the norm $\|T+\mathcal{S}\|_{\alpha} \equiv \alpha(T,\mathcal{S})$, it is complete.*

(f) *If $\{T_n\}$ is a sequence in $\mathcal{B}(\mathcal{H})$ such that $\alpha(T_n,\mathcal{S}) \to 0$, then there is a sequence $\{S_n\}$ in $\mathcal{S}$ such that $\|T_n - S_n\| \to 0$.*

If these conditions hold, then the constants a, b, and d that appear there can be chosen to satisfy $a = b^{-1} = d$.

In light of condition (c) and Theorem 56.9, it is observed that when $\mathcal{H}$ is finite dimensional, a subspace is reflexive if and only if it is hyperreflexive.

The proof of the theorem will require some preliminary work. The first requirement is a lemma from Banach space theory.

58.6 Lemma. *If $\mathcal{X}$ is a Banach space and $\mathcal{E}$ is a subset of $\mathcal{X}$ such that $\operatorname{cl}\mathcal{E} \supseteq \operatorname{ball}\mathcal{X}$, then for each x in $\mathcal{X}$ there is a sequence $\{e_n\}$ in $\mathcal{E}$ and a sequence of positive scalars $\{\lambda_n\}$ such that $\sum_n \lambda_n \le \|x\|$ and $x = \sum_n \lambda_n e_n$.*

Proof. Without loss of generality it can be assumed that $\|x\| = 1$. So $x \in \operatorname{cl}\mathcal{E}$ and, by hypothesis, there is a vector x_1 in $\mathcal{E}$ such that $\|x - x_1\| < 1/2$. Thus $x - x_1 \in \frac{1}{2}\operatorname{ball}\mathcal{X} \subseteq \frac{1}{2}\operatorname{cl}\mathcal{E}$ and so there is an x_2 in $\frac{1}{2}\mathcal{E}$ with $\|x - (x_1 + x_2)\| < 1/2^2$. Continue. For each $n \ge 1$ there is a vector x_n in $(\frac{1}{2})^{n-1}\mathcal{E}$ such that $\|x - (x_1 + \cdots + x_n)\| < 1/2^n$. Put $e_n = 2^{n-1}x_n$ and $\lambda_n = 1/2^{n-1}$. So $\sum_n \lambda_n \le 1 = \|x\|$ and $e_n \in \mathcal{E}$ and $x = \sum_n \lambda_n e_n$. ■

58.7 Lemma. *Let $\mathcal{S}$ be a norm closed subspace of $\mathcal{B}(\mathcal{H})$ and assume condition* (c) *in* Theorem 58.5 *is satisfied. If for each C in $\mathcal{S}_\perp$*

$$\|C\|_0 \equiv \inf\left\{ \sum_{n=1}^{\infty} \|g_n \otimes h_n\| : \text{each } g_n \otimes h_n \in \mathcal{S}_\perp, C = \sum_n g_n \otimes h_n, \right.$$
$$\left. \text{and } \sum_{n=1}^{\infty} \|g_n \otimes h_n\| < \infty \right\},$$

then $\|\cdot\|_0$ is a norm on $\mathcal{S}_\perp$ and $(\mathcal{S}_\perp, \|\cdot\|_0)$ is a Banach space. Moreover, for every T in $\mathcal{B}(\mathcal{H})$

$$\alpha(T, \mathcal{S}) = \sup\{|\operatorname{tr}(TC)| : C \in \mathcal{S}_\perp \text{ and } \|C\|_0 \le 1\}.$$

Proof. It is left as an exercise to check that $\|\cdot\|_0$ is a norm on $\mathcal{S}_\perp$ and $\|C\|_1 \le \|C\|_0$ for all C in $\mathcal{S}_\perp$. To show that $\mathcal{S}_\perp$ is complete with respect to this norm, it suffices to show that if $\{C_n\}$ is a sequence in $\mathcal{S}_\perp$ such that $\sum_n \|C_n\|_0 < \infty$, then there is a C in $\mathcal{S}_\perp$ such that $C = \sum_n C_n$, with convergence in the zero norm. But since the zero norm dominates the trace norm, there is a C in $\mathcal{S}_\perp$ such that $\|C - \sum_{k=1}^n C_k\|_1 \to 0$ as $n \to \infty$. For each $k \ge 1$ let $\{g_{kj} \otimes h_{kj}\}_j$ be a sequence of rank one functionals in $\mathcal{S}_\perp$ such that $C_k = \sum_j g_{kj} \otimes h_{kj}$ and $\sum_j \|g_{kj} \otimes h_{kj}\| < \|C_k\|_0 + 2^{-k}$. Thus

$$C - \sum_{k=1}^{n} C_k = \sum_{k=n+1}^{\infty} C_k = \sum_{k=n+1}^{\infty} \sum_{j=1}^{\infty} g_{kj} \otimes h_{kj}$$

and so

$$\left\| C - \sum_{k=1}^{n} C_k \right\|_0 \le \sum_{k=n+1}^{\infty} \sum_{j=1}^{\infty} \|g_{kj} \otimes h_{kj}\|$$

$$\leq \sum_{k=n+1}^{\infty}\left(\|C_k\|_0+\frac{1}{2^k}\right)$$
$$=\sum_{k=n+1}^{\infty}\|C_k\|_0+\frac{1}{2^n}$$

and this can be made as small as desired for sufficiently large n. Thus $(\mathcal{S}_\perp, \|\cdot\|_0)$ is a Banach space.

To prove the second part, fix an operator T and let σ be the supremum that appears in the lemma. If $g \otimes h \in \operatorname{ball}\mathcal{S}_\perp$, then $\|g\otimes h\|_0 \leq 1$ and so $|\langle Tg, h\rangle| \leq \sigma$. According to Proposition 58.2, this implies $\alpha(T,\mathcal{S}) \leq \sigma$. On the other hand, if $C \in \mathcal{S}_\perp$ and $\|C\|_0 < 1$, there are $g_n \otimes h_n$ in $\mathcal{S}_\perp$ such that $C = \sum_n g_n \otimes h_n$ and $\sum_n \|g_n \otimes h_n\| < 1$. Thus

$$|\operatorname{tr}(TC)| \leq \sum_{n=1}^{\infty}|\langle Tg_n, h_n\rangle|$$
$$\leq \sum_{n=1}^{\infty}\alpha(T,\mathcal{S})\|g_n\otimes h_n\|$$
$$\leq \alpha(T,\mathcal{S}).$$

Thus $\sigma \leq \alpha(T,\mathcal{S})$. ∎

Proof of Theorem 58.5. (a) *implies* (b). Let $\mathcal{X}$ be the collection of rank one operators in $\operatorname{ball}\mathcal{S}_\perp$ and let $\mathcal{E}$ be the closed convex hull of $\mathcal{X}$. Note that if $\lambda \in \mathbb{C}$ and $|\lambda| \leq 1$, then $\lambda\mathcal{X} \subseteq \mathcal{X}$; so $\mathcal{E} = (\mathcal{X}^\circ)_\circ$, the bipolar of $\mathcal{X}$ ([ACFA], V.1.8). Now from Proposition 58.1.e it follows that $T \in \mathcal{X}^\circ$ if and only if $\alpha(T,\mathcal{S}) \leq 1$. So if $T \in \mathcal{X}^\circ$, condition (a) implies $\operatorname{dist}(T,\mathcal{S}) \leq a$. That is, the norm of $a^{-1}T + \mathcal{S}$ in the quotient space $\mathcal{B}(\mathcal{H})/\mathcal{S}$ is at most 1. But from general duality theory, this is equivalent to the statement that $a^{-1}T \in (\operatorname{ball}\mathcal{S}_\perp)^\circ$. Hence $\mathcal{X}^\circ \subseteq a(\operatorname{ball}\mathcal{S}_\perp)^\circ$, so that $\mathcal{E} = (\mathcal{X}^\circ)_\circ \supseteq (a(\operatorname{ball}\mathcal{S}_\perp)^\circ)_\circ = a^{-1}\operatorname{ball}\mathcal{S}_\perp$. Take $b = a^{-1}$.

(b) *implies* (d). Let $\mathcal{E}$ be the closed convex hull of the rank one operators in $\operatorname{ball}\mathcal{S}_\perp$. Let $\mathcal{F} = \{\sum_{j=1}^n g_j \otimes h_j : n \geq 1, g_j \otimes h_j \in \mathcal{S}_\perp$, and $\sum_{j=1}^n \|g_j \otimes h_j\| \leq b^{-1}\}$. Note that $\mathcal{F}$ is a convex, bounded set and $\mathcal{F} \supseteq b^{-1}\mathcal{E}$. Thus (b) implies that $\operatorname{cl}\mathcal{F} \supseteq \operatorname{ball}\mathcal{S}_\perp$.

If $C \in \mathcal{S}_\perp$, Lemma 58.6 implies there is a sequence $\{C_n\}$ in $\mathcal{F}$ and a sequence of positive scalars $\{\lambda_n\}$ such that $\sum_n \lambda_n \leq \|C\|_1$ and $C = \sum_n \lambda_n C_n$. For each $n \geq 1$, put $C_n = \sum_{j=1}^{m_n} g_{nj} \otimes h_{nj}$, where each $g_{nj} \otimes h_{nj} \in$

$\mathcal{S}_\perp$ and $\sum_{j=1}^{m_n} \|g_{nj} \otimes h_{nj}\| \le b^{-1}$. Therefore,

$$C = \sum_{n=1}^{\infty} \sum_{j=1}^{m_n} \lambda_n g_{nj} \otimes h_{nj}$$

and

$$\sum_{n=1}^{\infty} \sum_{j=1}^{m_n} \|\lambda_n g_{nj} \otimes h_{nj}\| \le b^{-1} \sum_{n=1}^{\infty} \lambda_n \le b^{-1} \|C\|_1 .$$

This proves (d) with $d = b^{-1}$.

(d) *implies* (c). This is clear.

(c) *implies* (d). According to Lemma 58.7, $(\mathcal{S}_\perp, \|\cdot\|_0)$ is a Banach space. Moreover, the identity mapping $i\colon (\mathcal{S}_\perp, \|\cdot\|_0) \to (\mathcal{S}_\perp, \|\cdot\|_1)$ is a contractive bijection. Since both normed spaces are complete, the Open Mapping Theorem implies there is a constant d such that $\|C\|_0 < d\|C\|_1$ for all C in $\mathcal{S}_\perp$. This is precisely (d).

(d) *implies* (a). Since (d) implies (c), Lemma 58.7 is applicable and the norm $\|\cdot\|_0$ can be defined on $\mathcal{S}_\perp$. Moreover, the constant $d > 0$ in (d) shows that $\|C\|_0 \le d\|C\|_1$ for every C in $\mathcal{S}_\perp$. Using the last part of Lemma 58.7, it follows that if $T \in \mathcal{B}(\mathcal{H})$, $\operatorname{dist}(T, \mathcal{S}) = \sup\{|\operatorname{tr}(TC)| : C \in \mathcal{S}_\perp, \|C\|_1 \le 1\} \le \sup\{|\operatorname{tr}(TC)| : C \in \mathcal{S}_\perp, \|C\|_0 \le d\} = d^{-1}\alpha(T, \mathcal{S})$. This proves (a) with $a = d^{-1}$.

(a) *implies* (f). This is easy.

(f) *implies* (e). Note that it is always true the identity $i\colon (\mathcal{B}(\mathcal{H})/\mathcal{S}, \|\cdot\|) \to (\mathcal{B}(\mathcal{H})/\mathcal{S}, \|\cdot\|_\alpha)$ is a contraction. If $0 = \lim \|T_n + \mathcal{S}\|_\alpha = \lim \alpha(T_n, \mathcal{S})$, then part (f) implies there are operators $\{S_n\}$ in $\mathcal{S}$ such that $\|T_n - S_n\| \to 0$. But then $\|T_n + \mathcal{S}\| = \operatorname{dist}(T_n, \mathcal{S}) \le \|T_n - S_n\| \to 0$. That is, the inverse map i^{-1} is continuous. Part (e) now easily follows.

(e) *implies* (a). This is an easy consequence of the Open Mapping Theorem.

It remains to establish the relation between the constants. If the constant a is given as in (a), then the proof that (a) implies (b) shows explicitly that it is possible to take $b = a^{-1}$. The proof that (b) implies (d) shows that if b is given, it can be arranged that $d = b^{-1}$. Finally, the proof that (d) implies (a) shows that if d is given, $a = d^{-1}$ works. ■

Theorem 56.9 implies that if $\mathcal{S}$ is reflexive and $C \in \mathcal{S}_\perp$, there is a sequence $\{C_n\}$ in $\mathcal{S}_\perp$ such that $\|C - C_n\|_1 \to 0$ and each C_n is the finite linear combination of rank one elements of $\mathcal{S}_n$. By passing to a subsequence, it can be assumed that $\sum \|C_n - C_{n+1}\|_1 < \infty$. Thus for $D_n = C_n - C_{n-1}$, $C = \sum_n D_n = \sum_k g_k \otimes h_k$, where each $g_k \otimes h_k$ belongs to $\mathcal{S}_\perp$. The sequence of norms, $\{\, \|g_k \otimes h_k\| \,\}$, however, may not be summable. Paraphrasing this, it might be said that a subspace is reflexive if and only if every annihilating functional is the *conditionally convergent* sum of annihilating rank one functionals. On the other hand, as Arveson [1984] points out, it is legitimate to paraphrase Theorem 58.5 as saying that a subspace $\mathcal{S}$ is hyperreflexive if and only if every annihilating functional is the *absolutely convergent* sum of rank one annihilating functionals.

If $C \in \mathcal{S}_\perp$, put

$$\kappa(C, \mathcal{S}) = \inf \left\{ \sum_{n=1}^{\infty} \|h_n\| \, \|g_n\| : C = \sum_{n=1}^{\infty} h_n \otimes g_n \right.$$
$$\left. \text{and each } g_n \otimes h_n \in \mathcal{S}_\perp \right\}.$$

58.8 Proposition. *If $\mathcal{S}$ is a hyperreflexive subspace, then*

$$\kappa(\mathcal{S}) \equiv \sup \{\, \kappa(C, \mathcal{S}) : C \in \operatorname{ball} \mathcal{S}_\perp \,\}.$$

Proof. Let σ be the supremum defined in the statement of this proposition. If $d > \sigma$, then statement (d) in Theorem 58.5 is valid. Therefore that theorem says that $\mathcal{S}$ is hyperreflexive with constant d. Thus $\kappa(\mathcal{S}) \le d$; thus $\kappa(\mathcal{S}) \le \sigma$. Conversely, condition (a) of Theorem 58.5 is satisfied with $a = \kappa(\mathcal{S})$. Thus (d) holds with $d = \kappa(\mathcal{S})$; thus $\sigma \le \kappa(\mathcal{S})$. ■

Now to begin the construction of an example, due to Kraus and Larson [1985]. Let $\{e_1, e_2\}$ be the usual basis for $\mathbb{C}^2$ and fix ϵ with $0 < \epsilon < 1/3$. Put

$$u_1 = e_1, \quad u_2 = \frac{e_1 + \epsilon e_2}{\sqrt{1 + \epsilon^2}}.$$

Let G_1, G_2 be the rank one projections $G_j = u_j \otimes u_j$ and let $\mathcal{E}$ be the linear span of G_1 and G_2, thought of as being contained in $\mathcal{B}_1(\mathbb{C}^2)$. Set $\mathcal{S} = \mathcal{E}^\perp \subseteq \mathcal{B}(\mathbb{C}^2)$. Because $\mathcal{S}_\perp = \mathcal{E}$ is the linear span of its rank one elements, $\mathcal{S}$ is reflexive (56.9). Because the space is finite dimensional, $\mathcal{S}$ is hyperreflexive. It is claimed that the constant of hyperreflexivity is at least $1/3\epsilon$.

To see this, first observe that, relative to the standard basis for $\mathbb{C}^2$, G_1 and G_2 have the matrices

$$G_1 = \begin{bmatrix} 1 & 0 \\ 0 & 0 \end{bmatrix} \qquad G_2 = \frac{1}{1 + \epsilon^2} \begin{bmatrix} 1 & \epsilon \\ \epsilon & \epsilon^2 \end{bmatrix}.$$

Thus

$$\mathcal{E} = \left\{ \begin{bmatrix} \lambda + \mu & \mu\epsilon \\ \mu\epsilon & \mu\epsilon^2 \end{bmatrix} : \lambda, \mu \in \mathbb{C} \right\}.$$

From here, solving some equations shows that $\mathcal{S}$ is given by

$$\mathcal{S} = \left\{ \begin{bmatrix} 0 & \alpha \\ \beta & -\frac{\alpha+\beta}{\epsilon} \end{bmatrix} : \alpha, \beta \in \mathbb{C} \right\}.$$

Let $\mathcal{X}$ denote the convex hull of the rank 1 operators in ball $\mathcal{E}$ and let $r = \kappa(\mathcal{S})^{-1}$. So $r\,(\text{ball}\,\mathcal{E}) \subseteq \mathcal{X}$.

Assume that $F = \alpha_1 G_1 + \alpha_2 G_2$ has rank 1. So $Fe_1 = \alpha_1 u_1 + \alpha_2(1 + \epsilon^2)^{-1/2} u_2$ and $Fe_2 = \epsilon\alpha_2(1+\epsilon^2)^{-1/2}u_2$. Since u_1 and u_2 are linearly independent and F has rank 1, either $\alpha_1 = 0$ or $\alpha_2 = 0$. That is, all the rank one operators in $\mathcal{E}$ are multiples of either G_1 or G_2. Thus $\mathcal{X} = \{\, \lambda_1 G_1 + \lambda_2 G_2 : |\lambda_1| + |\lambda_2| \leq 1 \}$.

If $H = N^{-1}[(1+\epsilon^2)G_2 - G_1]$, where $N = \|(1+\epsilon^2)G_2 - G_1\|_1$, then $\|H\|_1 = 1$ and so $rH \in \mathcal{X}$. Hence $rH = \lambda_1 G_1 + \lambda_2 G_2$, where $|\lambda_1| + |\lambda_2| \leq 1$. Thus $\|rHe_2\| \leq \epsilon/\sqrt{1+\epsilon^2} < \epsilon$. That is, $\|He_2\| < \epsilon/r$. On the other hand, $He_2 = N^{-1}(1+\epsilon^2)G_2 e_2 = N^{-1}[\epsilon e_1 + \epsilon^2 e_2]$. Hence $\|He_2\| = N^{-1}\epsilon\sqrt{1+\epsilon^2}$. Now $N = \|(1+\epsilon^2)G_2 - G_1\|_1 = \|e_1 \otimes e_1 + \epsilon e_1 \otimes e_2 + \epsilon e_2 \otimes e_1 + \epsilon^2 e_2 \otimes e_2 - e_1 \otimes e_1\|_1 \leq 3\epsilon$. So $\|He_2\| \geq \epsilon(1+\epsilon^2)/3\epsilon > 1/3$. Therefore $1/3 < \|He_2\| < \epsilon/r$ so $r < 3\epsilon$. That is, $\kappa(\mathcal{S}) > 1/3\epsilon$.

Thus it has been shown that for every $\epsilon > 0$ there is a two dimensional hyperreflexive subspace $\mathcal{S}_\epsilon$ in $\mathcal{B}(\mathbb{C}^2)$ with the constant of hyperreflexivity at least $(3\epsilon)^{-1}$.

58.9 Example. (Kraus and Larson [1985]) There is an algebra that is reflexive but not hyperreflexive.

For each $n \geq 4$ let $\mathcal{S}_n$ be the linear subspace constructed above with $\epsilon = n^{-1}$. So $\mathcal{S}_n$ is hyperreflexive with constant at least $n/3$. Let $\mathcal{A}_n$ be the corresponding algebra in $\mathcal{B}(\mathbb{C}^4)$ as in (56.4). By Proposition 58.4, $\mathcal{A}_n$ is hyperreflexive with constant at least $n/3$. Put $\mathcal{A} = \bigoplus_{n=4}^{\infty} \mathcal{A}_n$.

Because each $\mathcal{A}_n$ is reflexive, $\mathcal{A}$ is reflexive. For each n there is an operator T_n on $\mathbb{C}^2$ such that $\text{dist}\,(T_n, \mathcal{A}_n) > \frac{n}{3}\,\alpha(T_n, \mathcal{A}_n)$. Let $\widehat{T}_n$ be the operator on $(\mathbb{C}^2)^{(\infty)}$ that equals T_n on the n-th coordinate and is zero elsewhere. It is easy to see that $\text{dist}\,(\widehat{T}_n, \mathcal{A}) = \text{dist}\,(T_n, \mathcal{A}_n)$. But $\text{Lat}\,\mathcal{A} = \bigoplus_{n=4}^{\infty} \text{Lat}\,\mathcal{A}_n$. So $\text{dist}\,(\widehat{T}_n, \mathcal{A}) = \text{dist}\,(T_n, \mathcal{A}_n) > \frac{n}{3}\,\alpha(T_n, \mathcal{A}_n) = \frac{n}{3}\,\alpha(\widehat{T}_n, \mathcal{A}_n)$. Therefore $\mathcal{A}$ is not hyperreflexive.

Exercises.

1. If $\mathcal{S}$ is a linear subspace of $\mathcal{B}(\mathcal{H})$ and P, Q are projections, show that $Q^\perp \mathcal{S} P = (0)$ if and only if $g \otimes h \in \mathcal{S}_\perp$ whenever $g \in P$ and $h \in Q^\perp$.

2. Assume that $\mathcal{S}$ is hyperreflexive, define the norm $\|\cdot\|_0$ on $\mathcal{S}_\perp$ as in Lemma 58.7, and define the norm $\|\cdot\|_\alpha$ as in Theorem 58.6.e. For $T+\mathcal{S}$ in $\mathcal{B}(\mathcal{H})/\mathcal{S}$ define $L_{T+\mathcal{S}}\colon \mathcal{S}_\perp \to \mathbb{C}$ by $L_{T+\mathcal{S}}(C) = \operatorname{tr}(TC)$. Show that the map $T+\mathcal{S} \to L_{T+\mathcal{S}}$ is an isometric isomorphism of $(\mathcal{B}(\mathcal{H})/\mathcal{S}, \|\cdot\|_\alpha)$ onto $(\mathcal{S}_\perp, \|\cdot\|_0)^*$.

§59. Reflexivity and duality

There is another concept that has a bearing on reflexivity. It is also the root of an entire arsenal of powerful techniques in operator theory.

59.1 Definition. A linear subspace $\mathcal{S}$ of $\mathcal{B}(\mathcal{H})$ is said to have *property* $\mathbb{A}_1$ if $\mathcal{S}$ is weak* closed and for every weak* continuous linear functional ϕ on $\mathcal{S}$ there are vectors h, g in $\mathcal{H}$ such that $\phi(S) = \langle Sh, g\rangle$ for all S in $\mathcal{S}$. If $r \geq 1$, say that $\mathcal{S}$ has *property* $\mathbb{A}_1(r)$ if $\mathcal{S}$ is weak* closed and for every $\epsilon > 0$ and every weak* continuous linear functional ϕ on $\mathcal{S}$ there are vectors h, g in $\mathcal{H}$ such that $\phi(S) = \langle Sh, g\rangle$ for all S in $\mathcal{S}$ and $\|h\|\,\|g\| < (r+\epsilon)\|\phi\|$.

These definitions, for the case that $\mathcal{S}$ is an algebra, appear in Bercovici, Foias, and Pearcy [1985] where the foundation for their study is explored. In that work they also define properties $\mathbb{A}_n$ and $\mathbb{A}_n(r)$, which are equivalent to the conditions that $M_n \otimes \mathcal{S} \subseteq \mathcal{B}(\mathcal{H}^{(n)})$ has property $\mathbb{A}_1$ or property $\mathbb{A}_1(r)$, respectively. Property $\mathbb{A}_1$ is also discussed in Larson [1982] where it is called property P_1; Hadwin and Nordgren [1982] discuss both conditions and call them D_σ and $\mathrm{D}_\sigma(r)$. The root of this notion lies in Brown [1978].

Note that for any $t > 0$, $(th) \otimes (t^{-1}g) = h \otimes g$. So in the definition of property $\mathbb{A}_1$ it can always be required that $\|h\| = \|g\|$.

For any weak* closed subspace $\mathcal{S}$ of $\mathcal{B}(\mathcal{H})$, there is a predual. Indeed, if $\mathcal{S}_* = \mathcal{B}_1(\mathcal{H})/\mathcal{S}_\perp$, $\mathcal{S}$ is the Banach space dual of $\mathcal{S}_*$. This can be identified with the weak* continuous linear functionals on $\mathcal{S}$. Thus $\mathcal{S}$ has property $\mathbb{A}_1$ if and only if for every trace class operator C there is a rank one operator $g \otimes h$ such that $C - g \otimes h \in \mathcal{S}_\perp$.

It is clear that if $\mathcal{S}$ is a weak* closed subspace of $\mathcal{B}(\mathcal{H})$ and $\mathcal{S}$ has property $\mathbb{A}_1$, then the relative weak and weak* topologies agree on $\mathcal{S}$. It is not clear, however, that $\mathcal{S}$ must be WOT closed. The next result first appeared in Olin and Thomson [1980], though in a more specialized setting.

59.2 Proposition. *If $\mathcal{S}$ is a weak* closed subspace of $\mathcal{B}(\mathcal{H})$ that has property $\mathbb{A}_1(r)$ for some $r \geq 1$, then $\mathcal{S}$ is* WOT *closed.*

Proof. Let $\{S_i\}$ be a net in $\mathcal{S}$ that converges to T in the WOT. Define a linear functional $\Lambda : \mathcal{S}_* \to \mathbb{C}$ as follows. If $\phi \in \mathcal{S}_*$, let $g, h \in \mathcal{H}$ such that $\phi(S) = \langle Sh, g\rangle$ for all S in $\mathcal{S}$ and $\|h\|\,\|g\| < (r+1)\|\phi\|$. Let $\Lambda(\phi) = \langle Th, g\rangle$.

It is routine to check that the hypotheses imply that Λ is well defined and $\|\Lambda\| \leq \|T\|(r+1)$. Thus there is an operator S in $\mathcal{S}$ such that $\Lambda(\phi) = \phi(S)$ for all ϕ in $\mathcal{S}_*$. But this implies that $\langle Th, g\rangle = \langle Sh, g\rangle$ for all g, h in $\mathcal{H}$ so that $T = S \in \mathcal{S}$. ∎

The interest in property $\mathbb{A}_1$ in the present context rests in the next two theorems.

59.3 Theorem. (Loginov and Sulman [1975]) *A subspace $\mathcal{S}$ of $\mathcal{B}(\mathcal{H})$ has the property that each of its weak* closed subspaces is reflexive if and only if it is reflexive and has property $\mathbb{A}_1$.*

Proof. Assume each weak* closed subspace of $\mathcal{S}$ is reflexive and let $C \in \mathcal{B}_1(\mathcal{H})$. Put $\mathcal{C} = \{S \in \mathcal{S} : \operatorname{tr}(CS) = 0\}$ and fix an operator B_0 in $\mathcal{S}$ such that $\operatorname{tr}(B_0) = 1$. So $\mathcal{C} = \operatorname{Ref}\mathcal{C}$ by assumption and, hence, $B_0 \notin \operatorname{Ref}\mathcal{C}$. Thus there is a positive number δ and a vector h such that $\|B_0h - Ah\| \geq \delta$ for all A in $\mathcal{C}$. Now if S is any operator in $\mathcal{S}$ and $\alpha = \operatorname{tr}(CS)$, then $A = \alpha B_0 - S \in \mathcal{C}$. Hence $\|Sh\| = \|\alpha B_0 h - Ah\| \geq |\alpha|\delta = \delta|\operatorname{tr}(CS)|$. Thus $|\operatorname{tr}(CS)| \leq \delta^{-1}\|Sh\|$ for all S in $\mathcal{S}$. In other words, $Sh \to \operatorname{tr}(CS)$ is a well defined, bounded linear functional on the linear manifold $\mathcal{S}h$ in $\mathcal{H}$. Hence there is a vector g such that $\operatorname{tr}(CS) = \langle Sh, g\rangle$ for all S in $\mathcal{S}$ and so $C - h \otimes g \in \mathcal{S}_\perp$.

Now assume that $\mathcal{S}$ has property $\mathbb{A}_1$. For any vectors h, g consider $\mathcal{F}_{h\otimes g} \equiv \{T \in \mathcal{B}(\mathcal{H}) : \langle Th, g\rangle = 0\}$. If $A \in \operatorname{Ref}\mathcal{F}_{h\otimes g}$, then $Ah \in [\mathcal{F}_{h\otimes g}h]$. But $g \perp [\mathcal{F}_{h\otimes g}h]$ so $g \perp Ah$. That is, $A \in \mathcal{F}_{h\otimes g}$. So $\mathcal{F}_{h\otimes g}$ is a reflexive subspace of $\mathcal{B}(\mathcal{H})$. Since $\mathcal{S}$ is reflexive, $\mathcal{S} \cap \mathcal{F}_{h\otimes g}$ is reflexive by Lemma 56.2. But the fact that $\mathcal{S}$ has property $\mathbb{A}_1$ implies that $\mathcal{S} \cap \mathcal{F}_{h\otimes g}$ is the generic weak* closed hyperplane in $\mathcal{S}$. Thus the Hahn–Banach Theorem implies that if $\mathcal{T}$ is any weak* closed subspace of $\mathcal{S}$, then

$$\mathcal{T} = \bigcap\{\mathcal{S} \cap \mathcal{F}_{h\otimes g} : \mathcal{T} \subseteq \mathcal{F}_{h\otimes g}\}.$$

Again Lemma 56.2 implies $\mathcal{T}$ is reflexive. ∎

Actually Loginov and Sulman [1975] proved a somewhat different result: if $\mathcal{S}$ is WOT closed, then every weakly closed subspace is reflexive if and only if for every finite rank operator C there are vectors h, g such that $C - h \otimes g \in \mathcal{S}_\perp$. It is unknown whether having the property that every WOT closed subspace is reflexive implies that every weak* closed subspace is reflexive, though one is tempted to believe it does not. Of course if $\mathcal{S}$ has property $\mathbb{A}_1$, then it is weak* closed and the weak* and WOT agree on $\mathcal{S}$.

One may wonder if just having property $\mathbb{A}_1$ guarantees reflexivity. This is not the case as the following example shows. Let $J = J_2$ be the 2×2 Jordan block operating on $\mathbb{C}^2$. That is, J has a 1 in the upper right hand corner and zeros elsewhere. Put $\mathcal{S} = \mathbb{C} + \mathbb{C}J$, the algebra generated by J and the identity. Let L be a linear functional on $\mathcal{S}$; so L is determined by the numbers $a = L(I)$ and $b = L(J)$. If x, y are the vectors $x = (a, b), y = (1, 0)$ in $\mathbb{C}^2$, it is easy to check that $L(S) = \langle Sx, y\rangle$ for all S in $\mathcal{S}$. So $\mathcal{S}$ has property $\mathbb{A}_1$ but is not reflexive (57.2).

59.4 Theorem. (Kraus and Larson [1986]) *If $\mathcal{S}$ is a hyperreflexive subspace of $\mathcal{B}(\mathcal{H})$, then every weak* closed subspace of $\mathcal{S}$ is hyperreflexive if and only if $\mathcal{S}$ has property $\mathbb{A}_1$. If $\mathcal{S}$ is hyperreflexive and has property $\mathbb{A}_1(r)$ for some $r \geq 1$, then for every weak* closed subspace $\mathcal{T}$ of $\mathcal{S}$,*

$$\kappa(\mathcal{T}) \leq r + (1 + r)\kappa(\mathcal{S}).$$

Proof. If every weak* closed subspace of $\mathcal{S}$ is hyperreflexive, then every weak* closed subspace is reflexive and so $\mathcal{S}$ has property $\mathbb{A}_1$ by Theorem 59.3. Now assume that $\mathcal{S}$ is hyperreflexive, $\mathcal{S}$ has property $\mathbb{A}_1$, and let $a > \kappa(\mathcal{S})$. If $\mathcal{T}$ is any weak* closed subspace of $\mathcal{S}$, let $C \in \mathcal{T}_\perp$. If ϕ is the linear functional defined on $\mathcal{S}$ by $\phi(S) = \operatorname{tr}(CS)$, then the assumption that $\mathcal{S}$ has property $\mathbb{A}_1$ implies there is a rank one operator $g \otimes h$ with $C - g \otimes h \in \mathcal{S}_\perp$. Note that this implies that $g \otimes h \in \mathcal{T}_\perp$. Since $\mathcal{S}$ is hyperreflexive, there are rank one operators $\{g_n \otimes h_n\}$ in $\mathcal{S}_\perp$ such that $C - g \otimes h = \sum_n g_n \otimes h_n$ and $\sum_n \|g_n\|\,\|h_n\| < \infty$. Thus $C = g \otimes h + \sum_n g_n \otimes h_n$ and so $\mathcal{T}$ is hyperreflexive by Theorem 58.5.c.

Now assume that $\mathcal{S}$ has property $\mathbb{A}_1(r)$. If $\epsilon > 0$, then the rank one operator $g \otimes h$ above can be chosen so that $\|g\|\,\|h\| < (r + \epsilon)\|C\|_1$. Thus

$$\begin{aligned}\kappa(C, \mathcal{T}) &\leq \|h\|\,\|g\| + \kappa(C - h \otimes g, \mathcal{S}_\perp)\\ &\leq \|h\|\,\|g\| + \kappa(\mathcal{S})\|C - h \otimes g\|\\ &< (r + \epsilon)\|C\|_1 + \kappa(\mathcal{S})(1 + r + \epsilon)\|C\|_1.\end{aligned}$$

Since ϵ was arbitrary, this implies that $\kappa(C, \mathcal{T}) \leq r + \kappa(\mathcal{S})(1 + r)$ and so $\kappa(\mathcal{T}) \leq r + \kappa(\mathcal{S})(1 + r)$. ∎

In light of the preceding two theorems, there is interest in finding a large collection of weak* closed subspaces of $\mathcal{B}(\mathcal{H})$ that have property $\mathbb{A}_1$ and property $\mathbb{A}_1(r)$. The next section will focus on von Neumann algebras. The rest of this section concentrates on other weak* closed subspaces.

59.5 Proposition. *Let $\{\mathcal{S}_n\}$ be a collection of weak* closed subspaces of $\mathcal{B}(\mathcal{H}_n)$.*

(a) *If $\{\mathcal{S}_n\}$ is a collection of weak* closed subspaces of $\mathcal{B}(\mathcal{H}_n)$, then each $\mathcal{S}_n$ has property $\mathbb{A}_1(r)$ if and only if $\bigoplus_n \mathcal{S}_n$ has property $\mathbb{A}_1(r)$.*

(b) *If $\mathcal{S}$ is a weak* closed subspace of $\mathcal{B}(\mathcal{H})$, $1 \leq n \leq \infty$, and $\mathcal{S}$ has property $\mathbb{A}_1(r)$, then $\mathcal{S}^{(n)}$ has property $\mathbb{A}_1(r)$.*

Proof. (a) Suppose L is a bounded linear functional on $\mathcal{S} = \bigoplus_n \mathcal{S}_n$, and, for each $n \geq 1$, let $L_n = L|\mathcal{S}_n$. Because $\mathcal{S}$ is an ℓ_∞-direct sum, $\|L\| = \sum_n \|L_n\|$. If each $\mathcal{S}_n$ has property $\mathbb{A}_1(r)$, $\epsilon > 0$, and L is a weak* continuous linear functional on $\mathcal{S}$, for each $n \geq 1$ let $g_n, h_n \in \mathcal{H}_n$ such that $L_n(S_n) = \langle S_n h_n, g_n\rangle$ for all S_n in $\mathcal{S}_n$ and $\|g_n\|, \|h_n\| \leq [(r+\epsilon)\|L_n\|]^{1/2}$. It follows that $\sum_n \|g_n\|^2 \leq (r+\epsilon)\sum_n \|L_n\| = (r+\epsilon)\|L\|$. Similarly, $\sum_n \|h_n\|^2 \leq (r+\epsilon)\|L\|$. Thus if $g = (g_n)$ and $h = (h_n)$, $L(\{S_n\}) = \langle \bigoplus S_n h, g\rangle$ and $\|g\|\,\|h\| \leq (r+\epsilon)\|L\|$. Therefore $\mathcal{S}$ has property $\mathbb{A}_1(r)$.

Conversely, assume that $\mathcal{S} = \bigoplus_n \mathcal{S}_n$ has property $\mathbb{A}_1(r)$. If $n \geq 1$ and L_n is a weak* continuous linear functional on $\mathcal{S}_n$, then $L(\bigoplus S_k) = L_n(S_n)$ defines a weak* continuous linear functional on $\mathcal{S}$. The ability to appropriately factor L implies that an appropriate factorization of L_n exists.

(b) Let $\epsilon > 0$. If L is a weak* continuous linear functional on $\mathcal{S}^{(n)}$, then $L_1(S) = L^{(n)}(S^{(n)})$ defines a weak* continuous linear functional on $\mathcal{S}$ and $\|L_1\| = \|L\|$. Thus there are vectors g and h in $\mathcal{H}$ such that $L_1(S) = \langle Sg, h\rangle$ and $\|g\|\,\|h\| \leq (r+\epsilon)\|L_1\|$. If $\widetilde{g} = (g, 0, \dots)$ and $\widetilde{h} = (h, 0, \dots) \in \mathcal{H}^{(n)}$, then $L(S^{(n)}) = \langle S^{(n)}\widetilde{g}, \widetilde{h}\rangle$ and $\|\widetilde{g}\|\,\|\widetilde{h}\| \leq (r+\epsilon)\|L_1\|$. ∎

The converse of part (b) is not true as the next proposition demonstrates. Indeed, this is a rephrasing of Proposition 21.3.

59.6 Proposition. *$\mathcal{B}(\mathcal{H})^{(\infty)}$ has property $\mathbb{A}_1(1)$.*

Proof. If $\phi\colon \mathcal{B}(\mathcal{H})^{(\infty)} \to \mathbb{C}$ is a weak* continuous linear functional, then $\widetilde{\phi}(T) = \phi(T^{(\infty)})$ defines a weak* continuous linear functional on $\mathcal{B}(\mathcal{H})$ and $\|\widetilde{\phi}\| = \|\phi\|$. Thus there is a K in $\mathcal{B}_1(\mathcal{H})$ such that $\widetilde{\phi}(T) = \operatorname{tr}(KT)$. Proposition 21.3 implies there are vectors g and h in $\mathcal{H}^{(\infty)}$ such that $\phi(T^{(\infty)}) = \langle T^{(\infty)}g, h\rangle$ and $\|g\|^2 = \|h\|^2 = \|K\|_1 = \|\widetilde{\phi}\| = \|\phi\|$. ∎

59.7 Corollary. *If $\mathcal{S}$ is any* weak* *closed subspace of $\mathcal{B}(\mathcal{H})$, then $\mathcal{S}^{(\infty)}$ is reflexive.*

Proof. Since $\mathcal{B}(\mathcal{H})^{(\infty)}$ has property $\mathbb{A}_1$, Theorem 59.3 implies it suffices to show that $\mathcal{B}(\mathcal{H})^{(\infty)}$ is reflexive. But this is a von Neumann algebra and therefore it is reflexive. ∎

59.8 Corollary. *$T^{(\infty)}$ is always a reflexive operator.*

Proof. Indeed, $W(T^{(\infty)}) = P^\infty(T^{(\infty)}) = P^\infty(T)^{(\infty)}$, and this is reflexive by Corollary 59.7. ∎

If $2 \le d < \infty$ and $\dim \mathcal{H} < \infty$, then it has been shown that $T^{(d)}$ is always reflexive (57.3). However when $\dim \mathcal{H} = \infty$ this is not necessarily the case as will be seen (59.14).

59.9 Proposition. *If $1 \le d \le \infty$ and $\mathcal{S}$ is a reflexive subspace of $\mathcal{B}(\mathcal{H})$, then $\mathcal{S}^{(d)}$ is a reflexive subspace of $\mathcal{B}(\mathcal{H}^{(d)})$.*

Proof. By Corollary 59.7 it can be assumed that $d < \infty$. If $B \in \operatorname{Ref} \mathcal{S}^{(d)}$, then $B = B_1^{(d)}$ for some operator B_1 in $\operatorname{Ref} \mathcal{S}$ (See Exercise 2). Since $\mathcal{S}$ is reflexive, $B \in \mathcal{S}^{(d)}$. ∎

59.10 Corollary. *If T is a reflexive operator, so is $T^{(d)}$ for any d.*

An ongoing discussion and investigation in operator theory is the concept of multiplicity and its correct definition. This discussion won't be entered into here, save for one small piece of it. The optimal multiplicity theory occurs in the study of normal operators. Here one associates with every normal operator a scalar-valued spectral measure and a multiplicity function. (See Exercise 51.7.) Two normal operators are unitarily equivalent if and only if they have equivalent spectral measures and equal multiplicity functions. A normal operator is said to have uniform infinite multiplicity if the multiplicity function is constantly equal to ∞ on its spectrum. It is not hard to see that a normal operator N has uniform infinite multiplicity if and only if $N \cong N^{(\infty)}$ (Exercise 51.9). Thus it is reasonable to say that an arbitrary operator T has *uniform infinite multiplicity* if there is an operator A such that $T \cong A^{(\infty)}$. It then follows that if T has uniform infinite multiplicity, then $T \cong T^{(2)} \cong T^{(\infty)}$. The next result says that the converse of this statement holds. If you will, it says another possible definition of uniform infinite multiplicity is equivalent to the given one.

59.11 Proposition. *If T is an operator such that $T \cong T^{(2)}$, then T has uniform infinite multiplicity.*

Proof. The hypothesis is equivalent to the statement that there are reducing subspaces $\mathcal{L}_0, \mathcal{H}_1$ for T such that $\mathcal{H} = \mathcal{L}_0 \oplus \mathcal{H}_1$ and $T \cong T|\mathcal{L}_0 \cong T|\mathcal{H}_1$. Let $U \colon \mathcal{H} \to \mathcal{H}_1$ be a unitary such that $UTU^{-1} = T|\mathcal{H}_1$. Considered as an operator from $\mathcal{H}$ into itself, U is an isometry and $UT = TU$. Examine the von Neumann–Wold decomposition of this isometry.

Put $\mathcal{H}_n = U^n\mathcal{H}$, $\mathcal{H}_\infty = \bigcap_n \mathcal{H}_n$. So $\mathcal{H}_\infty$ reduces U and $U|\mathcal{H}_\infty$ is unitary. It also follows that each $\mathcal{H}_n$ reduces T and, hence, so does $\mathcal{H}_\infty$. Let $\mathcal{L}_n = \mathcal{H}_n \ominus \mathcal{H}_{n+1}$ with $\mathcal{H}_0 = \mathcal{H}$. Clearly each $\mathcal{L}_n$ reduces T and $U\mathcal{L}_n = \mathcal{L}_{n+1}$ for all $n \geq 0$. Since $UT = TU$, $T \cong T|\mathcal{L}_n$ for all n. Thus

$$T \cong T|\mathcal{H}_\infty \oplus \bigoplus_{n=0}^{\infty} T|\mathcal{L}_n.$$

Now the fact that $T^{(2)} \cong T \cong T|\mathcal{L}_n$ implies that for $n \geq 0$ there are reducing subspaces $\mathcal{L}_n'$ and $\mathcal{L}_n''$ for T such that $\mathcal{L}_n = \mathcal{L}_n' \oplus \mathcal{L}_n''$ and $T|\mathcal{L}_n' \cong T|\mathcal{H}_\infty^\perp \cong T^{(\infty)}$ and $T|\mathcal{L}_n'' \cong T|\mathcal{H}_\infty$. This gives that

$$\begin{aligned}
T &\cong \bigoplus_{n=0}^{\infty} T|\mathcal{L}_n' \oplus T|\mathcal{H}_\infty \oplus \bigoplus_{n=0}^{\infty} T|\mathcal{L}_n'' \\
&\cong \bigoplus_{n=0}^{\infty} T|\mathcal{L}_n' \oplus T|\mathcal{H}_\infty \oplus (T|\mathcal{H}_\infty)^{(\infty)} \\
&\cong \bigoplus_{n=0}^{\infty} T|\mathcal{L}_n' \oplus (T|\mathcal{H}_\infty)^{(\infty)} \\
&\cong \bigoplus_{n=0}^{\infty} T|\mathcal{L}_n' \oplus \bigoplus_{n=0}^{\infty} T|\mathcal{L}_n'' \\
&\cong \bigoplus_{n=0}^{\infty} T|\mathcal{L}_n \\
&\cong T^{(\infty)}. \quad \blacksquare
\end{aligned}$$

59.12 Corollary. *If T is any operator such that $T \cong T^{(2)}$, then T is reflexive.*

59.13 Definition. Say that a weak* closed subspace $\mathcal{S}$ of $\mathcal{B}(\mathcal{H})$ is *d-reflexive* if $\mathcal{S}^{(d)}$ is reflexive. An operator T is *d-reflexive* if $T^{(d)}$ is reflexive.

It has been shown that a reflexive subspace is d-reflexive for all d. On the other hand, the operator $T = \begin{bmatrix} 0 & 1 \\ 0 & 0 \end{bmatrix}$ is 2-reflexive but not reflexive. Now for the analogue of Proposition 56.9. This result appeared in Larson [1982], Longstaff [1979], and Tsuji [1983].

59.14 Theorem. *If $\mathcal{S}$ is a* weak* *closed subspace of $\mathcal{B}(\mathcal{H})$, then $\mathcal{S}$ is d-reflexive if and only if $\mathcal{S}_\perp$ is the closed linear span of the operators in $\mathcal{S}_\perp$ having rank at most d.*

Proof. In light of Corollary 59.8 it can be assumed that $d < \infty$. First assume that $\mathcal{S}^{(d)}$ is reflexive and let $T \in \mathcal{B}(\mathcal{H})$ such that $\operatorname{tr}(TF) = 0$ whenever F is an element of $\mathcal{S}_\perp$ having rank at most d. It must be shown that $\operatorname{tr}(TK) = 0$ for every K in $\mathcal{S}_\perp$, or, equivalently, that $T \in \mathcal{S}$. But if $f, g \in \mathcal{H}^{(d)}$ such that $f \otimes g \in \mathcal{S}_\perp^{(d)}$, there is an operator F on $\mathcal{H}$ such that $\operatorname{rank} F \leq d$ and $\langle X^{(d)} f, g\rangle = \operatorname{tr}(XF)$ for every X in $\mathcal{B}(\mathcal{H})$. Thus $F \in \mathcal{S}_\perp$ so $0 = \operatorname{tr}(TF) = \langle T^{(d)} f, g\rangle$. Since $\mathcal{S}^{(d)}$ is reflexive, $\mathcal{S}_\perp^{(d)}$ is the norm closed linear span of the rank-one operators it contains. Thus $\operatorname{tr}(T^{(d)} C) = 0$ for all C in $\mathcal{S}_\perp^{(d)}$ and $T^{(d)} \in \mathcal{S}^{(d)}$. That is, $T \in \mathcal{S}$.

Now assume that $\mathcal{S}_\perp$ is the closed linear span of the operators it contains that have rank at most d. Let $T \in \operatorname{Ref} \mathcal{S}^{(d)}$. By Exercise 2, $T = T_1^{(d)}$ for some T_1 in $\operatorname{Ref} \mathcal{S}$. It must be shown that $T_1 \in \mathcal{S}$. Since $\mathcal{S}$ is weak* closed, this will be done if it is shown that $0 = \operatorname{tr}(KT_1)$ for every K in $\mathcal{S}_\perp$. By hypothesis it suffices to show this for the operators K in $\mathcal{S}_\perp$ having rank at most d. But in this case Proposition 21.3.a implies there are vectors f, g in $\mathcal{H}^{(d)}$ such that $\operatorname{tr}(KX) = \langle X^{(d)} f, g\rangle$ for all X in $\mathcal{B}(\mathcal{H})$. Since $K \in \mathcal{S}_\perp$, $f \otimes g \in \mathcal{S}_\perp^{(d)}$. This is equivalent to the statement that $g \perp [\mathcal{S}^{(d)} f]$. Hence $g \perp Tg = T_1^{(d)} f$, so $\operatorname{tr}(T_1 K) = \langle Tf, g\rangle = 0$. ∎

Put $\mathcal{H} = \bigoplus_n \mathcal{H}_n$, where $\dim \mathcal{H}_n = 2$ for all n. Let $\mathcal{L} = \{ (C_n) : C_n \in \mathcal{B}(\mathcal{H}_n) \text{ and } \sum_n \|C_n\|_1 < \infty \}$, considered as a subspace of $\mathcal{B}_1(\mathcal{H})$. If $\mathcal{S} = \mathcal{L}^\perp \subseteq \mathcal{B}(\mathcal{H})$, then the preceding theorem implies that $\mathcal{S}$ is 2-reflexive. It is not, however, reflexive.

You might wonder for a moment what happens if $\mathcal{S}_\perp$ has the property that it is the closed linear span of the finite rank operators it contains, but that is equivalent to the statement that $\mathcal{S}$ is WOT closed (20.10).

Exercises.

1. Which parts of Proposition 59.5 remain true if property $\mathbb{A}_1(r)$ is replaced by property $\mathbb{A}_1$?

2. Show that if $1 \leq d \leq \infty$ and $\mathcal{S}$ is a weak* closed subspace of $\mathcal{B}(\mathcal{H})$, then $\operatorname{Ref} \mathcal{S}^{(d)} \subseteq (\operatorname{Ref} \mathcal{S})^{(d)}$. Show that for $d = \infty$, equality occurs if and only if $\mathcal{S}$ is reflexive. What about when d is finite?

3. Let $\mathcal{S}$ be a weak* closed subspace of $\mathcal{B}(\mathcal{H})$ and for $d \geq 1$ let $\mathcal{E}_d$ be the closure of the operators in $\mathcal{S}_\perp$ having rank at most d. Show that $\mathcal{S}_d \equiv \mathcal{E}_d^\perp$ is the smallest d-reflexive subspace containing $\mathcal{S}$.

4. (Larson [1982]) Show that if $\mathcal{S}$ is a WOT closed subspace having property $\mathbb{A}_1$, then $\mathcal{S}$ is 3-reflexive.

5. (Larson [1982]) Show that if $\mathcal{S}$ is a weak* closed subspace, then every weak* closed subspace of $\mathcal{S}$ is d-reflexive if and only if $\mathcal{S}$ is d-reflexive and

for every trace class operator C on $\mathcal{H}$ there is an operator F having rank at most d such that $C - F \in \mathcal{S}_\perp$.

6. (Kraus and Larson [1986]) Suppose $\mathcal{A}$ is a weak* closed subalgebra of $\mathcal{B}(\mathcal{H})$ that contains the identity. Show that if $\mathcal{A}$ is generated as an algebra by operators having rank at most d, then $\mathcal{A}$ is d-reflexive and every weak* closed subspace of $\mathcal{B}(\mathcal{H})$ that is either a left or right module of $\mathcal{A}$ is also d-reflexive. (To say that $\mathcal{S}$ is a left module of $\mathcal{A}$ means that $\mathcal{A}\mathcal{S} \subseteq \mathcal{S}$. The definition of a right module is analogous.)

§60. Hyperreflexive von Neumann algebras

This section will explore hyperreflexivity for von Neumann algebras. But first some classical results are obtained by concentrating on the relations between reflexivity and duality developed in the preceding section, the first of which sets the stage.

60.1 Proposition. *An abelian von Neumann algebra has property* $\mathbb{A}_1(1)$.

Proof. First show this for a maximal abelian von Neumann algebra. But in this case, Theorem 14.5 implies there is a measure space (X, μ) such that $\mathcal{A}$ is spatially equivalent to $L^\infty(\mu)$ acting on $L^2(\mu)$. If F is a weak* continuous linear functional on $L^\infty(\mu)$, then there is an f in $L^1(\mu)$ such that $F(M_\phi) = \int \phi f \, d\mu$ for all ϕ in $L^\infty(\mu)$. Let u be a measurable function such that $|u| = 1$ a.e. and $f = u|f|$. Put $g = u|f|^{1/2}$ and $h = |f|^{1/2}$. This implies that $F(M_\phi) = \langle M_\phi g, h \rangle$ and $\|g\| \, \|h\| = \|F\|$. This proves the proposition in the case of a maximal abelian von Neumann algebra. But Theorem 51.4 together with Proposition 59.5 imply the result for any abelian von Neumann algebra. ∎

60.2 Corollary. *Normal operators are reflexive.*

Proof. It is known that a von Neumann algebra is reflexive (56.6). So Theorem 59.3 and the preceding proposition imply that every weak* closed subspace of an abelian von Neumann algebra is reflexive. In particular, if N is a normal operator , $P^\infty(N) \subseteq W^*(N)$ is reflexive. ∎

An additional lemma is needed to prove the next corollary.

60.3 Lemma. *If $\mathcal{M}$ is a linear manifold contained in $\mathcal{B}(\mathcal{H})$ that consists of normal operators, then $AB = BA$ for all A and B in $\mathcal{M}$.*

Proof. Note that

$$2(B^*A - AB^*) = (A+B)^*(A+B) - (A+B)(A+B)^* \\ + i[(A+iB)^*(A+iB) - (A+iB)(A+iB)^*].$$

If $A, B \in \mathcal{M}$ and $\mathcal{M}$ is a linear manifold, the right hand side is 0. So $B^*A = AB^*$. By the Fuglede-Putnam Theorem (12.5), $BA = AB$. ■

60.4 Corollary. (Sarason [1966]) *Every weak* closed linear space consisting of normal operators is reflexive.*

Proof. By the preceding lemma, each such linear subspace is contained in an abelian von Neumann algebra. The remainder of the proof follows the same line of reasoning as the proof of the last corollary. ■

60.5 Proposition. *The algebra of all analytic Toeplitz operators has property* $\mathbb{A}_1(1)$.

Proof. Let $\mathcal{A}$ be the algebra of all analytic Toeplitz operators. That is, $\mathcal{A} = \{T_\phi : \phi \in H^\infty\}$ acting on H^2. Since the predual of H^∞ is $L^1/H^\infty_\perp$, if $L\colon \mathcal{A} \to \mathbb{C}$ is a weak* continuous linear functional with $\|L\| = 1$ and $\epsilon > 0$, there is a function f in L^1 such that $L(T_\phi) = \int \phi f \, dm$ and $\|f\|_1 < 1 + \epsilon$. Now $(|f| + \epsilon)^{1/2}$ is bounded below and square integrable, so there is an outer function g in H^2 such that $|g| = (|f| + \epsilon)^{1/2}$ a.e. on $\partial\mathbb{D}$ (25.12). On the other hand, $h_1 = [|f|/(|f| + \epsilon)]^{1/2} |f|^{1/2}$ is the product of a bounded function and a function in L^2, so $h_1 \in L^2$. Since $h_1|g| = |f|$, there is a measurable function u such that $|u| \equiv 1$ and $f = uh_1g$. Let h be the projection of $\bar{u}h_1$ on H^2. So $\|h\|_2^2 \le \|\bar{u}h_1\|_2^2 \le \|f\|_1 < 1 + \epsilon$ and $\|g\|_2^2 = \|f\|_1 + \epsilon \le 1 + 2\epsilon$. So $L(T_\phi) = \int \phi g \bar{h} \, dm = \langle T_\phi g, h\rangle$ and $\|g\|_2 \, \|h\|_2$ can be made arbitrarily close to 1. ■

This proposition is extended in Bercovici [1988] where it is shown that any algebra of operators that is isometrically isomorphic and weak* homeomorphic with H^∞ has property $\mathbb{A}_1(1)$. By combining Proposition 59.5 with Proposition 56.7, the following result from Sarason [1966] follows.

60.6 Corollary. *Any weak* closed linear space of analytic Toeplitz operators is reflexive.*

60.7 Corollary. *Every analytic Toeplitz operator is reflexive.*

Analytic Toeplitz operators are subnormal, so the preceding corollary is a special case of the result from Olin and Thomson [1980] that subnormal operators are reflexive. This, however, is far more difficult to prove. (Also see Conway [1991], Theorem VII.8.5.) Another proof using Thomson's result on the existence of bounded point evaluations (Thomson [1991]) is in McCarthy [1993].

Davidson [1987] shows that any weak* closed subspace of Toeplitz operators is hyperreflexive with constant at most 39. This was recently generalized by Bercovici [preprint] where it is shown that a linear subspace of operators whose WOT continuous linear functionals satisfies a factorization condition related to $\mathbb{A}_1(r)$ is hyperreflexive.

Now to focus on hyperreflexive von Neumann algebras. The characterization of which von Neumann algebras are hyperreflexive is an unsloved problem that is equivalent to a classical problem that has persisted through the years. The most general result will not be proved here, but it will be shown that every abelian von Neumann algebra as well as every von Neumann algebra with an abelian commutant is hyperreflexive. As a consequence of this it will be shown that any WOT closed subspace of normal operators is hyperreflexive.

Begin with an alternative way to calculate $\alpha(T, \mathcal{A})$ when $\mathcal{A}$ is a von Neumann algebra.

60.8 Lemma. *If $\mathcal{A}$ is a von Neumann algebra, then for every operator T in $\mathcal{B}(\mathcal{H})$*

$$\alpha(T, \mathcal{A}) = \sup\{\|PT - TP\| : P \in \operatorname{Lat} \mathcal{A}\}.$$

Proof. let γ be the supremum appearing in the preceding equation. If $P \in \operatorname{Lat} \mathcal{A}$, then $\|P^\perp TP\| = \|P^\perp(TP - PT)\| \leq \|TP - PT\| \leq \gamma$. Thus $\alpha(T, \mathcal{A}) \leq \gamma$. Conversely, note that $TP - PT = P^\perp TP - PTP^\perp$. Now the operators $P^\perp TP$ and $PTP^\perp$ in this difference have orthogonal initial spaces and final spaces. Thus $\|TP - PT\| = \max\{\|P^\perp TP\|, \|PTP^\perp\|\} \leq \alpha(T, \mathcal{A})$, since $P^\perp \in \operatorname{Lat} \mathcal{A}$ whenever $P \in \operatorname{Lat} \mathcal{A}$. Thus $\gamma \leq \alpha(T, \mathcal{A})$. ■

If $\mathcal{A}$ is a C*-algebra and $a \in \mathcal{A}$, define $\Delta_a : \mathcal{A} \to \mathcal{A}$ by $\Delta_a(x) = ax - xa$. This is the *(inner) derivation* defined by a. A *derivation* on an algebra $\mathcal{A}$ is a linear map $\Delta : \mathcal{A} \to \mathcal{A}$ that satisfies $\Delta(xy) = x\Delta(y) + \Delta(x)y$ for all x, y in $\mathcal{A}$. It is the case that every derivation on a von Neumann algebra is inner (Sakai [1971], p. 156), but this is not true for every C*-algebra. The focus here will be on derivations on $\mathcal{B}(\mathcal{H})$.

60.9 Proposition. *If $\mathcal{A}$ is a von Neumann algebra and $T \in \mathcal{B}(\mathcal{H})$, then:*

(a) $\alpha(T, \mathcal{A}) \leq \|\Delta_T | \mathcal{A}'\|$;

(b) $\|\Delta_T | \mathcal{A}'\| \leq 4\alpha(T, \mathcal{A})$.

Proof. Part (a) is an immediate consequence of Lemma 60.8 and the fact that $\operatorname{Lat} \mathcal{A} \subseteq \mathcal{A}'$. Combining (60.8) and (54.2) gives that

$$\alpha(T, \mathcal{A}) = \sup\{\, \|TB - BT\| : B \in \mathcal{A}', 0 \leq B \leq 1\,\}.$$

But if $B \in \operatorname{ball} \mathcal{A}'$, $B = (B_1 - B_2) + i(B_3 - B_4)$, where each $B_i \in \mathcal{A}'$ and $0 \leq B_i \leq 1$. Thus $\|\Delta_T(B)\| \leq 4\alpha(T, \mathcal{A})$. This proves (b). ■

Thus a von Neumann algebra $\mathcal{A}$ is hyperreflexive if and only if there is a constant c such that $\operatorname{dist}(T, \mathcal{A}) \le c\|\Delta_T|\mathcal{A}'\|$ for every operator T in $\mathcal{B}(\mathcal{H})$. This says that hyperreflexivity for a von Neumann algebra is equivalent to the following 'almost commuting' statement.

60.10 Proposition. *A von Neumann algebra $\mathcal{A}$ is hyperreflexive if and only if for any sequence $\{T_n\}$ in $\mathcal{B}(\mathcal{H})$ such that* $\sup\{\,\|T_nB - BT_n\| : B \in \operatorname{ball}\mathcal{A}'\,\} \to 0$, *there is a sequence $\{A_n\}$ in $\mathcal{A}$ such that $\|T_n - A_n\| \to 0$.*

Proof. Assume $\mathcal{A}$ is hyperreflexive. So there is a constant a such that $\operatorname{dist}(T, \mathcal{A}) \le a\alpha(T, \mathcal{A})$ for all T in $\mathcal{B}(\mathcal{H})$. By the preceding proposition, $\operatorname{dist}(T, \mathcal{A}) \le a\|\Delta_T|\mathcal{A}'\|$. If $\{\,T_n\,\}$ is as described in the statement of the proposition, $\|\Delta_{T_n}\| \to 0$. Thus $\operatorname{dist}(T_n, \mathcal{A}) \to 0$ and the existence of the sequence $\{\,A_n\,\}$ in $\mathcal{A}$ is assured.

Now assume that the stated condition holds. Note that because $\mathcal{A}$ is a von Neumann algebra, if $T \in \mathcal{B}(\mathcal{H})$ and $\|\Delta_T|\mathcal{A}'\| = 0$, then $T \in \mathcal{A}$. Thus $p(T+\mathcal{A}) = \|\Delta_T|\mathcal{A}'\|$ defines a norm on $\mathcal{B}(\mathcal{H})/\mathcal{A}$. The condition in the proposition is equivalent to the condition that the identity map $(\mathcal{B}(\mathcal{H})/\mathcal{A}, p) \to (\mathcal{B}(\mathcal{H})/\mathcal{A}.\|\cdot\|)$ is continuous. But if this is the case, then there is a constant a such that $\operatorname{dist}(T, \mathcal{A}) = \|T + \mathcal{A}\| \le a\|\Delta_T|\mathcal{A}'\| \le 4a\alpha(T, \mathcal{A})$. ■

Now to begin the process of showing that a von Neumann algebra with an abelian commutant is hyperreflexive. To do this it is necessary to discuss some material on invariant means on an arbitrary group. If G is a group, $u \in G$, and $f \in \ell^\infty(G)$, define $f_u : G \to \mathbb{C}$ by $f_u(x) = f(ux)$ for all x in G. In the language of §43, $f_u = L_{u^{-1}}f$. An *invariant mean* for G is a positive linear functional $m: \ell^\infty(G) \to \mathbb{C}$ such that $m(1) = 1$ and $m(f_u) = m(f)$ for each f in $\ell^\infty(G)$ and each u in G. This is the same idea as a Banach limit on $\ell^\infty = \ell^\infty(\mathbb{N})$, except for the inconvenient fact that $\mathbb{N}$ is not a group. But of course the reader will see that the notion of an invariant mean makes sense in the setting of a semigroup, and, for the semigroup $\mathbb{N}$, the invariant means are precisely the Banach limits. Whenever G is an abelian group, an invariant mean exists. In fact this is an easy consequence of the Markov–Kakutani Fixed Point Theorem. (See [ACFA], Theorem V.10.1.) Now the plan is to use this to show the existence of an operator valued invariant mean for a special abelian group.

Let $\mathcal{A}$ be a von Neumann algebra whose commutant is abelian and let $\mathcal{G}$ be the group of hermitian unitary operators in $\mathcal{A}'$. If m is an invariant mean for $\mathcal{G}$, define an operator-valued invariant mean, $\widetilde{m}$, on $\ell^\infty(\mathcal{G}, \mathcal{B}(\mathcal{H}))$, the algebra of all bounded functions from $\mathcal{G}$ into $\mathcal{B}(\mathcal{H})$. Indeed, if $F \in \ell^\infty(\mathcal{G}, \mathcal{B}(\mathcal{H}))$ and $x, y \in \mathcal{H}$, $U \to \langle F(U)x, y\rangle$ is a function in $\ell^\infty(\mathcal{G})$ with norm at most $\|F\|_\infty \|x\|\,\|y\|$. Thus it makes sense to apply the invariant mean to the function $\langle F(\cdot)x,y\rangle$ to get $m(\langle F(\cdot)x,y\rangle)$. It is easily verified that this is a bounded sesquilinear form on $\mathcal{H}\oplus\mathcal{H}$ with bound $\|F\|_\infty$, and so there is an operator $\widetilde{m}(F)$ with $\|\widetilde{m}(F)\|\le\|F\|_\infty$ such that

60.11 $$\langle\widetilde{m}(F)x,y\rangle = m(\langle F(\cdot)x,y\rangle)$$

for all x,y in $\mathcal{H}$. It follows that $\widetilde{m}\colon \ell^\infty(\mathcal{G},\mathcal{B}(\mathcal{H}))\to\mathcal{B}(\mathcal{H})$ is a bounded linear transformation with $\|\widetilde{m}(F)\|\le\|F\|_\infty$ for all F. For U in $\mathcal{G}$ and F in $\ell^\infty(\mathcal{G},\mathcal{B}(\mathcal{H}))$, define F_U in $\ell^\infty(\mathcal{G},\mathcal{B}(\mathcal{H}))$ by $F_U(V)=F(UV)$. The properties of $\widetilde{m}$ are collected together in the next proposition.

60.12 Proposition. *If $\widetilde{m}\colon \ell^\infty(\mathcal{G},\mathcal{B}(\mathcal{H}))\to\mathcal{B}(\mathcal{H})$ is defined for the invariant mean m on $\mathcal{G}$ as in* (60.11) *and $F\in\ell^\infty(\mathcal{G},\mathcal{B}(\mathcal{H}))$, then the following hold.*

(a) *$\widetilde{m}$ is a contractive linear map.*

(b) *$\widetilde{m}(F)\ge 0$ when $F\ge 0$.*

(c) *$\widetilde{m}(F_U)=\widetilde{m}(F)$ for all U in $\mathcal{G}$.*

(d) *If $T\in\mathcal{B}(\mathcal{H})$ and $[T]\colon\mathcal{G}\to\mathcal{B}(\mathcal{H})$ is the function defined by $[T](U)=U^{-1}TU$, then $\widetilde{m}([T])\in\mathcal{A}$.*

(e) *If $X\in\mathcal{B}(\mathcal{H})$ and X is also used to denote the function in $\ell^\infty(\mathcal{G},\mathcal{B}(\mathcal{H}))$ that is constantly equal to X, then $\widetilde{m}(XF)=X\widetilde{m}(F)$.*

Proof. Part (a) has already been established. The proof of (b) is an immediate consequence of (60.11) and the definition of positivity in $\ell^\infty(\mathcal{G},\mathcal{B}(\mathcal{H}))$. The proof of (c) is left to the reader as training in the concepts.

To show (d), show that $\widetilde{m}([T])U=U\widetilde{m}([T])$ for every U in $\mathcal{G}$. Once this is done, it follows that T commutes with $2P-1$ for every P in $\operatorname{Lat}\mathcal{A}$; hence, T commutes with every projection in $\mathcal{A}'$. Since $\operatorname{Lat}\mathcal{A}$ generates $\mathcal{A}'$ as a Banach space, this will say that $T_0\in\{\mathcal{A}'\}'=\mathcal{A}$.

If U is fixed in $\mathcal{G}$ and vectors x,y in $\mathcal{H}$, then for any V in $\mathcal{G}$

$$\begin{aligned}\langle [T]_U(V)x,y\rangle &= \langle [T](UV)x,y\rangle\\ &= \langle UVTV^{-1}U^{-1}x,y\rangle\\ &= \langle U\left([T](V)\right)U^{-1}x,y\rangle.\end{aligned}$$

That is, $[T]_U(\cdot)=U\left([T](\cdot)\right)U^{-1}$. Thus

$$\begin{aligned}\langle\widetilde{m}([T])x,y\rangle &= \langle\widetilde{m}\left([T]_U\right)x,y\rangle\\ &= m\left(\langle [T]_U(\cdot)x,y\rangle\right)\\ &= m\left(\langle [T](\cdot)U^{-1}x,U^{-1}y\rangle\right)\end{aligned}$$

$$= \langle \widetilde{m}\,([T])\,U^{-1}x, U^{-1}y\rangle$$
$$= \langle U\widetilde{m}\,([T]U^{-1}x, y\rangle)\,.$$

But x and y were arbitrary vectors, so $\widetilde{m}\,([T]) = U\widetilde{m}\,([T])\,U^{-1}$.

Now to prove (e). If x, y are arbitrary vectors in $\mathcal{H}$, then

$$\langle \widetilde{m}(XF)x, y\rangle = m\,(\langle XF(\cdot)x, y\rangle) = m\,(\langle F(\cdot)x, X^*y\rangle)$$
$$= \langle \widetilde{m}(F)x, X^*y\rangle = \langle X\widetilde{m}(F)x, y\rangle. \;\blacksquare$$

Note that if $X \in \mathcal{B}(\mathcal{H})$ and X is considered as the constant function in $\ell^\infty(\mathcal{G}, \mathcal{B}(\mathcal{H}))$, then $\widetilde{m}(X) = X$.

60.13 Theorem. (Rosenoer [1982]) *If $\mathcal{A}$ is a von Neumann algebra with an abelian commutant, then $\mathcal{A}$ is hyperreflexive with $\kappa(\mathcal{A}) \leq 2$.*

Proof. Fix the operator T and let $\alpha = \alpha(T, \mathcal{A})$. It must be shown that $\operatorname{dist}(T, sA) \leq 2\alpha$. Let $\mathcal{G}$, $\widetilde{m}$, and $[T]$ be as in Proposition 60.12. Put $T_0 = \widetilde{m}([T])$. Considering T as the constant function in $\ell^\infty(\mathcal{G}, \mathcal{B}(\mathcal{H}))$, $\|(T - [T])(U)\| = \|T - U^{-1}TU\|$ for all U in $\mathcal{G}$. Now if $P \in \operatorname{Lat}\mathcal{A}$, $U = 2P - 1 \in \mathcal{G}$ and $\|T - U^{-1}TU\| = \|UT - TU\| = 2\|PT - TP\| \leq 2\alpha$. Thus $\|T - [T]\|_\infty \leq 2\alpha$. Hence $\|T - T_0\| = \|\widetilde{m}(T - [T])\| \leq 2\alpha$. Since $T_0 \in \mathcal{A}$ (60.12.d), $\operatorname{dist}(T, \mathcal{A}) \leq \|T - T_0\| \leq 2\alpha(T, \mathcal{A})$. ■

If this is combined with Theorem 59.4, the next result will be obtained. In particular, this says that an abelian von Neumann algebra is hyperreflexive, another result from Rosenoer [1982]. (Though it is shown there that for a von Neumann algebra $\mathcal{A}$, $\kappa(\mathcal{A}) \leq 2$ instead of the 5 obtained in the next theorem.)

60.14 Theorem. *Every weak* closed linear subspace $\mathcal{S}$ of $\mathcal{B}(\mathcal{H})$ consisting of normal operators is hyperreflexive with $\kappa(\mathcal{S}) \leq 5$.*

Proof. According to Lemma 60.3, the operators in $\mathcal{S}$ commute. Let $\mathcal{A}$ be a maximal abelian von Neumann algebra containing $\mathcal{S}$. Since $\mathcal{A}' = \mathcal{A}$, Theorem 60.13 implies $\mathcal{A}$ is hyperreflexive with $\kappa(\mathcal{A}) \leq 2$. But $\mathcal{A}$ has property $\mathbb{A}_1(1)$ (60.1), so Theorem 59.4 implies $\mathcal{S}$ is hyperreflexive with $\kappa(\mathcal{S}) \leq 1 + (1+1)\kappa(\mathcal{A}_1) \leq 5$. ■

60.15 Corollary. *Normal operators are hyperreflexive.*

For the problem of characterizing the hyperreflexive von Neumann algebras, the interested reader can see the comments in Arveson [1984] and the papers of Kadison and Kastler [1972], Christensen [1977a] and [1977b], and Johnson [1979].

§61. Some examples of operators

This section will present a technique for constructing examples of operators developed in Wogen [1987]. (Also see Wogen [1986].) These examples will answer many natural questions concerning operators and the algebras they generate.

Let $\mathcal{H}$ be a Hilbert space whose dimension may be finite or infinite and put $\mathcal{K} = \mathcal{H}^{(\infty)} = \mathcal{H}_0 \oplus \mathcal{H}_1 \oplus \cdots$, where each $\mathcal{H}_n = \mathcal{H}$. It is also possible to express $\mathcal{K}$ as $\mathcal{K} = \mathcal{K}_0 \oplus \mathcal{K}_\infty$, with $\mathcal{K}_0 = \mathcal{H}_0$ and $\mathcal{K}_\infty = \mathcal{H}_1 \oplus \mathcal{H}_2 \oplus \cdots$.

As usual, operators on $\mathcal{K}$ can be identified with certain operator-valued matrices (T_{ij}), where each $T_{ij} \in \mathcal{B}(\mathcal{H})$, $0 \le i, j < \infty$. Let P_i be the projection of $\mathcal{K}$ onto $\mathcal{H}_i$, $0 \le 1 < \infty$. For any subset $\mathcal{A}$ of $\mathcal{B}(\mathcal{K})$, let

$$\mathcal{A}_{ij} = P_i \mathcal{A} P_j \subseteq \mathcal{B}(\mathcal{K}).$$

It should be emphasized that $\mathcal{A}_{ij}$ is a subset of $\mathcal{B}(\mathcal{K})$ even though there is a natural way to identify this set with a subset of $\mathcal{B}(\mathcal{H})$. If $\mathcal{S} \subseteq \mathcal{B}(\mathcal{H})$, define another subset of $\mathcal{B}(\mathcal{K})$ by

$$[\mathcal{S}]_{ij} = \{\, S \in \mathcal{B}(\mathcal{K}) : S_{ij} \in \mathcal{S} \text{ and } S_{km} = 0 \text{ if } (k,m) \neq (i,j) \,\}.$$

Recall that for any operator T, $P(T)$ is the norm closed algebra generated by T. Similarly recall the definitions of $W(T)$ and $P^\infty(T)$. Now to prove the following.

61.1 Theorem. *If $\mathcal{C}$ is any norm closed, separable subspace of $\mathcal{B}(\mathcal{H})$, there is an operator T on $\mathcal{K}$ such that*

$$P(T)_{1,0} = [\mathcal{C}]_{1,0} \subseteq P(T).$$

Before proving the theorem, a few corollaries will be given.

61.2 Corollary. *If $\mathcal{S}$ is a* WOT *closed linear subspace of $\mathcal{B}(\mathcal{H})$, then there is an operator T on $\mathcal{K}$ such that*

$$W(T)_{1,0} = [\mathcal{S}]_{1,0} \subseteq W(T).$$

Proof. Since $\mathcal{H}$ is separable, there is a norm closed, separable subspace $\mathcal{C}$ of $\mathcal{B}(\mathcal{H})$ such that ball $\mathcal{C}$ is WOT dense in ball $\mathcal{S}$. Use this subspace $\mathcal{C}$ and construct T as in the theorem. If $S \in \mathcal{S}$, there is a sequence $\{C_n\}$ in $\mathcal{C}$ with $\|C_n\| \le \|S\|$ for all n such that $C_n \to S$ (WOT) in $\mathcal{B}(\mathcal{H})$. It follows that $[C_n]_{1,0} \to [S]_{1,0}$ (WOT) in $\mathcal{B}(\mathcal{K})$. Since the theorem implies that $[C_n]_{1,0} \in P(T) \subseteq W(T)$ for every $n \ge 1$, this gives that $[\mathcal{S}]_{1,0} \subseteq W(T)$. It follows that $[\mathcal{S}]_{1,0} \subseteq W(T)_{1,0}$. On the other hand, if $A \in W(T)$, there is a sequence $\{\, A_n \,\}$ in $P(T)$ such that $A_n \to A$ (WOT). But $[A_n]_{1,0} \in P(T)_{1,0} = [\mathcal{C}]_{1,0} \subseteq [\mathcal{S}]_{1,0}$, so that $[A]_{1,0} \in [\mathcal{S}]_{1,0}$. ■

A similar technique will prove the next corollary.

61.3 Corollary. *If $\mathcal{S}$ is a* weak* *closed linear subspace of $\mathcal{B}(\mathcal{H})$, then there is an operator T on $\mathcal{K}$ such that*

$$P^{\infty}(T)_{1,0} = [\mathcal{S}]_{1,0} \subseteq P^{\infty}(T).$$

There is another way to phrase the result of the theorem. For any subset $\mathcal{P}$ of $\mathcal{B}(\mathcal{K})$, let $\widetilde{\mathcal{P}} \equiv \{\, X - X_{1,0} : X \in \mathcal{P}\,\}$.

61.4 Corollary. *If $\mathcal{C}$ is any norm closed, separable subspace of $\mathcal{B}(\mathcal{H})$, there is an operator T on $\mathcal{K}$ such that*

$$P(T) = \widetilde{P(T)} \dotplus [\mathcal{C}]_{1,0}.$$

The direct sum $\dotplus$ that appears in this corollary is a Banach direct sum and does not imply anything about a formula for the norm on $P(T)$.

Proof of Theorem 61.1. Let $\{R_n\}$ be an enumeration of a dense subset of ball $\mathcal{C}$ such that each element appears infinitely often. Now take two decreasing sequences of positive numbers $\{b_n\}, \{w_n\}$ such that $\{b_n\} \in \ell^2$ and

$$\lim_{n\to\infty} \frac{w_n}{b_n} = 0.$$

Define the operator $T = (T_{ij})$ on $\mathcal{K}$ by setting

$$T_{ij} = \begin{cases} w_i & \text{for } j = i+1, 1 \le i < \infty; \\ b_i R_i & \text{for } 1 \le i < \infty \text{ and } j = 0; \\ 0 & \text{otherwise.} \end{cases}$$

It will be helpful to present the operator T in another way. Define the operator $W : \mathcal{K}_\infty \to \mathcal{K}_\infty$ as the operator-valued backward shift

$$W(h_1, h_2, \dots) = (w_1 h_2, w_2 h_3, \dots),$$

and define the operator $R\colon \mathcal{K}_0 \to \mathcal{K}_\infty$ by

$$Rh = (b_1 R_1 h, b_2 R_2 h, \dots).$$

An easy computation shows that

$$T = \begin{bmatrix} 0 & 0 \\ & \\ R & W \end{bmatrix} \qquad \mathit{on} \qquad \begin{matrix} \mathcal{K}_0 \\ \oplus \\ \mathcal{K}_\infty \end{matrix}$$

Matrix multiplication shows that

$$T^{n+1} = \begin{bmatrix} 0 & 0 \\ W^n R & W^{n+1} \end{bmatrix}.$$

Now because W is a backward shift, $W^n P_k = 0$ for $1 \le k \le n$. On the other hand, if $k > n$ and $h = (h_j) \in \mathcal{K}_\infty$, then

$$\begin{aligned} P_k W^n h &= w_k \cdots w_{k+n-1} h_{k+n} \\ &= \frac{\beta(k+n-1)}{\beta(k-1)} h_{k+n} \end{aligned}$$

where

$$\beta(0) = 1 \text{ and } \beta(m) = w_1 \cdots w_m$$

for all $m \ge 1$. Since $\{w_n\}$ is a decreasing sequence, this yields that

$$\begin{aligned} \|W^n h\|^2 &= \sum_{k=1}^{\infty} \left[\frac{\beta(k+n-1)}{\beta(k-1)}\right]^2 \|h_k\|^2 \\ &\le \beta(n)^2 \|h\|^2. \end{aligned}$$

From here it follows that $\|W^n\| = \beta(n)$ and so

61.5 $$\begin{aligned} \lim_{n\to\infty} \left\| \frac{1}{\beta(n) b_{n+1}} W^{n+1} \right\| &= \lim_{n\to\infty} \frac{\beta(n+1)}{\beta(n)} \frac{1}{b_{n+1}} \\ &= \lim_{n\to\infty} \frac{w_{n+1}}{b_{n+1}} \\ &= 0 \end{aligned}$$

by assumption.

Now using the definition of R, it follows that for h in $\mathcal{K}_0 = \mathcal{H}$ and $k > n \ge 1$

$$\begin{aligned} P_k W^n R h &= \frac{\beta(k+n-1)}{\beta(k-1)} P_{k+n}(Rh) \\ &= \frac{\beta(k+n-1)}{\beta(k-1)} b_{k+n} R_{k+n} h. \end{aligned}$$

Thus

61.6 $$\begin{aligned} \frac{1}{\beta(n) b_{n+1}} P_1 W^n R h &= \frac{\beta(n)}{\beta(n) b_{n+1} \beta(0)} b_{n+1} R_{n+1} h \\ &= R_{n+1} h. \end{aligned}$$

On the other hand, if $h \in \operatorname{ball} \mathcal{K}_0$, then

$$\begin{aligned} \|P_1^\perp W^n R h\|^2 &= \sum_{k=2}^{\infty} \left[\frac{\beta(k+n-1)}{\beta(k-1)}\right]^2 b_{k+n}^2 \|R_{k+n} h\|^2 \\ &\le \left[\frac{\beta(n+1)}{\beta(1)}\right]^2 \sum_{k=2}^{\infty} b_{k+n}^2 \|R_{k+n} h\|^2 \\ &\le \left[\frac{\beta(n+1)}{\beta(1)}\right]^2 \|R\|^2. \end{aligned}$$

Therefore

$$\textbf{61.7} \qquad \lim_{n\to\infty}\left\|\frac{1}{\beta(n)b_{n+1}}P_1^{\perp}W^nR\right\| \le \lim_{n\to\infty}\frac{w_{n+1}}{b_{n+1}}\frac{1}{w_1}\|R\| = 0.$$

Now let C be one of the operators in the sequence $\{R_j\}$. Recall that $C = R_j$ for infinitely many j. For every $n \ge 1$ it follows that

$$[C]_{1,0} - \frac{1}{\beta(n)b_{n+1}}T^{n+1} = \begin{bmatrix} 0 & 0 \\ P_1C - \frac{1}{\beta(n)b_{n+1}}W^nR & -\frac{1}{\beta(n)b_{n+1}}W^{n+1} \end{bmatrix}.$$

Equation (61.6) implies that

$$P_1C - \frac{1}{\beta(n)b_{n+1}}W^nR = P_1C - R_{n+1} - P_1^{\perp}\frac{1}{\beta(n)b_{n+1}}W^nR.$$

By (61.5) and (61.7), if $\epsilon > 0$, there is an integer $n \ge 1$ such that

$$\left\|-\frac{1}{\beta(n)b_{n+1}}W^{n+1}\right\| < \epsilon,$$

$$\left\|\frac{1}{\beta(n)b_{n+1}}W^nR\right\| < \epsilon,$$

$$R_{n+1} = C.$$

Therefore $[C]_{1,0}$ is included in the norm closure of $\{\,\alpha T^n : \alpha \in \mathbb{C}, n \ge 1\,\}$. Since C was arbitrary, $[C]_{1,0} \subseteq P(T)$. But for any $n \ge 1$, $(T^{n+1})_{1,0} = P_1T^{n+1}P_0 = P_1W^nR = \beta(n)b_{n+1}R_{n+1} \in [C]_{1,0}$. Hence, $P(T)_{1,0} \subseteq [C]_{1,0}$. ∎

The first illustration of this technique for building examples is the following. From a historical point of view it is interesting to note that in spite of the fact that the algebras $W(T)$ and $P^\infty(T)$ had been studied for a long time, it was unknown that they could be unequal before Wogen [1987].

61.8 Example. There is an operator T such that $W(T) \ne P^\infty(T)$.

Let $\mathcal{S}_1$ be a weak* closed subspace of $\mathcal{B}(\mathcal{H})$ that is not WOT closed. (For example, $\mathcal{S}_1$ could be the kernel of a weak* continuous linear functional on $\mathcal{B}(\mathcal{H})$ that is not WOT continuous.) Let $\mathcal{S}$ be the WOT closure of $\mathcal{S}_1$. Obtain the operator T as in Corollary 61.3 such that $[P^\infty(T)]_{1,0} = \mathcal{S}_1$. It is easy to check that $W(T)_{1,0} \supseteq \mathcal{S}$, so it must be that $W(T) \ne P^\infty(T)$.

Once a basic example like that in (61.8) is constructed, refinements can be added. For example, if T is any completely non-unitary contraction there is the Sz.-Nagy–Foias functional calculus $\phi \to \phi(T)$ defined for functions in H^∞. (See §35 and Sz.-Nagy and Foias [1970].) This is defined in several ways. For example, if $\phi \in H^\infty$ and $0 < r < 1$, let $\phi_r(z) = \phi(rz)$. So ϕ_r is analytic in a neighborhood of the closed unit disk and, hence, $\phi_r(T)$ can be defined by the Riesz functional calculus. It is then shown that since T is completely non-unitary, $\phi_r(T)$ converges in the SOT as $r \to 1-$. This strong limit is denoted by $\phi(T)$. This functional calculus is multiplicative, contractive, extends the Riesz functional calculus, and is unique.

In the study of this functional calculus it is often felt that the requirement that the unit circle, $\partial\mathbb{D}$, is contained in $\sigma(T)$, the spectrum of T, provides extra structure. This has been born out in many cases. (See Bercovici, Foias, and Pearcy [1985] for a discussion of these ideas.) The preceding example can be modified so that $\|T\| = 1$, $\sigma(T) \supseteq \partial\mathbb{D}$, and still $W(T) \neq P^\infty(T)$. To see this it is necessary to first establish the following.

61.9 Lemma. *If T is the operator obtained in Theorem* 61.1, *then T is quasinilpotent. That is, $\sigma(T) = \{0\}$.*

Proof. Retain the notation from the proof of Theorem 61.1. The fact that $w_n \to 0$ implies that $\sigma(W) = \{0\}$ (27.7.b). So if $\lambda \neq 0$, it is easy to check that the operator defined on $\mathcal{K} = \mathcal{K}_0 \oplus \mathcal{K}_\infty$ by the operator matrix

$$\begin{bmatrix} -\lambda^{-1} & 0 \\ \lambda^{-1}(W-\lambda)^{-1}R & (W-\lambda)^{-1} \end{bmatrix}$$

is the inverse on $(T - \lambda)$. ■

61.10 Example. There is an operator S such that $\|S\| = 1$, $\partial\mathbb{D} \subseteq \sigma(S)$, and $W(S) \neq P^\infty(S)$.

Let $\{\lambda_n\}$ be a Blaschke sequence such that $\lambda_n \neq 0$ for all n and $\partial\mathbb{D} \subseteq \operatorname{cl}\{\lambda_n\}$. (See Exercise VII.5.4 in Conway [1978].) Put $D = \operatorname{diag}(\lambda_1, \lambda_2, \dots)$. So $\sigma(D) = \{\lambda_n\} \cup \partial\mathbb{D}$. Letting T be as in Example 61.8, let $S = T \oplus D$. Clearly it can be arranged that $\|T\| = 1$, so S satisfies the first two requirements.

Now to show that $W(S) = W(T) \oplus W(D)$. In fact, let b be the Blaschke product with precisely the zeros $\{\lambda_n\}$. Thus $b(S) = b(T) \oplus 0 \in W(S)$. Since $b(0) \neq 0$, there is a sequence of polynomials $\{p_n\}$ that are uniformly bounded on some small disk Δ about the origin and such that $p_n(z) \to 1/b(z)$ uniformly of Δ. Thus $\|p_n(T)b(T) - 1\| \to 0$ and so $(p_n b)(S) \to 1 \oplus 0$. Thus $1 \oplus 0 \in P(S)$. It is now routine to show that $P(S) = P(T) \oplus P(D)$. It follows that $W(S) = W(T) \oplus W(D)$ and $P^\infty(S) = P^\infty(T) \oplus P^\infty(D)$.

But D is a normal operator and so $W(D) = P^\infty(D)$ (Proposition 60.1 and Proposition 59.2). Since $W(T) \neq P^\infty(T)$, it follows that $W(S) \neq P^\infty(S)$.

A linear space $\mathcal{S}$ in $\mathcal{B}(\mathcal{H})$ is called *transitive* if for each non-zero vector h in $\mathcal{H}$, $[\mathcal{S}h] = \mathcal{H}$. Of course $\mathcal{B}(\mathcal{H})$ is transitive, but there are proper transitive subspaces.

61.11 Proposition. *If $n \geq 2$ and $\mathcal{S}$ is the set of all S in $\mathcal{B}(\mathbb{C}^n)$ with $\operatorname{tr} S = 0$, then $\mathcal{S}$ is transitive.*

Proof. Let $e_1, \dots, e_n$ be the usual basis in $\mathbb{C}^n$ and fix $h = \alpha_1 e_1 + \cdots + \alpha_n e_n \neq 0$ in $\mathbb{C}^n$. So there is at least one index p such that $\alpha_p \neq 0$. Fix this integer p. For $k \neq p$ let A_k be defined by setting $A_k e_p = e_k$ and $A_k e_j = 0$ for $j \neq p$; $A_k \in \mathcal{S}$. Hence $A_k h = \alpha_p e_k$ and so $e_k \in [\mathcal{S}h]$ for all $k \neq p$. Now fix some integer k with $k \neq p$ and define U on $\mathbb{C}$ by $Ue_p = e_p$, $Ue_k = -e_k$, and $ue_j = 0$ for $j \neq k, p$. Again $U \in \mathcal{S}$ and so $Uh = \alpha_p e_p - \alpha_k e_k \in [\mathcal{S}h]$. Since it is already established that $e_k \in [\mathcal{S}h]$ and $\alpha_p \neq 0$, this implies $e_p \in [\mathcal{S}h]$ and so $\mathcal{S}$ is transitive. ∎

Recall (57.6) that if $\dim \mathcal{H} < \infty$, then $W(T) = \{T\}' \cap \operatorname{Alg} \operatorname{Lat} T$. This is not valid in infinite dimensional space as the following example from Wogen [1987] demonstrates.

61.12 Example. There is a trace class operator T such that $W(T) \neq \{T\}' \cap \operatorname{Alg} \operatorname{Lat} T$.

First it is shown that it is possible in the ambiance of Theorem 61.1 to choose T to be trace class. Retain the notation of the proof of that theorem.

61.13 Lemma. *If in the definition of the operator T in* Theorem 61.1 *the sequence $\{b_n\}$ is chosen in ℓ^1 and $\dim \mathcal{H} < \infty$, then $T \in \mathcal{B}_1(\mathcal{K})$.*

Proof. Note that since $\{b_n\} \in \ell^1$ and $w_n/b_n \to 0$, $\{w_n\} \in \ell^1$ as well. Thus $W^* \colon \mathcal{K}_\infty \to \mathcal{K}_\infty$ is given by

$$W^*(h_1, h_2, \dots) = (0, w_1 h_1, w_2 h_2, \dots).$$

From here it is easy to verify that

$$W^* W = \operatorname{diag}(0, w_1^2, w_2^2, \dots).$$

So if the hypotheses are fulfilled, $|W| \in \mathcal{B}_1(\mathcal{K}_\infty)$. Since R operates on a finite dimensional space, it follows that $T \in \mathcal{B}_1(\mathcal{K})$. ∎

Let $\mathcal{S}$ be any proper WOT closed transitive subspace of $\mathcal{B}(\mathcal{H})$ and let T be an operator as in Corollary 61.2 such that $[\mathcal{S}]_{1,0} = W(T)_{1,0} \subseteq W(T)$. Specify that $\dim \mathcal{H} < \infty$ and $\{b_n\} \in \ell^{-1}$ so that T is trace class.

Claim. $[\{T\}' \cap \operatorname{Alg Lat} T]_{1,0} = [\mathcal{B}(\mathcal{H})]_{1,0}$.

(Note that proving this claim will show that $W(T) \neq \{T\}' \cap \operatorname{Alg Lat} T$ and finish verifying the statement in the example.) Let $A \in [\mathcal{B}(\mathcal{H})]_{1,0}$. So there is an operator B on $\mathcal{H}$ such that $A(h_0, h_1, h_2, \dots) = (0, Bh_0, 0, \dots)$ for all $h = (h_0, h_1, \dots)$ in $\mathcal{K} = \mathcal{K}_0 \oplus \mathcal{K}_\infty = \mathcal{H}_0 \oplus \mathcal{H}_1 \oplus \mathcal{H}_2 \oplus \cdots$. An examination of the definition of T shows that for all $(h_0, h_1, \dots)$ in $\mathcal{H}_0 \oplus \mathcal{H}_1 \oplus \mathcal{H}_2 \oplus \cdots$,

61.14 $$T(h_0, h_1, \dots) = (0, b_1 R_1 h_0 + w_1 h_2, b_2 R_2 h_0 + w_2 h_3, \dots).$$

It is routine to verify that $AT = 0 = TA$, so $A \in \{T\}'$.

Now fix $h = (h_0, h_1, \dots)$ in $\mathcal{K}$ and assume that $h_0 \neq 0$. Since $[\mathcal{S}]_{1,0} = [W(T)]_{1,0} \subseteq W(T)$,

$$\begin{aligned} [W(T)h] &\supseteq [(W(T))_{1,0} h] \\ &= \operatorname{cl}\{ (0, Sh_0, 0, \dots) : S \in \mathcal{S} \} \\ &= (0) \oplus \mathcal{H}_1 \oplus (0) \oplus \cdots \\ &= P_1 \mathcal{K}. \end{aligned}$$

This shows that if $\mathcal{M} \in \operatorname{Lat} T$ and $P_0 \mathcal{M} \neq (0)$, then $\mathcal{M} \supseteq P_1 \mathcal{K}$. Now if $A \in [\mathcal{B}(\mathcal{H})]_{1,0}$, $\operatorname{ran} A \subseteq P_1 \mathcal{K}$. Therefore $A\mathcal{M} \subseteq \mathcal{M}$ for all $\mathcal{M}$ in $\operatorname{Lat} T$ for which $P_0 \mathcal{M} \neq (0)$. On the other hand, if $\mathcal{M} \in \operatorname{Lat} T$ and $P_0 \mathcal{M} = (0)$, then $\mathcal{M} \subseteq (0) \oplus \mathcal{K}_\infty \subseteq \ker A$, so $A\mathcal{M} \subseteq \mathcal{M}$. Thus the claim is verified and $W(T) \neq \{T\}' \cap \operatorname{Alg Lat} T$.

Bibliography

The numbers appearing in parentheses following each entry give the page number(s) on which reference is made to this entry.

[ACFA] J B Conway [1990], *A Course in Functional Analysis*, Springer-Verlag, New York.

D R Adams and L I Hedberg [1996], *Function Spaces and Potential Theory*, Springer–Verlag, Berlin. (153)

J Agler [1985], "Rational dilation on an annulus," *Ann Math* **121** 537–563. (197, 202)

J Agler [1988], "An abstract approach to model theory," *Surveys of some recent results in operator theory*, Vol. II, 1–23, Longman Sci. Tech., Harlow, 1988. (197)

P R Ahern and D N Clark [1970], "On functions orthogonal to invariant subspaces," *Acta Math* **124** 191–204. (146)

L V Ahlfors and A Beurling [1950], "Conformal invariants and function theoretic null sets," *Acta Math* **83** 101–129. (171)

C A Akemann [1967], "The dual space of an operator algebra," *Trans Amer Math Soc* **126** 286–302. (314)

C A Akemann and G K Pedersen [1977], "Ideal perturbations of elements of C*-algebras," *Math Scand* **41** 117–139. (206, 322)

J Akeroyd, D Khavinson, and H S Shapiro [1991], "Remarks concerning cyclic vectors in Hardy and Bergman spaces," *Michigan Math J* **38** 191–205. (153)

A Aleman [1996] , "Subnormal operators with compact selfcommutator," *Manuscripta Math* **91** 353–367. (177)

A Aleman, S Richter, and W T Ross [1996], "Bergman spaces on disconnected domains," *Canad J Math* **48** 225–243. (155)

A Aleman, S Richter, and C Sundberg [1996], "Beurling's theorem for the Bergman space," *Acta Math* **177** 275–310. (155)

H Alexander [1973], "Projections of polynomial hulls," *J Functional Analysis* **3** 13–19. (175)

C Apostol, H Bercovici, C Foias, and C Pearcy [1985], "Invariant subspaces, dilation theory, and the structure of the predual of a dual algebra, I," *J Functional Analysis* **63** 369–404. (154)

W B Arveson [1969], "Subalgebras of C*-algebras," *Acta Math* **123** 141–224. (196, 203)

W B Arveson [1975], "Interpolation problems in nest algebras," *J Functional Analysis* **20** 208–233. (323)

W B Arveson [1976], An Invitation to C^*-algebras, Springer–Verlag, New York. (181)

W B Arveson [1977], "Notes on extensions of C*-algebras," *Duke Math J* **44** 329–355. (206, 220, 229, 237)

W B Arveson [1984], "Ten Lectures on operator algebras," *CBMS Lecture Notes* **55** Amer Math Soc, Providence. (333, 347)

S Axler, J B Conway, and G McDonald [1982], "Toeplitz operators on Bergman spaces," *Canadian Math J* **34** 466–483. (148, 149)

S Axler and J Shapiro [1985], "Putnam's Theorem, Alexander's spectral area estimate, and VMO," *Math Ann* **271** 161–183. (176)

E A Azoff and M Ptak [1998], "A dichotomy for linear spaces of Toeplitz operators," *J Functional Anal* **156** 411–428. (322)

H Bercovici [1988], *Operator Theory and Arithmetic in H^∞*, Amer Math Soc, Providence. (146, 343)

H Bercovici [preprint], (344)

H Bercovici, C Foias, and C Pearcy [1985], *Dual algebras with applications to invariant subspaces and dilation theory*," *CBMS Lecture Notes* **56**, Amer Math Soc, Providence. (97, 154, 335, 352)

S K Berberian, [1959], "Note on a theorem of Fuglede and Putnam," *Proc Amer Math Soc* **10** 175–182. (58)

I D Berg [1971], "An extension of the Weyl-von Neumann theorem to normal operators," *Trans Amer Math Soc* **160** 365–371. (214)

C A Berger and L A Coburn [1996], "On Voiculescu's double commutant theorem," *Proc Amer Math Soc* **124** 3453–3457. (239)

C A Berger and B I Shaw [1973a], "Selfcommutators of multicyclic hyponormal operators are trace class," *Bull Amer Math Soc* **79** 1193–1199. (178)

C A Berger and B I Shaw [1973b], "Intertwining, analytic structure, and the trace norm estimate," *Proc Conf Operator Theory*, Springer–Verlag Lecture Notes vol 345, pp 1–6. (178)

C A Berger and B I Shaw [preprint], "Hyponormality: its analytic consequences." (178)

S Bergman [1947], *Sur les fonctions orthogonales de plusiers variables complexes avec les applications à la theorie des fonctions analytiques*, Gauthier–Villars, Paris. (153)

S Bergman [1950], *The kernel function and conformal mapping*, Math Surveys V, Amer Math Soc, Providence. (153)

A Beurling [1949], "On two problems concerning linear transformations in Hilbert space," *Acta Math* **81** 239–255. (123, 128)

P Billingsley [1965], *Ergodic Theory and Information*, Wiley, New York. (292)

J Bram [1955], "Subnormal operators," *Duke Math J* **22** 75–94. (53, 158, 164, 165, 197, 198)

L de Branges and J Rovnyak [1966a], "Canonical models in quantum scattering theory," in *Perturbation theory and its applications in quantum mechanics*, Wiley, New York, 295–392. (125)

L de Branges and J Rovnyak [1966b], *Square summable power series*, Holt, Reinhart and Winston, New York. (125)

J E Brennan [1977], "Approximation in the mean by polynomials on non-Carathédory domains," *Ark Math* **15** 117–168. (152, 153)

J E Brennan [1979], "Point evaluations, invariant subspaces and approximation in the mean by polynomials," *J Functional Analysis* **34** 407–420. (153)

L Brickman and P A Fillmore [1967], "The invariant subspace lattice of a linear transformation," *Canadian J Math* **19** 810–822. (326)

M S Brodskii [1957], "On a problem of I M Gelfand," *Uspekhi Mat Nauk* **12** 129–132. (143)

L G Brown, R G Douglas, and P A Fillmore [1973], "Unitary equivalence modulo the compact operators and extensions of C*-algebras," Lecture Notes vol **345**, 58–128, Springer–Verlag, Berlin. (170)

L G Brown, R G Douglas, and P A Fillmore [1977], "Extensions of C*-algebras and K-homology," *Ann Math* **105** 265–324. (170)

S W Brown [1978], "Some invariant subspaces for subnormal operators," *Integral Equations Operator Theory* **1** 310–333. (335)

J W Bunce and J A Deddens [1977], "On the normal spectrum of a subnormal operator," *Proc Amer Math Soc* **63** 107–110. (158)

E Christensen [1977a], "Perturbations of operator algebras," *Invent Math* **43** 1–13. (347)

E Christensen [1977b], "Perturbations of operator algebras, II," *Indiana U Math J* **26** 891–904. (347)

J Cima and A Matheson [1985], "Approximation in the mean by polynomials," *Rocky Mountain J Math* **15** 729–738. (152)

J B Conway [1977], "The direct sum of normal operators," *Indiana Univ Math J* **26** 277–289. (110)

J B Conway [1978], *Functions of One Complex Variable*, Springer–Verlag, New York. (124, 352)

J B Conway[1985], "Arranging the disposition of the spectrum," *Proc Royal Irish Acad* **85A**, 139–142. (211)

J B Conway [1991], *The Theory of Subnormal Operators*, Amer Math Soc Surveys and Monographs **36**, Providence. (53, 114, 124, 134, 154, 158, 169, 174, 175, 176, 178, 179, 201, 202, 322, 343)

J B Conway [1994], "On the fundamental problem for spectral sets," *Linear and Complex Analysis Problem Book 3*, vol 1, Springer–Verlag Lecture Notes **1573**, 373–377. (202)

J B Conway [1995], *Functions of One Complex Variable, II*, Springer–Verlag, New York. (124, 134, 144, 145, 150, 174, 203)

J B Conway and N S Feldman [1997], "The essential selfcommutator of a subnormal operator," *Proc Amer Math Soc* **125** 243–244. (178)

J B Conway and J Szücs [1973], "The weak sequential closure of certain sets of extreme points in a von Neumann algebra," *Indiana Univ Math J* **22** 763–768. (251)

J B Conway and P Y Wu [1977], "The splitting of $\mathcal{A}(T_1 \oplus T_2)$ and related questions," *Indiana Univ Math J* **26** 41–56. (108)

T Crimmins and P Rosenthal [1967], "On the decomposition of invariant subspaces," *Bull Amer Math Soc* **73** 97–99. (109)

A van Daele [1978], *Continuous Crossed Products and Type III von Neumann Algebras*, Cambridge University Press, Cambridge. (305)

K R Davidson [1987], "The distance to the analytic Toeplitz operators," *Ill J Math* **31** 265–273. (343)

K R Davidson [1996], *C^*-Algebras by Example*, Amer Math Soc, Fields Inst Monographs **6**, Providence. (170, 181)

J A Deddens and P A Fillmore [1975], "Reflexive linear transformations," *Lin Alg Appl* **10** 89–93. (323, 324, 326)

J Dixmier [1949], "Les operateurs permutables à l'opérateur integral," *Portugal Math* **8** 73–84. (143)

J Dixmier [1956], *Les Algèbres d'Operateurs dans l'Espace Hilbertien*, Gauthiers–Villars, Paris. (63, 241)

J Dixmier [1964], *Les C^*-Algèbras et leurs Représentations*, Gauthiers–Villars, Paris. (181)

W F Donoghue [1957], "The lattice of invariant subspaces of a quasi-nilpotent completely continuous transformation," *Pacific J Math* **7** 1031–1035. (140, 143)

R G Douglas [1966], "On majorization, factorization, and range inclusion of operators on Hilbert space," *Proc Amer Math Soc* **17** 413–415. (82)

R G Douglas [1969], "On the operator equation $S^*XT = X$ and related topics," *Acta Math Sci (Szeged)* **30** 19–32. (59)

R G Douglas [1972], *Banach Algebra Techniques in Operator Theory*, Academic Press, New York. (124, 125, 126)

P L Duren [1970], *H^p-spaces*, Academic Press, New York. (124)

P Duren, D Khavinson, H S Shapiro, and C Sundberg [1993], "Contractive zero-divisors in Bergman spaces," *Pacific J Math* **157** 37–56. (155)

P Duren, D Khavinson,H S Shapiro, and C Sundberg [1994], "Invariant subspaces in Bergman spaces and the biharmonic equation," *Michigan Math J* **41** 247–259. (155)

H A Dye [1953], "The unitary structure in finite rings of operators," *Duke math J* **20** 55–69. (251)

N Elias [1988], "Toeplitz operators on weighted Bergman spaces," *Integral Equations Operator Theory* **11** 310–331. (153)

O J Farrell [1934], "On approximation to an analytic function by polynomials," *Bull Amer Math Soc* **40** 908–914. (151)

N S Feldman [1997], "The self-commutator of a subnormal operator," PhD thesis, University of Tennessee. (239)

N S Feldman [1999], "Essentially subnormal operators," *Proc. Amer. Math. Soc.*, **127** 1171–1181. (169, 176)

P A Filmore and J P Williams [1971], "On operator ranges," *Adv Math* **7** 254–281. (82)

C Foias [1959], "Some applications of spectral sets I: harmonic spectral measure," *Acad R P Roumaine Stud Cerc Math* **10** 365–401. Also, *Amer Math Soc Translations* (2) **61** (1967) 25–62. (202)

B Fuglede [1950], "A commutativity theorem for normal operstors," *Proc nat Acad Sci* **36** 35–40. (58)

T W Gamelin and J Garnett [1971], "Pointwise bounded approximation and Dirichlet algebras," *J Functional Analysis* , **8** 360–404. (203)

T W Gamelin and D Khavinson [1989], "The isoperimetric inequality and rational approximation," *Amer Math Monthly* **96** 18–30. (171)

J Glimm [1960], "A Stone–Weierstrass Theorem for C*-algebras," *Ann Math* **72** 216–244. (221)

Don Hadwin [1994], "A general view of reflexivity," *Trans Amer Math Soc* **344** 325–360. (319)

D Hadwin and E Nordgren [1982], " Subalgebras of reflexive algebras," *J Operator Theory* **7** 3–23. (335)

P R Halmos, [1950], "Normal dilations and extensions of operators," *Summa Bras Math* **2** 125–134. (158)

P R Halmos, [1952], "Spectra and spectral manifolds," *Ann Soc Polon Math* **25** 43–49. (164)

P R Halmos [1961], "Shifts on Hilbert space," *J Reine Angew Math* **208** 102–112. (116, 125)

P R Halmos [1970], "Ten problems in Hilbert space," *Bull Amer Math Soc* **76** 887–933. (237)

P R Halmos [1972], "Continuous functions of hermitian operators," *Proc Amer Math Soc* **31** 130–132. (214)

P R Halmos [1982] *A Hilbert Space Problem Book*, Springer–Verlag, New York. (116, 138, 143)

P. R. Halmos, G. Lumer, and J. J. Schäffer [1953], "Square roots of operators," *Proc Amer Math Soc* **4** 142–149. (153)

P R Halmos and J von Neumann [1942], "Operator methods in classical mechanics, II," *Ann Math* **43** 332 – 350. (292)

L I Hedberg [1972a], "Approximation in the mean by analytic functions," *Trans Amer Math Soc* **163** 157–171. (153)

L I Hedberg [1972b], "Non-linear potentials and approximation in the mean by analytic functions," *Math Zeit* **129** 299–319. (153)

H Hedenmalm [1991], "A factorization theorem for square area-integrable analytic functions," *J Reine Angew Math* **422** 45–68. (155)

H Hedenmalm [1993], "An invariant subspace of the Bergman space having the codimension two property," *J Reine Angew Math* **443** 1–9. (155)

H Hedenmalm, B Korenblum, and K Zhu [1996], "Beurling type invariant subspaces of the Bergman spaces," *J London Math Soc* (2) **53** 601–614. (155)

H Hedenmalm, S Richter, and K Seip [1996], "Interpolating sequences and invariant subspaces of given index in Bergman spaces," *J Reine Angew Math* **477** 13–30. (155)

H Helson [1964], *Invariant Subspaces*, Academic Press, New York. (125)

H Helson and D Lowdenslager [1958], "Prediction thoery and Fourier series in several variables," *Acta Math* **99** 165–202. (125)

K Hoffman [1962], *Banach Spaces of Analytic Functions*, Prentice-Hall, Englewood Cliffs. (124)

K H Jin [1989], "On unbounded Bergman operators," PhD thesis, Indiana University. (153)

B E Johnson [1979], "Characterisation and norms of derivations on von Neumann algebras," Lecture Notes in Math, **725** 228–236, Springer–Verlag, New York. (347)

B E Johnson and S K Parrott [1972], "Operators commuting with a von Neumann algebra modulo the set of compact operators," *J Functional Analysis* **11** 39–61. (239)

R V Kadison [1956], "Operator algebras with a faithful weakly-closed representation," *Annals Math* **64** 175–181. (256)

R V Kadison [1957], "Irreducible operator algebras," *Proc Nat Acad Sci USA* **43** 273–276. (186)

R V Kadison [1985], "The von Neumann algebra characterization theorems," *Exp Math* **3** 193-227. (256)

R V Kadison and D Kastler [1972], "Perturbations of von Neumann algebras. I. Stability of type," *Amer J Math* **94** 38–54. (347)

R V Kadison and J R Ringrose [1983], *Fundamentals of the Theory of Operator Algebras, I*, Academic Press, New York. (81, 181, 186, 241)

R V Kadison and J R Ringrose [1986], *Fundamentals of the Theory of Operator Algebras, II*, Academic Press, New York. (181, 241)

G K Kalisch [1957], "On similarity, reducing manifolds, and unitary equivalence of certain Volterra operators," *Ann Math* **66** 481–494. (143)

T Kato [1957], "Perturbation of continuous spectra by trace class operators," *Proc Japan Acad* **33** 260–264. (212)

J L Kelley [1966], "Decomposition and representation theorems in measure theory," *Math Ann* **163** 89–94. (54)

G E Keough [1981], "Subnormal operators, Toeplitz operators, and spectral inclusion," *Trans Amer Math Soc* **263** 125–135. (169)

P Koosis [1980], *Introduction to H_p spaces*, London Math Soc Lecture Notes, vol 40, Cambridge Univ Press (Cambridge). (124)

J Kraus and D Larson [1985], "Some applications of a technique for constructing reflexive operator algebras," *J Operator Theory* **13** 227–236. (333, 344)

J Kraus and D Larson [1986], "Reflexivity and distance Formulae," *Proc London math Soc* **53** 340–356. (337, 342)

D R Larson [1982], "Annihilators of operator algebras," Operator Theory: Adv and Appl **6** 119–130. (335, 340, 341)

P Lax [1959], "Translation invariant subspaces," *Acta Math* **101** 163–178. (125)

P Lax [1961], "Translation invariant subspaces," *Proc Internat Symp Linear Spaces, Jerusalem, 1960*, 251–262, Macmillan, New York. (125)

A Lebow [1963], "On von Neumann's theory of spectral sets," *J Math Anal Appl* **7** 64–90. (202)

A Lebow [1968], "A Schroeder-Bernstein theorem for projections," *Proc Amer Math Soc* **19** 144–145. (269)

A I Loginov and V S Sulman [1975], "Hereditary and intermediate reflexivity of W*-algebras (Russian)," *Izv Akad Nauk SSSR Ser Mat* **39** 1260–1273; English trans.: *Math USSR-Izv* **9** 1189–1201. (39, 336)

W E Longstaff [1979], "On the operation AlgLat in finite dimensions," *Lin Alg Appl* **27** 27–29. (340)

A I Markusevic [1934], "Conformal mapping of regions with variable boundary and applications to the approximation of analytic functions by polynomials," Dissertation, Moskow. (151)

M Martin and M Putinar [1989], *Lectures on Hyponormal Operators*, Birkhäuser, Basel. (158)

J E McCarthy [1993], "Reflexivity of subnormal operators," *Pacific J. Math* **161** 359–370. (322, 343)

R Mercer [1986], "Dense G_δ's containing orthonormal bases," *Proc Amer Math Soc* **97** 449–452. (64)

S N Mergeljan [1953], "On the completeness of systems of analytic functions," *Uspeki Math Nauk* **8** 3–63. Also, *Amer Math Soc Translations* **19** (1962) 109–166. (152, 153)

B B Morrel [1973], "A decomposition for some operators," *Indiana Univ Math J* **23** 497–511. (178)

F J Murray and J von Neumann [1936], "On rings of operators," *Ann Math* **37** 116–229. (268, 311)

F J Murray and J von Neumann [1937], "On rings of operators, II," *Trans Amer Math Soc* **41** 208–248. (268, 311)

B Sz.-Nagy [1953], "Sur les contractions de l'espace de Hilbert," *Acta Sci Math* **15** 87–92. (125)

B Sz.-Nagy and C Foias [1970], *Harmonic Analysis of Operators on Hilbert Space*, North Holland, Amsterdam. (125, 200, 352)

M A Naimark [1943], "On a representation of additive operator set functions," (in Russian) *Doklady Acad Sci USSR* **41** 359–361. (195)

N K Nikolskii [1965], "Invariant subspaces of certain completely continuous operators," *Vestnik Leningrad Univ (Math 1)* **7** 68–77. (140)

N K Nikolskii [1985], *Treatise on the Shift Operator*, Springer–Verlag, Berlin. (137)

R F Olin and J E Thomson [1980], "Algebras of subnormal operators," *J Functional Analysis* **37** 271–301. (322, 335, 343)

V I Paulsen [1986], *Completely bounded maps and dilations*, Longman, Essex. (187, 196, 198, 203)

V I Paulsen [1988], "Toward a theory of K-spectral sets," *Surveys of Some Recent Results in Operator Theory*, vol 1, J B Conway and B B Morrel, Editors, 221–240, Research Notes in Mathematics, Longman, London. (198)

G K Pedersen [1972], "Monotone closures in operator algebras," *Amer J Math* **94** 955–962. (253)

G K Pedersen [1979], *C^*-Algebras and Their Automorphism Groups*, London Math Soc Monographs, vol 14, Academic Press, London. (181, 241)

S Popa [1987], "The commutant modulo the set of compact operators of a von Neumann algebra," *J Functional Anal* **71** 393–408. (239)

M Ptak [1997], "On the existence of invariant subspaces and reflexivity of N-tuples of operators," *Taiwanese J Math* **1** 231–290. (319)

C R Putnam [1951], "On normal operators in Hilbert space," *Amer J Math* **73** 357–362. (58, 59)

C R Putnam [1967], *Commutation properties of Hilbert Space Operators and Related Topics*, Springer–Verlag, New York. (212)

C R Putnam [1968], "The spectra of operators having resolvents of first order growth," *Trans Amer Math Soc* **133** 505–510. (210)

C R Putnam, [1970], "An inequality for the area of hyponormal spectra," *Math Zeit* **116** 323–330. (176)

H Radjavi and P Rosenthal [1969], "On invariant subspaces and reflexive algebras," *Amer J Math* **91** 683–692. (319)

H Radjavi and P Rosenthal [1973], *Invariant subspaces*, Springer–Verlag, Berlin. (58, 106, 146)

J R Ringrose [1971a], *Compact Non-self-adjoint Operators*, Van Nostrand-Reinhold, New York. (93)

J R Ringrose [1971b], "Lectures on the trace in a finite von-Neumann algebra," Springer–Verlag Lecture Notes vol 247. (311, 314)

J R Robertson [1965], "On wandering subspaces for unitary operators," *Proc Amer Math Soc* **16** 233–236. (116)

M Rosenblum [1957], "Perturbation of the continuous spectrum and unitary equivalence," *Pacific J Math* **7** 997–1010. (212)

M Rosenblum [1958], "On a theorem of Fuglede and Putnam," *J London Math Soc* **33** 376-377. (58)

M Rosenblum and J Rovnyak [1985], *Hardy Classes and Operator Theory*, Oxford Univ Press, New York. (125)

S Rosenoer [1982], "Distance estimates for von Neumann algebras," *Proc Amer Math Soc* **86** 248–252. (347)

L A Rubel and A L Shields [1964], "Bounded approximation by polynomials," *Acta Math* **112** 145–162. (152)

S Sakai [1971], *C^*-algebras and W^*-algebras*, Springer–Verlag, New York. (60, 241, 307, 344)

D Sarason [1965a], "On spectral sets having connected complements," *Acta Sci Math* (Szeged) **26** 289–299. (201)

D Sarason [1965b], "A remark on the Volterra operator," *J Math Anal Appl* **12** 244–246. (143, 146)

D Sarason [1966], "Invariant subspaces and unstarred operator algebras," *Pacific J Math* **17** 511–517. (343)

D Sarason [1967], "Generalized interpolation in H^∞," *Trans Amer Math Soc* **127** 179–203. (147)

D Sarason [1972], "Weak-star density of polynomials," *J Reine Angew Math* **252** 1–15. (110)

R Schatten [1960], *Norm Ideals of Completely Continuous Operators*, Springer–Verlag, Berlin. (93)

M Schecter [1965], "Invariance of the essential spectrum," *Bull Amer Math Soc* **71** 365–367. (210)

A L Shields [1974], "Weighted shift operators and analytic function theory," *Math Surveys*, Amer Math Soc **13** 49–128. (137, 139)

A L Shields and L J Wallen [1971], "The commutants of certain Hilbert space operators," *Indiana Univ Math J* **20** 777–788. (154)

J G Stampfli [1974], "Compact perturbations, normal eigenvalues, and a problem of Salinas," *J London Math Soc* **9** 165–175. (211)

W F Stinespring [1955], "Positive functionals on C*-algebras," *Proc Amer Math Soc* **6** 211–216. (192, 194)

S Stratila and L Zsido [1979], *Lectures on von Neumann Algebras*, Abacus Press, Tunbridge Wells. (241)

W Szymanski [1990], "The boundedness condition of dilation characterizes subnormals and contractions," *Rocky Mountain J Math* **20** 591–602. (158)

J E Thomson, [1991], "Approximation in the mean by polynomials," *Annals Math* **133** 477–507. (322, 343)

D Topping [1971], *Lectures on Von Neumann Algebras*, van Nostrand Reinhold, London. (241)

K Tsuji [1983], "Annihilators of operator algebras," *Mem Fac Sci Kochi Univ* **4** 9–21. (340)

D Voiculescu [1976], "A non-commutative Weyl–von Neumann Theorem," *Rev Roumaine math Pures Appl* **21** 97–113. (220, 229, 237)

D Voiculescu [1979], "Some results on norm-ideal perturbations of Hilbert space operators," *J Operator Theory* **2** 3–37. (217)

D Voiculescu [1981], 'Some results on norm-ideal perturbations of Hilbert space operators, II" *J Operator Theory* **5** 77–100. (217)

D Voiculescu [1990], "On the existence of quasicentral approximate units relative to normed ideals. Par I," *J Functional Analysis* 1–36. (217)

J von Neumann [1929], "Zur Algebra der funktional-operatoren und theorie der normalen operatoren," *Math Ann* **102** 370–427. (56)

J von Neumann [1935], "Charakterisierung des Spectrums eines Integraloperators," *Actualités Sci Ind* **229** 38–55. (212)

J Wermer [1955], "Report on subnormal operators," *Report on an International Conference on Operator Theory and Group Representations*, National Acad Sci-National Research Council, Harriman, New York, 1–3. (165)

H Weyl [1909], "Über beschrankte quadratischen Formen deren Differenz vollstetig ist," *Rend Circ Mat Palermo* **27** 373–392. (212)

W Wogen, [1986], "Counterexamples in the theory of nonselfadjoint operator algebras," *Bull Amer Math Soc* **15** 225–227. (322, 348)

W Wogen, [1987], "Some counterexamples in nonselfadjoint algebras," *Ann of Math* **126** 415–427. (108, 320, 332, 326, 348, 351, 353)

S Wright [1989], *Uniqueness of the injective III_1 factor,* Springer–Verlag Lecture Notes **1413**. (305)

F J Yeadon [1971], "A new proof of the existence of a trace in a finite von Neumann algebra," *Bull Amer Math Soc* **77** 257–260. (311)

K Zhu [1990], *Operator theory in function spaces*, Dekker, New York. (155)

Index

abelian C*-algebra, 7
abelian projection, 272, 275
abelian von Neumann algebra, 291, 342
absolute value, 15
additivity of equivalence, 267
adjoining an identity, 3
algebraic homomorphism, 5
all state representation, 34
analytic Toeplitz operator, 121
approximate identity, 18, 204
approximate point spectrum, 116
approximately equivalent representations, 229
approximately unitarily equivalent, 218
Arveson's Extension Theorem, 203
attached space, 39
Banach limit, 36, 345
Bergman operator, 147, 178, 198
Beurling's Theorem, 123
bilateral shift, 113
bilateral weighted shift, 137, 157
Blaschke product, 124
boundedly mutually absolutely continuous, 52
$C(X)$, 2, 10, 11, 12, 23, 30, 45, 189, 192, 194
$C_0(X)$, 2
Carathéodory region, 149
Cauchy transform, 173
Cauchy–Schwarz–Bunyakowski Inequality, 30
central cover, 245
central support, 245
commutant, 55
compact operator, 50, 71, 85, 93
Comparison Theorem, 268
completely bounded map, 192
completely positive map, 191
continuous algebra, 273, 283, 302
continuous projection, 273
convolution, 248
cornucopia, 150
C*-algebra, 1
C*-equivalent, 239
cyclic operator, 164
cyclic representation, 26
cyclic vector, 26, 165
d-fold inflation, 25
d-reflexive, 340
degenerate algebra, 71
decomposable measure space, 54
derivation, 344
diagonal algebra, 214
diagonalizable, 55
dilation, 196
direct sum of representations, 25
Dirichlet algebra, 202
discrete algebra, 272
discrete projection, 273
Donoghue shift, 140
double commutant, 55, 239
Double Commutant Theorem, 56
Douglas's Range Inclusion Theorem, 82
dual algebra, 97
dual algebra homomorphism, 97
dual algebra isomorphism, 97
equivalent projections, 265
equivalent representations, 25
essential boundary point, 148
essential spectrum, 116, 209
essentially normal, 170
extreme point, 145, 185, 306
F and M Riesz Theorem, 110, 127

factor, 249
faithful projection, 282
finite algebra, 270, 315, 316
finite projection, 270
Fourier transform, 119
Fredholm Alternative, 83
Fredholm index, 118, 139, 148, 170, 209
Fredholm operator, 207
Fuglede–Putnam Theorem, 58
functional calculus, 8, 162, 153, 166
functional calculus for a normal operator, 69
Gelfand transform, 7
Glimm's Lemma, 221
GNS Construction, 32
greatest common divisor, 131
Hardy spaces, 119
Hartogs–Rosenthal Theorem, 175
hereditary subalgebra, 21, 262
Herglotz's Theorem, 144
hermitian, 4
hermitian functional, 33, 308
Hilbert–Schmidt operator, 86, 92, 147
homogeneous algebra, 285
homomorphism, 2
*-homomorphism, 2, 5, 23
hyperreflexive, 328
hyponormal, 158, 169
ideal, 262
ideal (in C*-algebra), 21
ideal (in von Neumann algebra), 262
ideal (in $\mathcal{B}(\mathcal{H})$), 83
index of nilpotency, 324
infinite algebra, 270
infinite projection, 270, 280
inflation, 25, 99
inner function, 123
involution, 1
irreducible algebra, 73
irreducible representation, 46, 75, 181, 184, 236
isometry, 112, 114, 118, 253
isomorphism, 2
*-isomorphism, 2, 268
join, 105
Jordan canonical form, 107, 324
Jordan decomposition of functional, 35
Kaplansky Density Theorem, 251
Krein–Smulian Theorem, 96, 156
$L^\infty(\mu)$, 8, 10, 11, 57, 63, 97, 263, 292
lattice, 105
lattice of invariant subspaces, 106
least common multiple, 131
left essential spectrum, 116, 209
left Fredholm operator, 207
left spectrum, 116
M_ϕ, 10, 49, 53
maximal abelian von Neumann algebra, 63
maximal ideal space, 7
measurable (vector-valued function), 296
measurable (operator-valued function), 294
meet, 105
minimal normal extension, 163
minimal projection, 73
minimal unitary dilation, 203
multiplicity, 290
multiplicity function, 79, 294
mutually absolutely continuous, 52
n-th root, 13
Sz.-Nagy Dilation Theorem, 200
negative part, 13
non-degenerate representation, 27
normal, 4
normal functional, 259
normal map, 258, 261
normal operator, 48, 51, 53, 55, 58, 69, 293, 342, 347
normal spectrum, 164
operator system, 187
operator-valued measure, 195
orthogonal functionals, 308
outer function, 127
partial isometry, 15, 16, 306
Pedersen Up–Down Theorem, 254
polar decomposition, 15, 17, 50
polynomially convex, 109, 156
polynomially convex hull, 109
positive element, 12, 13
positive element of $\mathcal{B}(\mathcal{H})$, 14
positive linear functional, 29, 30
positive map, 187
positive part, 13
preannihilator, 98
properly infinite, 280, 302
property $\mathbb{A}_1$, 335
property $\mathbb{A}_1(r)$, 335
pure isometry, 113
pure state, 183, 185
purely infinite, 275
Putnam's Inequality, 176
quasicentral approximate identity, 205
quasinormal, 169
Radon–Nikodym Theorem, 182
rationally cyclic operator, 165
Rat(K) cyclic vector, 165
reflexive, 319
relatively dominates, 265
removable boundary point, 148
representation, 24, 181
reproducing kernel, 121
Riesz Functional Calculus, 8
right essential spectrum, 116, 209
right Fredholm operator, 207
scalar-valued spectral measure, 66, 69
Schatten p-class, 93

Schröder–Bernstein, 269
semi-Fredholm, 207
semi-invariant subspace, 201
separable measure space, 54
separating vector, 62
sequential approximate identity, 18
SOT (strong operator topology), 37, 40, 251
spatially isomorphic, 62
Spectral Mapping Theorem, 11, 69
spectral measure, 41
spectral set, 200
Spectral Theorem, 47, 50
spectrum of a C^*-algebra, 79
star-cyclic, 51
$*$-homomorphism, 2, 5, 23
$*$-isomorphism, 2, 261
state, 29
state space, 29
Stinespring's Theorem, 194
Stone–Čech compactification, 265
strong operator topology (SOT), 37, 40, 251
strong* topology, 41
subequivalent, 265
subnormal operator, 157, 176, 177, 197, 198
subrepresentation, 26, 75
support (of a functional), 308
support (of a measure), 9
The Gelfand–Naimark–Segal Construction, 31
Titchmarsh Convolution Theorem, 146
Toeplitz operator, 125, 322, 343
trace, 89, 286, 311
trace class, 86, 93, 94, 99, 101, 353
trace norm, 86
transitive, 353
trigonometric polynomial, 119
Type (of a von Neumann algebra), 275, 303
Type I factor, 284
Type I von Neumann algebra, 282, 283, 288, 290, 291
Type II_1 factor, 301
uniform infinite multiplicity, 339
unilateral shift, 40, 111, 120, 178, 198, 321
unilateral weighted shift, 136
unital, 1
unitary, 4
unitary operator, 50
universal representation, 34
Up–Down Theorem, 254
Voiculescu's Theorem, 220
Volterra operator, 143
von Neumann algebra, 60, 247, 353, 358
von Neumann's Inequality, 200
von Neumann–Wold decomposition, 112
weak operator topology (WOT), 37, 351
weak* topology, 95, 102, 351
weakly approximately equivalent representations, 232
weighted shift, 155, 169
Weyl spectrum, 210
Weyl–von Neumann Theorem, 211
Weyl–von Neumann–Berg Theorem, 216
WOT (weak operator topology), 37, 351

List of Symbols

cl A = the closure of a set A
int A = the interior of a set A
∂A = the boundary of the set A
$\mathbb{C}$ is the field of complex numbers, $\mathbb{R}$ the field of real numbers, $\mathbb{Z}$ the integers, $\mathbb{N}$ the natural numbers, and $\mathbb{D} = \{z \in \mathbb{C} : |z| < 1\}$
ball $X = \{x \in X : \|x\| \leq 1\}$
ran T and ker T denote the range and kernel of a linear transformation

$\mathcal{B} = \mathcal{B}(H)$, 1
$\mathcal{B}_0 = \mathcal{B}_0(\mathcal{H})$, 2
$C(X)$, 2
$C_0(X)$, 2
$\sigma(a)$, 4
$r(a)$, 4
Re $\mathcal{A}$, 4
$C_0(X, \mathcal{A})$, 6
$\bigoplus_0\{\mathcal{A}_i : i \in I\}$, 6
$C_0^*(S)$, 8
$C^*(S)$, 8
ess-ran (ϕ), 9
N_μ, 9
M_ϕ, 10
$C_b(X)$, 12
βX, 12
$\mathcal{A}_+$, 12
a_+, a_-, 13
$|a|$, 15
M_n, 24
π_μ, 25
$\mathcal{H}^d$, $A^{(d)}$, $\pi^{(d)}$, $\mathcal{H}^{(\infty)}$, $A^{(\infty)}$, and $\pi^{(\infty)}$, 25
$\pi_1 \cong \pi_2$, 25
$[S]$, 26
ϕ_+, ϕ_-, 34
LIM, 36
Ref $\mathcal{S}$, 39
$E_{g,h}$, 42
$\int \phi \, dE$, 43
$B(X)$, 44
$[\mu] = [\nu]$, 52
$\mathcal{S}'$, $\mathcal{S}''$, 55
$\mathcal{A}_\mu$, 57
$W^*(\mathcal{S})$, 61
$\mathcal{S}_1 \cong \mathcal{S}_2$, 62
$g \otimes h$, 72
$\mathcal{I}_\alpha$, 83
$\mathcal{B}_1 = \mathcal{B}_1(\mathcal{H})$, $\|A\|_1$, 86
$\mathcal{B}_2 = \mathcal{B}_2(\mathcal{H})$, $\|A\|_2$, 86
tr A, 89
$\mathcal{B}_{00} = \mathcal{B}_{00}(\mathcal{H})$, 89
$\mathcal{B}_p$, $\|A\|_p$, 93
$P^\infty(\mu)$, 97
$\mathcal{S}_\perp$, 98
$\mathcal{M} \vee \mathcal{N}$, $\mathcal{M} \wedge \mathcal{N}$, 105
Lat $\mathcal{A}$, Alg $\mathcal{L}$, 106
$P(\mathcal{S})$, $P^\infty(\mathcal{S})$, $W(\mathcal{S})$, 107
$\widehat{K}$, 109
$R(T)$, 111
$\sigma_{ap}(T)$, $\sigma_\ell(T)$, $\sigma_r(T)$, $\sigma_{\ell e}(T)$, $\sigma_{re}(T)$, $\sigma_e(T)$, 116
$\hat{f}$, 119 (also see page 133)
H^p, 119
k_λ, 120
$\hat{f}$, 133 (also see page 119)
P_z, 144
$d\mathcal{A}$, Area, 147
$L^p_a(G)$, 147
$P^2(G)$, $R^2(G)$, 149

$H^\infty(G)$, 153
$P^2(\mu)$, S_μ, 157
Rat K, 157
$R^2(K,\mu)$, 157
$\sigma_n(S)$, 164
$R(K)$, 166
$\|f\|_K$ 166
$[T^*,T]$, 170
$C^n(G)$, $C_c^n(G)$, 172
$\partial u \equiv \partial_z u \equiv \frac{\partial u}{\partial z}$, $\bar\partial u \equiv \partial_{\bar z} u \equiv \frac{\partial u}{\partial \bar z}$, 172
$\widehat{\mu}$, 173
$\Sigma(\mathcal{A})$, 185
$M_n(\mathcal{A})$, 190
ϕ_n, 191
$P(\lambda;T)$, 199
$\mathcal{F}_\ell$, $\mathcal{F}_r$, $\mathcal{SF}$, $\mathcal{F}$, 209
ind T, 209
$P_n(T)$, 210
$\sigma_w(T)$, 210
$\mathcal{D} = \mathcal{D}(\{e_k\})$, 214
$A \cong_a B$, 218
$A \cong_{sa} B$, 219
$\rho \cong_a \kappa$, $\rho \cong_{sa} \kappa$, 229
$\rho \cong_w \kappa$, 232
$C^*(A) \approx C^*(B)$, 239
$\mathcal{Z}$, 243
M_Φ, 244
$L^\infty(\mu :, M_d)$, 244
C_A, 245
cran A, 245
$\mathcal{A}_E$, 246
R_a, L_a, 247
$\mathcal{R}(G)$, $\mathcal{L}(G)$, 247
$f * g$, 248
$\mathcal{S}_\sigma$, $\mathcal{S}^\sigma$, 254
$E \sim F$, $E \sim_W F$, $E \precsim F$, 265
I_n, I_∞, II_1, II_∞, III, 275
$M_n(\mathcal{A})$, 285
$\kappa(\mu)$, 292
$L^p(\mu;H)$, 295
$L^\infty(\mu;\mathcal{B}(H))$, 297
M_Φ, 297
$\phi \perp \psi$, 308
$\widetilde{T}$, 321
J_n, 323
$\alpha(T,\mathcal{S})$, 327
$\kappa(\mathcal{S})$, 328
$\mathbb{A}_1$, $\mathbb{A}(r)_1$, 335
$[\mathcal{S}]_{ij}$, 348
$\widetilde{\mathcal{P}}$, 349